The use of neural networks in signal processing is becoming increasingly widespread, with applications in areas such as filtering, parameter estimation, signal detection, pattern recognition, signal reconstruction, system identification, signal compression, and signal transmission. *Applied Neural Networks for Signal Processing* is the first book to provide a comprehensive introduction to this broad field.

The book begins by covering the basic principles and models of neural networks in signal processing. The authors then discuss a number of powerful algorithms and architectures for a range of important problems and go on to describe practical implementation procedures. A key feature of the book is that many carefully designed examples are included to help guide the reader in the development of systems for new applications. The book will be an invaluable reference for scientists and engineers working in communications, control, or any other field related to signal processing. It can also be used as a textbook for graduate courses in electrical engineering and computer science.

Applied Neural Networks for Signal Processing

Applied Neural Networks for Signal Processing

FA-LONG LUO and ROLF UNBEHAUEN
University of Erlangen-Nuremberg, Erlangen, Germany

CAMBRIDGE
UNIVERSITY PRESS

PUBLISHED BY THE PRESS SYNDICATE OF THE UNIVERSITY OF CAMBRIDGE
The Pitt Building, Trumpington Street, Cambridge CB2 1RP, United Kingdom

CAMBRIDGE UNIVERSITY PRESS
The Edinburgh Building, Cambridge CB2 2RU, United Kingdom
40 West 20th Street, New York, NY 10011-4211, USA
10 Stamford Road, Oakleigh, Melbourne 3166, Australia

First published 1997

Printed in the United States of America

Typeset in Times Roman

Library of Congress Cataloging-in-Publication Data

Luo, Fa-Long,
Applied neural networks for signal processing / Fa-Long Luo, Rolf Unbehauen.
p. cm.
ISBN 0 521 56391 7
1. Signal processing. 2. Neural networks (Computer science)
I. Unbehauen, Rolf. II. Title.
TK5102.9.L85 1997
621.382′2′028563 – dc20 96-41384
CIP

A catalog record for this book is available from the British Library.

ISBN 0 521 56391 7 hardback

Contents

Preface

During the past decade neural networks have begun to find wide applicability in many diverse aspects of signal processing, for example, filtering, parameter estimation, signal detection, system identification, pattern recognition, signal reconstruction, time series analysis, signal compression, and signal transmission. The signals concerned include audio, video, speech, image, communication, geophysical, sonar, radar, medical, musical, and others. The key features of neural networks involved in signal processing are their asynchronous parallel and distributed processing, nonlinear dynamics, global interconnection of network elements, self-organization, and high-speed computational capability. With these features, neural networks can provide very powerful means for solving many problems encountered in signal processing, especially in nonlinear signal processing, real-time signal processing, adaptive signal processing, and blind signal processing.

From an engineering point of view, this book aims to provide a detailed treatment of neural networks for signal processing by covering basic principles, modeling, algorithms, architectures, implementation procedures, and well-designed simulation examples. This book is organized into nine chapters.

Chapter 1 presents basic models of neural networks for signal processing and related fundamentals such as stability theory, learning algorithms, and dynamics analysis. These basic models include mainly: the discrete-time Hopfield neural network, the continuous-time Hopfield neural network, cellular neural networks, multilayer-perceptron networks, self-organizing systems, radial basis function networks, and high-order neural networks.

Chapter 2 focuses on neural networks for filtering. This chapter includes essentially two parts. First, we show how to use neural networks to perform the computations required in various filtering algorithms such as the least-squares algorithm, the recursive least-squares algorithm, the constrained least-squares algorithm, and the total-least-squares algorithm. Next, we present the fundamentals of neural networks for the design and the implementation of nonlinear filters.

Chapter 3 is devoted to neural network approaches for maximum entropy spectral estimation, harmonic retrieval, two-dimensional spectral estimation, multichannel spectral estimation, and higher-order spectral estimation.

In Chapter 4, we discuss neural networks for signal detection. We deal in detail with applications of neural networks to the design, realization, preprocessing, and postprocessing of signal detection.

The main purpose of Chapter 5 is to present neural networks for signal reconstruction. We emphasize applications of neural networks to the implementation of optimal reconstruction algorithms and the development of nonlinear and blind reconstruction algorithms.

Chapter 6 is devoted to the adaptive extraction of the eigenvectors corresponding to the largest and smallest eigenvalues of the autocorrelation matrix of a signal. These two kinds of eigenvectors are referred to as the principal components and minor components, respectively. Their adaptive extraction is a preliminary requirement in adaptive signal processing. A detailed review of adaptive unsupervised learning algorithms for extracting the principal components and minor components is given in this chapter.

Chapter 7 deals with neural networks for array signal processing. We show how to use neural networks to perform the computations required for estimating the directions of arrival (DOA) of sources. The DOA estimation methods to be considered in this chapter include the maximum likelihood technique, the alternating projection maximum likelihood technique, the MUSIC method of Schmidt, and the propagator method of Marcos. In addition, another important problem of array signal processing – beamforming – is also considered in this chapter. The emphasis is put on neural networks for computing the optimal weights of beamformers in real time.

In Chapter 8, various neural network approaches for system identification are presented. Real-time system identification, nonlinear system identification, and blind system identification are all dealt with on the basis of key features of neural networks.

Chapter 9 is devoted to neural networks for signal compression. Neural networks for the real-time implementation of the optimal linear predictive coding algorithm, for nonlinear predictive coding systems, for the Karhunen–Loève transform coding, for wavelet transform coding, and for vector quantization are reported in this chapter.

We hope that this book serves not only as a reference for professional engineers and scientists (majoring in communications, radar, sonar, automatic control, and other fields related to signal processing) but also as a textbook for graduate students in electronics engineering and computer science.

We are deeply indebted to Professor G. O. Martens of the University of Manitoba for carefully reading the manuscript of this book. He made many of his insightful comments and contributed to very helpful discussions during his stay at the

Erlangen-Nürnberg University as a guest professor invited by the German Research Foundation (DFG). We are also very grateful to Professor Yan-Da Li at Tsinghua University for his help and many significant suggestions in writing this book.

We owe many thanks to Mr. H. Weglehner and Mrs. E. Orth who drew all the figures included in this book. We are also grateful to other staff members of the Lehrstul für Allgemeine and Theoretische Elektrotechnik at the Erlangen-Nürnberg University: Dipl.-Ing. B. Anhäupl, Mrs. H. Geisenfelder-Göhl, Dr.-Ing. W. Göttlicher, Mrs. S. He, Dipl.-Ing. H. Kicherer, Dr.-Ing. H. Rossmanith, Dipl.-Ing. M. Lendl, Dipl.-Phys. K. Reif, Mrs. B. Scholz, Dipl.-Ing. G. Triftshäuser, and Dipl.-Ing. A. Zell.

A number of experts in related fields have helped us in various ways, directly or indirectly. We especially thank Prof. S. Amari (University of Tokyo and RIKEN), Prof. Zheng Bao (Xidian University), Prof. R. Battiti (Trento University), Prof. A. Cichocki (RIKEN), Prof. V. David Sánchez A. (University of Miami), Prof. Hsin-Chia Fu (National Chiao-Tung University), Prof. N. E. Gough (University of Wolverhampton), Prof. Jenq-Neng Hwang (University of Washington), Dr. J. Karhunen (Helsinki University of Technology), Prof. A. Lansner (Sweden Royal Institute of Technology), Prof. Ping Liang (University of California at Riverside), Prof. W. Mecklenbräuker (Technical University of Vienna), Dr. D. W. Pearson (EMA-EERIE), Dr. N. Samardzija (DUPONT), Prof. Jun Wang (Chinese University of Hong Kong), Prof. Wei-Xin Xie (Shenzhen University), Dr. Jun Yang (Oldenburg University), and Dr. Ya-Qin Zhang (David Sarnoff Research Center).

Fa-Long Luo, Rolf Unbehauen
Erlangen, Germany
March 1996

1

Fundamental Models of Neural Networks for Signal Processing

Neural networks are information processing systems composed of a large number of simple processing elements called neurons. Many kinds of neural network models have been proposed and extensively used in signal processing. The key features of neural networks are asynchronous parallel and distributed processing, nonlinear dynamics, global interconnection of network elements, self-organization, and high-speed computational capability. The main objective of this chapter is to present the basic models of neural networks for signal processing and related fundamentals such as stability theory, learning algorithms, and dynamics analysis. These basic models mainly include: the discrete-time Hopfield neural network, the continuous-time Hopfield neural network, cellular neural networks, multilayer-perceptron networks, self-organizing systems, radial basis function networks, and high-order neural networks.

1.1 The Discrete-Time Hopfield Neural Network

The discrete-time Hopfield neural network [1], shown in Figure 1.1, generally consists of N neurons and is uniquely defined by a parameter set $\{\boldsymbol{T}, \boldsymbol{B}\}$. $\boldsymbol{T} = \{T_{ij}\}$ is an $N \times N$ matrix whose element T_{ij} (for $i, j = 1, 2, \ldots, N$) is the connection strength between the i'th neuron and the j'th neuron and specifies the contribution of the output of the j'th neuron to the potential acting on the i'th neuron. $\boldsymbol{B} = [b_1, b_2, \ldots, b_N]^T$ is a vector whose element b_i is the threshold applied externally to the i'th neuron.

Every neuron can be in one of two possible states: either 1 or -1. The state of the i'th neuron at time t is denoted by $v_i(t)$ (for $i = 1, 2, \ldots, N$); consequently, the state of the neural network at time t is the vector $\boldsymbol{V}(t) = [v_1(t), v_2(t), \ldots, v_N(t)]^T$. A state $\boldsymbol{V}(t)$ is called stable if and only if $\boldsymbol{V}(t + \Delta t) = \boldsymbol{V}(t)$ for all $\Delta t \geq 0$.

The time evolution and stability of the discrete-time Hopfield neural network depends not only on the parameter set $\{\boldsymbol{T}, \boldsymbol{B}\}$ but also on the mode of operation of

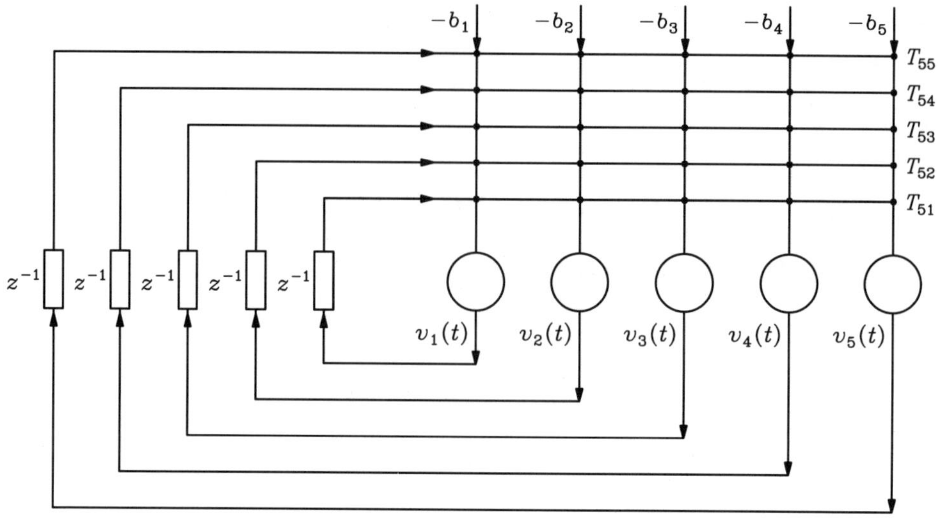

Figure 1.1. A discrete-time Hopfield neural network with five neurons.

the neural network. There are three modes of operation:

(1) a serial mode,
(2) a fully parallel mode, and
(3) a partially parallel mode.

In the serial mode the next state of each neuron is computed by

$$v_i(t+1) = \operatorname{sgn}\left(\sum_{j=1}^{N} T_{ij} v_j(t) - b_i\right) \quad \text{and}$$
$$v_j(t+1) = v_j(t), \quad j \neq i, \tag{1.1}$$

where sgn(·) is the signum function, that is,

$$\operatorname{sgn}(u) = \begin{cases} 1 & u \geq 0 \\ -1 & u < 0. \end{cases}$$

In this mode, only a single node in any time interval performs computation; the states of the remaining neurons are unchanged.

In the fully parallel mode, the network operates as follows:

$$v_i(t+1) = \operatorname{sgn}\left(\sum_{j=1}^{N} T_{ij} v_j(t) - b_i\right) \qquad (i = 1, 2, \ldots, N), \tag{1.2}$$

which means that the states of all neurons will vary in each time interval.

All the other cases will be called the partially parallel mode. In this mode, the next states of the neurons are calculated by

$$v_i(t+1) = \operatorname{sgn}\left(\sum_{j=1}^{N} T_{ij} v_j(t) - b_i\right), \qquad i = i_1, i_2, \ldots, i_k, \quad \text{and}$$
$$v_j(t+1) = v_j(t), \qquad j \neq i_1, i_2, \ldots, i_k. \tag{1.3}$$

We have the following theorem concerning the stability of the neural network operating in the above modes [2, 3, 4]:

Theorem 1.1 (1) *If the neural network is operating in a serial mode, and if the connection strength matrix $\boldsymbol{T}$ is symmetric and its diagonal elements are nonnegative, then the network will always converge to a stable state.*

(2) *If the neural network is operating in a fully parallel mode and the connection strength matrix $\boldsymbol{T}$ is symmetric, then the network will always converge to a stable state or to a cycle of length* 2.

(3) *If the neural network is operating in a fully parallel mode and the connection strength matrix $\boldsymbol{T}$ is an orthogonal projection matrix, then the network will always converge to a stable state.*

(4) *If the neural network is operating in a fully parallel mode, and if the connection strength matrix $\boldsymbol{T}$ is antisymmetric and its diagonal elements are all zeros, then the network will always converge to a cycle of length* 4.

The main idea in the proof of the four parts of the theorem is first to define an energy function and then to show that this energy function is bounded from below and is nonincreasing when the state of the network changes [2].

In the following, we will present the proof of the first part of the theorem; the proof of the other parts can be given in the same way according to the above idea.

Proof: An energy function is defined as

$$E(t) = -\frac{1}{2}\boldsymbol{V}^T(t)\boldsymbol{T}\boldsymbol{V}(t) + \boldsymbol{B}^T\boldsymbol{V}(t). \tag{1.4}$$

Let $\Delta E = E(t+1) - E(t)$ be the difference in the energy function associated with two consecutive states, and let Δv_i be the difference between the next state and the current state of the i'th neuron at time t. Clearly,

$$\Delta E = -\Delta v_i \frac{1}{2}\left(\sum_{j=1}^{N} T_{ij} v_j(t) + \sum_{k=1}^{N} T_{ki} v_k(t)\right) - T_{ii}(\Delta v_i)^2 + \Delta v_i b_i. \tag{1.5}$$

Because the connection strength matrix $\boldsymbol{T}$ is symmetric ($T_{ij} = T_{ji}$), Equation (1.5) can be written as

$$\Delta E = -\Delta v_i \left[\sum_{j=1}^{N} T_{ij} v_j(t) - b_i \right] - T_{ii}(\Delta v_i)^2. \tag{1.6}$$

Note that the term in brackets is exactly the argument of the signum function in (1.1); therefore, the sign of Δv_i and the sign of $\sum_{j=1}^{N} T_{ij} v_j(t) - b_i$ are the same. Together with the assumption that T_{ii} is nonnegative, it follows that $\Delta E \leq 0$ at any time t (i.e., $E(t)$ is nonincreasing).

Since $v_i(t)$ has only two states, the energy function $E(t)$ is bounded from below.

The fact that $E(t)$ is nonincreasing and is bounded from below guarantees the convergence of $E(t)$.

The second step in the proof is to show that the convergence of the energy function $E(t)$ implies the convergence of the network to a stable state.

From (1.6), we know

(1) if $\Delta v_i = 0$, then it follows that $\Delta E = 0$,
(2) if $\Delta v_i \neq 0$, then $\Delta E = 0$ only if the change in $v_i(t)$ is from -1 to 1 with $\sum_{j=1}^{N} T_{ij} v_j(t) - b_i = 0$.

Hence, once the energy function in the network has converged, the network will reach a stable state after at most N^2 time intervals.

This completes the proof of the first part of Theorem 1.1. QED

Bruck and Goodman [2] described a general result that enables transformation of a neural network whose connection matrix has nonnegative diagonal elements operating in a serial mode to an equivalent network whose connection matrix has zero diagonal elements. Their result also enables transformation of a neural network operating in a fully parallel mode to an equivalent network operating in a serial mode. The equivalence is in the sense that it is possible to derive the state of one network given the state of the other network, provided the two networks started from the same initial state. The transformation is accomplished by use of the parameter set

$$\mathbf{W}' = \begin{pmatrix} \mathbf{O} & \mathbf{W} \\ \mathbf{W} & \mathbf{O} \end{pmatrix}, \qquad \mathbf{B}' = \begin{pmatrix} \mathbf{B} \\ \mathbf{B} \end{pmatrix},$$

where $\mathbf{O}$ is an $N \times N$ zero matrix. This means that the new network has $2N$ neurons.

This result provides an efficient method to analyze the stability and performance of the discrete-time Hopfield neural network operating in different modes.

The main applications of the discrete-time Hopfield neural network are in signal classification and recognition. In these applications, there are three factors describing the network [5, 6, 7]: the number of the stable states, the convergence time (the time to approach stable states from the initial states), and the domain of attraction of the stable states.

1.2 The Continuous-Time Hopfield Neural Network

Hopfield has presented another type of neural network with continuous-time states represented by the following dynamic equations [8, 9, 10]:

$$C_i \frac{du_i(t)}{dt} = \sum_{j=1}^{N} T_{ij} v_j(t) - \frac{u_i(t)}{R_i} + b_i,$$
$$v_i(t) = g_i(u_i(t)) \qquad (i = 1, 2, \ldots, N), \tag{1.7}$$

where N is the number of neurons; $g_i(\cdot)$ is a nonlinear function representing the relationship between the input voltage $u_i(t)$ and output voltage $v_i(t)$ of the i'th neuron; $g_i(\cdot)$ is selected to be bounded, monotonically increasing, and continuous [generally, selected as the sigmoid function $g_i(u) = \frac{1}{2}(1+\tanh(\frac{u}{u_0}))$]; R_i, C_i, and b_i are the input resistor, capacitor, and the bias current of the i'th neuron, respectively; and T_{ij} is the connection strength between the i'th neuron and the j'th neuron. The continuous-time Hopfield neural network is shown in Figure 1.2.

In order to analyze the stability of this neural network, we first define an energy function

$$E(t) = -\frac{1}{2} \sum_{i=1}^{N} \sum_{j=1}^{N} T_{ij} v_i(t) v_j(t) - \sum_{i=1}^{N} v_i(t) b_i + \sum_{i=1}^{N} \frac{1}{R_i} \int_0^{v_i(t)} g_i^{-1}(v) dv \tag{1.8}$$

and then obtain the following theorem:

Theorem 1.2 *If the nonlinear function $g_i(\cdot)$ is bounded, monotonically increasing, and continuous, and $T_{ij} = T_{ji}$, then the energy function $E(t)$ is nonincreasing during time evolution and $\frac{dE(t)}{dt} = 0$ if and only if $\frac{dv_i(t)}{dt} = 0$, that is,*

$$\frac{dE(t)}{dt} \le 0, \qquad \frac{dE(t)}{dt} = 0 \quad \leftrightarrow \quad \frac{dv_i(t)}{dt} = 0$$
$$(i = 1, 2, \ldots, N). \tag{1.9}$$

Because the boundedness of the $g_i(\cdot)$ (for $i = 1, 2, \ldots, N$) guarantees that $E(t)$ is bounded from below, Theorem 1.2 shows that the time evolution of the network is a motion in the state space that seeks out the minimum of $E(t)$ and comes to a stop at such a point; that is, the neural network is stable. Moreover, any optimization

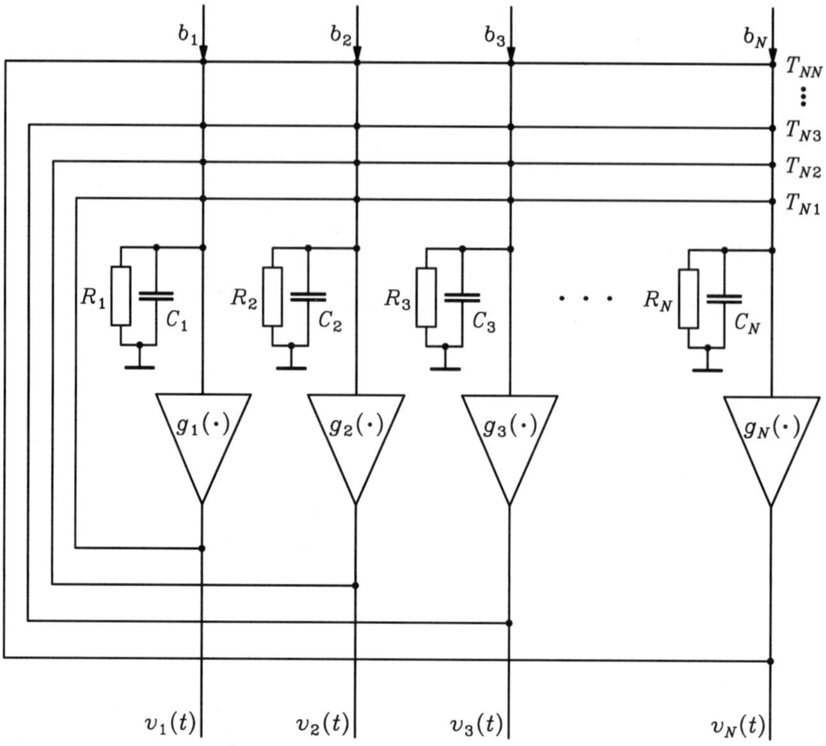

Figure 1.2. The continuous-time Hopfield neural network.

problem that is mapped to the energy function (1.8) can be solved by this neural network.

Proof: We have

$$\frac{dE(t)}{dt} = -\frac{1}{2}\sum_{i=1}^{N}\sum_{j=1}^{N}\left[T_{ij}v_i(t)\frac{dv_j(t)}{dt} + T_{ij}v_j(t)\frac{dv_i(t)}{dt}\right] - \sum_{i=1}^{N} b_i\frac{dv_i(t)}{dt} + \sum_{i=1}^{N}\frac{u_i(t)}{R_i}\frac{dv_i(t)}{dt}. \tag{1.10}$$

Using $T_{ij} = T_{ji}$ gives

$$\frac{dE(t)}{dt} = -\sum_{i=1}^{N}\sum_{j=1}^{N} T_{ij}v_j(t)\frac{dv_i(t)}{dt} - \sum_{i=1}^{N} b_i\frac{dv_i(t)}{dt} + \sum_{i=1}^{N}\frac{u_i(t)}{R_i}\frac{dv_i(t)}{dt}$$

$$= -\sum_{i=1}^{N}\frac{dv_i(t)}{dt}\left[\sum_{j=1}^{N} T_{ij}v_j(t) + b_i - \frac{u_i(t)}{R_i}\right]. \tag{1.11}$$

Substituting (1.7) into (1.11), we obtain

$$\frac{dE(t)}{dt} = -\sum_{i=1}^{N} C_i \frac{dv_i(t)}{dt}\frac{du_i(t)}{dt}$$
$$= -\sum_{i=1}^{N} C_i \left[g^{-1}(v_i(t))\right]' \left[\frac{dv_i(t)}{dt}\right]^2. \tag{1.12}$$

Since $[g^{-1}(v_i(t))]' \geq 0$, each term on the right-hand side of (1.12) is nonnegative and (1.9) holds. QED

Before a physical continuous-time Hopfield neural network is designed, it is necessary to know the dynamic range (i.e., the variable ranges of the input $u_i(t)$) of the network to guarantee that the realized network will satisfy the dynamic equations (1.7). The following theorem gives a quantitative analysis of the dynamic range.

Theorem 1.3 *If the nonlinear function $g_i(\cdot)$ (for $i = 1, 2, \ldots, N$) is bounded and continuous, then $u_i(t)$ (for $i = 1, 2, \ldots, N$) is bounded for $t \geq 0$, and a bound $u_{\max}$ can be computed using the expression*

$$u_{\max} = \max_i \left(|u_i(0)| + R_i|b_i| + R_i v_{i\max} \sum_{j=1}^{N} |T_{ij}| \right), \tag{1.13}$$

where $u_i(0)$ is the initial input voltage of the i'th neuron and $v_{i\max}$ is a bound of the function $g_i(\cdot)$, that is, $|g_i(u_i(t))| \leq v_{i\max}$.

In general, let $g_i(\cdot)$ be the sigmoid function $\frac{1}{2}(1 + \tanh(\frac{u}{u_0}))$ and $u_i(0)$ be zero for all i; hence,

$$u_{\max} = \max_i R_i \left(|b_i| + \sum_{j=1}^{N} |T_{ij}| \right). \tag{1.14}$$

Before the proof of this theorem is given, we prove that $u_i(t)$ can be written in the following integral form:

$$u_i(t) = u_i(0)e^{-\frac{t}{R_iC_i}} + \frac{1}{C_i}\int_0^t e^{-\frac{t-\tau}{R_iC_i}} \left(\sum_{j=1}^{N} T_{ij} v_j(\tau) + b_i \right) d\tau$$
$$(i = 1, 2, \ldots, N). \tag{1.15}$$

The derivatives of both sides of (1.15) with respect to t are

$$\begin{aligned}
\frac{du_i(t)}{dt} &= -\frac{u_i(0)}{R_iC_i}e^{-\frac{t}{R_iC_i}} + \frac{1}{C_i}\left(\sum_{j=1}^{N} T_{ij}v_j(t) + b_i\right) \\
&\quad -\frac{1}{R_iC_i}\frac{1}{C_i}\int_0^t e^{-\frac{t-\tau}{R_iC_i}}\left(\sum_{j=1}^{N} T_{ij}v_j(\tau) + b_i\right)d\tau \\
&= -\frac{1}{R_iC_i}\left[u_i(0)e^{-\frac{t}{R_iC_i}} + \frac{1}{C_i}\int_0^t e^{-\frac{t-\tau}{R_iC_i}}\left(\sum_{j=1}^{N} T_{ij}v_j(\tau) + b_i\right)d\tau\right] \\
&\quad +\frac{1}{C_i}\left(\sum_{j=1}^{N} T_{ij}v_j(\tau) + b_i\right) \\
&= -\frac{u_i(t)}{R_iC_i} + \frac{1}{C_i}\left(\sum_{j=1}^{N} T_{ij}v_j(\tau) + b_i\right) \qquad (i = 1, 2, \ldots, N),
\end{aligned} \tag{1.16}$$

which satisfies (1.7); that is, (1.15) is a solution of (1.7).

Proof: From (1.15), it follows that

$$\begin{aligned}
|u_i(t)| &\leq \left|u_i(0)e^{-\frac{t}{R_iC_i}}\right| + \frac{1}{C_i}\left|\int_0^t e^{-\frac{t-\tau}{R_iC_i}}\left(\sum_{j=1}^{N} T_{ij}v_j(\tau) + b_i\right)d\tau\right| \\
&\leq \left|u_i(0)e^{-\frac{t}{R_iC_i}}\right| + \frac{1}{C_i}\int_0^t e^{-\frac{t-\tau}{R_iC_i}}\left(\sum_{j=1}^{N} |T_{ij}v_j(\tau)| + |b_i|\right)d\tau \\
&\leq \left|u_i(0)e^{-\frac{t}{R_iC_i}}\right| + (h_i + \hat{b}_i)\int_0^t e^{-\frac{t-\tau}{R_iC_i}}d\tau \\
&\leq |u_i(0)| + R_iC_i(h_i + \hat{b}_i) \qquad (i = 1, 2, \ldots, N),
\end{aligned} \tag{1.17}$$

where

$$\hat{b}_i = \frac{|b_i|}{C_i} \tag{1.18}$$

and

$$\begin{aligned}
h_i &= \max_{\tau} \frac{1}{C_i}\left|\sum_{j=1}^{N} T_{ij}v_j(\tau)\right| \\
&\leq \frac{1}{C_i}\sum_{j=1}^{N} |T_{ij}| \max_{\tau} |v_j(\tau)|
\end{aligned} \tag{1.19}$$

$|v_j(\tau)| = |g_i(u_i(\tau))| \leq v_{i\,\max}$ gives

$$h_i \leq \frac{v_{i\,\max}}{C_i} \sum_{j=1}^{N} |T_{ij}| \qquad (i = 1, 2, \ldots, N). \tag{1.20}$$

Substituting (1.18) and (1.20) into (1.17), we have

$$\left|u_i(t)\right| \leq \left|u_i(0)\right| + R_i \left(v_{i\,\max} \sum_{j=1}^{N} |T_{ij}| + |b_i| \right) \qquad (i = 1, 2, \ldots, N). \tag{1.21}$$

Letting

$$u_{\max} = \max_i \left(\left|u_i(0)\right| + R_i \left(v_{i\,\max} \sum_{j=1}^{N} |T_{ij}| + |b_i| \right) \right), \tag{1.22}$$

we have

$$\max_t \left|u_i(t)\right| \leq u_{\max} \qquad (i = 1, 2, \ldots, N). \tag{1.23}$$

QED

This neural network has been widely applied to the solution of optimization problems in signal processing (e.g., in filtering, parameter estimation, and array signal processing). In fact, optimization problems in these applications can generally be mapped to

$$E(t) = -\frac{1}{2} \sum_{i=1}^{N} \sum_{j=1}^{N} T_{ij} v_i(t) v_j(t) - \sum_{i=1}^{N} v_i(t) b_i, \tag{1.24}$$

which is not exactly the same as (1.8). The third term of the energy function (1.8) may be neglected if the neurons have a very high gain or a steplike gain curve (that is, the steepness of $g_i(\cdot)$ approaches infinity). However, for a large but finite steepness of the nonlinear function $g_i(\cdot)$, the effect of the third term of (1.8) on the energy function $E(t)$ should be determined. This allows one to analyze the performance of this neural network in solving optimization problems of signal processing. Hopfield [8] has pointed out that this integral term becomes very large as $v_i(t)$ approaches its bounds ($\pm v_{i\,\max}$) because of the slowness with which $g_i^{-1}(\cdot)$ approaches its asymptotes. To give a quantitative analysis of the third term of (1.8), we will formulate a theorem that is useful for analyzing the performance and designing the hardware implementation of this neural network [11, 12].

For convenience, we split (1.8) into two parts as follows:

$$E(t) = E_1(t) + E_2(t), \tag{1.25}$$

where

$$E_1(t) = -\frac{1}{2}\sum_{i=1}^{N}\sum_{j=1}^{N} T_{ij}v_i(t)v_j(t) - \sum_{i=1}^{N} v_i(t)b_j \tag{1.26}$$

and

$$E_2(t) = \sum_{i=1}^{N} \frac{1}{R_i}\int_0^{v_i(t)} g_i^{-1}(v)dv. \tag{1.27}$$

Obviously, $E_1(t)$ also corresponds to the energy function (1.4) of the discrete-time Hopfield neuron network.

Theorem 1.4 *If the nonlinear function $g_i(\cdot)$ of the i'th neuron (for $i = 1, 2, \ldots, N$) is bounded and continuous, then the integral term $E_2(t)$ of the energy function is bounded for $t \geq 0$ and a bound $E_{2\max}$ can be computed with the expression*

$$E_{2\max} = \sum_{i=1}^{N} \frac{1}{R_i} u_{i\max} g_i(u_{i\max}), \tag{1.28}$$

where

$$u_{i\max} = |u_i(0)| + R_i\left(v_{i\max}\sum_{j=1}^{N}|T_{ij}| + |b_i|\right) \qquad (i = 1, 2, \ldots, N). \tag{1.29}$$

Proof: We have

$$\begin{aligned} |E_2(t)| &\leq \sum_{i=1}^{N} \frac{1}{R_i}\left|\int_0^{v_i(t)} g^{-1}(v)dv\right| \\ &\leq \sum_{i=1}^{N} \frac{1}{R_i}\int_0^{v_i(t)} |g^{-1}(v)|dv. \end{aligned} \tag{1.30}$$

In terms of the boundedness of $u_i(t)$ in Theorem 1.3, we know that

$$\begin{aligned} \left|g^{-1}(v_i(t))\right| &= |u_i(t)| \\ &\leq |u_i(0)| + R_i\left(v_{i\max}\sum_{j=1}^{N}|T_{ij}| + |b_i|\right) \\ &= u_{i\max} \end{aligned} \tag{1.31}$$

and

$$|v_i(t)| \leq g_i(u_{i\max}) \qquad (i = 1, 2, \ldots, N). \tag{1.32}$$

Substituting in (1.31) and (1.32) for (1.30), it follows that

$$\begin{aligned} |E_2(t)| &\le \sum_{i=1}^{N} \frac{1}{R_i} \int_0^{g_i(u_{i\,\max})} |g^{-1}(v)| dv \\ &\le \sum_{i=1}^{N} \frac{1}{R_i} u_{i\,\max} g_i(u_{i\,\max}). \end{aligned} \tag{1.33}$$

Letting

$$E_{2\,\max} = \sum_{i=1}^{N} \frac{1}{R_i} u_{i\,\max} g_i(u_{i\,\max}),$$

then we have

$$\max_t |E_2(t)| \le E_{2\,\max}.$$ QED

Because the values of T_{ij}, b_i, and R_i and the form of $g_i(\cdot)$ are known in analyzing the performance of the network, we can quantitatively determine the maximum effect of the integral term $E_2(t)$ on the energy function $E(t)$ by use of (1.28) and (1.29); subsequently, compute the maximum error between $E_1(t)$ (defined by optimization problems in signal processing) and $E(t)$ (defined by the neural network). In designing the hardware implementation of this neural network, Equations (1.28) and (1.29) are also useful because we can use them to determine the form (the steepness) of the nonlinear function $g_i(\cdot)$ and the other parameters such as R_i to satisfy specifications.

1.3 Cellular Neural Networks

Cellular neural networks (CNN) have been proposed by Chua and Yang and extensively analyzed [13, 14, 15]. This kind of neural network is very powerful in signal processing, particularly in real-time signal processing. A basic cellular neural network has a two-dimensional connection structure whose processing unit, called a cell, is connected only to its neighboring cells. Figure 1.3 shows the schematic of the cellular neural network with 3×3 cells. Each cell contains linear and nonlinear elements, such as linear capacitors, linear resistors, linear and nonlinear controlled sources, and independent sources, as shown in Figure 1.4. The adjacent cells can interact directly with each other; the cells not directly connected together may still affect each other indirectly mainly because of the propagation effects of the continuous-time dynamics of the network. Clearly, the structure of this kind of neural network resembles to that found in cellular automata.

For the cell $C(i, j)$ in Figure 1.4, let the suffices u, x, and y denote the input, state, and output, respectively; then, the node voltage $v_{xij}(t)$ is called the state of

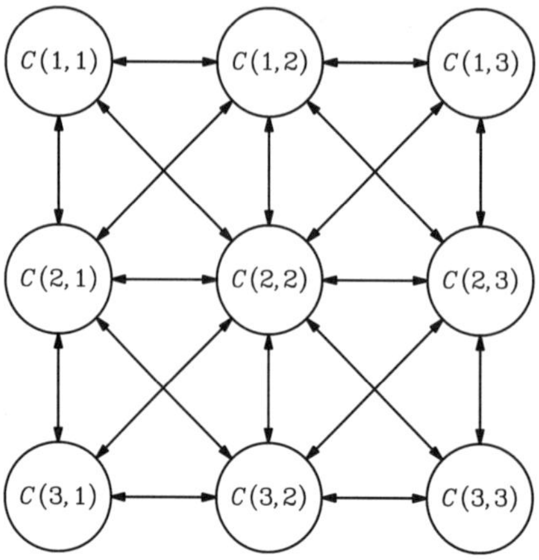

Figure 1.3. A cellular neural network with 3 × 3 cells.

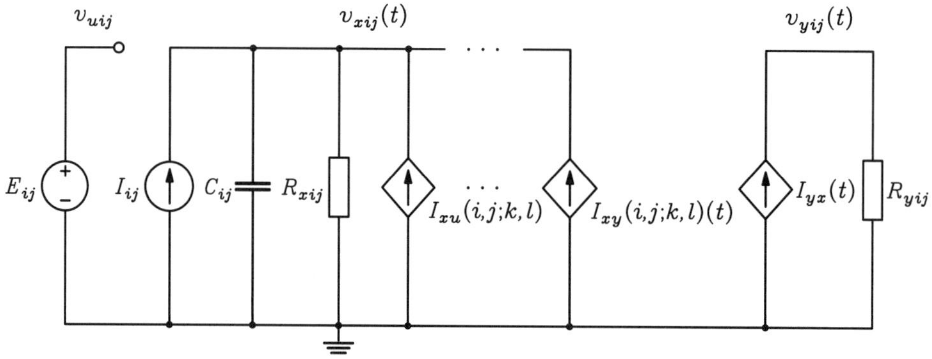

Figure 1.4. A cell of cellular neural networks.

the cell; the node voltage v_{uij} is the input of the cell and is assumed to be a constant; the node voltage $v_{yij}(t)$ is called the output of the cell $C(i, j)$.

Each cell in this network contains one independent voltage source E_{ij}, one independent current source I_{ij}, one linear capacitor C_{ij}, two linear resistors R_{xij} and R_{yij}, and at most $2m$ linear voltage-controlled current sources, which are coupled to its neighboring cells via the controlling input voltage v_{ukl} and the feedback from the output voltage $v_{ukl}(t)$ of each neighboring cell $C(k, l)$, where m is the number of neighboring cells. These two kinds of linear voltage-controlled current sources are characterized by $I_{xy}(i, j; k, l)(t) = A(i, j; k, l)v_{ykl}(t)$ and $I_{xu}(i, j; k, l) = B(i, j; k, l)v_{ukl}$ for all $C(k, l) \in N_r(i, j)$, where $N_r(i, j)$ is called the r-neighborhood of the cell $C(i, j)$ and is defined by

$$N_r(i, j) = \Big\{C(k, l)\big|\max\{|k - i|, |l - j|\} \leq r, 1 \leq k \leq M; 1 \leq l \leq N\Big\}.$$

M and N are the cell numbers arranged in rows and columns, respectively; r is a positive integer number. In general, the only nonlinear element in each cell is a voltage-controlled current source with the characteristic

$$I_{yx}(t) = \frac{f\big(v_{xij}(t)\big)}{R_{yij}}.$$

The function $f(\cdot)$ is selected to be bounded and monotonically increasing (e.g., the sigmoid function or a piecewise-linear function).

In terms of Kirchhoff's laws, we can obtain the dynamic equations for each cell as

$$\begin{aligned} C_{ij}\frac{dv_{xij}(t)}{dt} &= -\frac{v_{xij}(t)}{R_{xij}} + \sum_{C(k,l)\in N_r(i,j)} A(i,j;k,l)v_{ykl}(t) \\ &\quad + \sum_{C(k,l)\in N_r(i,j)} B(i,j;k,l)v_{ukl} + I_{ij} \end{aligned} \tag{1.34}$$

and

$$v_{yij}(t) = f\big(v_{xij}(t)\big) \qquad (i = 1, 2, \ldots, M;\ j = 1, 2, \ldots, N). \tag{1.35}$$

For mathematical convenience, but without loss of generality, we assume for all cells that

(1) $R_{xij} = R_x$, $R_{yij} = R_y$, $C_{ij} = C$, and $I_{ij} = I$;
(2) the absolute value of the initial condition $v_{xij}(0)$ is less than or equal to unity;
(3) the absolute value of the input voltage v_{uij} is less than or equal to unity;
(4) $A(i,j;k,l) = A(k,l;i,j)$;
(5) $f(v) = \frac{1}{2}(|v+1| - |v-1|)$, that is, f is the simplest piecewise-linear function.

Then (1.34) and (1.35) become

$$\begin{aligned} C\frac{dv_{xij}(t)}{dt} &= -\frac{v_{xij}(t)}{R_x} + \sum_{C(k,l)\in N_r(i,j)} \big[A(i,j;k,l)v_{ykl}(t) \\ &\quad + B(i,j;k,l)v_{ukl}\big] + I \end{aligned} \tag{1.36}$$

and

$$\begin{aligned} v_{yij}(t) &= \frac{1}{2}\Big(\big|v_{xij}(t)+1\big| - \big|v_{xij}(t)-1\big|\Big) \\ &\qquad (i = 1, 2, \ldots, M;\ j = 1, 2, \ldots, N). \end{aligned} \tag{1.37}$$

A cellular neural network is completely characterized by the nonlinear differential equations (1.36) and (1.37).

For the purpose of analyzing the stability, we first define an energy function $E(t)$ as

$$E(t) = -\frac{1}{2}\sum_{(i,j)}\sum_{(k,l)} A(i,j;k,l)v_{yij}(t)v_{ykl}(t) + \frac{1}{2R_x}\sum_{(i,j)}\left(v_{yij}(t)\right)^2$$
$$-\sum_{(i,j)}\sum_{(k,l)} B(i,j;k,l)v_{yij}(t)v_{ukl} - \sum_{(i,j)} v_{yij}(t)I. \tag{1.38}$$

Then, we have a theorem similar to Theorem 1.2.

Theorem 1.5 *The energy function $E(t)$ of (1.38) is nonincreasing during time evolution, and*

$$\lim_{t\to\infty}\frac{dE(t)}{dt} = 0, \tag{1.39}$$

$$\lim_{t\to\infty}\frac{dv_{yij}(t)}{dt} = 0 \qquad (i = 1, 2, \ldots, M; j = 1, 2, \ldots, N). \tag{1.40}$$

This theorem means that the cellular neural network is stable. That is, after the transient has settled down, the cellular network always approaches one of the stable equilibrium points and seeks out the minimum of $E(t)$. Moreover, any optimization problem that may be mapped to the energy function (1.38) can be solved by this cellular neural network.

Proof: First, we prove that $E(t)$ is nonincreasing during time evolution, that is, $\frac{dE(t)}{dt} \le 0$ for $t \ge 0$.

It follows that

$$\frac{dE(t)}{dt} = -\sum_{(i,j)}\sum_{(k,l)} A(i,j;k,l)\frac{dv_{yij}(t)}{dv_{xij}(t)}\frac{dv_{xij}(t)}{dt}v_{ykl}(t)$$
$$+\frac{1}{R_x}\sum_{(i,j)}\frac{dv_{yij}(t)}{dv_{xij}(t)}\frac{dv_{xij}(t)}{dt}v_{yij}(t)$$
$$-\sum_{(i,j)}\sum_{(k,l)} B(i,j;k,l)\frac{dv_{yij}(t)}{dv_{xij}(t)}\frac{dv_{xij}(t)}{dt}v_{ukl}$$
$$-\sum_{(i,j)} I\frac{dv_{yij}(t)}{dv_{xij}(t)}\frac{dv_{xij}(t)}{dt}. \tag{1.41}$$

Here we have used the symmetry assumption $A(i,j;k,l) = A(k,l;i,j)$.

According to (1.37), we have

$$\frac{dv_{yij}(t)}{dv_{xij}(t)} = \begin{cases} 1 & |v_{xij}(t)| < 1 \\ 0 & |v_{xij}(t)| \ge 1 \end{cases} \tag{1.42}$$

and

$$v_{yij}(t) = v_{xij}(t), \quad |v_{xij}(t)| < 1 \quad (i = 1, 2, \ldots, M; j = 1, 2, \ldots, N). \tag{1.43}$$

In addition, we have

$$A(i, j; k, l) = 0, \quad B(i, j; k, l) = 0 \quad (\text{for} \quad C(k, l) \notin N_r(i, j)). \tag{1.44}$$

Substituting (1.42)–(1.44) into (1.41) yields

$$\frac{dE(t)}{dt} = -\sum_{(i,j)} \frac{dv_{yij}(t)}{dv_{xij}(t)} \frac{dv_{xij}(t)}{dt} \Bigg[\sum_{C(k,l)\in N_r(i,j)} A(i, j; k, l) v_{ykl}(t) - \frac{1}{R_x} v_{xij}(t) + \sum_{C(k,l)\in N_r(i,j)} B(i, j; k, l) v_{ukl} + I \Bigg]. \tag{1.45}$$

Using (1.36), Equation (1.45) becomes

$$\frac{dE(t)}{dt} = -\sum_{(i,j)} C \frac{dv_{yij}(t)}{dv_{xij}(t)} \left(\frac{dv_{xij}(t)}{dt} \right)^2. \tag{1.46}$$

Because (1.42) shows

$$\frac{dv_{yij}(t)}{dv_{xij}(t)} \geq 0$$

(a sufficient condition for this inequality is that $f(\cdot)$ is a monotonically increasing function), we have $\frac{dE(t)}{dt} \leq 0$ for $t \geq 0$.

Moreover, the boundedness of the nonlinear function $f(\cdot)$ (that is, $v_{yij}(t)$ is bounded for all i and j) guarantees that the energy function $E(t)$ is bounded from below. As a result, we have

$$\lim_{t\to\infty} \frac{dE(t)}{dt} = 0. \tag{1.47}$$

Because each term on the right-hand side of (1.46) is nonnegative, (1.47) means

$$\lim_{t\to\infty} \frac{dv_{yij}(t)}{dt} = 0 \quad (i = 1, 2, \ldots, M; j = 1, 2, \ldots, N). \tag{1.48}$$

QED

Chua and Yang have also proved that if $A(i, j; i, j) > 1/R_x$, then

$$\lim_{t\to\infty} |v_{xij}(t)| \geq 1, \tag{1.49}$$

or equivalently,

$$\lim_{t\to\infty} |v_{yij}(t)| = \pm 1 \qquad (i = 1, 2, \ldots, M; j = 1, 2, \ldots, N), \tag{1.50}$$

which means that the stationary outputs of the cellular neural network are binary values. This property proves to be very useful in signal classification problems.

Concerning the dynamic ranges of each cell in a cellular neural network, we have the following theorem, which is similar to Theorem 1.3.

Theorem 1.6 *All states $v_{xij}(t)$ in a cellular neural network are bounded and a bound $v_{\max}$ can be computed using the following expression:*

$$v_{\max} = 1 + R_x|I| + R_x \times \max_{1\le i\le M, 1\le j\le N} \left(\sum_{C(k,l)\in N_r(i,j)} \left(|A(i, j; k, l)| + |B(i, j; k, l)| \right) \right). \tag{1.51}$$

The proof of this theorem can be given in the same way as for Theorem 1.3.

It is worth noting that if the differential equations (1.34) and (1.35) are transformed into difference equations (that is, approximated in discrete time) by some methods such as the Euler method or Runge-Kutta method, we can use the difference equations to construct corresponding digital cellular neural networks. Applications and a VLSI design of digital cellular neural networks are reported in [16].

1.4 Multilayer Perceptron Networks

Multilayer perceptron (MLP) networks comprise a large class of feedforward neural networks with one or more layers of neurons, called hidden neurons, between the input and output neurons. In general, all neurons in a layer are connected to all neurons in the adjacent layers through unidirectional links. These links are represented by connection weights.

Figure 1.5 shows the architectural graph of a multilayer perceptron network with one hidden layer. The first layer is the input layer, which simply feeds input signals to the second layer (hidden layer) without modification. The last layer is the output layer, where the response of the network comes from.

Let $W_{ij}^{(k-1)}$ denote the connection weight between the i'th neuron in the $(k-1)$th layer and the j'th neuron in the k'th layer; let $y_j^{(k)}$, $f_j^{(k)}(\cdot)$, and $\theta_j^{(k)}$ be the output, activation function, and threshold of the j'th neuron in the k'th layer, respectively; N_k is the number of the neurons in the k'th layer; M is the number of the layers including the input layer and the output layer. Then we have the following input–

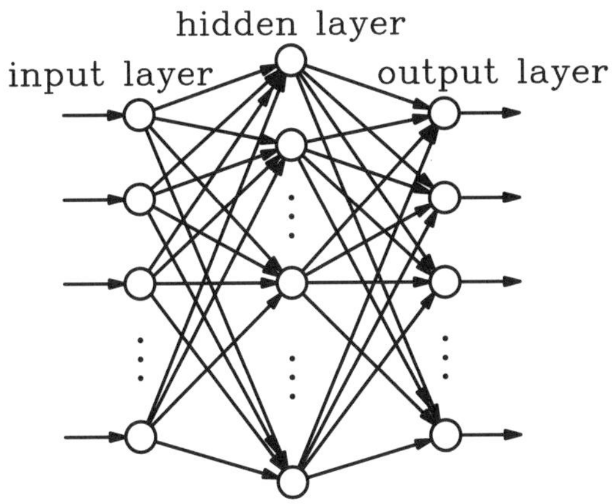

Figure 1.5. An MLP network with one hidden layer.

output relationship for each neuron:

$$y_j^{(k)} = f_j^{(k)}\left(\sum_{i=1}^{N_{k-1}} W_{ij}^{(k-1)} y_i^{(k-1)} - \theta_j^{(k)}\right)$$

$$(j = 1, 2, \ldots, N_k; k = 1, 2, \ldots, M). \qquad (1.52)$$

Clearly, the inputs of the network are denoted by $y_1^{(0)}, y_2^{(0)}, \ldots, y_{N_0}^{(0)}$; the outputs of the network are $y_1^{(M)}, y_2^{(M)}, \ldots, y_{N_M}^{(M)}$; N_0 and N_M are the numbers of the neurons in the input layer and output layer, respectively.

The key function of MLP networks is the implementation of a nonlinear input–output mapping of a general nature. The following two theorems guarantee this mapping property [17, 18].

Theorem 1.7 *Given any $\epsilon > 0$ and any L_2-function $f : [0, 1]^{N_0} \to \boldsymbol{R}^{N_M}$, there exists a three-layer perceptron network that can approximate f to within ϵ mean squared error accuracy.*

Theorem 1.8 *Any continuous function $f : [0, 1]^{N_0} \to \boldsymbol{R}^{N_M}$ can be implemented exactly by a three-layer perceptron network having $2N_0 + 1$ neurons in the hidden layer.*

The proof of these two theorems are in [17, 18, 19, 20].

It should be noted that, although these two theorems show that three layers are always enough, in practical cases one often uses four, five, or even more layers [17]. This is because for many problems an approximation with three layers would require an impractically large number of hidden neurons, whereas an adequate solution can be obtained with a practical network size using more than three layers.

Moreover, the implementation of the network with more than three layers is more flexible than that with only three layers.

It is this mapping capability of the MLP network that lays a mathematical foundation for its application to a wide variety of signal processing problems such as nonlinear filtering and signal compression.

With these two theorems, we can implement the desired nonlinear mapping by selecting suitable connection weights, thresholds, and activation functions. However, in many signal processing applications, the desired nonlinear mapping is not available but there exist the input–output sample sets as $\{\boldsymbol{Y}_1^{(0)}, \hat{\boldsymbol{Y}}_1^{(M)}\}$, $\{\boldsymbol{Y}_2^{(0)}, \hat{\boldsymbol{Y}}_2^{(M)}\}$, ..., $\{\boldsymbol{Y}_I^{(0)}, \hat{\boldsymbol{Y}}_I^{(M)}\}$, where $\{\boldsymbol{Y}_i^{(0)}, \hat{\boldsymbol{Y}}_i^{(M)}\}$ is the i'th input–output sample set,

$$\boldsymbol{Y}_i^{(0)} = [y_{i,1}^{(0)}, y_{i,2}^{(0)}, \ldots, y_{i,N_0}^{(0)}]^T$$

is the i'th input vector, and

$$\hat{\boldsymbol{Y}}_i^{(M)} = [\hat{y}_{i,1}^{(M)}, \hat{y}_{i,2}^{(M)}, \ldots, \hat{y}_{i,N_M}^{(M)}]^T$$

is the i'th output vector (for $i = 1, 2, \ldots, I$, where I is the number of available sample sets). In this case, the connection weights, thresholds, and activation functions can be obtained by solving the optimization problem

$$\min \sum_{i=1}^{I} \|\boldsymbol{Y}_i^{(M)} - \hat{\boldsymbol{Y}}_i^{(M)}\|^2, \tag{1.53}$$

where $\boldsymbol{Y}_i^{(M)} = [y_{i,1}^{(M)}, y_{i,2}^{(M)}, \ldots, y_{i,N_M}^{(M)}]^T$ is the output vector of the network computed using (1.52) with $\boldsymbol{Y}_i^{(0)}$ as the input vector.

There have been many methods proposed to solve this optimization problem. The back-propagation (BP) learning algorithm is the simplest and most common one [21, 22]. In the BP algorithm, all the activation functions and thresholds are fixed a priori and the updating of the connection weights is summarized as follows:

(1) Initialize randomly all connection weights.

(2) Compute by proceeding forward the output vector $\boldsymbol{Y}_n^{(k)}$ of each layer (for $k = 1, 2, \ldots, M$) using (1.52) and the available input $\boldsymbol{Y}_n^{(0)} = [y_{n,1}^{(0)}, y_{n,2}^{(0)}, \ldots, y_{n,N_0}^{(0)}]^T$ (for $n = 1, 2, \ldots, I$).

(3) Compute by proceeding backward the error propagation terms $\delta_{nj}^{(k)}$ (for $k = 1, 2, \ldots, M$; $n = 1, 2, \ldots, I$ and $j = 1, 2, \ldots, N_k$) as

$$\delta_{nj}^{(M)} = (\hat{y}_{n,j}^{(M)} - y_{n,j}^{(M)})(f_j^{(M)}(x_{nj}^{(M)}))' \tag{1.54}$$

for the output layer ($k = M$) and

$$\delta_{nj}^{(k)} = (f_j^{(k)}(x_{nj}^{(k)}))' \sum_{l=1}^{N_{k+1}} \delta_{nl}^{(k+1)} W_{jl}^{(k)}(t) \tag{1.55}$$

for the other layers ($k < M$), where

$$(f_j^{(k)}(x_{nj}^{(k)}))' = \left. \frac{\partial f_j^{(k)}(u)}{\partial u} \right|_{u=x_{nj}^{(k)}} \tag{1.56}$$

and

$$x_{nj}^{(k)} = \sum_{i=1}^{N_{k-1}} W_{ij}^{(k-1)}(t) y_i^{(k-1)} - \theta_j^{(k)}. \tag{1.57}$$

Note that $W_{jl}^{(k)}(t)$ and $W_{ij}^{(k-1)}(t)$ stand for the related weights at the t'th iteration.

(4) Adjust the connection weights according to

$$W_{ij}^{(k-1)}(t+1) = W_{ij}^{(k-1)}(t) + \gamma \sum_{n=1}^{I} \delta_{nj}^{(k)} y_{ni}^{(k-1)}, \tag{1.58}$$

where γ is the learning-rate parameter.

(5) Compute the total error ϵ as

$$\epsilon = \sum_{n=1}^{I} \|\boldsymbol{Y}_n^{(M)} - \hat{\boldsymbol{Y}}_n^{(M)}\|^2 = \sum_{n=1}^{I} \sum_{j=1}^{N_M} (y_{n,j}^{(M)} - \hat{y}_{n,j}^{(M)})^2 \tag{1.59}$$

and iterate the computation by returning to Step (2) until this error is less than the specified one.

This is a kind of batch-processing; that is, all I sample sets are available when the training of the connection weights is initiated. In some cases, each new sample set becomes available during the training. With this condition, (1.58) and (1.59) of the BP algorithm become, respectively,

$$W_{ij}^{(k-1)}(n+1) = W_{ij}^{(k-1)}(n) + \gamma \delta_{nj}^{(k)} y_{ni}^{(k-1)} \tag{1.60}$$

and

$$\epsilon = \|\boldsymbol{Y}_n^{(M)} - \hat{\boldsymbol{Y}}_n^{(M)}\|^2 = \sum_{j=1}^{N_M} (\hat{y}_{n,j}^{(M)} - y_{n,j}^{(M)})^2. \tag{1.61}$$

$W_{ij}^{(k-1)}(n+1)$ indicates the connection weight to be computed by using the n'th sample set; consequently, $W_{ij}^{k-1}(n)$ indicates that obtained by using the $(n-1)$th sample set.

Because the BP algorithm is based on the gradient descent method, it suffers from the following drawbacks: slow convergence speed and a local minimum problem. However, many methods have been proposed [23, 24, 25, 26] to make the BP algorithm perform better. These improvements mainly include

(1) initializing the connection weights of the network so that they are uniformly distributed inside a small range;
(2) varying the learning-rate parameter during the training and assigning a smaller value in the last layers than in the front-end layers; and
(3) using prior information about the desired nonlinear mapping as much as possible.

In addition, there are different ways to improve the performance of the BP algorithm for different applications. We will deal with these in more detail in the next few chapters.

1.5 Self-Organizing Systems

The MLP network and the corresponding BP algorithm are focused on the external and supervised adjustment of the system parameters (such as the connection weights and thresholds). In many fields of signal processing such as adaptive signal classification and blind identification, however, an unsupervised learning or a self-organized learning is desired mainly because the supervised information (or say, the teacher) is unavailable. The neighboring units in self-organizing systems compete in their activities by means of mutual lateral interactions, and they develop adaptively into specific detectors of different signal patterns, which are similar to those encountered in the brain [27, 28, 29, 30, 31]. This property of self-organizing systems can be used to discover and extract features from the input data without the supervised information. The structures of self-organizing systems may take on a variety of different forms. As an example, a simple two-dimensional self-organizing system is briefly introduced in this chapter. For a more detailed discussion see Chapters 6 and 9.

A simple two-dimensional self-organizing system is depicted in Figure 1.6. The neurons can be arranged in any planar configuration (e.g., in a rectangular or hexagonal lattice).

Let $\boldsymbol{X}(n) = [x_1(n), x_2(n), \ldots, x_N(n)]^T$ be the input vector and be connected in parallel to all neurons in the network. The connection weight vector between the input vector and the i'th neuron is denoted by $\boldsymbol{W}_i(n) = [W_{i1}(n), W_{i2}(n), \ldots, W_{iN}(n)]^T$.

For the purpose of making the weight vectors in the stable state approximate the input vectors in an orderly fashion, the following steps can be used to update the weight vectors.

(1) Initialize randomly all connection weight vectors and select the size (width

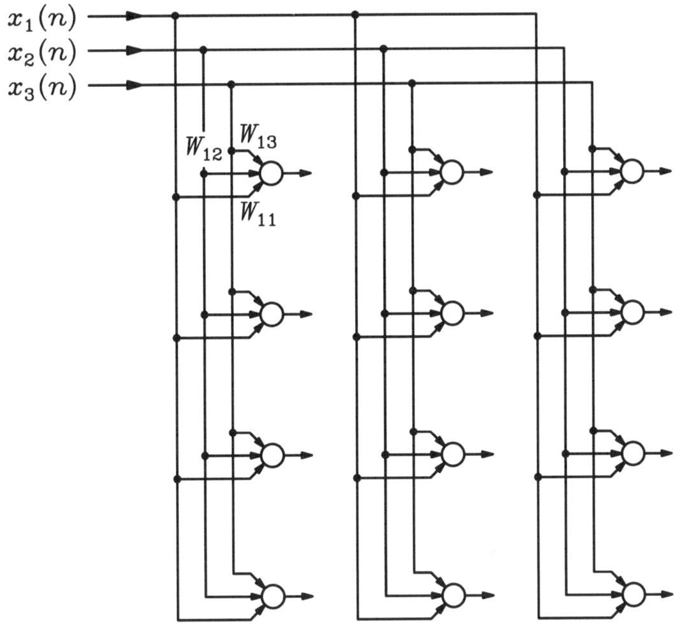

Figure 1.6. A two-dimensional self-organizing system.

or radius) of the neighborhood set N_c (which is defined by all the neurons neighboring a so-called central neuron).

The size N_c can also vary in time. Experimentally, it turns out to be advantageous to let N_c be very wide in the beginning and to shrink it monotonically with time.

(2) Compute the distance d_i ($i \in N_c$) from

$$d_i = \|\boldsymbol{X}(n) - \mathbf{W}_i(n)\|. \tag{1.62}$$

The distance is a measure for the match of $\boldsymbol{X}(n)$ with the weight vector $\mathbf{W}_i(n)$, which may take another form of measure such as the inner product of two vectors.

(3) Find the central neuron i^* that satisfies

$$\|\boldsymbol{X}(n) - \mathbf{W}_{i^*}(n)\| = \min_i \|\boldsymbol{X}(n) - \mathbf{W}_i(n)\|. \tag{1.63}$$

The central neuron i^* is also called the winning neuron whose weight vector is the best match with the input vector $\boldsymbol{X}(n)$.

(4) Adjust the connection weights by

$$\mathbf{W}_i(n+1) = \begin{cases} \mathbf{W}_i(n) + \gamma(n)\big(\boldsymbol{X}(n) - \mathbf{W}_i(n)\big), & i \in N_c, \\ \mathbf{W}_i(n), & i \notin N_c. \end{cases} \tag{1.64}$$

$\gamma(n)$ is an adaptive parameter that should satisfy $0 < \gamma(n) < 1, \gamma(n) \to 0$, $\sum_n \gamma^p(n) < \infty$ for some p, and $\sum_n \gamma(n) = \infty$.

An alternative notation of the above updating is

$$\boldsymbol{W}_i(n+1) = \boldsymbol{W}_i(n) + h_i(n)\big(\boldsymbol{X}(n) - \boldsymbol{W}_i(n)\big). \tag{1.65}$$

Clearly,

$$h_i(n) = \begin{cases} \gamma(n), & i \in N_c, \\ h_i(n) = 0, & i \notin N_c. \end{cases} \tag{1.66}$$

The advantage of this alternative notation is that we can define $h_i(n)$ in a more general way than by (1.66) [30].

(5) Iterate the computation by presenting a new input vector and returning to Step (2) until the weight vectors stabilize their values.

It has been shown by simulation results that the point density function of the weight vectors tends to approximate the probability density function (PDF) of the input vectors and the weight vectors tend to be ordered according to their mutual similarity. In other words, each neuron in this system becomes maximally sensitized to a particular input vector, but for different neurons this sensitization occurs in an orderly fashion corresponding to the PDF of the input vectors and the distance measure.

This property is the reason that the above self-organizing system finds a wide use in signal processing as will be seen in the following chapters of this book.

1.6 Radial Basis Function Networks

An alternative network to the MLP network for many applications of signal processing is the radial basis function (RBF) network, which has been proposed by different authors [32, 33, 34]. An RBF is a multidimensional function that depends on the distance between the input vector and a center vector. RBFs provide a powerful tool for multidimensional approximation or fitting that essentially does not suffer from the problem of proliferation of the adjustable parameters as the dimensionality of the problem increases [33].

Figure 1.7 shows the basic structure of the RBF network. The input layer has neurons with a linear function that simply feeds the input signals to the hidden layer. Moreover, the connection between the input layer and the hidden layer are not weighted, that is, each hidden neuron receives each corresponding input value unaltered. The hidden neurons are processing units that perform the radial basis function. In contrast to the MLP network, the RBF network usually has only one hidden layer. The transfer function of the hidden neurons in the RBF network can be, for example, the

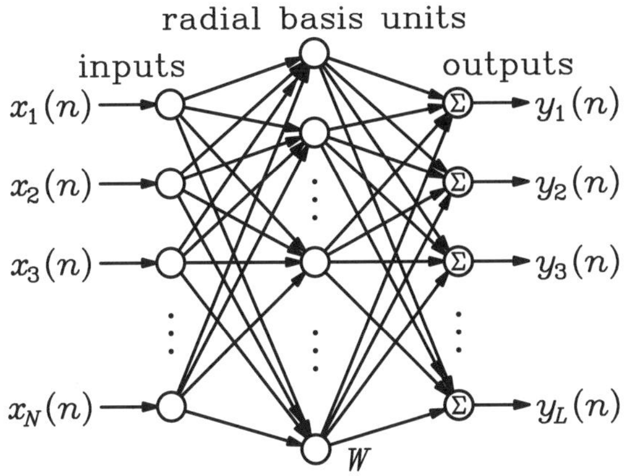

Figure 1.7. Schematic of a radial basis function network.

(1) Gaussian function: $f(r) = e^{-\frac{r^2}{\sigma^2}}$,
(2) multiquadratic function: $f(r) = (r^2 + \sigma^2)^{\frac{1}{2}}$,
(3) inverse multiquadratic function: $f(r) = (r^2 + \sigma^2)^{-\frac{1}{2}}$,
(4) thin-plate-spline function : $f(r) = r^2 \log(r)$,
(5) piece-wise linear function: $f(r) = \frac{1}{2}(|r + 1| - |r - 1|)$, or the
(6) cubic approximation function: $f(r) = \frac{1}{2}(|r^3 + 1| - |r^3 - 1|)$,

where σ is a real parameter (called a scaling parameter) and r is the distance between the input vector and the center vector. The distance is usually measured by the Euclidean norm.

Let the input vector at time n be denoted by $\boldsymbol{X}(n) = [x_1(n), x_2(n), \ldots, x_N(n)]^T$, and let the center vector of each hidden neuron be denoted by $\boldsymbol{C}_i$ (for $i = 1, 2, \ldots, H$) (H is the neuron number of the hidden layer). Then the output of each neuron in the hidden layer is

$$h_i(n) = f_i(\|\boldsymbol{X}(n) - \boldsymbol{C}_i\|) \qquad (i = 1, 2, \ldots, H). \tag{1.67}$$

The connections between the hidden layer and the output layer are weighted. Each neuron of the output layer has a linear input–output relationship so that they perform simple summations; that is, the output of the i'th neuron in the output layer at time n is

$$y_i(n) = \sum_{j=1}^{H} W_{ij} h_j(n) = \sum_{j=1}^{H} W_{ij} f_j\left(\|\boldsymbol{X}(n) - \boldsymbol{C}_j\|\right)$$

$$(i = 1, 2, \ldots, L), \tag{1.68}$$

where L is the neuron number of the output layer and W_{ij} is the connection weight between the j'th neuron in the hidden layer and i'th neuron in the output layer.

It has been shown experimentally that if a sufficient number of hidden neurons is used and the center vectors are suitably distributed in the input domain, then the RBF network is able to approximate a wide class of nonlinear multidimensional functions. Moreover, the choice of the nonlinearity of the RBF is not crucial for the approximation performance of the network. For example, this holds for the nonlinearity of the thin-plate-spline function, $f(r) \to \infty$ as $r \to \infty$, and for the nonlinearity of the Gaussian function, $f(r) = 0$ as $r \to \infty$. Although these two nonlinearities have quite different properties, both the resulting RBF networks have good approximation capabilities. However, the approximation performance of an RBF network critically depends on the choice of the centers [35, 36, 37].

From Figure 1.7 and Equations (1.67) and (1.68), we know that in general an RBF network is specified by two sets of parameters: the connection weights and the center vectors. These parameters can in principle be determined from the available sample vectors (training data) by solving an optimization problem similar to (1.53), that is,

$$E = \sum_{n=1}^{M} \left\| \boldsymbol{Y}(n) - \hat{\boldsymbol{Y}}(n) \right\|^2, \tag{1.69}$$

where M is the number of the available sample vectors,

$$\boldsymbol{Y}(n) = \left[y_1(n), y_2(n), \ldots, y_L(n)\right]^T$$

is the output vector computed from the sample input vector by using (1.67) and (1.68), and $\hat{\boldsymbol{Y}}(n)$ is the corresponding desired output vector. Clearly, the gradient descent procedure may be used to solve this optimization problem. However, in practice, an alternative approach is used to find the parameters of an RBF network. This approach essentially includes the following two steps:

(1) determining the center vectors $\boldsymbol{C}_i$ (for $i = 1, 2, \ldots, H$) and
(2) determining the connection weights.

For the center vectors the simplest technique involves choosing these vectors randomly from a subset of the available sample vectors. However, in such a case the number of hidden neurons needs to be relatively large to cover the entire input domain. An improved approach is to apply the so-called k-means clustering algorithm [37]. The basic idea of this algorithm is to distribute the center vectors according to the natural measure of the attractor (i.e., if the density of the data points is high, so is the density of the centers). The k-means clustering algorithm finds a set of cluster centers and partitions the training samples into subsets. Each cluster center is associated with one of the H hidden neurons in the RBF network. The data are partitioned in such a way that the training points are assigned to the cluster with the nearest center. The cluster center corresponds to one of the minima of the following optimization problem:

$$E_{k\text{-means}} = \sum_{j=1}^{H} \sum_{n=1}^{M} B_{jn} \left\| \boldsymbol{X}(n) - \boldsymbol{C}_j \right\|^2, \tag{1.70}$$

where B_{jn} is the cluster partition or membership function forming an $H \times M$ matrix. Each column represents an available sample vector and each row represents a cluster. Each column has a single "1" in the row corresponding to the cluster nearest to that training point and zeros elsewhere.

The center of each cluster is initialized to a different randomly chosen training point. Then each training example is assigned to the neuron nearest to it. When all training points have been assigned, the average position of the training points for each cluster is found and the cluster center is moved to that point. Once all the clusters have been updated, the procedure is repeated until it converges. The cluster centers become the desired centers of the hidden neurons.

It should be noted that for some RBF functions such as the Gaussian function it is necessary to determine the scaling parameter σ before determining the connection weights. The goal in setting this parameter is to cover the training points to allow a smooth fit of the desired network outputs, which means that any point within the convex hull of the neuron centers must significantly activate more than one neuron. To achieve this goal, each hidden neuron must activate at least one other hidden neuron to a significant degree. An appropriate method to determine the scaling parameter σ is based on the P-nearest neighbor heuristic, that is,

$$\sigma_i = \frac{1}{P} \sum_{j=1}^{P} \| \boldsymbol{C}_j - \boldsymbol{C}_i \|^2 \qquad (i = 1, 2, \ldots, H), \tag{1.71}$$

where the $\boldsymbol{C}_j$ (for $j = 1, 2, \ldots, P$) are the P-nearest neighbors of $\boldsymbol{C}_i$.

We now turn to determining the connection weights.

Because an RBF network has only one layer of weighted connections and the output neurons are simple summation units, Equation (1.69) will become a linear least-squares problem once the center vectors and the scaling parameter have been determined. That is,

$$\min_{\boldsymbol{W}} \sum_{n=1}^{M} \| \boldsymbol{Y}(n) - \hat{\boldsymbol{Y}}(n) \|^2 = \min_{\boldsymbol{W}} \| \boldsymbol{W}\boldsymbol{F} - \hat{\boldsymbol{Y}} \|^2, \tag{1.72}$$

where $\boldsymbol{W} = \{W_{ij}\}$ is the $L \times H$ matrix of the connection weights, $\boldsymbol{F}$ is an $H \times M$ matrix consisting of the outputs of the hidden neurons and whose elements are computed with

$$F_{in} = f_i\left(\left\| \boldsymbol{X}(n) - \boldsymbol{C}_i \right\| \right) \qquad (i = 1, 2, \ldots, H; n = 1, 2, \ldots, M), \tag{1.73}$$

and $\hat{Y} = [\hat{Y}(1), \hat{Y}(2), \ldots, \hat{Y}(M)]$ is the $L \times M$ matrix of the desired outputs. We can find the connection weight matrix W from (1.72) in an explicit form as

$$W = \hat{Y}F^{+} = \hat{Y} \lim_{\alpha \to 0} F^{T}(FF^{T} + \alpha I)^{-1}, \tag{1.74}$$

where F^{+} is the pseudoinverse of F.

This is a batch-processing version to determine the connection weights of the RBF network, and it is applied in the case where all sample sets are available at one time. In some cases, each new sample set becomes available recursively. For these cases, the following recursive procedure can be used to determine the right connection weights:

(1) Initialize randomly all connection weights.
(2) Compute the output vector $Y(n)$ by use of (1.68).
(3) Compute the error term $e_i(n)$ of each output neuron

$$e_i(n) = y_i(n) - \hat{y}_i(n) \qquad (i = 1, 2, \ldots, L). \tag{1.75}$$

(4) Adjust the connection weights according to

$$W_{ij}(n+1) = W_{ij}(n) + \gamma e_i(n) f_j\left(\|X(n) - C_i\|\right) \qquad (i = 1, 2, \ldots, L;\ j = 1, 2, \ldots, H), \tag{1.76}$$

where γ is the learning-rate parameter.

(5) Compute the total error

$$\epsilon = \|Y(n) - \hat{Y}(n)\|^2 \tag{1.77}$$

and iterate the computation by returning to Step (2) until this error is less than the specified one.

This recursive procedure is derived from (1.72) by use of the gradient descent method.

Comparisons of the RBF network and the MLP network are given in [23] and [38].

1.7 High-Order Neural Networks

High-order neuron networks are another class of mapping networks [40, 41, 42, 43]. A basic high-order neuron network consists of three layers, called the input layer, the high-order layer, and the output layer, respectively, as shown in Figure 1.8. Similar to the input layer of the MLP network, the input layer of the high-order neural network simply feeds input signals to the second layer (the high-order layer) without modification. The units of the high-order layer are in effect multipliers whose outputs are computed as follows:

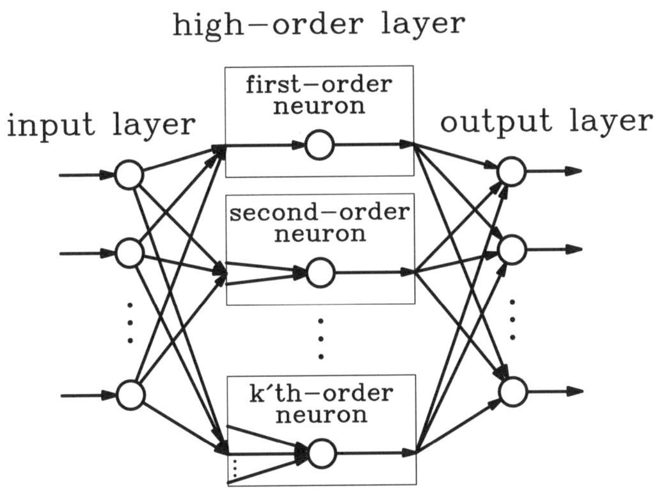

Figure 1.8. Schematic of a high-order neuron network.

(1) The outputs of the first-order neurons are

$$q_i^{(1)}(n) = x_i(n) \qquad (i = 1, 2, \ldots, N_1), \tag{1.78}$$

where N_1 is the number of the first-order neurons. $x_i(n)$ (for $i = 1, 2, \ldots, N$) are the inputs at time n, $1 \le N_1 \le N$.

(2) The outputs of the second-order neurons are determined by

$$q_{ij}^{(2)}(n) = x_i(n)x_j(n) \qquad (i, j = 1, 2, \ldots, N_2), \tag{1.79}$$

where N_2 is the number of the second-order neurons, $1 \le N_2 \le N$.

(3) The outputs of the k'th-order neurons are determined by

$$q_{i_1 i_2 \ldots i_k}^{(k)}(n) = x_{i_1}(n)x_{i_2}(n)\ldots x_{i_k}(n) \qquad (i_j = 1, 2, \ldots, N_k; j = 1, 2, \ldots, k), \tag{1.80}$$

where N_k is the number of the k'th-order neurons, $1 \le N_k \le N$.

The connections between the high-order layer and the output layer are weighted. The output of each neuron in the output layer is

$$y_l(n) = f_l\left(\theta_l + \sum_{i=1}^{N_1} W_l(i)q_i^{(1)}(n) + \sum_{i=1}^{N_2}\sum_{j=1}^{N_2} W_l(ij)q_{ij}^{(2)}(n) + \cdots + \sum_{i_1=1}^{N_k}\sum_{i_2=1}^{N_k}\cdots\sum_{i_k=1}^{N_k} W_l(i_1 i_2 \ldots i_k)q_{i_1 i_2 \ldots i_k}^{(k)}(n)\right) \qquad (l = 1, 2, \ldots, L), \tag{1.81}$$

where L is the neuron number of the output layer, $f_l(\cdot)$ and θ_l are the activation function and the threshold of the l'th neuron in the output layer, respectively, and $W_l(\cdot)$ are the corresponding connection weights.

The mapping between the input and the output is defined by the activation functions, thresholds, and the connection weights. However, the activation functions and thresholds are usually fixed. The activation functions can be the sigmoid function, a piecewise-linear function, the unit-step function, or a linear function. All the connection weights can be obtained from the training data by solving the optimization problem

$$\min_{\boldsymbol{W}} \sum_{n=1}^{M} \|\boldsymbol{Y}(n) - \hat{\boldsymbol{Y}}(n)\|^2, \tag{1.82}$$

where M is the number of the available sample vectors, $\boldsymbol{W}$ denotes all the connection weights, which can be written as an $L \times (N_1 + N_2 + \cdots + N_k)$ matrix, $\boldsymbol{Y}(n) = [y_1(n), y_2(n), \ldots, y_L(n)]^T$ is the output vector computed with (1.81), and $\hat{\boldsymbol{Y}}(n)$ is the corresponding desired output vector.

Clearly, the simplest way to solve (1.82) uses the gradient descent procedure. Details of the procedure are as follows:

(1) Initialize randomly all connection weights.
(2) Compute each element of the output vector $\boldsymbol{Y}(n)$ by use of (1.81) and the available input $\boldsymbol{X}(n)$ (for $n = 1, 2, \ldots, M$).
(3) Compute the error term $\delta_l(n)$ of each neuron in the output layer

$$\delta_l(n) = \left[\hat{y}_l(n) - y_l(n)\right]\left\{f_l\left[u_l(n)\right]\right\}' \qquad (l = 1, 2, \ldots, L), \tag{1.83}$$

where

$$\left\{f_l\left[u_l(n)\right]\right\}' = \left.\frac{\partial f_l(u)}{\partial u}\right|_{u=u_l(n)} \tag{1.84}$$

and

$$u_l(n) = \sum_{i=1}^{N_1} W_l(i)q_i^{(1)}(n) + \sum_{i=1}^{N_2}\sum_{j=1}^{N_2} W_l(ij)q_{ij}^{(2)}(n) + \cdots + \sum_{i_1=1}^{N_k}\sum_{i_2=1}^{N_k}\cdots\sum_{i_k=1}^{N_k} W_l(i_1 i_2 \ldots i_k)q_{i_1 i_2 \ldots i_k}^{(k)}(n) \tag{1.85}$$

(i.e., $u_l(n)$ is the input of the l'th neuron in the output layer).
(4) Update the connection weights by the correction term

$$\Delta W_l(i_1 i_2 \ldots i_k) = \gamma \sum_{n=1}^{M} \delta_l(n) q_{i_1 i_2 \ldots i_k}^{(k)}(n), \tag{1.86}$$

where γ is the learning-rate parameter.

(5) Compute the total error

$$\epsilon = \sum_{n=1}^{M} \left\| \boldsymbol{Y}(n) - \hat{\boldsymbol{Y}}(n) \right\|^2 \tag{1.87}$$

and iterate the computation by returning to Step (2) until this error is less than a specified one.

Similarly to Equations (1.60) and (1.61), for the case in which a new sample set becomes available during the above training, (1.86) and (1.87) will become

$$\Delta W_l(i_1 i_2 \ldots i_k) = \gamma \delta_l(n) q^{(k)}_{i_1 i_2 \ldots i_k}(n) \tag{1.88}$$

and

$$\epsilon = \left\| \boldsymbol{Y}(n) - \hat{\boldsymbol{Y}}(n) \right\|^2. \tag{1.89}$$

In addition, if a linear function is taken as the activation function of the output neuron (i.e., the output neuron performs a simple summation of its inputs), (1.82) will be a linear least-squares problem whose solution can be obtained in an explicit form in the same way as for (1.72). Moreover, in this case, (1.81) has the form of Volterra series expansions. As a result, the theories related to Volterra series can be used to analyze the various performances of the network.

Bibliography

[1] Hopfield, J. J., "Neural Networks and Physical Systems with Emergent Collective Computational Abilities," *Proc. of the National Academy of Sciences, USA*, Vol. 79, 1982, pp. 2554–58.

[2] Bruck, J. and Goodman, J. W., "A Generalized Convergence Theorem for Neural Networks," *IEEE Trans. on Information Theory*, Vol. 34, No. 5, 1988, pp. 1089–93.

[3] Bruck, J. and Roychowdhury, V. P., "On the Number of Spurious Memories in the Hopfield Model," *IEEE Trans. on Information Theory*, Vol. 36, No. 2, 1990, pp. 393–97.

[4] Goles, E., "Antisymmetrical Neural Networks," *Discrete Appl. Math.*, Vol. 13, 1986, pp. 97–100.

[5] Piret, P., "Analysis of a Modified Hebbian Rule," *IEEE Trans. on Information Theory*, Vol. 36, No. 6, 1990, pp. 1391–97.

[6] Abu-Mostafa, Y. S. and St. Jacques, J. M., "Information Capacity of the Hopfield Model," *IEEE Trans. on Information Theory*, Vol. 31, No. 4, 1985, pp. 461–64.

[7] Kuh, A. and Dickinson, B. W., "Information Capacity of Associative Memories," *IEEE Trans. on Information Theory*, Vol. 35, No. 1, 1989, pp. 59–67.

[8] Hopfield, J. J., "Neurons with Graded Response Have Collective Computational Properties Like Those Two-State Neurons," *Proc. of the National Academy of Sciences, USA*, Vol. 81, 1984, pp. 3088–92.

[9] Li, J. H., Michael, A. N., and Porod, W., "Analysis and Synthesis of a Class of Neural Networks: Linear Systems Operating on a Closed Hypercube," *IEEE Trans. on Circuits and Systems*, Vol. 36, 1989, pp. 1405–22.

[10] Hopfield, J. J. and Tank, D. W., "Neural Computation of Decision Optimization Problems," *Biol. Cybernetics*, Vol. 52, 1985, pp. 141–52.

[11] Luo, F. L. and Yang, J., "Bound on Inputs to Neurons of Hopfield Continuous-Variable Neural Network," *IEE Proc.*, Pt. G, Vol. 138, 1991, pp. 671–72.

[12] Luo, F. L. and Li, Y. D., "A Theorem on the Energy Function of Hopfield Neural Network," *Int. J. Electronics*, Vol. 76, No. 3, 1994, pp. 443–46.

[13] Chua, L. O. and Yang, L., "Cellular Neural Networks: Theory," *IEEE Trans. on Circuits and Systems*, Vol. 35, No. 10, 1988, pp. 1257–72.

[14] Chua, L. O. and Yang, L., "Cellular Neural Networks: Applications," *IEEE Trans. on Circuits and Systems*, Vol. 35, No. 10, 1988, pp. 1273–90.

[15] Nossek, J., *Cellular Neural Networks*, Special issue of *IEEE Trans. on Circuits and Systems*, II, Vol. 39, No. 3, 1993, pp. 129–231.

[16] Wen, K. A., Su, J. Y., and Lu, C. Y., "VLSI Design of Digital Cellular Neural Networks for Image Processing," *J. Visual Commun. Image Representation*, Vol. 5, No. 2, 1994, pp. 117–26.

[17] Hecht-Nielsen, R., *Neurocomputing*, Addison-Wesley, Reading, MA, 1989.

[18] Hecht-Nielsen, R., "Counterpropagation Networks," *Proc. Int. Conf. on Neural Networks*, Vol. II, New York, 1987, pp. 19–32.

[19] Hecht-Nielsen, R., "Theory of the Backpropagation Neural Networks," *Proc. Int. Joint Conf. on Neural Networks*, Vol. II, 1989, pp. 593–611.

[20] Kolmogorov, A. N., "On the Representation of Continuous Functions of Many Variables by Superposition of Continuous Functions of One Variable and Addition," *Dokl. Akad. Nauk, USSR*, Vol. 114, 1957, pp. 953–56.

[21] Rumelhart, D. E., "Learning Representation by Error Propagation," *Nature*, Vol. 323, 1986, pp. 533–36.

[22] Rumelhart, D. E., McClelland, J.L., and the PDP Research Group, *Parallel Distributed Processing*, MIT Press, Cambridge, MA, 1986.

[23] Haykin, S., *Neural Networks, A Comprehensive Foundation*, IEEE Press, New York, 1994.

[24] Kollias, S. and Anastassion, D., "An Adaptive Least Squares Algorithm for the Efficient Training of Artificial Neural Networks," *IEEE Trans. on Circuits and Systems*, Vol. 36, No. 8, 1989, pp. 1092–101.

[25] Guyon, I., "Applications of Neural Networks to Character Recognition," *Int. J. Pattern Recognition and Artificial Intelligence*, Vol. 5, 1991, pp. 353–82.

[26] Abu-Mostafa, Y. S., "Learning from Hints in Neural Networks," *J. Complexity*, Vol. 6, 1990, pp. 192–98.

[27] Kohonen, T., *Self-Organization and Associative Memory*, Springer-Verlag, New York, 1989.

[28] Kohonen, T., "An Introduction to Neural Networks," *Neural Networks*, Vol. 1, 1988, pp. 3–16.

[29] Kohonen, T., "Learning Vector Quantization," *Neural Networks*, Vol. 1, suppl.1, 1988, p. 303.

[30] Kohonen, T., "The Self-Organizing Map," *Proc. of IEEE*, Vol. 78, No. 9, 1990, pp. 1464–80.

[31] Kohonen, T., Raivio, K., Simula, O, Ventae, O., and Henriksson, J., "An Adaptive

Discrete-Signal Detector Based on Self-Organizing Maps," *Proc. of Int. Joint Conf. on Neural Networks*, Vol. II, 1990, pp. 249–52.

[32] Powell, M. J. D., "Radial Basis Functions for Multivariable Interpolation: A Review", in *Proc. of IMA Conf. on Algorithms for Approximation*, Mason, J. C. and Cox, M. G. (eds.), Oxford, 1987, pp. 143–67.

[33] Broomhead, D. S. and Lowe, D.,"Multivariable Functional Interpolation and Adaptive Networks," *Complex Systems*, Vol. 2, 1988, pp. 321–55.

[34] Renals, S. and Rohwer, R., "Phoneme Classification Experiments Using Radial Basis Functions", *Proc. of Int. Joint Conf. on Neural Networks*, Vol. I, 1989, pp. 461–67.

[35] Chen, S., Covan, C. F. N., and Grant, P. M., "Orthogonal Least Squares Learning Algorithm for Radial Basis Function Networks," *IEEE Trans. on Neural Networks*, Vol. 2, 1991, pp. 302–9.

[36] Chen, S., Gibson, G. J., Covan, C. F. N., and Grant, P. M., "Reconstruction of Binary Signals Using an Adaptive Radial-Basis-Function Equalizer," *Signal Processing*, Vol. 22, 1991, pp. 77–93.

[37] Moody, J. E. and Darken, C. J., "Fast Learning in Networks of Locally Tuned Processing Units", *Neural Computation*, Vol. 1, 1989, pp. 281–294.

[38] Leonard, J. A. and Kramer, M. A., "Radial Basis Function Networks for Classifying Process Faults," *IEEE Control Systems Magazine*, April, 1991, pp. 31–38.

[39] Sanger, T. D., "A Tree-Structured Adaptive Network for Function Approximation in High-Dimensional Space," *IEEE Trans. on Neural Networks*, Vol. 2, 1991, pp. 285–293.

[40] Giles, C. L. and Maxwell, T., "Learning, Invariance, and Generalization in High-Order Neural Networks," *Appl. Optics*, Vol. 26, No. 23, 1987, pp. 4972–78.

[41] Maxwell, T., Giles, C. L., Lee, Y. C., and Chen, H. H., "Transformation Invariance Using High-Order Correlations in Neural Net Architecture," *IEEE Trans. on Systems, Man and Cybernetics*, 1986, pp. 627–29.

[42] Maxwell, T., Giles, C. L., Lee, Y. C., and Chen, H. H., "Nonlinear Dynamics of Artificial Neural Systems," *AIP Conf. Proc.*, Vol. 151, 1986, pp. 299–304.

[43] Chen, H. H., Lee, Y. C., Maxwell, T., Sun, G. Z., Lee, H. Y. and Giles, C. L., "High-Order Correlation Model for Associative Memory," *AIP Conference Proc.*, Vol. 151, 1986, pp. 86–92.

2

Neural Networks for Filtering

Filtering theory and technique play a very important role in signal processing. Filtering is, in effect, performing some mapping from the input signal to the output signal [1]. The key problem of filtering is the design and implementation of the mapping relationships. There are many kinds of filters. Of all of them, linear filters with time-invariant weight coefficients are the simplest. A filter is referred to as an adaptive filter if its weight coefficients are varied in some way according to a specific criterion as new information becomes available [2]. We say that a filter is nonlinear if the output signal is the result of a nonlinear operation on the input signal [3]. Current research emphasis is on adaptive filtering and nonlinear filtering, which deal mainly with the following problems:

(1) The real-time computation of the weight coefficients. In adaptive filtering, it is absolutely necessary to compute the weight coefficients of the filter in real time so as to make the corresponding transform function trace the characteristics of the input signals in the desired way. However, the computational complexity of the available algorithms for computing the weight coefficients is very intensive. For example, in an adaptive Wiener filter, one has to solve in real time the Wiener–Hopf equations, which involve a matrix inversion. Although many methods have been proposed to decrease the computational complexity, it is still difficult to deliver the desired real-time performance if conventional digital and sequential methods are used.

(2) The design and implementation of nonlinear filters. Unlike linear filters, nonlinear filters usually have complicated mapping relationships that make it difficult to design and to implement nonlinear filters in algorithms and structures.

In this chapter, we will show how to use neural networks to attack the above problems. There are two general motivations behind the use of neural networks in filtering. The first motivation is based on the high-speed computational capability of neural networks. Neural networks can be considered as specialized computers and can perform in real time the necessary computations of the weight coefficients. The second motivation is based on the nonlinear dynamics and mapping ability of the MLP network. According to Theorem 1.7, we are able to guarantee that any nonlinear filter can be realized by an MLP network.

This chapter is organized as follows. From Section 2.1 to Section 2.4, we will deal with the architectures and dynamics of neural networks for filtering with the least-squares (LS) algorithm, recursive least-squares (RLS) algorithm, constrained least-squares algorithm, and total-least-squares (TLS) algorithm, respectively. Sections 2.5–2.7 present the fundamentals of neural networks for the design and implementation of nonlinear filters.

2.1 Neural Networks for the Least-Squares Algorithm

The principle of an adaptive linear filter is shown in Figure 2.1. The least-squares (LS) algorithm for computing the weight coefficients of the adaptive filter can be regarded as the following optimization problem:

$$\min_{\boldsymbol{W}} \sum_{n=1}^{M} \left| d(n) - \boldsymbol{X}^T(n)\boldsymbol{W} \right|^2, \tag{2.1}$$

where $d(1), d(2), \ldots, d(M)$ are a set of known quantities and

$$\boldsymbol{X}(n) = \left[x_1(n), x_2(n), \ldots, x_N(n)\right]^T \qquad (n = 1, 2, \ldots, M)$$

are a set of vectors that stand for the input samples of the filter. $\boldsymbol{W} = [W_1, W_2, \ldots, W_N]^T$ is the unknown weight coefficient vector of the filter.

The problem (2.1) can be written in a matrix form:

$$\min_{\boldsymbol{W}} \|\boldsymbol{d} - \boldsymbol{X}\boldsymbol{W}\|^2, \tag{2.2}$$

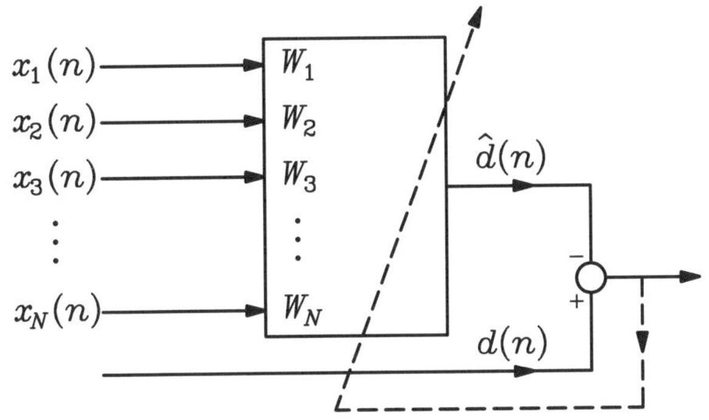

Figure 2.1. The principle of an adaptive linear filter.

where

$$X = \begin{pmatrix} x_1(1) & x_2(1) & \cdots & x_N(1) \\ x_1(2) & x_2(2) & \cdots & x_N(2) \\ \vdots & \vdots & \ddots & \vdots \\ x_1(M) & x_2(M) & \cdots & x_N(M) \end{pmatrix}$$

and $\boldsymbol{d} = [d(1), d(2), \ldots, d(M)]^T$. The dimensions of $\boldsymbol{X}$, $\boldsymbol{d}$, and $\boldsymbol{W}$ are $M \times N$, $M \times 1$, and $N \times 1$, respectively.

According to [4], the solution of (2.2) is

$$\boldsymbol{W}_{LS} = \boldsymbol{X}^+ \boldsymbol{d} = \lim_{\alpha \to 0} (\boldsymbol{X}^T \boldsymbol{X} + \alpha \boldsymbol{I})^{-1} \boldsymbol{X}^T \boldsymbol{d}, \tag{2.3}$$

where $\boldsymbol{X}^+$ is the pseudoinverse of $\boldsymbol{X}$.

Because (2.3) involves a matrix pseudoinversion, the computational complexity is intensive. In the following, we show how to use neural networks to solve this problem.

Takeda and Goodman [5] first used the continuous-time Hopfield neural network in Figure 1.2 to solve (2.1), but they encountered the "programming complexity" problem. In other words, although the Hopfield neural network can provide the solution of (2.1) during an elapsed time of a few characteristic time constants of the network, it is time consuming to compute from the given matrix $\boldsymbol{X}$ and vector $\boldsymbol{d}$ the parameters such as the interconnected conductance strengths and the bias currents of the network before the network begins to work. In [5], it was shown that the computational complexity invested in finding the right neural network parameters nearly equals the computational complexity involved in directly solving (2.1) without the neural network. In addition, the Hopfield neural network can provide only a local optimum solution of (2.1) but not the global one (2.3). As a result, the Hopfield neural network of Figure 1.2 is unsuitable for solving the LS problem (2.1) in real time.

An alternative neural network for solving (2.1), which is derived from the linear programming neural network model [6], is presented in Figure 2.2.

The input–output relationships of the neurons in the left and right parts of this network are denoted by $g(u)$ and $f(u)$, respectively. The left part has N neurons and the right part has M neurons. R_i and C_i (for $i = 1, 2, \ldots, N$) are the input resistance and capacitance of the i'th neuron in the left part, respectively (for mathematical convenience, here we let $R_i = R$ and $C_i = C$). $\boldsymbol{T} = \{T_{ji}\}$ (for $j = 1, 2, \ldots, M; i = 1, 2, \ldots, N$) is the connection strength matrix. $\boldsymbol{B} = [b_1, b_2, \ldots, b_M]^T$ is the bias current vector of the right part. $v_i(t)$ and $q_j(t)$ (for $i = 1, 2, \ldots, N; j = 1, 2, \ldots, M$) are the neuron outputs of the left and right parts of the network, respectively. $u_i(t)$ (for $i = 1, 2, \ldots, N$) is the neuron input

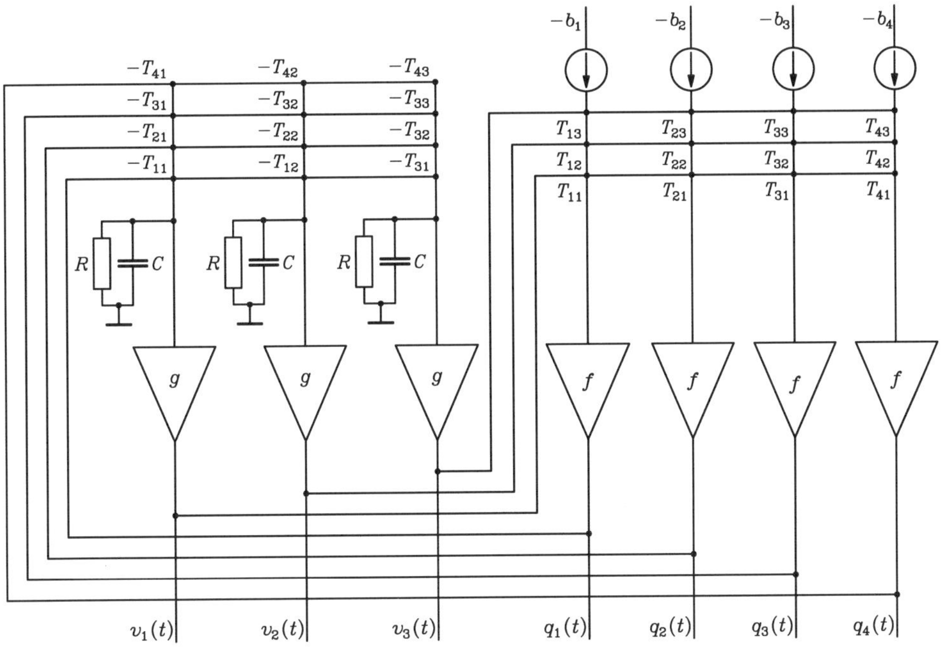

Figure 2.2. A neural network for solving the LS problem. $M = 4$ and $N = 3$.

of the left part. In terms of Kirchhoff's laws, we have the following relationship:

$$C\frac{du_i(t)}{dt} = -\sum_{j=1}^{M} T_{ji} f\left(\sum_{k=1}^{N} T_{jk}v_k(t) - b_j\right) - \frac{u_i(t)}{R}. \tag{2.4}$$

For this neural network, we define an energy function

$$E(t) = \sum_{j=1}^{M} F\left(\sum_{i=1}^{N} T_{ji}v_i(t) - b_j\right) + \sum_{i=1}^{N} \frac{1}{R}\int_0^{v_i(t)} g^{-1}(v)dv, \tag{2.5}$$

where $F(\cdot)$ is the indefinite integral of the function $f(u)$. The following theorem describes the dynamics and the stability of this neural network.

Theorem 2.1 *If $E(t)$ is bounded from below and $g(u)$ is a monotonically increasing function, then this neural network is asymptotically stable.*

Proof: The time derivative of $E(t)$ is

$$\begin{aligned}\frac{dE(t)}{dt} &= \sum_{j=1}^{M}\sum_{i=1}^{N} T_{ji}\frac{dv_i(t)}{dt} f\left(\sum_{k=1}^{N} T_{jk}v_k(t) - b_j\right) + \sum_{i=1}^{N}\frac{u_i(t)}{R}\frac{dv_i(t)}{dt} \\ &= \sum_{i=1}^{N}\frac{dv_i(t)}{dt}\left[\sum_{j=1}^{M} T_{ji} f\left(\sum_{k=1}^{N} T_{jk}v_k(t) - b_j\right) + \frac{u_i(t)}{R}\right].\end{aligned} \tag{2.6}$$

Substituting (2.4) into (2.6), we obtain

$$\begin{aligned}\frac{dE(t)}{dt} &= -\sum_{i=1}^{N} C\frac{dv_i(t)}{dt}\frac{du_i(t)}{dt}\\ &= -\sum_{i=1}^{N} C\left\{g^{-1}[v_i(t)]\right\}'\left(\frac{dv_i(t)}{dt}\right)^2. \end{aligned} \tag{2.7}$$

Since each term on the right-hand side of (2.7) is nonnegative, together with the assumption that $E(t)$ is bounded from below, we have

$$\frac{dE(t)}{dt} \le 0, \qquad \frac{dE(t)}{dt} = 0 \to \frac{dv_i(t)}{dt} = 0 \qquad (i = 1, 2, \ldots, N). \tag{2.8}$$

Equation (2.8) shows that the time evolution of the network is a motion in the state space that seeks out the minimum of $E(t)$ and comes to a stop at such a point; that is, this neural network is asymptotically stable. QED

To use the proposed neural network to solve (2.1), we select

(1) $\boldsymbol{T} = \boldsymbol{X}$ and $\boldsymbol{B} = \boldsymbol{d}$. That is, the available matrix $\boldsymbol{X}$ and vector $\boldsymbol{d}$ are taken directly as the connection strength matrix and bias current vector of the network, respectively.

(2) $f(u) = K_1 u$ (where u has the unit of current and K_1 has the unit of resistance) and $g(u) = K_2 u$ (where u has the unit of voltage and K_2 is the voltage gain).

Under these conditions, Equation (2.5) becomes

$$\begin{aligned}E(t) &= \frac{1}{2}K_1\|\boldsymbol{d} - \boldsymbol{X}^T\boldsymbol{V}(t)\|^2 + \frac{1}{2RK_2}\|\boldsymbol{V}(t)\|^2\\ &= \frac{1}{2}K_1\left(\|\boldsymbol{d} - \boldsymbol{X}^T\boldsymbol{V}(t)\|^2 + \frac{1}{RK_1K_2}\|\boldsymbol{V}(t)\|^2\right), \end{aligned} \tag{2.9}$$

where $\boldsymbol{V}(t) = [v_1(t), v_2(t), \ldots, v_N(t)]^T$ is the output vector of the left part of the network.

Equation (2.9) shows that $E(t)$ is nonnegative (bounded from below). Together with the fact that $g(u)$ is a linear function with positive gains, we know that such a constructed neural network satisfies the sufficient conditions of Theorem 2.1; hence, this neural network is stable.

The stability of this neural network can also be proved in another way. Let us write (2.4) in a matrix form

$$C\frac{d\boldsymbol{U}(t)}{dt} = -\frac{\boldsymbol{U}(t)}{R} - \boldsymbol{X}^T\boldsymbol{Q}(t) \tag{2.10}$$

$$\boldsymbol{Q}(t) = K_1(\boldsymbol{X}\boldsymbol{V}(t) - \boldsymbol{d}) = K_1(K_2\boldsymbol{X}\boldsymbol{U}(t) - \boldsymbol{d}). \tag{2.11}$$

Moreover,

$$\frac{d\boldsymbol{U}(t)}{dt} = -\left(\frac{1}{RC}\boldsymbol{I} + \frac{K_1K_2}{C}\boldsymbol{X}^T\boldsymbol{X}\right)\boldsymbol{U}(t) + \frac{K_1}{C}\boldsymbol{X}^T\boldsymbol{d}, \tag{2.12}$$

where $\boldsymbol{U}(t) = [u_1(t), u_2(t), \ldots, u_N(t)]^T$ is the input voltage vector of the left part and $\boldsymbol{Q}(t) = [q_1(t), q_2(t), \ldots, q_M(t)]^T$ is the output voltage vector of the right part. Because $\boldsymbol{X}^T\boldsymbol{X}$ is nonnegative definite, $\frac{1}{RC}\boldsymbol{I} + \frac{K_1K_2}{C}\boldsymbol{X}^T\boldsymbol{X}$ is positive definite, which means that the neural network is stable.

The input vector $\boldsymbol{U}_f$ and output vector $\boldsymbol{V}_f$ in the stationary state can be obtained by letting $du_i(t)/dt = 0$ for all i:

$$\boldsymbol{U}_f = \lim_{t\to\infty}\boldsymbol{U}(t) = K_1\left(\frac{1}{R}\boldsymbol{I} + K_1K_2\boldsymbol{X}^T\boldsymbol{X}\right)^{-1}\boldsymbol{X}^T\boldsymbol{d}, \tag{2.13}$$

$$\boldsymbol{V}_f = \lim_{t\to\infty}\boldsymbol{V}(t) = \left(\frac{1}{RK_1K_2}\boldsymbol{I} + \boldsymbol{X}^T\boldsymbol{X}\right)^{-1}\boldsymbol{X}^T\boldsymbol{d}. \tag{2.14}$$

Comparing (2.14) with (2.3) shows that $\boldsymbol{V}_f$ is an approximation of the LS solution $\boldsymbol{W}_{LS}$. The error is

$$\|\boldsymbol{V}_f - \boldsymbol{W}_{LS}\| \le \frac{\|(\boldsymbol{X}^T\boldsymbol{X})^{-1}\|\,\|\boldsymbol{W}_{LS}\|}{RK_1K_2}. \tag{2.15}$$

If $\boldsymbol{X}^T\boldsymbol{X}$ is not of full rank, then (2.15) can be written as

$$\begin{aligned}\|\boldsymbol{V}_f - \boldsymbol{W}_{LS}\| &= \frac{1}{RK_1K_2}\left\|\sum_{i=1}^{r}\frac{1}{\lambda_i + \frac{1}{RK_1K_2}}\frac{1}{\lambda_i}\boldsymbol{\Lambda}_i\boldsymbol{\Gamma}_i\right\| \\ &\le \frac{1}{1 + RK_1K_2\lambda_{\min}}\|\boldsymbol{W}_{LS}\|,\end{aligned} \tag{2.16}$$

where r is the rank of the matrix $\boldsymbol{X}$, λ_i (for $i = 1, 2, \ldots, r$) are the nonzero eigenvalues of $\boldsymbol{X}^T\boldsymbol{X}$, $\boldsymbol{\Lambda}_i$ and $\boldsymbol{\Gamma}_i$ are the right and left singular vectors corresponding to λ_i, and $\lambda_{\min}$ is the smallest nonzero eigenvalue of $\boldsymbol{X}^T\boldsymbol{X}$.

In deriving the above inequalities, we used the following approximation

$$\begin{aligned}\left(\frac{1}{RK_1K_2}\boldsymbol{I} + \boldsymbol{X}^T\boldsymbol{X}\right)^{-1} &= \left(\boldsymbol{I} + \frac{1}{RK_1K_2}(\boldsymbol{X}^T\boldsymbol{X})^{-1}\right)^{-1}(\boldsymbol{X}^T\boldsymbol{X})^{-1} \\ &\approx \left(\boldsymbol{I} - \frac{1}{RK_1K_2}(\boldsymbol{X}^T\boldsymbol{X})^{-1}\right)(\boldsymbol{X}^T\boldsymbol{X})^{-1}.\end{aligned} \tag{2.17}$$

Obviously, the difference between $\boldsymbol{V}_f$ and $\boldsymbol{W}_{LS}$ can be made arbitrarily small by appropriately selecting R, K_1, and K_2.

We make several comments concerning this neural network for computing the weight coefficient vector of a filter under the LS criterion.

Comment 1 The convergence time (the time to approach closely the stationary state) of this neural network depends on the eigenvalues of the positive definite matrix $(\frac{1}{RC}\boldsymbol{I} + \frac{K_1K_2}{C}\boldsymbol{X}^T\boldsymbol{X})$. The possible smallest eigenvalue is $1/RC$, which corresponds to the time constant RC. As a result, the convergence time is on the order of $5RC$. For example, if $R = 10$ KΩ and $C = 10$ pF, then $5RC = 500$ ns, which means that no matter what the initial value is (as long as it is nonzero), this neural network can provide the desired LS solution during an elapsed time on the order of hundreds of nanoseconds. As we know, it is impossible for any available digital and sequential scheme to provide the desired LS solution during hundreds of nanoseconds. Therefore, for the LS problem, this neural network is much more powerful than any available digital and sequential scheme.

Comment 2 Although the difference between $\boldsymbol{V}_f$ and $\boldsymbol{W}_{LS}$ may be decreased by increasing R, this makes the convergence time increase. To satisfy the specified error, we should adjust the parameters K_1 and K_2 but not R.

Comment 3 Because the available matrix $\boldsymbol{X}$ and vector $\boldsymbol{d}$ are taken directly as the connection strength matrix and the bias current vector, respectively, without any computation, we do not encounter the problem of "programming complexity" as pointed out in [5].

Comment 4 Because $(\frac{1}{RC}\boldsymbol{I} + \frac{K_1K_2}{C}\boldsymbol{X}^T\boldsymbol{X})$ is a positive definite matrix, $E(t)$ in (2.9) has only one minimum.

Comment 5 This neural network can be generalized to the case in which $\boldsymbol{X}, \boldsymbol{d}$, and $\boldsymbol{W}$ take complex values. In this case, we write $\boldsymbol{X} = \boldsymbol{X}_r + j\boldsymbol{X}_i$, $\boldsymbol{d} = \boldsymbol{d}_r + j\boldsymbol{d}_i$, and $\boldsymbol{W} = \boldsymbol{W}_r + j\boldsymbol{W}_i$, where $\boldsymbol{X}_r, \boldsymbol{d}_r$, and $\boldsymbol{W}_r$ and $\boldsymbol{X}_i, \boldsymbol{d}_i$, and $\boldsymbol{W}_i$ are the real and imaginary parts of $\boldsymbol{X}, \boldsymbol{d}$, and $\boldsymbol{W}$, respectively. Equation (2.2) becomes

$$\begin{aligned}\min_{\boldsymbol{W}} \|\boldsymbol{d} - \boldsymbol{X}\boldsymbol{W}\|^2 &= \min_{\boldsymbol{W}_r, \boldsymbol{W}_i} \|\boldsymbol{d}_r + j\boldsymbol{d}_i - (\boldsymbol{X}_r + j\boldsymbol{X}_i)(\boldsymbol{W}_r + j\boldsymbol{W}_i)\|^2 \\ &= \min_{\boldsymbol{W}_r, \boldsymbol{W}_i} \Big\|\boldsymbol{d}_r - (\boldsymbol{X}_r\boldsymbol{W}_r - \boldsymbol{X}_i\boldsymbol{W}_i) \\ &\qquad + j(\boldsymbol{d}_i - (\boldsymbol{X}_r\boldsymbol{W}_i + \boldsymbol{X}_i\boldsymbol{W}_r))\Big\|^2 \\ &= \min_{\boldsymbol{W}_c} \|\boldsymbol{d}_c - \boldsymbol{X}_c\boldsymbol{W}_c\|^2, \end{aligned} \tag{2.18}$$

where

$$\boldsymbol{X}_c = \begin{pmatrix} \boldsymbol{X}_r & -\boldsymbol{X}_i \\ \boldsymbol{X}_i & \boldsymbol{X}_r \end{pmatrix}, \qquad \boldsymbol{d}_c = \begin{pmatrix} \boldsymbol{d}_r \\ \boldsymbol{d}_i \end{pmatrix}, \qquad \boldsymbol{W}_c = \begin{pmatrix} \boldsymbol{W}_r \\ \boldsymbol{W}_i \end{pmatrix}$$

are all of real values. The solution of (2.18) is

$$\boldsymbol{W}_{CLS} = \boldsymbol{X}_c^+ \boldsymbol{d}_c = \lim_{\alpha \to 0} (\boldsymbol{X}_c^T \boldsymbol{X}_c + \alpha \boldsymbol{I})^{-1} \boldsymbol{X}_c^T \boldsymbol{d}_c. \tag{2.19}$$

If we select $\boldsymbol{T} = \boldsymbol{X}_c$, $\boldsymbol{B} = \boldsymbol{d}_c$ and take the other parameters the same as for the real-valued case, the neural network will provide

$$\boldsymbol{V}_f = \lim_{t \to \infty} \boldsymbol{V}(t) = \left(\frac{1}{RK_1K_2} \boldsymbol{I} + \boldsymbol{X}_c^T \boldsymbol{X}_c \right)^{-1} \boldsymbol{X}_c^T \boldsymbol{d}_c, \tag{2.20}$$

which approximates the desired solution (2.19) with an arbitrarily small error. From the above, we know that the number of neurons needed in the complex-valued case is twice as many as in the real-valued case; consequently, the dimension of the connection strength matrix in the complex-valued case is four times as many as in the real-valued case. This manner of generalization to the complex-valued case can also be used in the next sections. For the sake of space, we will only deal with the real-valued case in the next sections unless there are some differences.

Comment 6 Because M (the number of samples) is fixed, this network is only suitable for batch-processing or window-shift processing. The case in which the number of samples increases as time progresses will be discussed in the next section.

For further illustrations, we present two sets of simulation results about this neural network for solving the LS problem (2.1). In these simulation results, we used $R = 10$ KΩ, $C = 10$ pF, $K_1 = 1$, and $K_2 = 1$.

In these two examples, $\boldsymbol{X}$ is the given matrix, $\boldsymbol{d}$ is the given vector, $\boldsymbol{W}_{LS}$ is the exact solution, and $\boldsymbol{V}_f$ is the solution provided by the neural network in the steady state. Tables 2.1 and 2.2 and Figures 2.3 and 2.4 show the time evolution of the neural network.

Table 2.1. *The time evolution of the output vector $V(t)$*

t (10^{-12} seconds)	$v_1(t)$	$v_2(t)$	$v_3(t)$	$v_4(t)$
0	0.000000	0.000000	0.000000	0.000000
5	−0.054771	−0.270287	0.141488	−1.327200
10	−0.651601	0.271680	0.424670	−0.633752
⋮	⋮	⋮	⋮	⋮
200	−0.362038	0.071155	0.254689	−0.619635

Table 2.2. *The time evolution of the output vector $V(t)$*

t (10^{-12} seconds)	$v_1(t)$	$v_2(t)$	$v_3(t)$	$v_4(t)$
0	0.000000	0.000000	0.00000	0.000000
5	0.246472	0.058609	0.063853	0.136686
10	0.183092	−0.006556	0.093263	0.158946
⋮	⋮	⋮	⋮	⋮
200	0.066749	−0.098957	0.155229	0.209371

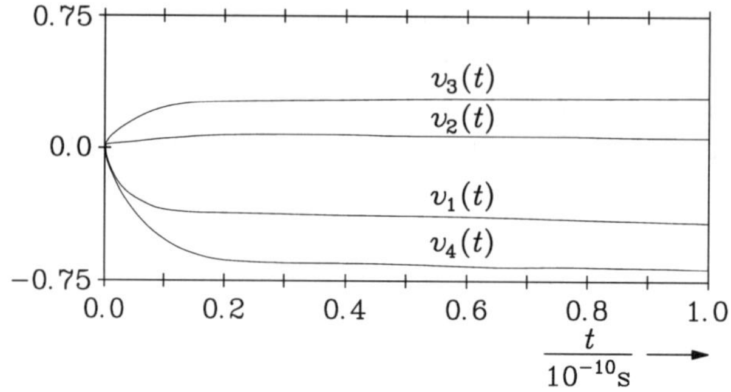

Figure 2.3. The dynamic curves of the output vector $\boldsymbol{V}(t)$ in Table 2.1.

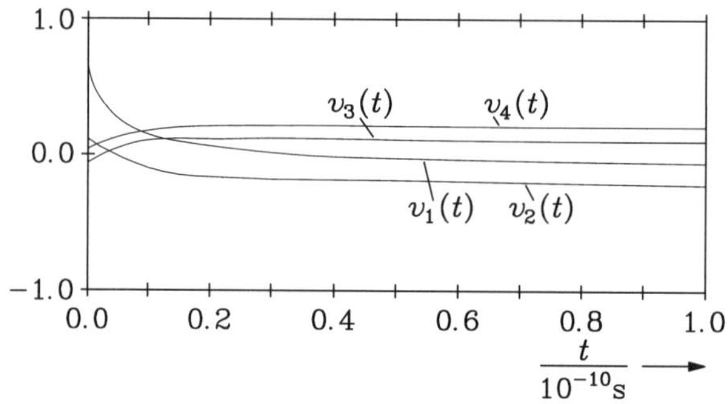

Figure 2.4. The dynamic curves of the output vector $\boldsymbol{V}(t)$ in Table 2.2.

Example 2.1

$$\boldsymbol{X} = \begin{pmatrix} -8.7279 & 6.1566 & 3.5920 & 3.2134 \\ 0.4227 & -3.9608 & -3.4599 & -2.6540 \\ -4.4573 & -4.3388 & 1.2256 & 2.6654 \\ 2.5976 & 4.1698 & 0.3388 & -0.2568 \\ -3.1500 & 1.6686 & -5.1235 & 2.2546 \\ -0.4573 & 3.4679 & 1.6581 & 3.5255 \\ 0.6486 & 1.2176 & -3.3869 & -1.7768 \\ 0.3466 & -0.3468 & 2.2663 & 2.0782 \end{pmatrix}, \qquad \boldsymbol{d} = \begin{pmatrix} 2.8436 \\ 3.9436 \\ 2.0460 \\ 3.0530 \\ -2.5474 \\ 1.1648 \\ -0.2557 \\ -2.3469 \end{pmatrix}$$

$$W_{LS} = \begin{pmatrix} -0.362038 \\ 0.071154 \\ 0.254689 \\ -0.619636 \end{pmatrix}, \quad V_f = \begin{pmatrix} -0.362038 \\ 0.071154 \\ 0.254689 \\ -0.619635 \end{pmatrix}$$

Example 2.2

$$X = \begin{pmatrix} -2.4573 & 7.1533 & 1.9728 & 6.2136 \\ 4.5976 & -1.9708 & 1.4599 & -0.7670 \\ 1.1500 & -2.3358 & 3.2117 & -2.6214 \\ -2.4573 & 2.1898 & 2.3358 & -3.1068 \\ 2.6486 & 3.6496 & -7.1533 & 0.2916 \\ 0.3466 & 1.4599 & 3.6496 & 0.5825 \\ -3.3751 & 3.2117 & -1.3869 & -3.1068 \\ 2.0036 & -2.3358 & 4.2864 & 0.0212 \end{pmatrix}, \quad d = \begin{pmatrix} 0.8544 \\ 1.9417 \\ 0.0000 \\ 1.0680 \\ -0.4854 \\ 3.1068 \\ -2.7007 \\ -0.3649 \end{pmatrix}$$

$$W_{LS} = \begin{pmatrix} 0.066675 \\ -0.098957 \\ 0.155229 \\ 0.209371 \end{pmatrix}, \quad V_f = \begin{pmatrix} 0.066675 \\ -0.098956 \\ 0.155229 \\ 0.209371 \end{pmatrix}$$

All of the above examples deal only with the ideal case. In practice, there are many departures from ideality in the hardware implementation of this network. We can expect two complications: that the connection strength matrix $\boldsymbol{T}$ will not exactly match the matrix $\boldsymbol{X}$ and that the neurons will have a time delay.

In the following we will analyze the effect of these two nonideal conditions on the dynamic and stationary performance of the network, respectively.

The mismatched connection matrices on the left and right sides of the network can be written as

$$\boldsymbol{T}_g = \boldsymbol{X}^T + \boldsymbol{\Delta}_g \tag{2.21}$$

and

$$\boldsymbol{T}_f = \boldsymbol{X} + \boldsymbol{\Delta}_f. \tag{2.22}$$

In this case, (2.12) becomes

$$\frac{d\boldsymbol{U}(t)}{dt} = -\left(\frac{1}{RC}\boldsymbol{I} + \frac{K_1 K_2}{C}\boldsymbol{T}_g\boldsymbol{T}_f\right)\boldsymbol{U}(t) + \frac{K_1}{C}\boldsymbol{T}_g\boldsymbol{d}. \tag{2.23}$$

Obviously, a sufficient condition on the stability is that the smallest eigenvalue of $K_1 K_2 \boldsymbol{T}_g \boldsymbol{T}_f$ is larger than $-1/R$. The input and output vectors in the stationary state are

$$\boldsymbol{U}_f = \lim_{t\to\infty} \boldsymbol{U}(t) = K_1\left(\frac{1}{R}\boldsymbol{I} + K_1 K_2 \boldsymbol{T}_g \boldsymbol{T}_f\right)^{-1} \boldsymbol{T}_g \boldsymbol{d} \tag{2.24}$$

and

$$V_f = \lim_{t\to\infty} V(t) = \left(\frac{1}{RK_1K_2}I + T_gT_f\right)^{-1} T_g d. \tag{2.25}$$

Substituting (2.21) and (2.22) into (2.25), we obtain

$$V_f = \left(\frac{1}{RK_1K_2}I + X^TX + X^T\Delta_f + \Delta_g X + \Delta_g\Delta_f\right)^{-1} (X^T + \Delta_g)d. \tag{2.26}$$

Using an approximation similar to (2.17) gives

$$V_f = \left[(X^TX)^{-1} - (X^TX)^{-1}\left(\frac{1}{RK_1K_2}I + X^T\Delta_f + \Delta_g X + \Delta_g\Delta_f\right) \times (X^TX)^{-1}\right](X^T + \Delta_g)d. \tag{2.27}$$

If we let $\Delta = X^T\Delta_f + \Delta_g X$ and ignore the higher-order terms of Δ_f and Δ_g, then (2.27) becomes

$$\begin{aligned} V_f &= (X^TX)^{-1}X^Td + (X^TX)^{-1}\Delta_g d \\ &\quad - (X^TX)^{-1}\left(\frac{1}{RK_1K_2}I + \Delta\right)(X^TX)^{-1}X^Td \\ &= (X^TX)^{-1}X^Td - \frac{(X^TX)^{-1}(X^TX)^{-1}X^Td}{RK_1K_2} \\ &\quad - (X^TX)^{-1}\Delta(X^TX)^{-1}X^Td + (X^TX)^{-1}\Delta_g d. \end{aligned}$$

The error between V_f and the exact solution W_{LS} by (2.3) is

$$\begin{aligned} \|V_f - W_{LS}\| &= \left\| \frac{(X^TX)^{-1}W_{LS}}{RK_1K_2} - (X^TX)^{-1}\Delta W_{LS} + (X^TX)^{-1}\Delta_g d \right\| \\ &\le \frac{\|(X^TX)^{-1}\|\|W_{LS}\|}{RK_1K_2} + \|(X^TX)^{-1}\|\|\Delta\|\|W_{LS}\| \\ &\quad + \|(X^TX)^{-1}\|\|\Delta_g\|\|d\| \\ &= \|(X^TX)^{-1}\|\left(\frac{\|W_{LS}\|}{RK_1K_2} + \|\Delta\|\|W_{LS}\| + \|\Delta_g\|\|d\|\right). \end{aligned} \tag{2.28}$$

This error consists of three terms. The first term is equal to (2.15), which exists even in the ideal case but can be decreased by selecting reasonable values for the parameters R, K_1, and K_2. The second and third terms are caused by the mismatched connection strength matrices and are independent of R, K_1, and K_2. The only way to reduce these two error terms is to reduce both $\boldsymbol{\Delta}_g$ and $\boldsymbol{\Delta}_f$.

The sufficient condition mentioned above for the stability of the network and the error by (2.28) in the mismatched connection strength case are very useful in designing the hardware implementation of the network.

The time delay of the neurons can be described by an additional input resistance and capacitance. Assuming that R_f and C_f stand for the total resistance and capacitance of the neurons on the right side of the network when considering the time delay and that R_g and C_g are those on the left side of the network, the dynamic equations of the network can be written as

$$\begin{aligned} C_g \frac{d\boldsymbol{U}(t)}{dt} &= -\frac{\boldsymbol{U}(t)}{R_g} - \boldsymbol{X}^T \boldsymbol{Q}(t) \\ &= -\frac{\boldsymbol{U}(t)}{R_g} - K_1 \boldsymbol{X}^T \boldsymbol{I}_f(t) \end{aligned} \tag{2.29}$$

and

$$\begin{aligned} R_f C_f \frac{d\boldsymbol{I}_f(t)}{dt} &= -\boldsymbol{I}_f(t) + \boldsymbol{X}\boldsymbol{V}(t) - \boldsymbol{d} \\ &= -\boldsymbol{I}_f(t) + K_2 \boldsymbol{X}\boldsymbol{U}(t) - \boldsymbol{d}, \end{aligned} \tag{2.30}$$

where $\boldsymbol{I}_f(t) = [i_{f1}(t), i_{f2}(t), \ldots, i_{fM}(t)]^T$ and $i_{fi}(t)$ (for $i = 1, 2, \ldots, M$) is the current through the input resistor R_f of the i'th neuron in the right part of the network.

Moreover, it follows that

$$\begin{pmatrix} \frac{d\boldsymbol{U}(t)}{dt} \\ \frac{d\boldsymbol{I}_f(t)}{dt} \end{pmatrix} = -\begin{pmatrix} \frac{1}{R_g C_g}\boldsymbol{I} & \frac{K_1 \boldsymbol{X}^T}{C_g} \\ -\frac{K_2}{R_f C_f}\boldsymbol{X} & \frac{1}{R_f C_f}\boldsymbol{I} \end{pmatrix} \begin{pmatrix} \boldsymbol{U}(t) \\ \boldsymbol{I}_f(t) \end{pmatrix} + \begin{pmatrix} \boldsymbol{0} \\ -\frac{1}{R_f C_f}\boldsymbol{d} \end{pmatrix}, \tag{2.31}$$

where $\boldsymbol{O}$ is a zero vector. Let $\tau_1 = 1/R_g C_g$, $\tau_2 = 1/R_f C_f$, $K_3 = K_1/C_g$, and $K_4 = K_2/R_f C_f$. Then (2.31) becomes

$$\begin{pmatrix} \frac{d\boldsymbol{U}(t)}{dt} \\ \frac{d\boldsymbol{I}_f(t)}{dt} \end{pmatrix} = -\begin{pmatrix} \tau_1 \boldsymbol{I} & K_3 \boldsymbol{X}^T \\ -K_4 \boldsymbol{X} & \tau_2 \boldsymbol{I} \end{pmatrix} \begin{pmatrix} \boldsymbol{U}(t) \\ \boldsymbol{I}_f(t) \end{pmatrix} + \begin{pmatrix} \boldsymbol{0} \\ -\frac{1}{R_f C_f}\boldsymbol{d} \end{pmatrix}. \tag{2.32}$$

Set

$$\boldsymbol{A} = \begin{pmatrix} \tau_1 \boldsymbol{I} & K_3 \boldsymbol{X}^T \\ -K_4 \boldsymbol{X} & \tau_2 \boldsymbol{I} \end{pmatrix}.$$

We will now show that the real parts of all the eigenvalues of the matrix $\boldsymbol{A}$ are positive. Let λ be an eigenvalue of $\boldsymbol{A}$ and

$$\boldsymbol{S} = \begin{pmatrix} \boldsymbol{S}_g \\ \boldsymbol{S}_f \end{pmatrix}$$

be the corresponding eigenvector; that is,

$$\boldsymbol{A}\begin{pmatrix}\boldsymbol{S}_g\\ \boldsymbol{S}_f\end{pmatrix}=\lambda\begin{pmatrix}\boldsymbol{S}_g\\ \boldsymbol{S}_f\end{pmatrix}$$

or

$$\begin{pmatrix}\tau_1\boldsymbol{I} & K_3\boldsymbol{X}^T\\ -K_4\boldsymbol{X} & \tau_2\boldsymbol{I}\end{pmatrix}\begin{pmatrix}\boldsymbol{S}_g\\ \boldsymbol{S}_f\end{pmatrix}=\lambda\begin{pmatrix}\boldsymbol{S}_g\\ \boldsymbol{S}_f\end{pmatrix}$$

or

$$\begin{cases}\tau_1\boldsymbol{S}_g+K_3\boldsymbol{X}^T\boldsymbol{S}_f=\lambda\boldsymbol{S}_g\\ -K_4\boldsymbol{X}\boldsymbol{S}_g+\tau_2\boldsymbol{S}_f=\lambda\boldsymbol{S}_f.\end{cases}\tag{2.33}$$

Moreover,

$$-K_3K_4\boldsymbol{X}^T\boldsymbol{X}\boldsymbol{S}_g=(\lambda-\tau_1)(\lambda-\tau_2)\boldsymbol{S}_g,\tag{2.34}$$

which means that $(\lambda-\tau_1)(\lambda-\tau_2)$ is an eigenvalue of the matrix $-K_3K_4\boldsymbol{X}^T\boldsymbol{X}$. Because $\boldsymbol{X}^T\boldsymbol{X}$ is nonnegative definite, $(\lambda-\tau_1)(\lambda-\tau_2)$ has a nonpositive real value, that is,

$$(\lambda-\tau_1)(\lambda-\tau_2)\le 0.\tag{2.35}$$

If λ is real, we obtain from (2.35)

$$0\le\tau_1\le\lambda\le\tau_2\quad\text{or}\quad 0\le\tau_2\le\lambda\le\tau_1.\tag{2.36}$$

If λ has a complex value ($\lambda=\lambda_r+j\lambda_i$), using (2.35) gives

$$(\lambda_r+j\lambda_i-\tau_1)(\lambda_r+j\lambda_i-\tau_2)\le 0,\tag{2.37}$$

$$\lambda_i(\lambda_r-\tau_2)+\lambda_i(\lambda_r-\tau_1)=0,\tag{2.38}$$

and

$$\lambda_r=\frac{\tau_2+\tau_1}{2}>0.\tag{2.39}$$

Expressions (2.36) and (2.39) show that the real parts of all the eigenvalues of the matrix $\boldsymbol{A}$ are positive; consequently, the network is asymptotically stable. The input and output vectors in the stationary state are obtained by letting the left term of (2.31) be zero:

$$-\frac{\boldsymbol{U}(t)}{R_g}-K_1\boldsymbol{X}^T\boldsymbol{I}_f(t)=\boldsymbol{0},\tag{2.40}$$

$$-\boldsymbol{I}_f(t)+K_2\boldsymbol{X}\boldsymbol{U}(t)=\boldsymbol{d}\tag{2.41}$$

and

$$U_f = K_1\left(\frac{1}{R_g}I + K_1K_2X^TX\right)^{-1}X^Td, \tag{2.42}$$

$$V_f = \left(\frac{1}{R_gK_1K_2}I + X^TX\right)^{-1}X^Td. \tag{2.43}$$

Equations (2.42) and (2.43) are identical to (2.13) and (2.14), respectively, which means that the time delay of the neurons has no effect on the stationary performance of the network.

To account for more nonideal conditions, one may refer to [7].

2.2 Neural Networks for the Recursive Least-Squares Algorithm

Unlike the LS algorithm, the recursive least-squares (RLS) algorithm is generally used to minimize recursively the time-indexed sum of squares (2.1). This means that the available samples in the RLS algorithm are time-increasing. The LS solution of (2.1) at time n is

$$W_{LS}(n) = X_n^+d_n = \lim_{\alpha\to 0}(X_n^TX_n + \alpha I)^{-1}X_n^Td_n, \tag{2.44}$$

where

$$X_n = \begin{pmatrix} x_1(1) & x_2(1) & \cdots & x_N(1) \\ x_1(2) & x_2(2) & \cdots & x_N(2) \\ \vdots & \vdots & \ddots & \vdots \\ x_1(n) & x_2(n) & \cdots & x_N(n) \end{pmatrix}, \qquad d_n = \begin{pmatrix} d(1) \\ d(2) \\ \vdots \\ d(n) \end{pmatrix}. \tag{2.45}$$

At time $n+1$, the LS solution of (2.1) is

$$W_{LS}(n+1) = X_{n+1}^+d_{n+1} = \lim_{\alpha\to 0}(X_{n+1}^TX_{n+1} + \alpha I)^{-1}X_{n+1}^Td_{n+1} \tag{2.46}$$

and the corresponding sample matrix and vector become

$$X_{n+1} = \begin{pmatrix} x_1(1) & x_2(1) & \cdots & x_N(1) \\ x_1(2) & x_2(2) & \cdots & x_N(2) \\ \vdots & \vdots & \ddots & \vdots \\ x_1(n) & x_2(n) & \cdots & x_N(n) \\ x_1(n+1) & x_2(n+1) & \ldots & x_N(n+1) \end{pmatrix},$$

$$\boldsymbol{d}_n = \begin{pmatrix} d(1) \\ d(2) \\ \vdots \\ d(n) \\ d(n+1) \end{pmatrix}. \tag{2.47}$$

We can write (2.47) in the following form

$$\boldsymbol{X}_{n+1} = \begin{pmatrix} \boldsymbol{X}_n \\ z_{n+1} \end{pmatrix}, \qquad \boldsymbol{d}_{n+1} = \begin{pmatrix} \boldsymbol{d}_n \\ d(n+1) \end{pmatrix}, \tag{2.48}$$

where $z_{n+1} = [x_1(n+1), x_2(n+1), \ldots, x_N(n+1)]$.

Based on (2.48), several fast and efficient RLS algorithms similar to the Kalman filtering algorithm have been proposed [8, 9]. These algorithms can provide the LS solution $\boldsymbol{W}_{LS}(n+1)$ at time $n+1$ by using only $\boldsymbol{W}_{LS}(n)$, z_{n+1}, and $d(n+1)$. Although these RLS algorithms have greatly reduced the computational complexity of solving Equation (2.46), the computation required in these RLS algorithms is still intensive, and hence it is difficult to provide $\boldsymbol{W}_{LS}(n+1)$ in real time.

As pointed out in the previous section, the neural network in Figure 2.2 is not suitable for the RLS algorithm. The neuron number of the neural network for the RLS algorithm has to be independent of the sample number. Let us consider the continuous-time Hopfield neural network in Figure 1.2, except that we select the neurons with a linear input–output relationship function $g(u) = K_1 u$ instead of the sigmoid function $\frac{1}{2}(1+\tanh(\frac{u}{u_0}))$. In addition, at time n, the connection strengths and bias currents of the network are computed from the expressions

$$\begin{aligned} \boldsymbol{T}_n &= \boldsymbol{X}_n^T \boldsymbol{X}_n \quad \text{and} \\ \boldsymbol{B}_n &= -\boldsymbol{X}_n^T \boldsymbol{d}_n. \end{aligned} \tag{2.49}$$

In order to prove the stability of this neural network, we first define an energy function

$$E(t) = \frac{1}{2}\boldsymbol{V}^T(t)\boldsymbol{X}_n^T\boldsymbol{X}_n\boldsymbol{V}(t) - (\boldsymbol{X}_n^T\boldsymbol{d}_n)^T\boldsymbol{V}(t) + \frac{1}{2RK_1}\|\boldsymbol{V}(t)\|^2, \tag{2.50}$$

where $\boldsymbol{V}(t) = [v_1(t), v_2(t), \ldots, v_N(t)]^T$ is the output vector of the network. It follows that

$$\begin{aligned} E(t) &= \frac{1}{2}\|\boldsymbol{d}_n - \boldsymbol{X}_n^T\boldsymbol{V}(t)\|^2 + \frac{1}{2RK_1}\|\boldsymbol{V}(t)\|^2 - \frac{1}{2}\|\boldsymbol{d}_n\|^2 \\ &\geq -\frac{1}{2}\|\boldsymbol{d}_n\|^2. \end{aligned} \tag{2.51}$$

Moreover,

$$\frac{dE(t)}{dt} = \boldsymbol{V}^T(t)\boldsymbol{X}_n^T\boldsymbol{X}_n\frac{d\boldsymbol{V}(t)}{dt} - (\boldsymbol{X}_n^T\boldsymbol{d}_n)^T\frac{d\boldsymbol{V}(t)}{dt} + \frac{1}{RK_1}\boldsymbol{V}^T(t)\frac{d\boldsymbol{V}(t)}{dt}. \tag{2.52}$$

In terms of Kirchhoff's current law, we have

$$C\left(\frac{d\boldsymbol{U}(t)}{dt}\right)^T = -\boldsymbol{V}^T(t)\boldsymbol{X}_n^T\boldsymbol{X}_n + (\boldsymbol{X}_n^T\boldsymbol{d}_n)^T - \frac{1}{RK_1}\boldsymbol{V}^T(t), \tag{2.53}$$

where $\boldsymbol{U}(t) = [u_1(t), u_2(t), \ldots, u_N(t)]^T$ is the input voltage vector of the network.

Substituting (2.53) into (2.52) yields

$$\frac{dE(t)}{dt} = -C\left(\frac{d\boldsymbol{U}(t)}{dt}\right)^T\frac{d\boldsymbol{V}(t)}{dt} = -\frac{C}{K_1}\left(\frac{d\boldsymbol{V}(t)}{dt}\right)^T\frac{d\boldsymbol{V}(t)}{dt} \leq 0. \tag{2.54}$$

Equations (2.51) and (2.54) prove this network to be stable. The output vector $\boldsymbol{V}_{n,f}$ in the stationary state is computed by letting $d\boldsymbol{U}(t)/dt = \boldsymbol{O}$, that is,

$$\boldsymbol{V}_{n,f} = \lim_{t\to\infty}\boldsymbol{V}(t) = \left(\frac{1}{RK_1}\boldsymbol{I} + \boldsymbol{X}_n^T\boldsymbol{X}_n\right)^{-1}\boldsymbol{X}_n^T\boldsymbol{d}_n, \tag{2.55}$$

which is an approximation of (2.44).

Consequently, at time $n+1$, the connection strengths and bias currents of the network are computed by

$$\boldsymbol{T}_{n+1} = \boldsymbol{X}_{n+1}^T\boldsymbol{X}_{n+1} \quad \text{and}$$
$$\boldsymbol{B}_{n+1} = -\boldsymbol{X}_{n+1}^T\boldsymbol{d}_{n+1}, \tag{2.56}$$

and the network provides the solution

$$\boldsymbol{V}_{n+1,f} = \lim_{t\to\infty}\boldsymbol{V}(t) = \left(\frac{1}{RK_1}\boldsymbol{I} + \boldsymbol{X}_{n+1}^T\boldsymbol{X}_{n+1}\right)^{-1}\boldsymbol{X}_{n+1}^T\boldsymbol{d}_{n+1}, \tag{2.57}$$

which approximates the exact solution (2.46).

Although the computation complexity invested in computing the connection strengths and bias currents by use of (2.49) and (2.56) is $O(N^2n^2)$, which is much less than the complexity required in (2.44) and (2.46), the computational complexity needed in (2.49) and (2.56) is on the same order of complexity as that of the fast RLS algorithm [9]. As a result, it is necessary to decrease the complexity for computing the connection strengths and bias currents before this neural network is applied to real-time fields.

If we compare (2.56) with (2.49), we have

$$\boldsymbol{T}_{n+1} = \boldsymbol{X}_{n+1}^T \boldsymbol{X}_{n+1} = \begin{pmatrix} \boldsymbol{X}_n \\ \boldsymbol{z}_{n+1} \end{pmatrix}^T \begin{pmatrix} \boldsymbol{X}_n \\ \boldsymbol{z}_{n+1} \end{pmatrix}$$
$$= \boldsymbol{X}_n^T \boldsymbol{X}_n + \boldsymbol{z}_{n+1}^T \boldsymbol{z}_{n+1} = \boldsymbol{T}_n + \boldsymbol{z}_{n+1}^T \boldsymbol{z}_{n+1} \tag{2.58}$$

and

$$\boldsymbol{B}_{n+1} = -\boldsymbol{X}_{n+1}^T \boldsymbol{d}_{n+1} = -\begin{pmatrix} \boldsymbol{X}_n \\ \boldsymbol{z}_{n+1} \end{pmatrix}^T \begin{pmatrix} \boldsymbol{d}_n \\ d(n+1) \end{pmatrix}$$
$$= -\boldsymbol{X}_n^T \boldsymbol{d}_n - \boldsymbol{z}_{n+1}^T d(n+1) = \boldsymbol{B}_n - \boldsymbol{z}_{n+1}^T d(n+1). \tag{2.59}$$

From (2.58) and (2.59), we know that the complexity for recursively computing the connection strengths and bias currents is $O(N^2)$, which is less than the complexity of the available fast RLS algorithms. In addition, no division computations are needed if this proposed neural network is used to solve the RLS problem.

It should be noted that for temporal processing the complexity of computing the connection strengths and bias currents may be reduced to $O(N)$ because, in this case, the input sample vector $\boldsymbol{X}(n)$ becomes a scalar variable $x(n)$, and consequently, the matrices $\boldsymbol{X}_n$ in (2.45) and $\boldsymbol{X}_{n+1}$ in (2.47) become

$$\boldsymbol{X}_n = \begin{pmatrix} x(N) & x(N-1) & \cdots & x(1) \\ x(N+1) & x(N) & \cdots & x(2) \\ \vdots & \vdots & \ddots & \vdots \\ x(n) & x(n-1) & \cdots & x(n+1-N) \end{pmatrix} \tag{2.60}$$

and

$$\boldsymbol{X}_{n+1} = \begin{pmatrix} x(N) & x(N-1) & \cdots & x(1) \\ x(N+1) & x(N) & \cdots & x(2) \\ \vdots & \vdots & \ddots & \vdots \\ x(n) & x(n-1) & \cdots & x(n+1-N) \\ x(n+1) & x(n) & \cdots & x(n+2-N) \end{pmatrix}, \tag{2.61}$$

which means that only one sample $(x(n+1))$ but not a vector $(\boldsymbol{z}_{n+1})$ with N elements is involved in recursively computing the connection strengths and bias currents by (2.58) and (2.59).

The above discussion demonstrates the suitability of this neural network approach for the RLS algorithm.

2.3 Neural Networks for the Constrained Least-Squares Algorithm

In some cases [10, 11], the weight coefficients of an adaptive filter are the solutions of the following constrained optimization problem:

$$\begin{cases} \min_{W} & \sum_{n=1}^{M} \left| \boldsymbol{X}^T(n)\boldsymbol{W} \right|^2 \\ s.t. & \boldsymbol{W}^T\boldsymbol{d} = 1, \end{cases} \tag{2.62}$$

where the vector $\boldsymbol{d}$ is called the constraint vector and the other variables are the same as those in (2.2).

Using Lagrange's method we can define a cost function

$$E(\boldsymbol{W}) = \sum_{n=1}^{M} \left| \boldsymbol{X}^T(n)\boldsymbol{W} \right|^2 + \lambda(1 - \boldsymbol{W}^T\boldsymbol{d}), \tag{2.63}$$

where λ is an arbitrary constant. Differentiating with respect to the weight vector $\boldsymbol{W}$ and equating to zero give the optimum weight vector $\boldsymbol{W}_{\text{opt}}$ of (2.62) as

$$\boldsymbol{W}_{\text{opt}} = \frac{\boldsymbol{R}_x^{-1}\boldsymbol{d}^*}{\boldsymbol{d}^T\boldsymbol{R}_x^{-1}\boldsymbol{d}^*}, \tag{2.64}$$

where "$*$" stands for the conjugate and

$$\boldsymbol{R}_x = \sum_{n=1}^{M} \boldsymbol{X}^*(n)\boldsymbol{X}^T(n). \tag{2.65}$$

In practical applications, we can use the constant λ instead of $1/(\boldsymbol{d}^T\boldsymbol{R}_x^{-1}\boldsymbol{d}^*)$, which has no effect on the filtering performance, that is,

$$\boldsymbol{W}_{\text{opt}} = \lambda\boldsymbol{R}_x^{-1}\boldsymbol{d}^*. \tag{2.66}$$

For simplicity, we let $\lambda = 1$, and then (2.66) becomes

$$\boldsymbol{W}_{\text{opt}} = \boldsymbol{R}_x^{-1}\boldsymbol{d}^*. \tag{2.67}$$

Equation (2.67) involves both a matrix inversion and a matrix multiplication and hence is of very intensive computational complexity.

In the following, we show how to use the neural network in Figure 2.2 to solve this problem.

Equation (2.67) can be written in real-value form as

$$\boldsymbol{W}_{c\,\text{opt}} = \boldsymbol{R}_{cx}^{-1}\boldsymbol{d}_c, \tag{2.68}$$

where

$$\boldsymbol{W}_{c\,\mathrm{opt}} = \begin{pmatrix} \boldsymbol{W}_{r\,\mathrm{opt}} \\ \boldsymbol{W}_{i\,\mathrm{opt}} \end{pmatrix}, \qquad \boldsymbol{R}_{cx} = \begin{pmatrix} \boldsymbol{R}_{rx} & -\boldsymbol{R}_{ix} \\ \boldsymbol{R}_{ix} & \boldsymbol{R}_{rx} \end{pmatrix}, \qquad \boldsymbol{d}_c = \begin{pmatrix} \boldsymbol{d}_r \\ -\boldsymbol{d}_i \end{pmatrix},$$

and $\boldsymbol{W}_{\mathrm{opt}} = \boldsymbol{W}_{r\,\mathrm{opt}} + j\boldsymbol{W}_{i\,\mathrm{opt}}$, $\boldsymbol{R}_x = \boldsymbol{R}_{rx} + j\boldsymbol{R}_{ix}$, and $\boldsymbol{d}^* = \boldsymbol{d}_r - j\boldsymbol{d}_i$.

Let the known matrix $\boldsymbol{R}_{cx}$ be the connection strength matrix $\boldsymbol{T}$ and the constraint vector $\boldsymbol{d}_c$ be the bias current vector $\boldsymbol{B}$ of the network, respectively; the other parameters of the network are selected as for (2.1). According to the analysis given in Section 2.1, this network gives the solution

$$\boldsymbol{V}_f = \lim_{t\to\infty} \boldsymbol{V}(t) = \left(\frac{1}{RK_1K_2}\boldsymbol{I} + \boldsymbol{R}_{cx}^T\boldsymbol{R}_{cx}\right)^{-1} \boldsymbol{R}_{cx}^T\boldsymbol{d}_c \approx \boldsymbol{R}_{cx}^{-1}\boldsymbol{d}_c. \tag{2.69}$$

That is, it is an approximation of the optimum solution (2.68).

In order to reduce the complexity of computing $\boldsymbol{R}_{cx}$ (or say, programming $\boldsymbol{T}$) from the sample vector $\boldsymbol{X}(n)$ (for $n = 1, 2, \ldots, M$), we can use the recursive algorithm as shown in (2.58).

In addition, using the penalty function method [12] for (2.62), it is sufficient to solve the following unconstrained problem of the form

$$\min_{\boldsymbol{W}} \left[\sum_{n=1}^{M} |\boldsymbol{X}^T(n)\boldsymbol{W}|^2 + k(\boldsymbol{W}^T\boldsymbol{d} - 1)^2\right], \tag{2.70}$$

where k is a positive constant.

From [12], we know that (2.62) and (2.70) have the same solution if k tends to infinity. However, if k is large enough, the solution of (2.70) can approximate that of (2.62).

Equation (2.70) is, in effect, the LS problem shown in (2.1). As a result, we can also use the neural network in Figure 2.2 to solve (2.70) as long as we select the connection strength matrix $\boldsymbol{T}$ and bias current vector $\boldsymbol{B}$ as

$$\boldsymbol{T} = \begin{pmatrix} x_1(1) & x_2(1) & \cdots & x_N(1) \\ x_1(2) & x_2(2) & \cdots & x_N(2) \\ \vdots & \vdots & \ddots & \vdots \\ x_1(M) & x_2(M) & \cdots & x_N(M) \\ \sqrt{k}d_1 & \sqrt{k}d_2 & \cdots & \sqrt{k}d_N \end{pmatrix}, \qquad \boldsymbol{B} = \begin{pmatrix} 0 \\ 0 \\ \vdots \\ 0 \\ \sqrt{k} \end{pmatrix},$$

which means that the programming complexity of this network from the known samples $\boldsymbol{X}(n)$ and the constraint vector $\boldsymbol{d}$ is zero. Figure 2.5 shows the construction of the neural network with the above parameters. However, this network cannot be applied to the case in which the number M of samples is time-increasing.

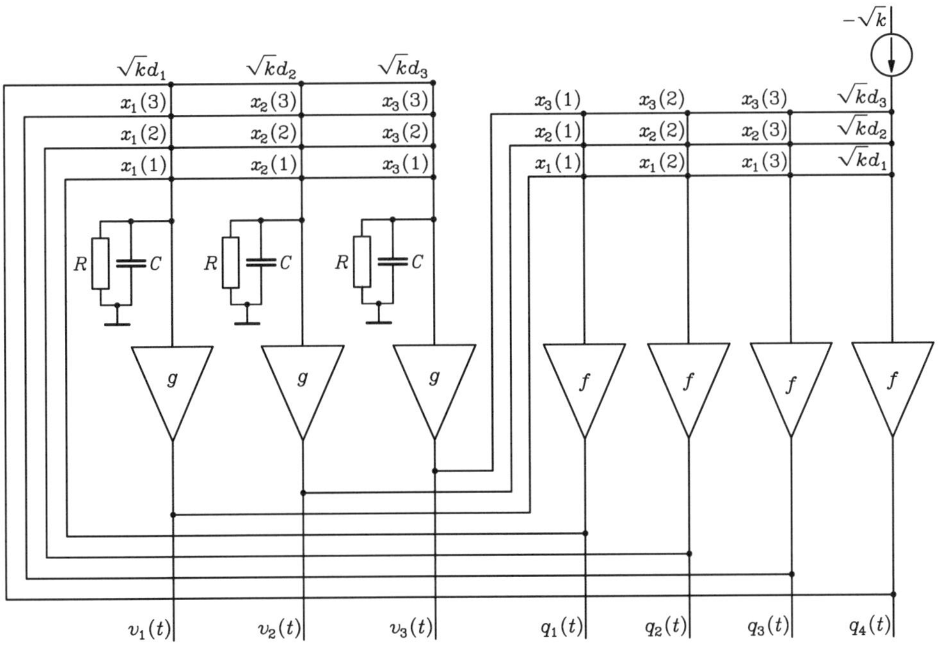

Figure 2.5. A neural network for solving (2.70). $M = 3$ and $N = 3$.

2.4 Neural Networks for the Total-Least-Squares Algorithm

The total-least-squares (TLS) algorithm for computing the weight coefficients of an adaptive filter can be formulated as seeking a solution such that

$$\|\boldsymbol{D}\|_F = \text{minimum} \tag{2.71}$$

and

$$\boldsymbol{d} + \Delta\boldsymbol{d} \in \text{range}(\boldsymbol{X} + \Delta\boldsymbol{X}), \tag{2.72}$$

where $\Delta\boldsymbol{d}$ and $\Delta\boldsymbol{X}$ denote the perturbation of the vector $\boldsymbol{d}$ and the matrix $\boldsymbol{X}$ in (2.2), respectively, $\boldsymbol{D} = [\Delta\boldsymbol{X}, \Delta\boldsymbol{d}]$ whose dimension is $M \times (N+1)$, and $\|\cdot\|_F$ is the Frobenius norm given by

$$\|\boldsymbol{D}\|_F = \left(\sum_{i=1}^{M}\sum_{j=1}^{N+1} D_{ij}^2\right)^{\frac{1}{2}}. \tag{2.73}$$

Let

$$\boldsymbol{\Lambda}_{N+1} = [\Lambda_{N+1,1}, \Lambda_{N+1,2}, \ldots \Lambda_{N+1,N+1}]^T \tag{2.74}$$

be the right singular vector corresponding to the smallest singular value σ_{N+1} of the expanded matrix $\boldsymbol{Y} = [\boldsymbol{X}, \boldsymbol{d}]$. Then the TLS solution of (2.71) with (2.72) is

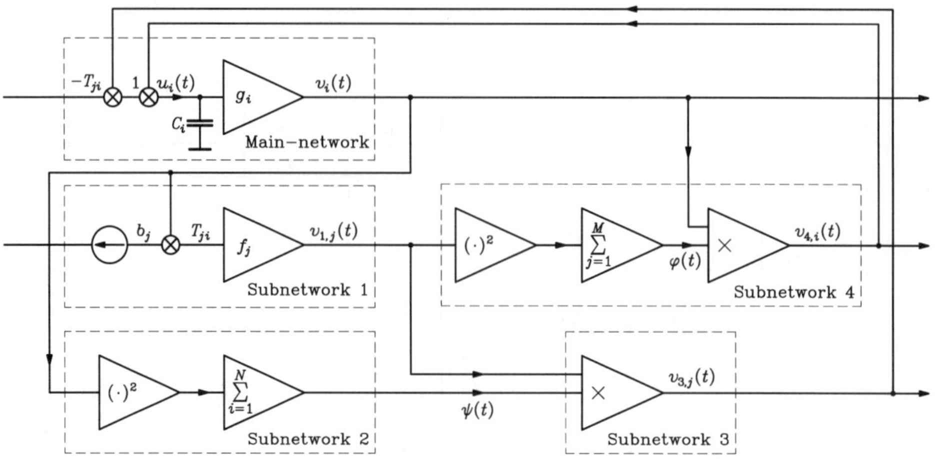

Figure 2.6. A neural network for solving the TLS problem.

obtained as

$$\boldsymbol{W}_{TLS} = -\frac{1}{\Lambda_{N+1,N+1}}[\Lambda_{N+1,1}, \Lambda_{N+1,2}, \ldots, \Lambda_{N+1,N}]^T. \tag{2.75}$$

If the singular value is not single but multiple, Equation (2.75) becomes

$$\boldsymbol{W}_{TLS} = -\sum_{l=L+1}^{N+1}\left(\frac{\Lambda_{l,N+1}^*}{\sum_{i=L+1}^{N+1}|\Lambda_{i,N+1}|^2}\right)[\Lambda_{l,1}, \Lambda_{l,2}, \ldots, \Lambda_{l,N}]^T, \tag{2.76}$$

where $\boldsymbol{\Lambda}_l$ (for $l = L+1, L+2, \ldots, N+1$) is the l'th right singular vector corresponding to the smallest singular value σ_{N+1} and where $N - L + 1$ is the multiplicity of σ_{N+1}. As the singular value σ_{N+1} tends to zero, the LS solution by (2.3) and the TLS solution by (2.75) (or (2.76)) approach each other [13, 14].

Obtaining the TLS solution $\boldsymbol{W}_{TLS}$ involves the singular value decomposition of the expanded matrix $\boldsymbol{Y} = [\boldsymbol{X}, \boldsymbol{d}]$. Hence, it is generally quite burdensome and very time consuming, and this prevents the TLS algorithm from being widely used, especially in real-time applications. To tackle this problem, we present a neural network approach as follows.

The architecture of the neural network for the TLS problem is shown in Figure 2.6, which consists of five parts, called the main-network, Subnetwork 1, Subnetwork 2, Subnetwork 3, and Subnetwork 4, respectively. The neurons in the main-network and subnetworks are interconnected with one another.

The main-network has N neurons. The i'th neuron is modeled as an amplifier with a nonlinear input–output relationship $g_i(\cdot)$, and has an input capacitor C_i. The terms $v_i(t)$ and $u_i(t)$ stand for the output and input voltages of the i'th neuron. The relationship $g_i(\cdot)$ is selected to be a monotonically increasing function.

Subnetwork 1 has M neurons with an input–output relationship $f_j(z) = K_2 z$; that is, all neurons in this subnetwork are transresistance amplifiers (for convenience in the following descriptions, we let $K_2 = 1$). The output voltage of the j'th neuron is expressed as $v_{1,j}(t)$. b_j is the bias current of the j'th neuron.

The j'th neuron in Subnetwork 1 receives the current from the i'th neuron of the main-network by an amount proportional to T_{ji}; $\boldsymbol{T} = \{T_{ji}\}$ is called the connection strength matrix. Obviously, for each neuron in Subnetwork 1, we have

$$v_{1,j}(t) = \sum_{i=1}^{N} T_{ji} v_i(t) - b_j. \tag{2.77}$$

Subnetwork 2 consists of two layers. The first layer has N neurons with the input $v_i(t)$ and output $v_i^2(t)$. The second layer has only one neuron whose output $\psi(t)$ is the sum of the neuron outputs in the first layer and is biased by unity, that is,

$$\psi(t) = \sum_{i=1}^{N} v_i^2(t) + 1. \tag{2.78}$$

Subnetwork 3 consists of M neurons. The j'th neuron has two inputs, $\psi(t)$ and $v_{1,j}(t)$. Therefore, the output $v_{3,j}(t)$ of the j'th neuron is

$$v_{3,j}(t) = v_{1,j}(t)\psi(t), \tag{2.79}$$

where the multiplication can be implemented by the CMOS analog multipliers given in [15].

Subnetwork 4 has three layers. The output of the j'th neuron in the first layer is $v_{1,j}^2(t)$. The second layer has only one neuron whose output is $\phi(t) = \sum_{j=1}^{M} v_{1,j}^2(t)$. The third layer consists of N neurons, and the i'th neuron of this layer has the output

$$v_{4,i}(t) = v_i(t)\phi(t) \qquad (i = 1, 2, \ldots, N), \tag{2.80}$$

where the multiplication can also be implemented by the CMOS analog multipliers [15].

The i'th neuron in the main-network receives the currents from the output $v_{3,j}(t)$ of the j'th neuron in Subnetwork 3 by an amount proportional to $-T_{ji}$ and from the i'th neuron output $v_{4,i}(t)$ of Subnetwork 4 by a unity connection strength. In terms of Kirchhoff's law, we have

$$C_i \frac{du_i(t)}{dt} = v_{4,i}(t) - \sum_{j=1}^{M} T_{ji} v_{3,j}(t) \qquad (i = 1, 2, \ldots, N). \tag{2.81}$$

To guarantee that this neural network is stable, we define an energy function as

$$E(t) = \frac{1}{2}\frac{\phi(t)}{\psi(t)}. \tag{2.82}$$

Because $\psi(t) = \sum_{i=1}^{N} v_i^2(t) + 1 > 0$ and $\phi(t) = \sum_{j=1}^{M} v_{1,j}^2(t) \geq 0$, we have $E(t) \geq 0$, that is, $E(t)$ is bounded from below.

Moreover, from (2.82), we have

$$\frac{dE(t)}{dt} = \frac{1}{2} \frac{\frac{d\phi(t)}{dt}\psi(t) - \frac{d\psi(t)}{dt}\phi(t)}{[\psi(t)]^2}, \tag{2.83}$$

where

$$\begin{aligned} \frac{d\phi(t)}{dt} &= 2\sum_{j=1}^{M} v_{1,j}(t)\frac{dv_{1,j}(t)}{dt} \\ &= 2\sum_{j=1}^{M} v_{1,j}(t)\sum_{i=1}^{N} T_{ji}\frac{dv_i(t)}{dt} \end{aligned} \tag{2.84}$$

and

$$\frac{d\psi(t)}{dt} = 2\sum_{i=1}^{N} v_i(t)\frac{dv_i(t)}{dt}. \tag{2.85}$$

Substituting (2.84) and (2.85) into (2.83), we have

$$\begin{aligned} \frac{dE(t)}{dt} &= \frac{\sum_{j=1}^{M} \psi(t)v_{1,j}(t)\sum_{i=1}^{N} T_{ji}\frac{dv_i(t)}{dt} - \sum_{i=1}^{N} \phi(t)v_i(t)\frac{dv_i(t)}{dt}}{[\psi(t)]^2}. \\ &= \frac{\sum_{i=1}^{N}[\sum_{j=1}^{M} \psi(t)v_{1,j}(t)T_{ji} - \phi(t)v_i(t)]\frac{dv_i(t)}{dt}}{[\psi(t)]^2}. \end{aligned} \tag{2.86}$$

Using (2.79) and (2.80), the above expression becomes

$$\frac{dE(t)}{dt} = \frac{\sum_{i=1}^{N}[\sum_{j=1}^{M} v_{3,j}(t)T_{ji} - v_{4,i}(t)]\frac{dv_i(t)}{dt}}{[\psi(t)]^2}. \tag{2.87}$$

Substituting (2.81) into the bracketed expression in the numerator of (2.87) gives

$$\begin{aligned} \frac{dE(t)}{dt} &= -\frac{\sum_{i=1}^{N} C_i\frac{du_i(t)}{dt}\frac{dv_i(t)}{dt}}{[\psi(t)]^2} \\ &= -\frac{\sum_{i=1}^{N} C_i[g^{-1}(v_i(t))]'(\frac{dv_i(t)}{dt})^2}{[\psi(t)]^2}. \end{aligned} \tag{2.88}$$

Since C_i is positive and $g^{-1}(v_i(t))$ is a monotonically increasing function, each term on the right-hand side of (2.88) is nonnegative and

$$\frac{dE(t)}{dt} \leq 0, \tag{2.89}$$

$$\frac{dE(t)}{dt} = 0 \quad \rightarrow \quad \frac{dv_i(t)}{dt} = 0 \qquad (i = 1, 2, \ldots, N). \tag{2.90}$$

Together with the fact that $E(t)$ is bounded from below, (2.89) and (2.90) show that the time evolution of this network is a motion in the state space that seeks out the minimum to $E(t)$ and comes to a stop at such a point. Hence, any optimization problem that can be mapped to the energy function (2.82) can be solved by this neural network.

For mathematical convenience in the following, we express the energy function $E(t)$ in a matrix form

$$\begin{aligned} E(t) &= \frac{1}{2}\frac{\phi(t)}{\psi(t)} = \frac{1}{2}\frac{\sum_{j=1}^{M} v_{1,j}^2(t)}{\sum_{i=1}^{N} v_i^2(t)+1} \\ &= \frac{1}{2}\frac{\sum_{j=1}^{M}(\sum_{i=1}^{N} T_{ji}v_i(t)-b_j)^2}{\sum_{i=1}^{N} v_i^2(t)+1} \\ &= \frac{1}{2}\frac{\|\boldsymbol{T}\boldsymbol{V}(t)-\boldsymbol{B}\|^2}{\boldsymbol{V}^T(t)\boldsymbol{V}(t)+1}, \end{aligned} \tag{2.91}$$

where $\boldsymbol{V}(t) = [v_1(t), v_2(t), \ldots, v_N(t)]^T$ is the output vector of the main-network and $\boldsymbol{B} = [b_1, b_2, \ldots, b_N]^T$ is the bias current vector of Subnetwork 1.

It should be noted that this network can be generalized to cases in which the matrix $\boldsymbol{T}$ and vector $\boldsymbol{B}$ take complex values. In such cases, two neurons are needed to provide the real and imaginary parts, respectively, of each complex value, and the energy function $E(t)$ becomes

$$E(t) = \frac{1}{2}\frac{\|\boldsymbol{T}\boldsymbol{V}(t)-\boldsymbol{B}\|^2}{\boldsymbol{V}^H(t)\boldsymbol{V}(t)+1}, \tag{2.92}$$

where "H" denotes the conjugate transpose.

In order to use the proposed neural network to solve the TLS problem, we first seek the mapping relationship between the TLS problem (2.71) and the energy function (2.92). According to matrix theory [14], it is easy to verify that

$$\frac{\|\boldsymbol{Y}\boldsymbol{h}\|_2}{\|\boldsymbol{h}\|_2} \geq \sigma_{N+1}, \qquad \boldsymbol{h} \neq \boldsymbol{O} \tag{2.93}$$

and that the equality holds for a nonzero vector $\boldsymbol{h}$ if and only if the vector $\boldsymbol{h}$ is in the subspace spanned by the right singular vectors corresponding to the smallest singular value σ_{N+1} of the matrix $\boldsymbol{Y}$, where $\|\cdot\|_2$ is the l_2 norm. Combining this fact with Equation (2.75), we see that the TLS problem (2.71) amounts to finding a vector $\boldsymbol{W}$ such that

$$\frac{\left\|\boldsymbol{Y}\begin{pmatrix}\boldsymbol{W}\\-1\end{pmatrix}\right\|_2}{\left\|\begin{pmatrix}\boldsymbol{W}\\-1\end{pmatrix}\right\|_2} = \sigma_{N+1}. \tag{2.94}$$

Equation (2.94) can be written in another form:

$$\min_{W} \frac{1}{2} \frac{\|XW - d\|^2}{W^H W + 1}. \tag{2.95}$$

If we compare the TLS problem by (2.95) with the energy function by (2.92), we know that they have the same minimum as long as we choose $T = X$, $B = d$, and $V(t) = W$, which means that the output vector $V(t)$ of the main-network is used to provide the vector W, and the available data matrix X and vector d are taken directly as the connection strength matrix T and bias current vector B of the network, respectively. If the other parameters of this network are selected as for (2.82), the above analyses show that this network is stable and provides the minimum of the TLS problem by (2.95). In other words, the output vector $V_f = \lim_{t\to\infty} V(t)$ of the main-network in the stationary state is the TLS solution W_{TLS}.

The smallest singular value σ_{N+1}, easily obtained from the energy function $E_f = \lim_{t\to\infty} E(t)$ in the stationary state of the network, is

$$\sigma_{N+1} = 2E_f. \tag{2.96}$$

In addition, the initial state vector $V(0)$ of the main-network should be nonzero so as to avoid a zero solution.

Because this network is based on the analog circuit architecture (which has continuous-time dynamics), the convergence time (the time to approach closely the stationary state) is within an elapsed time of only a few characteristic time constants of the network, which shows that this network can provide the TLS solutions in real time with an arbitrarily small error.

Two sets of simulation results are given as follows. In the simulations, we used $g_i(u) = u$ (for $i = 1, 2, \ldots, N$), $K_2 = 1$, and $C = 100$ pF. V_f is the output vector of the main-network in the stationary state, σ_{N+1} is the exact smallest singular value of the expanded matrix Y, and $\sigma_f = 2E_f$; obviously, $\sigma_{N+1} \approx \sigma_f$. Tables 2.3 and 2.4 show the time evolution of the outputs in the main-network. Figures 2.7 and 2.8 are the dynamic curves of the first output neuron in the main-network.

Example 2.3

$$Y^T Y = \begin{pmatrix} 1.6500 & -2.0000 & 0.5000 \\ -2.0000 & 4.6500 & -1.0000 \\ 0.5000 & -1.0000 & 0.9000 \end{pmatrix}$$

$$\sigma_{N+1} = 0.8062, \qquad \sigma_f = 0.8063$$

$$V_f = \begin{pmatrix} 1.106019 \\ 0.821224 \end{pmatrix}$$

Table 2.3. *The time evolution of the outputs in the main-netowrk*

t (10^{-10} seconds)	$v_1(t)$	$v_2(t)$
0	1.000000	1.00000
1	1.017499	0.973549
2	1.032095	0.950667
⋮	⋮	⋮
22	1.105154	0.822897
23	1.105622	0.821991

Table 2.4. *The time evolution of the outputs in the main-network*

t (10^{-10} seconds)	$v_1(t)$	$v_2(t)$	$v_3(t)$
0	1.000000	1.000000	1.000000
1	1.001824	0.999194	1.014673
2	1.003504	0.998287	1.028989
⋮	⋮	⋮	⋮
120	0.977810	0.873353	1.488217
121	0.977669	0.873099	1.488628

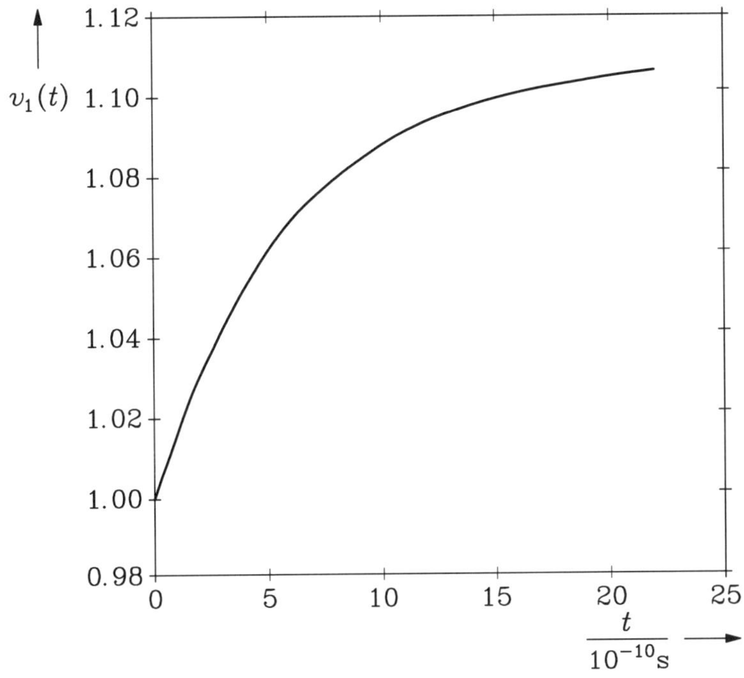

Figure 2.7. The dynamic curve of the output of the first neuron in the main-network (corresponding to Table 2.3).

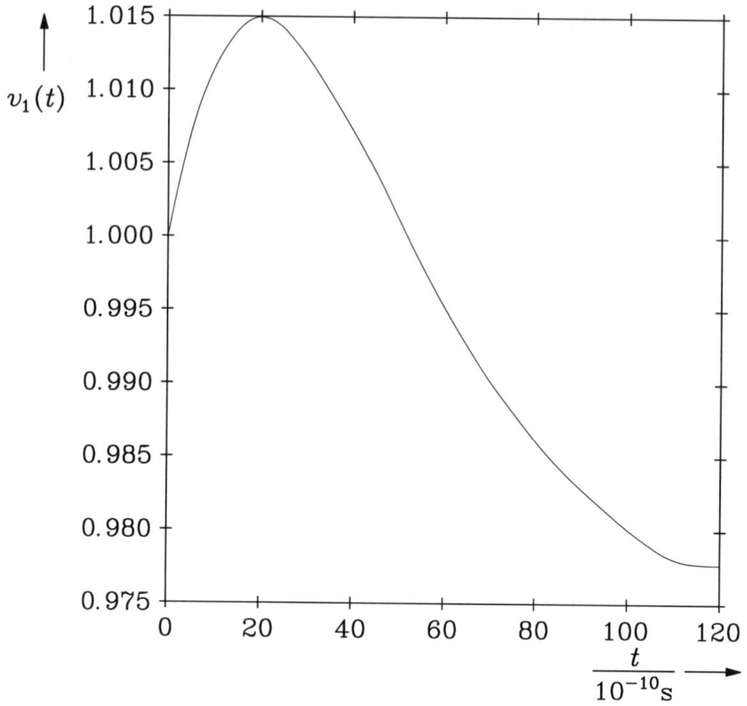

Figure 2.8. The dynamic curve of the output of the first neuron in the main-network (corresponding to Table 2.4).

Example 2.4

$$\boldsymbol{Y}^T\boldsymbol{Y} = \begin{pmatrix} 0.7644 & 0.0252 & -0.0384 & 0.0864 \\ 0.0252 & 0.7941 & -0.0672 & 0.1512 \\ -0.0384 & -0.0672 & 0.8524 & -0.2304 \\ 0.0864 & 0.1512 & -0.2304 & 1.2684 \end{pmatrix}$$

$$\sigma_{N+1} = 0.8660, \qquad \sigma_f = 0.8662$$

$$\boldsymbol{V}_f = \begin{pmatrix} 0.977532 \\ 0.872851 \\ 1.489027 \end{pmatrix}$$

2.5 Neural Networks for a Class of Nonlinear Filters

Linear filters have a poor performance in cases where the noise is non-Gaussian or not additive. Recently, a growing interest has been devoted to the applications and synthesis of nonlinear filters. Many different classes of adaptive filters have been developed [3, 16]. Palmieri and Boncelet [17] described a large class of nonlinear

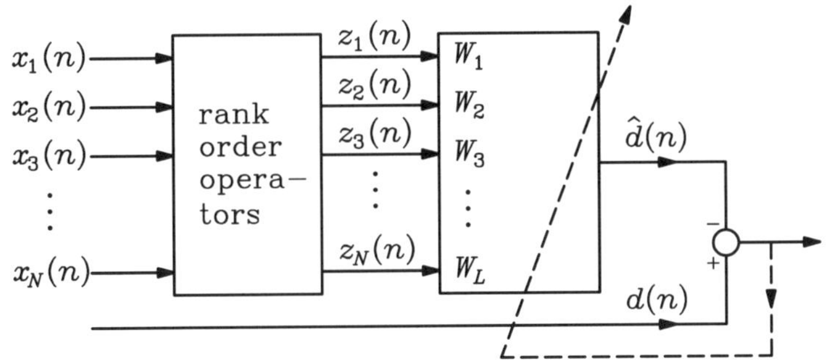

Figure 2.9. A class of nonlinear filters.

filters that includes order statistic filters [18] (OSF or L-filters), C filters (Ll-filters) [19], Volterra filters [20], and ZNL-LTI filters [21]. This class of filters is based on the compositions of linear filters and rank-order operators (mappings) as shown in Figure 2.9.

The N-dimensional input vector $\boldsymbol{X}(n) = [x_1(n), x_2(n), \ldots, x_N(n)]^T$ is first transformed to an L-dimensional vector $\mathbf{Z}(n) = [z_1(n), z_2(n), \ldots, z_L(n)]^T$ (for $n = 1, 2, \ldots, M$) by an operator ζ, which is fixed a priori, and then $\mathbf{Z}(n)$ is taken as the input of a linear digital FIR filter with weight vector $\boldsymbol{W} = [W_1, W_2, \ldots, W_L]^T$. The variable $d(n)$ (for $n = 1, 2, \ldots, M$) is known and $\hat{d}(n)$ is

$$\hat{d}(n) = \sum_{i=1}^{L} z_i(n) W_i \qquad (n = 1, 2, \ldots, M). \tag{2.97}$$

Different nonlinear filters are defined by different operators ζ as follows.

Volterra Filters

For Volterra filters, ζ forms a subset of all possible cross products of the elements of $\boldsymbol{X}(n)$. Namely, the generic element of $\mathbf{Z}(n)$ may take the form

$$z_i(n) = x_{i_1}(n) x_{i_2}(n) \cdots x_{i_k}(n) \qquad (i = 1, 2, \ldots, L), \tag{2.98}$$

where the vector of the indexes $(i_1, i_2, \ldots, i_k)$ with $i_k \leq N$ selects for each i a subset of the elements of $\boldsymbol{X}(n)$.

ZNL-LTI Filters

Each element of the vector $\mathbf{Z}(n)$ has the form

$$z_i(n) = \zeta_i\big(x_i(n)\big) \qquad (i = 1, 2, \ldots, L), \tag{2.99}$$

where ζ_i is a zero-memory, nonlinear and Borel-measurable function.

L-Filters

For L-filters, the vector $\boldsymbol{X}(n)$ forms the vector $\mathbf{Z}(n)$ as

$$\mathbf{Z}(n) = \left[x_{(1)}(n), x_{(2)}(n), \ldots, x_{(N)}(n)\right]^T, \tag{2.100}$$

which means that the elements of $\mathbf{Z}(n)$ are the elements of $\boldsymbol{X}(n)$ arranged in increasing order: $x_{(1)}(n) \leq x_{(2)}(n) \leq \ldots \leq x_{(N)}(n)$.

Ll-Filters

The vector $\mathbf{Z}(n)$ has N^2 elements and can be denoted as

$$\begin{aligned}\mathbf{Z}(n) = \big[&x_{(1)1}(n), x_{(1)2}(n), \ldots, x_{(1)N}(n)| \ldots | \\ &x_{(N)1}(n), x_{(N)2}(n), \ldots, x_{(N)N}(n)\big]^T\end{aligned} \tag{2.101}$$

with

$$x_{(i)j}(n) = \begin{cases} x_{(j)}(n), & x_{(i)}(n) \leftrightarrow x_j(n), \\ 0, & \text{else} \end{cases} \qquad (i, j = 1, 2, \ldots, N), \tag{2.102}$$

where the notation $x_{(i)}(n) \leftrightarrow x_j(n)$ means that the j'th element in the sample vector $\boldsymbol{X}(n)$ occupies the i'th location in the ranked vector by (2.100).

This class of nonlinear filters is composed of two distinct parts and involves two estimation procedures: a filter, which estimates the output $\hat{d}(n)$ from the sample vector $\mathbf{Z}(n)$, and a block, which estimates the optimum weight coefficient vector. The appealing feature of this class of nonlinear filters is that the weight estimation problem can be formulated as a linear problem on the transformed sample space. Thus, the optimum coefficient weight vector under the LS criterion is the solution of the following optimization problem:

$$\min_{\mathbf{W}} \sum_{n=1}^{M} \left|d(n) - \mathbf{Z}^T(n)\mathbf{W}\right|^2. \tag{2.103}$$

Comparing (2.1) and (2.103) shows that any algorithm for solving (2.1) can be immediately extended to solving (2.103) as long as we replace $\boldsymbol{X}(n)$ with $\mathbf{Z}(n)$. However, the computational complexity of (2.103) is usually more intensive than that of (2.1), mainly because $L > N$.

For the real-time applications of this class of nonlinear filters, we use the neural network in Figure 2.2 to solve (2.103) by selecting the connection matrix $\boldsymbol{T}$ as

$$\boldsymbol{T} = \begin{pmatrix} z_1(1) & z_2(1) & \cdots & z_L(1) \\ z_1(2) & z_2(2) & \cdots & z_L(2) \\ \vdots & \vdots & \ddots & \vdots \\ z_1(M) & z_2(M) & \cdots & z_L(M) \end{pmatrix}.$$

If the other parameters of the network are the same as for (2.1), then according to the analysis in Section 2.1, this network can solve (2.103) during an elapsed time of a few characteristic time constants of the network with an arbitrarily small error.

Because $L > N$, the neurons needed on the right side of the network for solving (2.103) are more than those needed for (2.1), but this has almost no effect on the computational time (the time to approach closely the stationary state) of the network.

In addition, the neural network approaches proposed in Sections 2.2 and 2.4 for the RLS and the TLS algorithms can also be applied to the weight coefficient vector estimation of this class of nonlinear filters.

One of the disadvantages of these neural network approaches for this class of nonlinear filters is that they cannot perform the nonlinear operator (mapping) ζ. The realization of the mapping of nonlinear filters will be dealt with in the next two sections.

2.6 Neural Networks for General Nonlinear Filters

2.6.1 Fundamentals

As shown in the preceding five sections, filtering is composed of two distinct estimation (computation) procedures. One is the estimation of the mapping (transformation) from the available samples. The other is the estimation of the output of the filter from the input by using this mapping (or say, the realization of the mapping). For a linear filter, it is not difficult to realize the mapping once the mapping is available. This is the reason we emphasized in the previous sections how to use neural networks to estimate the mapping (weight coefficients) of a linear filter. For a nonlinear filter, the realization of the mapping is not as easy as that of linear filters. How to estimate and to realize effectively the mapping of nonlinear filters is the current research focus in this field.

The nonlinear mapping capability and the corresponding learning algorithm of the MLP network provide us with a new approach to attack the above problems of nonlinear filters.

A general nonlinear filter is shown in Figure 2.10, where $\boldsymbol{X}(n) = [x_1(n), x_2(n), \ldots, x_N(n)]^T$ is the input vector, $d(n)$ is the known variable, $g(\cdot)$ stands for the mapping of this nonlinear filter, $\hat{d}(n)$ is the output

$$\hat{d}(n) = g\big(\boldsymbol{X}(n)\big), \tag{2.104}$$

and

$$\varepsilon(n) = d(n) - \hat{d}(n). \tag{2.105}$$

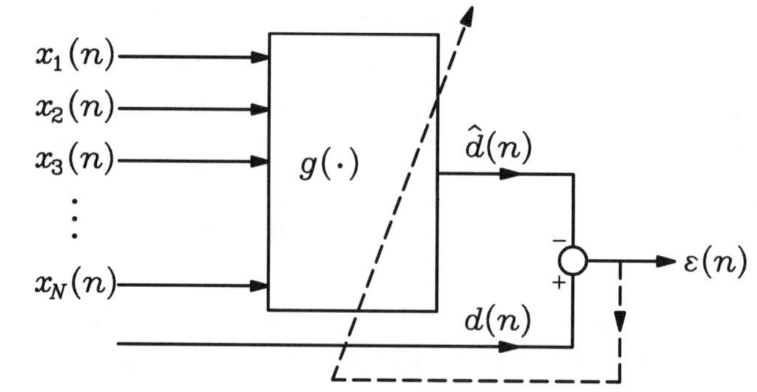

Figure 2.10. A general nonlinear filter.

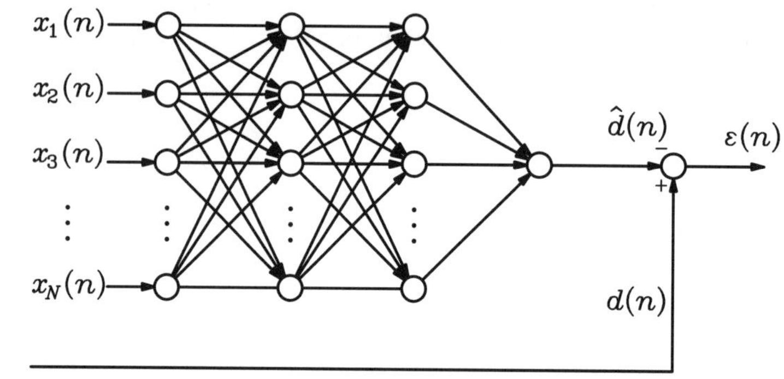

Figure 2.11. An MLP network for nonlinear filters ($i = 4$).

The MLP network shown in Figure 2.11 can be used to approximate this nonlinear filter; that is,

$$\hat{d}(n) = f_1\left(\sum_{j=1}^{N_{i-1}} W_{1j}^{(i-1)} y_j^{(i-1)}(n) - \theta_1\right), \tag{2.106}$$

where

$$y_j^{(i-1)}(n) = f_j^{(i-1)}\left(\sum_{k=1}^{N_{i-2}} W_{jk}^{(i-2)} y_k^{(i-2)}(n) - \theta_j^{(i-1)}\right). \tag{2.107}$$

The i'th layer is the output layer that has one neuron with the function $f_1(\cdot)$ and threshold θ_1; $y_j^{(i-1)}(n)$, $f_j^{(i-1)}$, and $\theta_j^{(i-1)}$ are the output, activation function, and threshold of the j'th neuron in the $(i-1)$th layer, respectively; $W_{1j}^{(i-1)}$ is the connection weight between the output neuron and the j'th neuron in the $(i-1)$th layer; $W_{jk}^{(i-2)}$ is the connection weight between the j'th neuron in the $(i-1)$th layer and the k'th neuron in the $(i-2)$'th layer; N_{i-1} and N_{i-2} are the number of the neurons of the $(i-1)$th layer and the $(i-2)$'th layer; and $y_k^{(0)}(n) = x_k(n)$ (the input sample).

Different neuron activation functions, connection weights, and thresholds will define different mappings $g(\cdot)$. In other words, the estimation of the mapping $g(\cdot)$ of nonlinear filters can be accomplished by selecting suitable neuron functions, connection weights, and thresholds. The criterion employed to select these parameters may be the least-squares so that

$$\min_{\boldsymbol{W},\boldsymbol{F},\boldsymbol{\Theta}} \sum_{n=1}^{M} \left(d(n) - \hat{d}(n)\right)^2 \tag{2.108}$$

or the least-mean-square so that

$$\min_{\boldsymbol{W},\boldsymbol{F},\boldsymbol{\Theta}} E\left[\left(d(n) - \hat{d}(n)\right)^2\right] \tag{2.109}$$

or the least-mean-absolute value so that

$$\min_{\boldsymbol{W},\boldsymbol{F},\boldsymbol{\Theta}} E\left[\left|d(n) - \hat{d}(n)\right|\right], \tag{2.110}$$

where $E(\cdot)$ denotes the mathematical expectation and $\boldsymbol{W}$, $\boldsymbol{F}$, and $\boldsymbol{\Theta}$ denote the sets of connection weights, activation functions, and thresholds, respectively. Obviously, to estimate the mapping $g(\cdot)$, we have to solve the complicated nonlinear programming problem (2.108) (or (2.109), (2.110)). However, in practical applications, the form of the activation functions (such as the sigmoid function, the piecewise-linear function, or the unit-step function) and the thresholds are usually fixed a priori. As a result, we can employ the BP algorithm described in Section 1.4 to estimate adaptively the connection weights. If a fast convergence speed is desired, the other available learning algorithms [22, 23] can be used.

We have illustrated the fundamentals of the neural network approach to nonlinear filtering. For further illustration, we will present two examples: one for signal prediction (in Subsection 2.6.2), the other for generalized stack filters (in Section 2.7).

2.6.2 An Application Example: Signal Prediction

Signal prediction from past observations is a basic signal processing operation by use of filters. It has applications in many diverse fields such as target tracking, weather forecasting, and financial market forecasting. Conventional parametric approaches to this problem involve mathematical modeling of the signal characteristics, which is then used to carry out the prediction. In a general case, this is a rather complex task involving many steps such as model hypothesis, identification and estimation of model parameters, and their verification. Moreover, it will give rise to a very poor prediction performance if the modeling of signal is not suitable [24]. Using the MLP network presented in Subsection 2.6.1, we can bypass the modeling phase and perform a nonlinear and nonparametric signal prediction [25].

Let us consider the prediction of a signal denoted by $x(n)$ and concentrate on the one-step-ahead prediction (only for simplicity) based on the past N observations, that is,

$$\hat{x}(n+1) = g\big(x(n), x(n-1), \ldots, x(n-N+1)\big), \tag{2.111}$$

where $g(\cdot)$ is a nonlinear mapping that can obviously be performed by the MLP network of Figure 2.11. We usually select three layers for (2.111). The neurons in the input layer and output layer are all linear amplifiers with unity gains. The second layer has N neurons with a sigmoid activation function. The connection weights among the neurons in the input layer and the second layer are all unity. The thresholds of all neurons are zeros. As a result, the unknown variables for performing (2.111) are only the connection weights $W_{1j}^{(2)}$ (for $j = 1, 2, \ldots, N$) between the output neuron and the j'th neuron in the second layer, which can be trained by the available sample set $\{x(1), x(2), \ldots, x(L)\}$ (where $L \gg N$, usually). If we divide this set into M overlapping subsets of length $N+1$, the first N points of each subset are used as inputs to the network and the $(N+1)$th point is designated as the corresponding desired output ($d(n)$ in Figure 2.11). The error during the training phase is

$$\sum_{m=1}^{M} x(mN+1) - \sum_{j=1}^{N} W_{1j}^{(2)} f\big(x(mN+1-j)\big), \tag{2.112}$$

where $f(\cdot)$ denotes the sigmoid function.

Once the weights $W_{1j}^{(2)}$ are obtained by some algorithm (such as the BP algorithm), the prediction of (2.111) can be given by

$$\hat{x}(n+1) = \sum_{j=1}^{N} W_{1j}^{(2)} f\big(x(n+1-j)\big). \tag{2.113}$$

It should be noted that the weight vector $\boldsymbol{W}^{(2)} = \{W_{1j}^{(2)}\}$ under the LS criterion can be obtained in an explicit form; that is,

$$\min_{\boldsymbol{W}^{(2)}} \sum_{m=1}^{M} \left| (x(mN+1)) - \sum_{j=1}^{N} W_{1j}^{(2)} f\big(x(mN+1-j)\big) \right|^2 \tag{2.114}$$

and

$$\boldsymbol{W}^{(2)} = \boldsymbol{Z}^{+}\boldsymbol{d} = \lim_{\alpha \to 0} (\boldsymbol{Z}^T \boldsymbol{Z} + \alpha \boldsymbol{I})^{-1} \boldsymbol{Z}^T \boldsymbol{d}, \tag{2.115}$$

where

$$\boldsymbol{Z} = \begin{pmatrix} f(x(N)) & f(x(N-1)) & \cdots & f(x(1) \\ f(x(2N)) & f(x(2N-1)) & \cdots & f(x(N+1)) \\ \vdots & \vdots & \ddots & \vdots \\ f(x(MN)) & f(x(MN-1)) & \cdots & f(x((M-1)N+1)) \end{pmatrix}$$

and $\boldsymbol{d} = [x(N+1), x(2N+1), \ldots, x(MN+1)]^T$.

Obviously, Equation (2.115) can be computed by the network of Figure 2.2 with the connection strength matrix $\boldsymbol{T} = \boldsymbol{Z}$ and the bias current vector $\boldsymbol{B} = \boldsymbol{d}$.

In summary, we have used two kinds of neural networks to perform the computations required in the nonlinear and nonparametric signal prediction: The MLP network is used to realize the nonlinear mapping, and the analog neural network of Figure 2.2 is used to estimate the connection weight vector (the nonlinear mapping).

2.7 Neural Networks for Generalized Stack Filters

Generalized stack (GS) filters comprise an important class of nonlinear filters that includes stack filters and discrete-time morphological filters [26]. The definition of GS filters is based on the threshold decomposition and Boolean operators [27]. In other words, any filter that possesses both the threshold decomposition property and the consistency property (the stacking property) is known as a stack filter. The threshold decomposition first maps the M-level input signal into $M-1$ binary signals by thresholding the original signal at each of the allowable levels. The set of $M-1$ binary output signals is then filtered by $M-1$ Boolean operators, which are constrained to have the stacking property. The output is finally obtained as the sum of the $M-1$ binary output signals. The computational complexity for the optimal GS filter is very intensive mainly because of the truth table representation of Boolean functions. For a Boolean function of M variables, there are 2^M entries in the truth table that have to be determined in the optimal GS filtering algorithms. In the following, we will show that the MLP network in Figure 2.11 with a unit step activation function can effectively realize the GS filters [28]; however, we first give some necessary definitions concerning GS filters.

Definition of the Threshold Decomposition

The threshold decomposition of an M-level quantized signal vector $\boldsymbol{X}(n) = [x_1(n), x_2(n), \ldots, x_N(n)]$ at time n is the set of $(M-1)$ binary sequences, called threshold signals $\boldsymbol{x}^m(n)$, whose i'th element is defined by

$$x_i^m(n) = T^m(x_i(n)) = \begin{cases} 1 & x_i(n) \geq m \\ 0 & \text{else} \end{cases} \qquad (m = 1, 2, \ldots, M-1). \tag{2.116}$$

Obviously, $\boldsymbol{X}(n)$ is the sum of its binary threshold signals

$$\boldsymbol{X}(n) = \sum_{m=1}^{M-1} T^m(\boldsymbol{X}(n)) = \sum_{m=1}^{M-1} \boldsymbol{x}^m(n) \qquad (m = 1, 2, \ldots, M-1). \tag{2.117}$$

In addition,

$$\begin{aligned}\boldsymbol{x}^m(n) &= \left[x_1^m(n), x_2^m(n), \ldots, x_N^m(n)\right] \\ &= \left[T^m\left(x_1(n)\right), T^m\left(x_2(n)\right), \ldots, T^m\left(x_N(n)\right)\right] \\ &\qquad (m = 1, 2, \ldots, M-1).\end{aligned} \tag{2.118}$$

By stacking $2I + 1$ threshold vectors in (2.118), we obtain a $(2I+1) \times N$ matrix at the threshold level m:

$$\boldsymbol{X}^m(n) = \begin{pmatrix} \boldsymbol{x}^{m+I}(n) \\ \boldsymbol{x}^{m+I-1}(n) \\ \vdots \\ \boldsymbol{x}^{m-I}(n) \end{pmatrix} = \begin{pmatrix} x_1^{m+I}(n) & x_2^{m+I}(n) & \cdots & x_N^{m+I}(n) \\ x_1^{m+I-1}(n) & x_2^{m+I-1}(n) & \cdots & x_N^{m+I-1}(n) \\ \vdots & \vdots & \ddots & \vdots \\ x_1^{m-I}(n) & x_2^{m-I}(n) & \cdots & x_N^{m-I}(n) \end{pmatrix}$$

$$(m = 1, 2, \ldots, M-1), \tag{2.119}$$

where I is an integer. Based on (2.116), we can write $\boldsymbol{X}^m(n)$ in another form:

$$\boldsymbol{X}^m(n) = T^m\left(\bar{\boldsymbol{X}}(n)\right), \tag{2.120}$$

where

$$\bar{\boldsymbol{X}}(n) = \begin{pmatrix} x_1(n) - I & x_2(n) - I & \cdots & x_N(n) - I \\ x_1(n) - I + 1 & x_2(n) - I + 1 & \cdots & x_N(n) - I + 1 \\ \vdots & \vdots & \ddots & \vdots \\ x_1(n) + I & x_2(n) + I & \cdots & x_N(n) + I \end{pmatrix}. \tag{2.121}$$

To account for the upper I and lower I thresholded binary array we let $\boldsymbol{x}^m(n)$ be a zero-valued vector for $m > M - 1$ and a unity-valued vector for $m < 1$.

Definition of the Stacking Property

Let $\boldsymbol{X} = \{x_{i,j}\}$ and $\boldsymbol{Y} = \{y_{i,j}\}$ be binary arrays with the same dimension. Then $\boldsymbol{X}$ is said to stack on $\boldsymbol{Y}$ and is denoted as $\boldsymbol{X} \leq \boldsymbol{Y}$ if and only if $x_{i,j} \leq y_{i,j}$ for all i and j. In addition, we say $\boldsymbol{X} = \boldsymbol{Y}$ if and only if $x_{i,j} = y_{i,j}$ for all i and j. Consequently, we have $\boldsymbol{X} < \boldsymbol{Y}$ if $\boldsymbol{X} \leq \boldsymbol{Y}$ and $\boldsymbol{X} \neq \boldsymbol{Y}$ [28].

With the above definitions, the stacking property holds for the binary thresholded arrays in (2.119). That is,

$$\boldsymbol{X}^1(n) \geq \boldsymbol{X}^2(n) \geq \cdots \geq \boldsymbol{X}^{M-1}(n) \tag{2.122}$$

because each element of $\boldsymbol{X}^{m_1}(n)$ is greater than or equal to the corresponding element of $\boldsymbol{X}^{m_2}(n)$ for $1 \leq m_1 \leq m_2 \leq M-1$.

Definition of the Stacking Set of Boolean Functions

The ordered set of $M-1$ Boolean functions $\{f^1(\cdot), f^2(\cdot), \ldots, f^{M-1}(\cdot)\}$ is called a stacking set of Boolean functions if

$$f^{m+1}\left(\boldsymbol{X}^{m+1}(n)\right) \leq f^m\left(\boldsymbol{X}^m(n)\right) \qquad (m = 1, 2, \ldots, M-2). \tag{2.123}$$

Definition of GS Filters

An M-valued generalized stack filter $F(\cdot)$ with a window width N is a stacking set of $M-1$ Boolean functions [26].

If the input of a GS filter is $\boldsymbol{X}(n)$, the output $\hat{d}(n)$ of this GS filter is defined by

$$\hat{d}(n) = F\left(\boldsymbol{X}(n)\right) = \sum_{n=1}^{M-1} f^m\left(\boldsymbol{X}^m(n)\right). \tag{2.124}$$

An example of a GS filter is shown in Figure 2.12.

Besides the Boolean functions, $f^m(\cdot)$ in (2.124) could be any operation such as a linear or a nonlinear operation. If a linear operation is used on each threshold level, the corresponding filters are called microstatistics filters [29].

In order to use the MLP network shown in Figure 2.11 to perform the operation of a GS filter by (2.124), we must select the three layers and connection weight matrices as follows:

(1) The input layer has $(2I+1) \times (MN-N)$ neurons with a linear activation function; $\boldsymbol{x}^m(n)$ (for $m = 1, 2, \ldots, M-1$) is the input array corresponding to the threshold level m.

(2) The second layer has $M-1$ neurons with unit step activation function $U(\cdot)$. The threshold of the m'th neuron is denoted by $\theta^m(n)$.

(3) The output layer performs the summation of the $M-1$ outputs of the second layer (i.e., the connection weight between the output neuron and any neuron in the second layer is unity).

(4) The connection weight matrix between the input array $\boldsymbol{x}^m(n)$ and the m'th

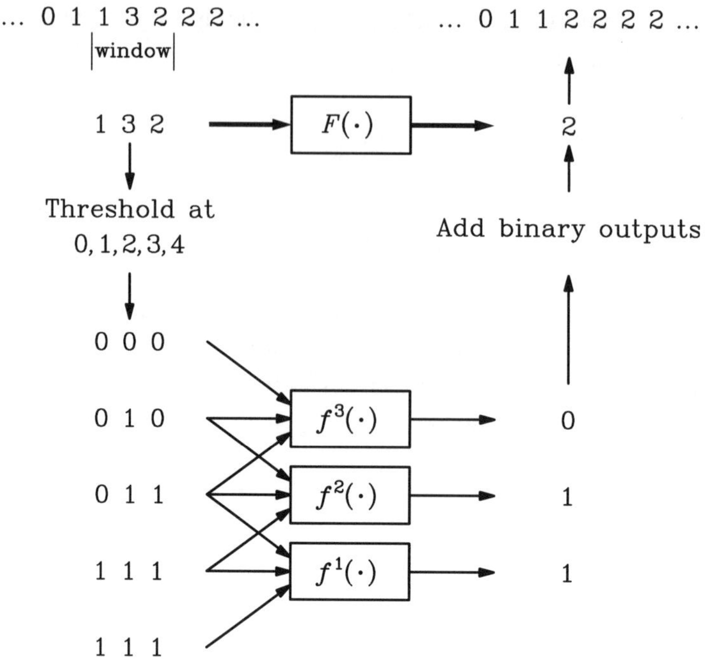

Figure 2.12. A GS filter with $N = 3$ and $2I + 1 = 3$. From Reference [28].

neuron in the second layer is written as $\boldsymbol{W}^m(n)$ and defined as

$$\boldsymbol{W}^m(n) = \begin{pmatrix} W^m_{-I,1}(n) & W^m_{-I,2}(n) & \cdots & W^m_{-I,N}(n) \\ W^m_{-I+1,1}(n) & W^m_{-I+1,2}(n) & \cdots & W^m_{-I+1,N}(n) \\ \vdots & \vdots & \ddots & \vdots \\ W^m_{I,1}(n) & W^m_{I,2}(n) & \cdots & W^m_{I,N}(n) \end{pmatrix}. \tag{2.125}$$

The connection weights between the input array $\boldsymbol{x}^{m_1}(n)$ and the m_2'th neuron in the second layer are all zeros for any $m_1 \neq m_2$, (for $m_1, m_2 = 1, 2, \ldots, M-1$).

As a result, we have the output $\hat{d}(n)$ of this network as

$$\hat{d}(n) = \sum_{m=1}^{M-1} U\big(\boldsymbol{W}^m(n) \odot \boldsymbol{x}^m(n) - \theta^m(n)\big), \tag{2.126}$$

where $\odot$ denotes the sum of the products of the corresponding elements in two matrices.

From (2.126), we know that this MLP network can be viewed as an extension of a GS filter by dropping the stacking constraints. In other words, a GS filter can be realized by this MLP network as long as the stacking property is satisfied. The following theorem concerns the conditions under which the stacking property is satisfied.

Theorem 2.2 *If*

$$W_{i,j}^{m+1}(n) \le W_{i,j}^{m}(n) \tag{2.127}$$

and

$$\theta^m(n) \le \theta^{m+1}(n)$$
$$(i = -I, -I+1, \ldots, I; \quad j = 1, 2, \ldots, N; \quad m = 1, 2, \ldots, M-2) \tag{2.128}$$

then the stacking condition of GS filters holds, that is,

$$U\left(\mathbf{W}^{m+1}(n) \odot \boldsymbol{x}^{m+1}(n) - \theta^{m+1}(n)\right) \le U\left(\mathbf{W}^{m}(n) \odot \boldsymbol{x}^{m}(n) - \theta^{m}(n)\right). \tag{2.129}$$

The proof of this theorem is straightforward according to the definition of the stacking property of GS filters.

This MLP network for GS filters requires much fewer parameters than the corresponding truth tables, which greatly reduces the computational complexity for finding out the optimal GS filter. The adaptive algorithms for the optimal GS filters under the LMS criterion by (2.109) and LMA criterion by (2.110) with the stacking constraints are as follows:

Under the LMS criterion we have

$$\begin{cases} \min E\left|d(n) - \hat{d}(n)\right|^2 \\ s.t. \quad W_{i,j}^{m+1}(n) \le W_{i,j}^{m}(n) \\ s.t. \quad \theta^m(n) \le \theta^{m+1}(n) \end{cases} \tag{2.130}$$

$$(i = -I, -I+1, \ldots, I; \quad j = 1, 2, \ldots, N; \quad m = 1, 2, \ldots, M-2).$$

Using the gradient descent method, we have

$$\begin{aligned} \mathbf{W}^m(n) = \mathbf{W}^m(n-1) &+ 2\gamma(n)\left|d(n) - \hat{d}(n)\right|\left(d^m(n) - \hat{d}^m(n)\right) \\ &\times \hat{d}^m(n)\left(1 - \hat{d}^m(n)\right)\boldsymbol{x}^m(n), \end{aligned} \tag{2.131}$$

$$\begin{aligned} \theta^m(n) = \theta^m(n-1) &- 2\gamma(n)\left|d(n) - \hat{d}(n)\right|\left(d^m(n) - \hat{d}^m(n)\right) \\ &\times \hat{d}^m(n)\left(1 - \hat{d}^m(n)\right), \end{aligned} \tag{2.132}$$

and

$$W_{i,j}^m(n) = W_{i,j}^{m+1}(n), \quad \text{if} \quad W_{i,j}^{m+1}(n) > W_{i,j}^m(n), \tag{2.133}$$

$$\theta^m(n) = \theta^{m+1}(n), \quad \text{if} \quad \theta^m(n) > \theta^{m+1}(n)$$
$$(i = -I, -I+1, \ldots, I; \quad j = 1, 2, \ldots, N; \quad m = 1, 2, \ldots, M-2). \tag{2.134}$$

Under the LMA criterion we have

$$\begin{cases} \min E\left|d(n) - \hat{d}(n)\right| \\ s.t \quad W_{i,j}^{m+1}(n) \leq W_{i,j}^m(n) \\ s.t \quad \theta^m(n) \leq \theta^{m+1}(n) \end{cases} \tag{2.135}$$

$$(i = -I, -I+1, \ldots, I; \quad j = 1, 2, \ldots, N; \quad m = 1, 2, \ldots, M-2).$$

Similarly to (2.131)–(2.134), it follows that

$$\boldsymbol{W}^m(n) = \boldsymbol{W}^m(n-1) + 2\gamma(n)\big(d^m(n) - \hat{d}^m(n)\big)\hat{d}^m(n)\big(1 - \hat{d}^m(n)\big)\boldsymbol{x}^m(n), \tag{2.136}$$

$$\theta^m(n) = \theta^m(n-1) - 2\gamma(n)\big(d^m(n) - \hat{d}^m(n)\big)\hat{d}^m(n)\big(1 - \hat{d}^m(n)\big), \tag{2.137}$$

$$W_{i,j}^m(n) = W_{i,j}^{m+1}(n), \quad \text{if} \quad W_{i,j}^{m+1}(n) > W_{i,j}^m(n), \tag{2.138}$$

and

$$\theta^m(n) = \theta^{m+1}(n), \quad \text{if} \quad \theta^m(n) > \theta^{m+1}(n) \tag{2.139}$$
$$(i = -I, -I+1, \ldots, I; \quad j = 1, 2, \ldots, N; \quad m = 1, 2, \ldots, M-2),$$

where $\hat{d}^m(n)$ is the output of the mth neuron in the second layer, that is,

$$\hat{d}^m(n) = U\big(\boldsymbol{W}^m(n) \odot \boldsymbol{x}^m(n) - \theta^m(n)\big) \qquad (m = 1, 2, \ldots, M-1), \tag{2.140}$$

$d^m(n)$ is the desired value corresponding to $\hat{d}^m(n)$, and $\gamma(n)$ is the convergence factor. For simplicity, $\gamma(n)$ is usually not changed during the learning phase.

Equations (2.133), (2.134), (2.138), and (2.139) are to ensure that, no matter when the learning process is terminated, the resulting filters always satisfy the stacking constraints. In practice, the procedure of enforcing the stacking constraints can be further simplified. For details, see Reference [28].

Because these algorithms are based on the gradient descent method, they have the same drawbacks as the BP algorithm, that is, a slow convergence speed and the local minimum problem. Although these drawbacks can be overcome to some extent by using the methods proposed in References [22] and [23], this remains an unsolved problem.

In addition, these algorithms are presented here only in the discrete-time form. We can also use the nonlinear programming neural networks [30] to find out the continuous-time dynamic equations and corresponding analog neural networks for solving the stacking-constrained optimization problems (by (2.130) and (2.135)). These analog neural networks will provide the connection weight matrix $\boldsymbol{W}^m(n)$ and threshold $\theta^m(n)$ during an elapsed time of a few characteristic time constants of the network. However, the local minimum problem still exists in these analog neural network approaches.

Finally, it should be mentioned that References [31–39] can serve as sources for further reading concerning this chapter.

Bibliography

[1] Haykin, S., *Adaptive Filter Theory*, Prentice-Hall, Englewood Cliffs, NJ, 1991.

[2] Bellanger, M. G., *Adaptive Digital Filters and Signal Analysis*, Marcel Dekker, New York, 1987.

[3] Pitas, I. and Venetsanopoulos, A. N., *Nonlinear Digital Filters*, Kluwer, Dordrecht, 1989.

[4] Golub, G. H. and Van Loan, C. T., *Matrix Computation*, Johns Hopkins Univ. Press, Baltimore, MD, 1983.

[5] Takeda, M. and Goodman, J. W., "Neural Networks for Computation: Numerical Representation and Programming Complexity," *Appl. Optics*, Vol. 25, 1986, pp. 3033–52.

[6] Tank, D. W. and Hopfield, J. J., "Simple 'Neural' Optimization Networks: A/D Converter, Signal Decision Circuit, and Linear Programming Circuit," *IEEE Trans. on Circuits and Systems*, Vol. 33, 1986, pp. 533–41.

[7] Culhane, A. D., Peckerar, M. C., and Marrian, C. R. K., "A Neural Net Approach to Discrete Hartley and Fourier Transforms," *IEEE Trans. on Circuits and Systems*, Vol. 36, No. 5, 1989, pp. 695–703.

[8] Gioffi, J. M. and Kailath, T., "Fast RLS Transversal Filters for Adaptive Filtering," *IEEE Trans. on Acoustics, Speech and Signal Processing*, Vol. 32, No. 2, 1984, pp. 304–37.

[9] Fabre, M. and Gueguen, C., "Fast RLS Algorithms," *Proc. IEEE Int. Conf. on Acoustics, Speech and Signal Processing*, Tampa, FL, 1985, pp. 1149–52.

[10] Griffiths, J. W. R., "Adaptive Array Processing, a Tutorial," *IEE Proc.*, Pt. F, Vol. 130, 1983, pp. 3–10.

[11] Klemm, R., "Adaptive Airborne MTI: An Auxiliary Channel Approach," *IEE Proc.*, Pt. F, Vol. 134, 1987, pp. 269–76.

[12] Luenberger, D. G., *Linear and Nonlinear Programming*, Addision-Wesley, Reading, MA, 1978, pp. 366–69.

[13] Xu, L., Oja, E., and Suen, C. Y., "Modified Hebbian Learning for Curve and Surface Fitting," *Neural Networks*, Vol. 5, 1992, pp. 441–57.

[14] Golub, G. H. and Van Loan, C. T., "An Analysis of the Total Least Squares Problem," *SIAM J. Numer. Anal.*, Vol. 17, No. 6, 1980, pp. 883–93.

[15] Unbehauen, R. and Cichocki, A., *MOS Switched-Capacitor and Continuous-Time Integrated Circuits and Systems: Analysis and Design*, Springer-Verlag, New York, 1989.

[16] Coyle, E. J., Lin, J. H., and Gabbouj, M., "Optimal Stack Filtering and the Estimation and Structural Approaches to Image Processing," *IEEE Trans. on Acoustics, Speech and Signal Processing*, Vol. 37, No. 12, 1989, pp. 2037–66.

[17] Palmieri, F. and Boncelet Jr., C. G., "A Class of Nonlinear Adaptive Filters," *Proc. IEEE Int. Conf. on Acoustics, Speech and Signal Processing*, 1988, pp. 1483–86.

[18] Bovik, A. C., Huang, T. S., and Munson Jr., D. C., "A Generalization of Median Filtering Using Linear Combination of Order Statistics," *IEEE Trans. on Acoustics, Speech and Signal Processing*, Vol. 31, 1983, pp. 1342–50.

[19] Palmieri, F. and Boncelet Jr., C. G., "Design of Order Statistics Filters with Given Spectral Behavior," *Proc. Conf. on Information Science and Systems*, Johns Hopkins Univ. Press, Baltimore, MD, 1987.

[20] Sicuranza, G. L. and Ramponi, G. L., "Theory and Realization of M-D Nonlinear Digital Filters," *Proc. IEEE Int. Conf. on Acoustics, Speech and Signal Processing*, pp. 1061–64, Tokyo, Japan, 1986.

[21] McCannon, T. E., "On the Design of Nonlinear Discrete-Time Predictors," *IEEE Trans. on Information Theory*, Vol. 28, No. 2, 1982.

[22] Haykin, S., *Neural Networks, A Comprehensive Foundation*, IEEE Press, New York, 1994.

[23] Kollias, S. and Anastassion, D., "An Adaptive Least Squares Algorithm for the Efficient Training of Artificial Neural Networks," *IEEE Trans. on Circuits and Systems*, Vol. 36, No. 8, 1989, pp. 1092–101.

[24] Marple, S. L., *Digital Spectral Analysis with Applications*, Prentice-Hall, Englewood Cliffs, NJ, 1987.

[25] Khotanzad, A. and Lu, J. H., "Non-Parametric Prediction of AR Processes Using Neural Networks," *Proc. IEEE Int. Conf. on Acoustics, Speech and Signal Processing*, 1990, pp. 2551–54.

[26] Lin, J. H. and Coyle, E. J., "Minimum Mean Absolute Error Estimation Over the Class of Generalized Stack Filters," *IEEE Trans. on Acoustics, Speech and Signal Processing*, Vol. 38, No. 4, 1990, pp. 663–78.

[27] Wendt, P. D., Coyle, E. J., and Gallagher, N. C., "Stack Filter," *IEEE Trans. on Acoustics, Speech and Signal Processing*, Vol. 34, No. 4, 1986, pp. 898–911.

[28] Yin, L., Astola, J., and Neuvo, Y., "A New Class of Nonlinear Filters–Neural Filters," *IEEE Trans. on Signal Processing*, Vol. 41, No. 3, 1993, pp. 1201–22.

[29] Arce, G. R., "Microstatistics in Signal Decomposition and the Optimal Filtering Problem," *IEEE Trans. on Signal Processing*, Vol. 40, No. 11, 1992, pp. 2669–82.

[30] Kennedy, M. P. and Chua, L. O., "Neural Networks for Nonlinear Programming," *IEEE Trans. on Circuits and Systems*, Vol. 35, 1988, pp. 554–62.

[31] Cichocki, A. and Unbehauen, R., *Neural Networks for Optimization and Signal Processing*, Wiley-Teubner Verlag, Chichester, England, 1993.

[32] Cichocki, A. and Unbehauen, R., "Neural Networks for Solving Systems of Linear Equations and Related Problems," *IEEE Trans. on Circuits and Systems II*, Vol. 39, 1992, pp. 124–38.

[33] Cichocki, A. and Unbehauen, R., "Simplified Neural Networks for Solving Squares and Total Least Squares Problems in Real-Time," *IEEE Trans. on Neural Networks*, Vol. 5, No. 6, 1994, pp. 910–23.

[34] Luo, F. L. and Bao, Z., "A Neural Network Approach to the Implementation of LS

Algorithm," *Proc. IEEE Int. Conf. on Information Theory*, 1991, p. 81, Budapest, Hungary.

[35] Luo, F. L. and Bao, Z., "Neural Network Approach to Adaptive FIR Filtering and Deconvolution Problems," *Proc. IEEE Int. Conf. on Industrial Electronics* (IECON'91), 1991, pp. 1449–53, Kobe, Japan.

[36] Luo, F. L., Li, Y. D., and Bao, Z., "A Modified Hopfield Neural Network and its Applications," *Int. J. Neural Networks*, Vol. 3, No. 4, 1992, pp. 135–41.

[37] Luo, F. L. and Li, Y. D., "Neural Networks for the Exact Adaptive RLS Algorithm," *Appl. Math. and Computation*, Vol. 60, No. 2, 1994, pp. 103–12.

[38] Luo, F. L., Li, Y. D., and He, C. X., "Neural Network Approach to Total Least Square Linear Prediction Frequency Estimation Problem," *Neurocomputing*, Vol. 11, 1996.

[39] Luo, F. L. and Bao, Z., "Neural Network for a Class of Nonlinear Adaptive Filters," *Adv. Model. Simulation*, Vol. 31, No. 1, 1992, pp. 45–54.

3

Neural Networks for Spectral Estimation

Spectral estimation techniques can be split into two groups: nonparametric (conventional) methods and parametric (modern) methods. Because many fast algorithms such as the fast Fourier transform (FFT) have been used in nonparametric spectral methods, it is not too difficult to implement nonparametric methods in real time. As a result, the conventional nonparametric methods have found wide use in advanced radar, sonar, communication, speech, biomedical, geophysical, and other data processing systems. However, two problems plague nonparametric spectral estimation methods, namely, high estimation variances and low resolution, particularly, in the cases that the data are short and the signal-to-noise ratio is low. As an alternative, parametric spectral estimation methods have been proposed and extensively analyzed during the past two decades. Parametric spectral estimation methods first estimate the parameters of an underlying data-generating model and then use the model and the estimated parameters to compute the spectrum. Unfortunately, parametric spectral estimation methods are usually computationally intensive, which prevents parametric methods from being widely used, especially in real-time applications. The conflict between the estimation performance and computational complexity is the key problem in this field and has attracted great research attention [1, 2, 3]. This chapter will show how to use neural networks to attack the above problem. We will present neural network approaches for maximum entropy spectral estimation, harmonic retrieval, multichannel spectral estimation, two-dimensional spectral estimation, and higher-order spectral estimation.

3.1 Maximum Entropy Spectral Estimation by Neural Networks

Consider a zero-mean stationary time series $x(n)$ whose spectrum is defined as

$$s_x(f) = \sum_{m=-\infty}^{+\infty} R(m)e^{-j2\pi mf\Delta t}, \tag{3.1}$$

where $R(m) = E[x^*(n+m)x(n)]$ is the autocorrelation function of $x(n)$ and Δt is the unit time interval. For mathematical convenience, we let $\Delta t = 1$ in this chapter. The estimation $\hat{s}_x(f)$ of $s_x(f)$ under the maximum entropy criterion [4] is

$$\hat{s}_x(f) = \frac{\sigma_P^2}{\left|1 + \sum_{k=1}^{P} a_k e^{-j2\pi kf}\right|^2}, \tag{3.2}$$

where the a_k (for $k = 1, 2, \ldots, P$) are a set of model coefficients through which $x(n)$ can be modeled as an AR process, that is,

$$x(n) = -\sum_{k=1}^{P} a_k x(n-k) + \sigma_P^2 \delta(n). \tag{3.3}$$

P is the order of the AR process and σ_P^2 is the variance of the white noise $\sigma_P^2\delta(n)$; $\delta(n)$ is the white noise whose variance is unity.

In some cases, only the normalized spectral estimation is needed, which means that (3.2) becomes

$$\hat{s}_x(f) = \frac{1}{\left|1 + \sum_{k=1}^{P} a_k e^{-j2\pi kf}\right|^2}. \tag{3.4}$$

Equations (3.2) and (3.4) show that the key to the spectral estimation is to estimate the model coefficients a_k (for $k = 1, 2, \ldots, P$) that satisfy

$$\sum_{k=0}^{P} a_k R(m-k) = \begin{cases} \sigma_P^2, & m = 0, \\ 0, & m = 1, 2, \ldots, P, \end{cases} \tag{3.5}$$

where $a_0 = 1$. This can be written in the matrix form as

$$\begin{pmatrix} R(0) & R(-1) & \cdots & R(-P) \\ R(1) & R(0) & \cdots & R(1-P) \\ \vdots & \vdots & \ddots & \vdots \\ R(P) & R(P-1) & \cdots & R(0) \end{pmatrix} \begin{pmatrix} 1 \\ a_1 \\ a_2 \\ \vdots \\ a_P \end{pmatrix} = \begin{pmatrix} \sigma_P^2 \\ 0 \\ \vdots \\ 0 \end{pmatrix}, \tag{3.6}$$

which is called the Yule–Walker equation.

Clearly, once the $P+1$ autocorrelation values $R(k)$ (for $k = 0, 1, \ldots, P$) are known (note that $R(-k) = R^*(k)$), the model coefficients a_k (for $k = 1, 2, \ldots, P$) and the variance σ_P^2 can be determined by solving the above Yule–Walker equation. Then, using (3.4) (or (3.2)) the desired spectral estimation can be computed. However, in most of practical uses, only the samples $x(0), x(1), \ldots, x(N-1)$ are available. In these cases, the procedure to estimate the spectrum of $x(n)$ is as follows:

(1) Estimate the $P + 1$ autocorrelation values from the available samples $x(0), x(1), \ldots, x(N-1)$:

$$\hat{R}(k) = \frac{1}{N} \sum_{n=0}^{N-k-1} x^*(n+k)x(n) \qquad (k = 0, 1, \ldots, P). \tag{3.7}$$

(2) Construct the Yule–Walker equation using (3.7) and the relationship $\hat{R}(-k) = \hat{R}^*(k)$, that is,

$$\sum_{k=0}^{P} \hat{a}_k \hat{R}(m-k) = \begin{cases} \hat{\sigma}_P^2, & m = 0, \\ 0, & m = 1, 2, \ldots, P. \end{cases} \tag{3.8}$$

(3) Find $\hat{a}_k$ (for $k = 1, 2, \ldots, P$) and $\hat{\sigma}_P^2$ by solving (3.8).

(4) Compute the spectrum from $\hat{a}_k$ (for $k = 1, 2, \ldots, P$) and $\hat{\sigma}_P^2$ by using (3.2).

Note that the computational complexity in the above steps is very intensive (e.g., the multiplication complexity required from Step (1) to Step (3) is $O(P^3)$). There are, however, some properties of the Yule–Walker equation that can be used to reduce the computational complexity [5, 6]. The most successful example is the Levinson–Durbin algorithm (Figure 3.1 shows the flow chart), which employs the Toeplitz property of the matrix in (3.6) and reduces the computational complexity to $O(P^2)$. However, it remains difficult to deliver the desired real-time spectral estimation even if the Levinson–Durbin algorithm is used.

In the following, we will show how to use the neural network in Figure 2.2 to perform, in real time, all the computations required from Step (1) to Step (3). Without loss of generality, we consider only the normalized spectral estimation.

Equation (3.8) can be written as

$$\sum_{k=1}^{P} \hat{a}_k \hat{R}(m-k) = -\hat{R}(m) \qquad (m = 1, 2, \ldots, P) \tag{3.9}$$

and

$$\begin{pmatrix} \hat{R}(0) & \hat{R}(-1) & \cdots & \hat{R}(1-P) \\ \hat{R}(1) & \hat{R}(0) & \cdots & \hat{R}(2-P) \\ \vdots & \vdots & \ddots & \vdots \\ \hat{R}(P-1) & \hat{R}(P-2) & \cdots & \hat{R}(0) \end{pmatrix} \begin{pmatrix} \hat{a}_1 \\ \hat{a}_2 \\ \vdots \\ \hat{a}_P \end{pmatrix} = \begin{pmatrix} \hat{R}(1) \\ \hat{R}(2) \\ \vdots \\ \hat{R}(P) \end{pmatrix}. \tag{3.10}$$

Substituting (3.7) into (3.10) yields

$$\boldsymbol{X}^H \boldsymbol{X} \boldsymbol{A} = \boldsymbol{X}^H \boldsymbol{X}_1, \tag{3.11}$$

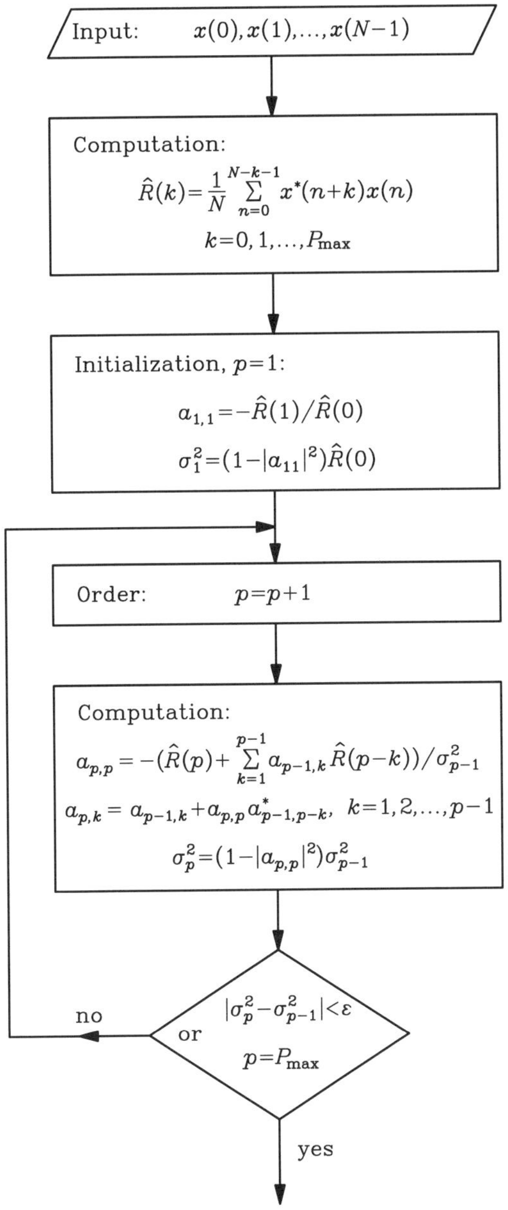

Figure 3.1. The flowchart of the Levinson–Durbin algorithm. $P_{\max}$ is assumed a priori and ε is specified.

where

$$
\boldsymbol{X} = \begin{pmatrix}
x(0) & 0 & \cdots & 0 \\
x(1) & x(0) & \cdots & 0 \\
\vdots & \vdots & \ddots & \vdots \\
x(P-1) & x(P-2) & \cdots & x(0) \\
\vdots & \vdots & \ddots & \vdots \\
x(N-1) & x(N-2) & \cdots & x(N-P) \\
\vdots & \vdots & \ddots & \vdots \\
0 & \cdots & x(0) & x(1) \\
0 & \cdots & 0 & x(0)
\end{pmatrix}_{(N+P)\times P}, \tag{3.12}
$$

$$
\boldsymbol{A} = \begin{pmatrix} \hat{a}_1 \\ \hat{a}_2 \\ \vdots \\ \hat{a}_P \end{pmatrix}_{P\times 1}, \quad \text{and} \quad \boldsymbol{X}_1 = \begin{pmatrix} x(1) \\ x(2) \\ \vdots \\ x(N-1) \\ 0 \\ \vdots \\ 0 \end{pmatrix}_{(N+P)\times 1}. \tag{3.13}
$$

$\boldsymbol{X}$ and $\boldsymbol{X}_1$ are referred to as the data matrix and data vector, respectively. $\boldsymbol{A}$ is the coefficient vector.

The data matrix $\boldsymbol{X}$ and data vector $\boldsymbol{X}_1$ have three other forms if we use different data windows. These different forms will give different estimation performances. The other three forms of $\boldsymbol{X}$ and $\boldsymbol{X}_1$ are:

(1)

$$
\boldsymbol{X} = \begin{pmatrix}
x(P-1) & x(P-2) & \cdots & x(0) \\
x(P) & x(P-1) & \cdots & x(1) \\
\vdots & \vdots & \ddots & \vdots \\
x(N-1) & x(N-2) & \cdots & x(N-P)
\end{pmatrix}_{(N-P+1)\times P},
$$

$$
\boldsymbol{X}_1 = \begin{pmatrix} x(P) \\ x(P+1) \\ \vdots \\ x(N-1) \\ 0 \end{pmatrix}_{(N-P+1)\times 1},
$$

(2)

$$\boldsymbol{X} = \begin{pmatrix} x(0) & 0 & \cdots & 0 \\ x(1) & x(0) & \cdots & 0 \\ \vdots & \vdots & \ddots & \vdots \\ x(P-1) & x(P-2) & \cdots & x(0) \\ \vdots & \vdots & \ddots & \vdots \\ x(N-1) & x(N-2) & \cdots & x(N-P) \end{pmatrix}_{N\times P},$$

$$\boldsymbol{X}_1 = \begin{pmatrix} x(1) \\ x(2) \\ \vdots \\ x(N-1) \\ 0 \end{pmatrix}_{N\times 1},$$

and

(3)

$$\boldsymbol{X} = \begin{pmatrix} x(P-1) & x(P-2) & \cdots & x(0) \\ \vdots & \vdots & \ddots & \vdots \\ x(N-1) & x(N-2) & \cdots & x(N-P) \\ \vdots & \vdots & \ddots & \vdots \\ 0 & \cdots & x(N-1) & x(N-2) \\ 0 & \cdots & 0 & x(N-1) \end{pmatrix}_{N\times P},$$

$$\boldsymbol{X}_1 = \begin{pmatrix} x(P) \\ x(P+1) \\ \vdots \\ x(N-1) \\ 0 \\ \vdots \\ 0 \end{pmatrix}_{N\times 1}.$$

The solution of (3.11) is

$$\boldsymbol{A} = (\boldsymbol{X}^H \boldsymbol{X})^{-1} \boldsymbol{X}^H \boldsymbol{X}_1, \tag{3.14}$$

which can be written in the real-valued form

$$\boldsymbol{A}_c = (\boldsymbol{X}_c^T \boldsymbol{X}_c)^{-1} \boldsymbol{X}_c^T \boldsymbol{X}_{1c}, \tag{3.15}$$

where

$$A_c = \begin{pmatrix} A_r \\ A_i \end{pmatrix}, \qquad X_c = \begin{pmatrix} X_r & -X_i \\ X_i & X_r \end{pmatrix}, \qquad X_{1c} = \begin{pmatrix} X_{1r} \\ X_{1i} \end{pmatrix}$$

and

$$A = A_r + jA_i, \qquad X = X_r + jX_i, \qquad X_1 = X_{1r} + jX_{1i}. \tag{3.16}$$

Equations (3.16) signify that A_r, X_r, and X_{1r} and A_i, X_i, and X_{1i} are the real and imaginary parts of A, X, and X_1, respectively.

Comparing (3.15) with (2.3) and (2.14), we see that if we let the data matrix X_c and vector X_{1c} be taken as the connection strength matrix T and the bias current vector B of the neural network of Figure 2.2, respectively, then such a constructed network is stable and provides the solution

$$V_f = \left(\frac{1}{RK_1K_2} I + X_c^T X_c \right)^{-1} X_c^T X_{1c}, \tag{3.17}$$

which is an approximation of (3.15). The error can be made arbitrarily small by appropriately selecting the parameters K_1, K_2, and R. Together with the high-speed computational capability of the neural network shown in Section 2.1, this means that the neural network can perform the computation, required from Step (1) to Step (3) of the maximum entropy spectral estimation method, during an elapsed time of only a few characteristic time constants of the network with arbitrarily small error. As a result, this neural network approach is suitable in real-time applications of the maximum entropy spectral estimation method.

The solution (3.14) is, in effect, the LS solution of the equations

$$XA = X_1. \tag{3.18}$$

If the neural network approach proposed in Section 2.4 is used, we can solve (3.18) and find the coefficients a_k (for $k = 1, 2, \ldots, P$) in real time under the TLS criterion. Using the coefficients under the TLS criterion may improve the spectral estimation performance as shown in [7]. However, the architecture of the network for the TLS solution is more complicated than that for the LS solution (3.14). As a result, one has to make a trade-off in practical situations.

3.2 Harmonic Retrieval by Neural Networks

Estimating the frequencies of multiple sinusoids in the presence of noise is a special case of spectral estimation that arises in many applications. How to increase the resolution of closely spaced frequencies is the emphasis in this field. Using the above maximum entropy spectral estimation can provide a high resolution, but the method suffers from spurious-spike and line-splitting problems. To overcome these two problems, many other high-resolution frequency estimation techniques have been developed and extensively analyzed. Of all of them, the methods based on the eigenstructure (Pisarenko's method and the MUSIC method are two representative examples) are the most appealing [8, 9]. These eigenstructure-based methods utilize the orthogonal relationship between the signal vectors and the eigenvectors of the covariance matrix of the underlying data. However, it is not easy to implement these eigenstructure-based methods in real time mainly because of the costly computational complexity of the eigendecomposition.

Neural networks with high-speed computational capability provide us with an alternative tool to implement these eigenstructure-based frequency estimation methods in real time. In this section, we concentrate on applying neural networks to performing the eigendecomposition required in Pisarenko's method. Neural network approaches to other eigenstructure-based methods will be dealt with in Chapter 7 by combining them with the high-resolution bearing estimation.

Let us consider a received signal consisting of P sinusoids corrupted by white noise:

$$x(n) = \sum_{i=1}^{P} \alpha_i \cos(\omega_i n + \theta_i) + \epsilon(n) \qquad (n = 1, 2, \ldots, L), \tag{3.19}$$

where $x(n)$ and $\epsilon(n)$ are the measured samples and noise samples, respectively. The parameters α_i, ω_i, and θ_i denote the amplitude, frequency (normalized), and initial phase (uniformly distributed over $[0, 2\pi]$) of the i'th sinusoid. $\epsilon(n)$ is zero mean and has variance σ^2.

To estimate the frequencies of the P sinusoids from the available data samples, Pisarenko's frequency estimation method includes essentially the following steps [10, 11, 12]:

(1) Estimate the covariance matrix $\boldsymbol{R}$ of size $N \times N$ from the measured samples. The exact covariance matrix $\boldsymbol{R}$ is given by

$$\boldsymbol{R} = \begin{pmatrix} R(0) & R(1) & \cdots & R(N-1) \\ R(-1) & R(0) & \cdots & R(N-2) \\ \vdots & \vdots & \ddots & \vdots \\ R(-N+1) & R(-N+2) & \cdots & R(0), \end{pmatrix}, \tag{3.20}$$

where

$$R(k) = \sum_{l=1}^{P} \frac{\alpha_l^2}{2} \cos(\omega_l k) + \sigma^2 \delta(k) \qquad (k = -N+1, -N+2, \ldots, N-1) \tag{3.21}$$

and $N \geq 2P + 1$.

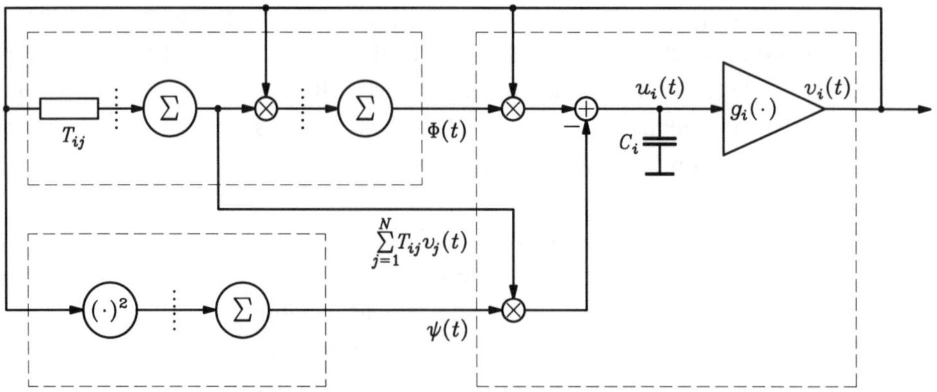

Figure 3.2. The neural network for computing the desired eigenvector in Pisarenko's frequency estimation method.

(2) Compute the eigenvector corresponding to the smallest eigenvalue of the estimated covariance matrix $\boldsymbol{R}$.

(3) Compute the roots of the polynomial formed by the elements of the above eigenvector. This polynomial will have $2P$ roots located at $\exp(\pm j\omega_i)$ (for $i = 1, 2, \ldots, P$).

Note that the second step involves a matrix eigendecomposition, and therein lies the major computational burden of Pisarenko's frequency estimation method, making it very difficult to implement the method in real time. To tackle this problem, we present a new neural network computational model that will be shown to be a very powerful means for computing, in real time, the eigenvector corresponding to the smallest eigenvalue of the estimated covariance matrix $\boldsymbol{R}$.

The proposed neural network, shown in Figure 3.2, can be represented by the continuous-time differential equations

$$C_i \frac{du_i(t)}{dt} = \phi(t) v_i(t) - \psi(t) \sum_{j=1}^{N} T_{ij} v_j(t) \quad \text{and}$$

$$v_i(t) = g_i\big(u_i(t)\big) \qquad (i = 1, 2, \ldots, N), \tag{3.22}$$

where $u_i(t)$ and $v_i(t)$ stand for the input and output voltages of the i'th neuron, respectively; $v_i(t) = g_i(u_i(t))$; and $g_i(\cdot)$ is selected to be a monotonically increasing function. C_i is the input capacitor of the i'th neuron. Furthermore,

$$\psi(t) = \sum_{i=1}^{N} \big(v_i(t)\big)^2 \tag{3.23}$$

and

$$\phi(t) = \sum_{i=1}^{N} \sum_{j=1}^{N} T_{ij} v_i(t) v_j(t). \tag{3.24}$$

The matrix $\boldsymbol{T} = \{T_{ij}\}$ (for $i = 1, 2, \ldots, N$; $j = 1, 2, \ldots, N$) is called the connection strength matrix and is selected to be positive definite.

In order to guarantee that this neural network is stable, we define an energy function as

$$E(t) = \frac{\phi(t)}{\psi(t)}. \tag{3.25}$$

It follows that

$$\begin{aligned}\frac{dE(t)}{dt} &= \frac{2}{\psi(t)^2}\left(\psi(t)\sum_{i=1}^{N}\frac{dv_i(t)}{dt}\sum_{j=1}^{N}T_{ij}v_j(t) - \phi(t)\sum_{i=1}^{N}v_i(t)\frac{dv_i(t)}{dt}\right)\\ &= \frac{2}{\psi(t)^2}\left(\sum_{i=1}^{N}\frac{dv_i(t)}{dt}\left[\psi(t)\sum_{j=1}^{N}T_{ij}v_j(t) - \phi(t)v_i(t)\right]\right).\end{aligned} \tag{3.26}$$

Comparing (3.22) with (3.26) gives

$$\begin{aligned}\frac{dE(t)}{dt} &= -\frac{2}{\psi(t)^2}\sum_{i=1}^{N}C_i\frac{du_i(t)}{dt}\frac{dv_i(t)}{dt}\\ &= -\frac{2}{\psi(t)^2}\sum_{i=1}^{N}C_i\left[g_i^{-1}(v_i(t))\right]'\left(\frac{dv_i(t)}{dt}\right)^2.\end{aligned} \tag{3.27}$$

Since C_i is positive and $g_i^{-1}(v_i(t))$ is a monotonically increasing function, each term on the right-hand side of (3.27) is nonnegative and

$$\frac{dE(t)}{dt} \le 0, \quad \frac{dE(t)}{dt} = 0 \rightarrow \frac{dv_i(t)}{dt} = 0 \qquad (i = 1, 2, \ldots, N). \tag{3.28}$$

Together with the fact that $E(t)$ is also nonnegative (hence, $E(t)$ is bounded from below), (3.28) shows that the time evolution of this neural network is a motion in the state space that seeks out the minimum to $E(t)$ and comes to a stop at such a point. Thus, any optimization problem with an objective function that is mapped to (3.25) can be solved by this proposed neural network. One of the minimum points of $E(t)$ is provided by

$$\frac{du_i(t)}{dt} = 0 \quad \text{or} \quad \frac{dv_i(t)}{dt} = 0 \qquad (i = 1, 2, \ldots, N). \tag{3.29}$$

We next wish to use the proposed neural network to compute the desired eigenvector corresponding to the smallest eigenvalue of the estimated covariance matrix $\boldsymbol{R}$ in Pisarenko's frequency estimation method. We choose

(1) $\boldsymbol{R} = \boldsymbol{T}$, that is, the covariance matrix $\boldsymbol{R}$ is directly taken as the connection strength matrix of the network;

(2) $C_i = C$ (for $i = 1, 2, \ldots, N$); and

(3) $g_i(u) = K_1 u$ (for $i = 1, 2, \ldots, N$), which means that each neuron is modeled as a linear amplifier with voltage gain K_1.

According to the above relationships, we can write (3.22) in the matrix form

$$\frac{d\boldsymbol{V}(t)}{dt} = K\{\phi(t)\boldsymbol{V}(t) - \psi(t)\boldsymbol{R}\boldsymbol{V}(t)\}, \tag{3.30}$$

where $\boldsymbol{V}(t) = [v_1(t), v_2(t), \ldots, v_N(t)]^T$ and $K = K_1/C$ is a positive constant.

We also have the following relationships in matrix form:

$$\psi(t) = \boldsymbol{V}^T(t)\boldsymbol{V}(t), \tag{3.31}$$

$$\phi(t) = \boldsymbol{V}^T(t)\boldsymbol{R}\boldsymbol{V}(t). \tag{3.32}$$

The following theorem concerns the dynamics of this network for Pisarenko's frequency estimation method.

Theorem 3.1 *The norm of the output vector $\boldsymbol{V}(t)$ of the network is invariant during the time evolution and is equal to the norm of the initial state $\boldsymbol{V}(0)$; that is,*

$$\|\boldsymbol{V}(t)\|^2 = \|\boldsymbol{V}(0)\|^2, \qquad t \geq 0 \tag{3.33}$$

$$\psi(t) = \psi(0), \qquad t \geq 0. \tag{3.34}$$

Proof: Multiplying (3.30) by $\boldsymbol{V}^T(t)$ on the left yields

$$\frac{d\|\boldsymbol{V}(t)\|^2}{dt} = -2K\{\psi(t)\boldsymbol{V}^T(t)\boldsymbol{R}\boldsymbol{V}(t) - \phi(t)\boldsymbol{V}^T(t)\boldsymbol{V}(t)\}. \tag{3.35}$$

Substituting (3.31) and (3.32) into (3.35) gives

$$\frac{d\|\boldsymbol{V}(t)\|^2}{dt} = -2K\{\psi(t)\phi(t) - \phi(t)\psi(t)\} = 0, \tag{3.36}$$

which shows that (3.33) and (3.34) hold and thus concludes the proof.

QED

Because the norm of the desired eigenvector in Pisarenko's frequency estimation method is unity, we have to select $\psi(0) = \|\boldsymbol{V}(0)\|^2 = 1$.

If we set

$$V(t) = \sum_{i=1}^{N} y_i(t) S_i \tag{3.37}$$

then Theorem 3.1 shows that

$$0 \le y_i^2(t) \le 1, \qquad t \ge 0 \tag{3.38}$$

and

$$\psi(t) = 1, \qquad t \ge 0, \tag{3.39}$$

where $0 < \lambda_1 \le \lambda_2 \le \cdots \le \lambda_N$ and $S_1, S_2, \ldots, S_N$ are the eigenvalues and the corresponding orthonormal eigenvectors of the matrix $\boldsymbol{R}$.

Theorem 3.2 *If the initial state of the network satisfies $V^T(0)S_1 \ne 0$, the network is asymptotically stable and*

$$\lim_{t\to\infty} V(t) = S_1, \tag{3.40}$$

$$\lim_{t\to\infty} \phi(t) = \lambda_1. \tag{3.41}$$

The stability of this proposed network has been proved in the previous part of this section (3.28). We next prove that (3.40) and (3.41) hold.

We substitute (3.37) into (3.30) and obtain

$$\frac{dy_i(t)}{dt} = K\big(- \lambda_i y_i(t) + \phi(t) y_i(t)\big) \qquad (i = 1, 2, \ldots, N). \tag{3.42}$$

According to the assumption of the initial state $V(0)$ of the network, we may define (see [13] and [14])

$$z_i(t) = \frac{y_i(t)}{y_1(t)} \qquad (i = 2, 3, \ldots, N). \tag{3.43}$$

Moreover,

$$\frac{dz_i(t)}{dt} = \frac{y_1(t)\frac{dy_i(t)}{dt} - y_i(t)\frac{dy_1(t)}{dt}}{(y_1(t))^2} \qquad (i = 2, 3, \ldots, N). \tag{3.44}$$

Combining (3.42) and (3.44) gives the differential equation

$$\frac{dz_i(t)}{dt} = K(\lambda_1 - \lambda_i) z_i(t) \tag{3.45}$$

with the solution

$$z_i(t) = K_{i1} \exp\big(K(\lambda_1 - \lambda_i)t\big) \qquad (i = 2, 3, \ldots, N), \tag{3.46}$$

where K_{i1} is a constant depending on K, the initial value, and the eigenvalues of the matrix $\boldsymbol{R}$.

Using (3.43) yields

$$y_i(t) = K_{i1} y_1(t) \exp\big(K(\lambda_1 - \lambda_i)t\big) \qquad (i = 2, 3, \ldots, N). \tag{3.47}$$

Because of $y_1^2(t) \leq 1$, we have

$$y_i^2(t) \leq M_{i1} \exp\big(2K(\lambda_1 - \lambda_i)t\big), \tag{3.48}$$

where $M_{i1} = K_{i1}^2$.

If $\lambda_i > \lambda_1$, we know

$$\lim_{t\to\infty} y_i(t) = 0 \qquad (i = 2, 3, \ldots, N). \tag{3.49}$$

Together with $\|V_1(t)\|^2 = 1$, Equations (3.37) and (3.49) lead to

$$\lim_{t\to\infty} y_1(t) = \pm 1 \tag{3.50}$$

and

$$\lim_{t\to\infty} \boldsymbol{V}(t) = \pm \boldsymbol{S}_1. \tag{3.51}$$

Since $-\boldsymbol{S}_1$ is also a unit eigenvector corresponding to the eigenvalue λ_1 of the matrix $\boldsymbol{R}$, Equation (3.51) can be written in the unified form

$$\lim_{t\to\infty} \boldsymbol{V}(t) = \boldsymbol{S}_1. \tag{3.52}$$

For the case in which the smallest eigenvalue λ_1 is not single but multiple (i.e., $\lambda_1 = \lambda_2 = \cdots = \lambda_M = \lambda$, $M \leq N$), we have

$$\lim_{t\to\infty} \boldsymbol{V}(t) = \sum_{i=1}^{M} \lim_{t\to\infty} y_i(t) \boldsymbol{S}_i. \tag{3.53}$$

Multiplying the above equation by $\boldsymbol{R}$ on the left yields

$$\begin{aligned} \lim_{t\to\infty} \boldsymbol{R}\boldsymbol{V}(t) &= \sum_{i=1}^{M} \lim_{t\to\infty} y_i(t) \boldsymbol{R}\boldsymbol{S}_i = \sum_{i=1}^{M} \lim_{t\to\infty} \lambda_i y_i(t) \boldsymbol{S}_i \\ &= \lambda \sum_{i=1}^{M} \lim_{t\to\infty} y_i(t) \boldsymbol{S}_i = \lambda \lim_{t\to\infty} \boldsymbol{V}(t), \end{aligned} \tag{3.54}$$

which means that $\lim_{t\to\infty} \boldsymbol{V}(t)$ is still the eigenvector corresponding to the multiple eigenvalue λ of the matrix $\boldsymbol{R}$.

This concludes the proof of Theorem 3.2.

Theorems 3.1 and 3.2 guarantee that the output of the network in the stationary state provides the eigenvector corresponding to the smallest eigenvalue of the covariance matrix $\boldsymbol{R}$.

Table 3.1. *The time evolution of the network outputs*

t (ps)	$v_1(t)$	$v_2(t)$	$v_3(t)$	$v_4(t)$	$v_5(t)$
0	1.00000	0.00000	0.00000	0.00000	0.00000
10	1.00000	−0.06566	−0.06466	−0.06300	−0.06069
20	0.98726	−0.11814	−0.11591	−0.11250	−0.10794
30	0.97051	−0.15686	−0.15327	−0.14812	−0.14145
⋮	⋮	⋮	⋮	⋮	⋮
6140	0.65767	−0.63531	−0.32024	−0.00193	0.31635
6150	0.65760	−0.63534	−0.32024	−0.00190	0.31642
6160	0.65754	−0.63537	−0.32024	−0.00187	0.31648

Table 3.2. *The time evolution of the network outputs*

t (ps)	$v_1(t)$	$v_2(t)$	$v_3(t)$	$v_4(t)$	$v_5(t)$	$v_6(t)$	$v_7(t)$
0	1.00000	0.00000	0.00000	0.00000	0.00000	0.00000	0.00000
10	1.00000	−0.06264	−0.06154	−0.05974	−0.05725	−0.05410	−0.05032
20	0.98640	−0.10729	−0.10458	−0.10066	−0.09559	−0.08941	−0.08220
30	0.97228	−0.13674	−0.13210	−0.12593	−0.11831	−0.10993	−0.09909
⋮	⋮	⋮	⋮	⋮	⋮	⋮	⋮
6140	0.75351	−0.48864	−0.34300	−0.19427	−0.04427	−0.10519	0.25234
6150	0.75351	−0.48864	−0.34300	−0.19427	−0.04427	−0.10519	0.25234
6160	0.75351	−0.48864	−0.34300	−0.19427	−0.04427	−0.10519	0.25234

For further illustration, we will now present two simulation examples (Examples 3.1 and 3.2) for computing the desired eigenvector in Pisarenko's frequency estimation method by use of the proposed neural network. In these two examples, we used $K_1 = 1$ and $C = 10^4$ pF. In the first example, two sinusoids are considered with the frequencies 0.1 and 0.11, respectively; their associated signal-to-noise ratios are 10 dB and 0 dB, respectively. In the second example, three sinusoids are considered with the frequencies 0.1, 0.11, and 0.15, and their associated signal-to-noise ratios are 10 dB, 10 dB, and 0 dB, respectively.

In these simulations, $\boldsymbol{R}$ is the covariance matrix, $\boldsymbol{V}_f$ and $\boldsymbol{V}(0)$ are the outputs of the network in the stationary state and the initial state, respectively, $\phi(f)$ is the eigenvalue provided by the output $\phi(t)$ in the stationary state, and $\lambda(a)$ is the exact eigenvalue of the matrix $\boldsymbol{R}$. Obviously, $\phi(f) = \lambda(a)$. Tables 3.1 and 3.2 and Figures 3.3 and 3.4 describe the dynamics of the output vector $\boldsymbol{V}(t)$ of the network.

These simulation results demonstrate the accuracy of the above analysis and the effectiveness of this neural network approach.

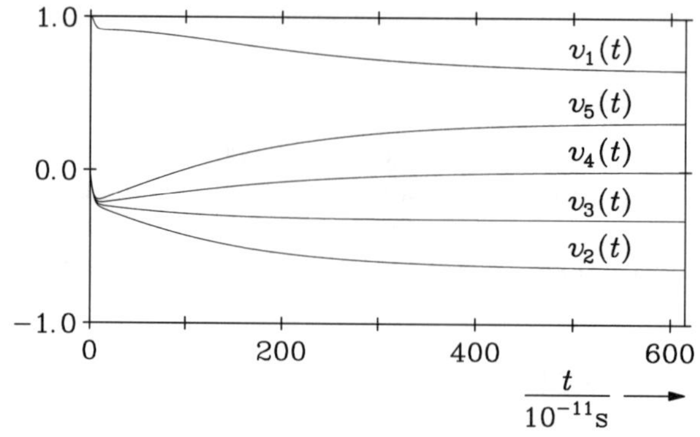

Figure 3.3. The dynamic curves of the output vector $\boldsymbol{V}(t)$ in Table 3.1.

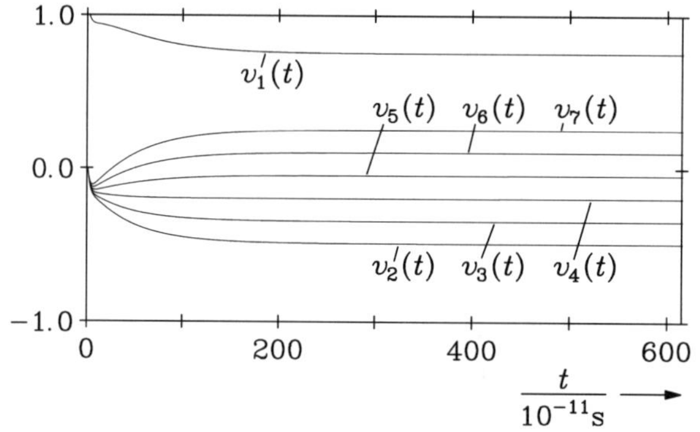

Figure 3.4. The dynamic curves of the output vector $\boldsymbol{V}(t)$ in Table 3.2.

Example 3.1

$P = 2, N = 5$

$$\boldsymbol{R} = \begin{pmatrix} 12.0000 & 10.9440 & 10.7766 & 10.4994 & 10.1154 \\ 10.9440 & 12.0000 & 10.9440 & 10.7766 & 10.4994 \\ 10.7766 & 10.9440 & 12.0000 & 10.9440 & 10.7766 \\ 10.4994 & 10.7766 & 10.9440 & 12.0000 & 10.9440 \\ 10.1154 & 10.4994 & 10.7766 & 10.9440 & 12.0000 \end{pmatrix}$$

$$\boldsymbol{V}_f = \begin{pmatrix} 0.6575 \\ -0.6354 \\ -0.3202 \\ -0.0018 \\ 0.3156 \end{pmatrix}, \qquad \boldsymbol{V}(0) = \begin{pmatrix} 1.0000 \\ 0.0000 \\ 0.0000 \\ 0.0000 \\ 0.0000 \end{pmatrix}$$

$\phi(f) = 1.0002, \qquad \lambda(a) = 1.0000$

Example 3.2

$P = 3, N = 7$

$$\boldsymbol{R} = \begin{pmatrix} 22.0000 & 20.8784 & 20.5150 & 19.9142 & 19.0835 & 18.0328 & 16.7749 \\ 20.8784 & 22.0000 & 20.8784 & 20.5150 & 19.9142 & 19.0835 & 18.0328 \\ 20.5150 & 20.8784 & 22.0000 & 20.8784 & 20.5150 & 19.9142 & 19.0835 \\ 19.9142 & 20.5150 & 20.8784 & 22.0000 & 20.8784 & 20.5150 & 19.9142 \\ 19.0835 & 19.9142 & 20.5150 & 20.8784 & 22.0000 & 20.8784 & 20.5150 \\ 18.0328 & 19.0835 & 19.9142 & 20.5150 & 20.8784 & 22.0000 & 20.8784 \\ 16.7749 & 18.0328 & 19.0835 & 19.9142 & 20.5150 & 20.8784 & 22.0000 \end{pmatrix}$$

$$\boldsymbol{V}_f = \begin{pmatrix} 0.7535 \\ -0.4886 \\ -0.3430 \\ -0.1943 \\ -0.0443 \\ 0.1052 \\ 0.2523 \end{pmatrix}, \qquad \boldsymbol{V}(0) = \begin{pmatrix} 1.0000 \\ 0.0000 \\ 0.0000 \\ 0.0000 \\ 0.0000 \\ 0.0000 \\ 0.0000 \end{pmatrix}$$

$\phi(f) = 1.0017, \qquad \lambda(a) = 1.0000$

We can summarize the key features of this neural network approach for computing the eigenvector required in Pisarenko's frequency estimation method as follows:

(1) This approach is based on an analog circuit architecture with continuous-time dynamics. As a result, the convergence time of the network (the time to approach closely the stationary state) is within an elapsed time of only a few characteristic time constants of the network, which means that this approach can provide the desired eigenvector in real time.

(2) This approach can still provide the desired eigenvector even if the smallest eigenvalue of the covariance matrix is not single. That is, we make no assumption for the smallest eigenvalue of the covariance matrix.

(3) The norm of the output vector of the network is invariant and equal to the norm of the initial vector during the time evolution, so that we can provide the desired unit eigenvector by letting the norm of the initial vector be unity. This property is also useful in designing a hardware implementation of the proposed approach.

(4) The covariance matrix is directly taken as the network parameters without any computation, so we do not encounter the "programming complexity" problem pointed out in [15].

(5) No division computation is needed in this neural network approach.

(6) The initial vector $\boldsymbol{V}(0)$ of the network can be selected almost arbitrarily because we only have the assumption that $\boldsymbol{V}(0)$ is not orthogonal to

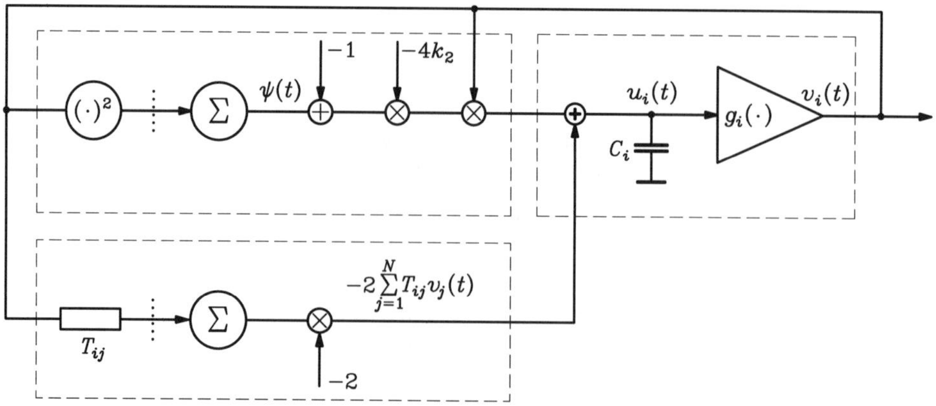

Figure 3.5. An alternative neural network for computing the desired eigenvector in Pisarenko's frequency estimation method.

the eigenvector corresponding to the smallest eigenvalue λ_1 (i.e., $y_1(0) = \boldsymbol{V}^T(0)\boldsymbol{S}_1 \neq 0$). From (3.42)–(3.51), we know that the output vector of the network will converge to the eigenvector $\boldsymbol{S}_2$ corresponding to the eigenvalue λ_2 (the second smallest eigenvalue) if $y_1(0) = \boldsymbol{V}^T(0)\boldsymbol{S}_1 = 0$ and $y_2(0) = \boldsymbol{V}^T(0)\boldsymbol{S}_2 \neq 0$. We can also conclude that the output vector $\boldsymbol{V}(t)$ of the network will converge to the eigenvector $\boldsymbol{S}_{i+1}$ corresponding to the eigenvalue λ_{i+1} if $y_1(0) = y_2(0) = \cdots = y_i(0) = 0$ and $y_{i+1}(0) \neq 0$.

Figure 3.5 shows another neural network (proposed in [16]) for computing the eigenvector required in Pisarenko's frequency estimation method. This network can be represented by the continuous-time differential equations

$$C_i \frac{du_i(t)}{dt} = -4K_2\big(\psi(t) - 1\big)v_i(t) - 2\sum_{j=1}^{N} T_{ij}v_j(t)$$
$$v_i(t) = g_i\big(u_i(t)\big) \qquad (i = 1, 2, \ldots, N), \tag{3.55}$$

where K_2 is a positive constant and the other quantities are the same as those described in Figure 3.2.

By defining and analyzing the energy function

$$E(t) = \phi(t) + K_2\big(\psi(t) - 1\big)^2 \tag{3.56}$$

we can prove that this neural network is stable. The proof may be divided into two steps.

First, Equation (3.56) indicates that $E(t)$ is nonnegative, and hence, it is bounded from below. Second, it follows that

$$\frac{dE(t)}{dt} = 2\sum_{i=1}^{N}\sum_{j=1}^{N} T_{ij}v_j(t)\frac{dv_i(t)}{dt} + 4K_2\sum_{i=1}^{N} v_i(t)(\psi(t)-1)\frac{dv_i(t)}{dt}$$
$$= \sum_{i=1}^{N}\frac{dv_i(t)}{dt}\left(2\sum_{j=1}^{N} T_{ij}v_j(t) + 4K_2 v_i(t)(\psi(t)-1)\right). \tag{3.57}$$

Substituting (3.55) into (3.57) gives

$$\frac{dE(t)}{dt} = -\sum_{i=1}^{n} C_i \frac{du_i(t)}{dt}\frac{dv_i(t)}{dt}$$
$$= -\sum_{i=1}^{n} C_i \left[g_i^{-1}(v_i(t))\right]' \left(\frac{dv_i(t)}{dt}\right)^2, \tag{3.58}$$

which tells us that $E(t)$ is nonincreasing during the time evolution. These two facts imply

$$\lim_{t\to\infty}\frac{dE(t)}{dt} = 0, \tag{3.59}$$

or equivalently,

$$\lim_{t\to\infty}\frac{du_i(t)}{dt} = 0, \quad \text{or} \quad \lim_{t\to\infty}\frac{dv_i(t)}{dt} = 0 \qquad (i = 1, 2, \ldots, N). \tag{3.60}$$

Thus, the output vector $\boldsymbol{V}(t) = [v_1(t), v_2(t), \ldots, v_N(t)]^T$ of this network will always approach a stationary point of $E(t)$ and provides one of the minima of $E(t)$.

In the following, we will present two theorems [16] that show that each minimizer of $E(t)$ is the eigenvector corresponding to the smallest eigenvalue of the matrix $\boldsymbol{R}$ if we let the covariance matrix $\boldsymbol{R}$ be taken directly as the connection strength matrix $\boldsymbol{T}$ of the network and if K_2 satisfies $K_2 > \lambda_1/2$.

Theorem 3.3 *A vector $\boldsymbol{V}_f$ is a stationary point (minimizer) of $E(t)$ if and only if $\boldsymbol{V}_f$ is an eigenvector corresponding to the eigenvalue λ of the matrix $\boldsymbol{R}$ with norm*

$$\beta = \sqrt{1 - \frac{\lambda}{2K_2}}.$$

This result immediately follows from the gradient vector $\boldsymbol{g}(\boldsymbol{V}(t))$ of $E(t)$ given by

$$\boldsymbol{g}(\boldsymbol{V}(t)) = 2\boldsymbol{R}\boldsymbol{V}(t) + 4K_2\boldsymbol{V}(t)(\psi(t)-1). \tag{3.61}$$

Theorem 3.4 *A vector V_f is a global minimizer of $E(t)$ if and only if V_f is the eigenvector corresponding to the smallest eigenvalue λ_1 with norm*

$$\beta = \sqrt{1 - \frac{\lambda_1}{2K_2}}.$$

Proof: (from [16]*) for the "if" part:* Clearly, from Theorem 3.3, V_f is a minimizer of $E(t)$. To prove that V_f is a global minimizer, we have to show that $E(V(t)) - E(V_f) \geq 0$ for any $V(t)$.

Let $V(t) = V_f + \Delta V(t)$; then we have

$$\begin{aligned} E(V(t)) - E(V_f) &= 2(\Delta V(t))^T R V_f + (\Delta V(t))^T R \Delta V(t) \\ &\quad + K_2\Big[(\Delta V(t))^T V_f + (\Delta V(t))^T \Delta V(t)\Big]^2 \\ &\quad + 2K_2(\beta^2 - 1)\Big[(\Delta V(t))^T V_f + (\Delta V(t))^T \Delta V(t)\Big]. \end{aligned} \tag{3.62}$$

From the hypothesis, we have

$$R V_f = \lambda_1 V_f \tag{3.63}$$

and

$$\lambda_1 = 2K_2(1 - \beta^2). \tag{3.64}$$

Moreover,

$$\begin{aligned} (\Delta V(t))^T R \Delta V(t) &= \sum_{i=1}^{N} \lambda_i (\Delta V(t))^T S_i S_i^T \Delta V(t) \\ &\geq \lambda_1 (\Delta V(t))^T \Delta V(t). \end{aligned} \tag{3.65}$$

Substituting (3.63)–(3.65) into (3.62), we obtain

$$E(V(t)) - E(V_f) = K_2\Big[2(\Delta V(t))^T V_f + (\Delta V(t))^T \Delta V(t)\Big]^2 \geq 0, \tag{3.66}$$

which means that V_f is a global minimizer of $E(t)$. QED

Proof: (from [16]*) for the "only if" part:* If the vector V_f is a global minimizer, then from Theorem 3.3, it follows that

$$R V_f = \lambda_i V_f = \lambda_i \beta S_i \tag{3.67}$$

and

$$\lambda_i = 2K_2(1 - \beta^2) \quad \text{for some} \quad i \in \{1, 2, \ldots, N\}. \tag{3.68}$$

Substituting (3.67) and (3.68) into the Hessian matrix $\boldsymbol{H}(\boldsymbol{V}(t))$ of $E(t)$ given by

$$\boldsymbol{H}(\boldsymbol{V}(t)) = 2\boldsymbol{R} + 8K_2\boldsymbol{V}(t)\boldsymbol{V}^T(t) + 4K_2(\psi(t) - 1)\boldsymbol{I} \tag{3.69}$$

we get

$$\boldsymbol{H}(\boldsymbol{V}_f) = \sum_{j=1}^{N} h_j \boldsymbol{S}_j \boldsymbol{S}_j^T, \tag{3.70}$$

where

$$h_j = 2(\lambda_j - \lambda_i) + 8K_2\beta^2\delta_{ij} \tag{3.71}$$

is the j'th eigenvalue of the Hessian matrix $\boldsymbol{H}(\boldsymbol{V}_f)$ with $\boldsymbol{S}_j$ as the corresponding eigenvector and δ_{ij} is the Kronecker delta function. Because the vector $\boldsymbol{V}_f$ is a global minimizer, $\boldsymbol{H}(\boldsymbol{V}_f)$ is nonnegative definite. Hence, its eigenvalues h_j (for $j = 1, 2, \ldots, N$) are all nonnegative, that is,

$$2(\lambda_j - \lambda_i) + 8K_2\beta^2\delta_{ij} \geq 0 \qquad (j = 1, 2, \ldots, N), \tag{3.72}$$

which is true only if $\lambda_i = \lambda_1$. QED

From the proof of Theorem 3.4, we know that every local minimizer of $E(t)$ is also a global minimizer because the Hessian matrix $\boldsymbol{H}(\boldsymbol{V}_f)$ is at least nonnegative definite [17].

The above theorem guarantees that if $K_2 > \lambda_1/2$ is satisfied, the neural network will certainly converge to the eigenvector corresponding to the smallest eigenvalue λ_1 of the matrix $\boldsymbol{R}$.

The neural network in Figure 3.5 may serve as an alternative to the neural network in Figure 3.2 in some applications of Pisarenko's frequency estimation method, because the structure of Figure 3.5 is simpler than that of Figure 3.2. However, in the neural network of Figure 3.5, it is necessary to know the value of λ_1 for choosing K_2 so as to satisfy $K_2 > \lambda_1/2$. Because λ_1 will not be known a priori, we suggest the following practical lower bound

$$K_2 > \frac{\mathrm{trace}\{\boldsymbol{R}\}}{2N}, \tag{3.73}$$

which means that an additional computation has to be invested to estimate $(\mathrm{trace}\{\boldsymbol{R}\})/2N$ before the network can be used. Moreover, because the norm of the eigenvector provided by the neural network of Figure 3.5 is not unity but depends on the values of λ_1 and K_2, an additional division is also needed so as to provide the unit eigenvector required in Pisarenko's frequency estimation method. As a result, in practical uses, one has to trade off the features of both neural networks.

Finally, we should mention that the generalization of the above neural network approaches for the case in which $\boldsymbol{R}$ takes complex values remains an open problem.

3.3 Neural Networks for Multichannel Spectral Estimation

Multichannel spectral estimation is being widely used, particularly in the sonar, radar, and geoseismic communities. The success of unichannel high-resolution spectral estimation techniques such as the maximum entropy spectral estimation method in Section 3.1 has encouraged researchers to develop multichannel extensions. Most of the available multichannel spectral estimation methods are based on linear prediction (LP) concepts and the AR model (or ARMA and MA models), and hence they are referred to as the multichannel LP spectral estimation methods [18, 19]. Although these multichannel LP spectral estimation methods have satisfactory performance, it is difficult to implement them. In this section, we will discuss neural network approaches for performing the computation required in the multichannel LP spectral estimation methods. To set the stage, we first give an overview of the multichannel LP spectral estimation methods.

If we define $X(n)$ as the vector of L channel samples at time n from a stationary zero-mean multichannel process,

$$\boldsymbol{X}(n) = \left[x_1(n), x_2(n), \ldots, x_L(n)\right]^T, \tag{3.74}$$

then the covariance function at the time lag k is given by

$$\boldsymbol{R}(k) = E\left[\boldsymbol{X}(n+k)\boldsymbol{X}^H(n)\right]. \tag{3.75}$$

Clearly,

$$\boldsymbol{R}(k) = \boldsymbol{R}^H(-k). \tag{3.76}$$

Define the forward linear prediction error as

$$\boldsymbol{e}_a(n) = \sum_{k=0}^{P} \boldsymbol{A}_k^{(P)} \boldsymbol{X}(n-k) \tag{3.77}$$

and the backward linear prediction error as

$$\boldsymbol{e}_b(n) = \sum_{k=0}^{P} \boldsymbol{B}_k^{(P)} \boldsymbol{X}(n-P+k), \tag{3.78}$$

where $\boldsymbol{A}_k^{(P)}$ and $\boldsymbol{B}_k^{(P)}$ (for $k = 1, 2, \ldots, P$) are, respectively, the forward and backward linear prediction coefficient matrices of dimension $L \times L$; $\boldsymbol{A}_0^{(P)} = \boldsymbol{B}_0^{(P)} = \boldsymbol{I}$, the identity matrix.

Minimization of the forward and backward LP mean square errors, $tr\{E[\boldsymbol{e}_a(n)\,\boldsymbol{e}_a^H(n)]\}= tr\{\boldsymbol{E}_a^{(P)}\}$ and $tr\{E[\boldsymbol{e}_b(n)\boldsymbol{e}_b^H(n)]\}= tr\{\boldsymbol{E}_b^{(P)}\}$ ("tr" denotes the matrix trace), respectively, will yield coefficient matrices ($\boldsymbol{A}_k^{(P)}$ and $\boldsymbol{B}_k^{(P)}$) that satisfy the block-Toeplitz normal equations

$$\begin{pmatrix} \boldsymbol{I} & \boldsymbol{A}_1^{(P)} & \cdots & \boldsymbol{A}_{P-1}^{(P)} & \boldsymbol{A}_P^{(P)} \\ \boldsymbol{B}_P^{(P)} & \boldsymbol{B}_{P-1}^{(P-1)} & \cdots & \boldsymbol{B}_1^{(P)} & \boldsymbol{I} \end{pmatrix} \begin{pmatrix} \boldsymbol{R}(0) & \boldsymbol{R}^H(1) & \cdots & \boldsymbol{R}^H(P) \\ \boldsymbol{R}(1) & \boldsymbol{R}(0) & \cdots & \boldsymbol{R}^H(P-1) \\ \vdots & \vdots & \ddots & \vdots \\ \boldsymbol{R}(P) & \boldsymbol{R}(P-1) & \cdots & \boldsymbol{R}(0) \end{pmatrix}$$

$$= \begin{pmatrix} \boldsymbol{E}_a^{(P)} & \boldsymbol{0} & \cdots & \boldsymbol{0} & \boldsymbol{0} \\ \boldsymbol{0} & \boldsymbol{0} & \cdots & \boldsymbol{0} & \boldsymbol{E}_b^{(P)} \end{pmatrix}. \tag{3.79}$$

Once all the coefficient matrices are obtained from (3.79), the multichannel spectral estimation is given by

$$\begin{aligned} \boldsymbol{S}(f) &= \left(\boldsymbol{A}(e^{j2\pi f})\right)^{-1} \boldsymbol{E}_a^{(P)} \boldsymbol{A}(e^{j2\pi f})^{-H} \\ &= \left(\boldsymbol{B}(e^{j2\pi f})\right)^{-1} \boldsymbol{E}_b^{(P)} \boldsymbol{B}(e^{j2\pi f})^{-H}, \end{aligned} \tag{3.80}$$

where

$$\boldsymbol{A}(e^{j2\pi f}) = \sum_{k=0}^{P} \boldsymbol{A}_k^{(P)} e^{-j2\pi kf} \tag{3.81}$$

and

$$\boldsymbol{B}(e^{j2\pi f}) = \sum_{k=0}^{P} \boldsymbol{B}_k^{(P)} e^{-j2\pi kf}. \tag{3.82}$$

Note that the computational complexity for solving (3.79) is very intensive. To reduce the computational complexity, we can use the Levinson–Wiggins–Robinson algorithm to solve (3.79). This algorithm uses the block-Toeplitz property and relates the coefficient matrices of the order p to those of the order $p-1$ according to the recursions

$$\begin{aligned} \boldsymbol{E}^{(p)} &= \sum_{k=0}^{p-1} \boldsymbol{A}_k^{(p-1)} \boldsymbol{R}(p-k), \\ \boldsymbol{A}_p^{(p)} &= -\boldsymbol{E}^{(p)} (\boldsymbol{E}_b^{(p-1)})^{-1}, \\ \boldsymbol{B}_p^{(p)} &= -(\boldsymbol{E}^{(p)})^H (\boldsymbol{E}_a^{(p-1)})^{-1}, \\ \boldsymbol{E}_a^{(p)} &= (\boldsymbol{I} - \boldsymbol{A}_p^{(p)} \boldsymbol{B}_p^{(p)}) \boldsymbol{E}_a^{(p-1)}, \\ \boldsymbol{E}_b^{(p)} &= (\boldsymbol{I} - \boldsymbol{B}_p^{(p)} \boldsymbol{A}_p^{(p)}) \boldsymbol{E}_b^{(p-1)}, \\ \boldsymbol{A}_k^{(p)} &= \boldsymbol{A}_k^{(p-1)} + \boldsymbol{A}_p^{(p)} \boldsymbol{B}_{p-k}^{(p-1)}, \\ \boldsymbol{B}_k^{(p)} &= \boldsymbol{B}_k^{(p-1)} + \boldsymbol{B}_p^{(p)} \boldsymbol{A}_{p-k}^{(p-1)} \qquad (k = 1, 2, \ldots, p; \quad p = 1, 2, \ldots, P), \end{aligned}$$

with initial conditions

$$E_a^{(0)} = E_b^{(0)} = R(0)$$

and

$$A_k^{(0)} = B_k^{(0)} = I \qquad (k = 1, 2, \ldots, p; \quad p = 1, 2, \ldots, P).$$

In practical cases, there are only $N + 1$ data sample vectors $X(n)$ available, but no covariance values are known at any lags. The covariance values can be estimated from the available data vectors by

$$R_{ij} = \sum_{n=P+1}^{N} X(n-i)X^H(n-j) \qquad (i, j = 1, 2, \ldots, P). \tag{3.83}$$

Note that (3.76) does not hold in this case.

The corresponding coefficient matrices satisfy

$$\begin{pmatrix} I & A_1^{(P)} & \cdots & A_{P-1}^{(P)} & A_P^{(P)} \\ B_P^{(P)} & B_{P-1}^{(P)} & \cdots & B_1^{(P)} & I \end{pmatrix} \begin{pmatrix} R_{0,0} & R_{0,1} & \cdots & R_{0,P} \\ R_{1,0} & R_{1,1} & \cdots & R_{1,P} \\ \vdots & \vdots & \ddots & \vdots \\ R_{P,0} & R_{P,1} & \cdots & R_{P,P} \end{pmatrix}$$

$$= \begin{pmatrix} E_a^{(P)} & 0 & \cdots & 0 & 0 \\ 0 & 0 & \cdots & 0 & E_b^{(P)} \end{pmatrix} \tag{3.84}$$

so as to minimize both

$$tr\{E_a^{(P)}\} = tr\left\{ \sum_{n=P+1}^{N} e_a(n) e_a^H(n) \right\} \tag{3.85}$$

and

$$tr\{E_b^{(P)}\} = tr\left\{ \sum_{n=P+1}^{N} e_b(n) e_b^H(n) \right\}. \tag{3.86}$$

Although a recursive algorithm similar in structure to that of the Levinson–Wiggins–Robinson algorithm can solve (3.84) [18], the computational complexity in this recursive algorithm is still very intensive. In the following, we present the neural network approach for solving (3.84).

First, we combine (3.85) with (3.86) and obtain

$$tr\{FF^H\} = tr\{(Y - DX)(Y - DX)^H\}, \tag{3.87}$$

where

$$F = \begin{pmatrix} e_a(P+1) & e_a(P+2) & \cdots & e_a(N) \\ e_b(P) & e_b(P+1) & \cdots & e_b(N-1) \end{pmatrix}_{2L \times (N-P)},$$

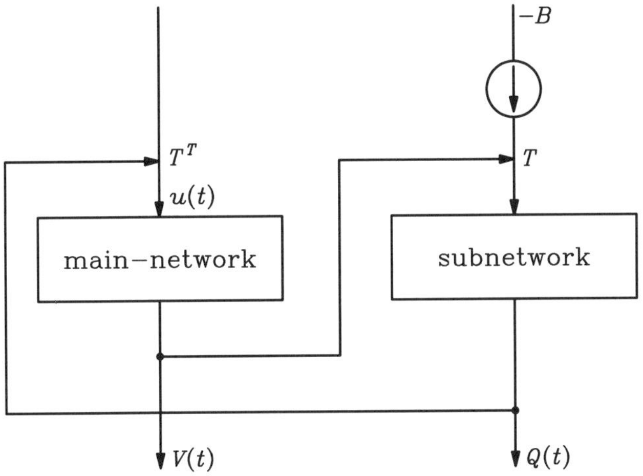

Figure 3.6. A 2-D neural network with the main-network and subnetwork.

$$\boldsymbol{Y} = \begin{pmatrix} \boldsymbol{X}(P+1) & \boldsymbol{X}(P+2) & \cdots & \boldsymbol{X}(N) \\ \boldsymbol{X}(0) & \boldsymbol{X}(1) & \cdots & \boldsymbol{X}(N-P-1) \end{pmatrix}_{2L\times(N-P)},$$

$$\boldsymbol{X} = \begin{pmatrix} \boldsymbol{X}(P) & \boldsymbol{X}(P+1) & \cdots & \boldsymbol{X}(N-1) \\ \boldsymbol{X}(P-1) & \boldsymbol{X}(P) & \cdots & \boldsymbol{X}(N-2) \\ \vdots & \vdots & \ddots & \vdots \\ \boldsymbol{X}(1) & \boldsymbol{X}(2) & \cdots & \boldsymbol{X}(N-P) \end{pmatrix}_{PL\times(N-P)},$$

and

$$\boldsymbol{D} = \begin{pmatrix} \boldsymbol{A}_1^{(P)} & \boldsymbol{A}_2^{(P)} & \cdots & \boldsymbol{A}_P^{(P)} \\ \boldsymbol{B}_P^{(P)} & \boldsymbol{B}_{P-1}^{(P)} & \cdots & \boldsymbol{B}_1^{(P)} \end{pmatrix}_{2L\times PL}.$$

The solution of (3.87) can be given in an explicit form,

$$\boldsymbol{D} = \boldsymbol{Y}\boldsymbol{X}^{+}. \tag{3.88}$$

Note that although (3.88) has the same form as (2.3), $\boldsymbol{D}$ is a $2L \times PL$ dimensional matrix but not a vector. As a result, the neural network of Figure 2.2 cannot be used to compute (3.88).

A two-dimensional (2-D) neural network for solving (3.87) is shown in Figure 3.6. It consists of two parts: the main-network and the subnetwork. The neurons between the main-network and the subnetwork are interconnected by one another.

The main-network has $p \times n$ neurons; each neuron denoted by (i, j) (for $i = 1, 2, \ldots, p$; $j = 1, 2, \ldots, n$) is modeled as an amplifier with an input–output

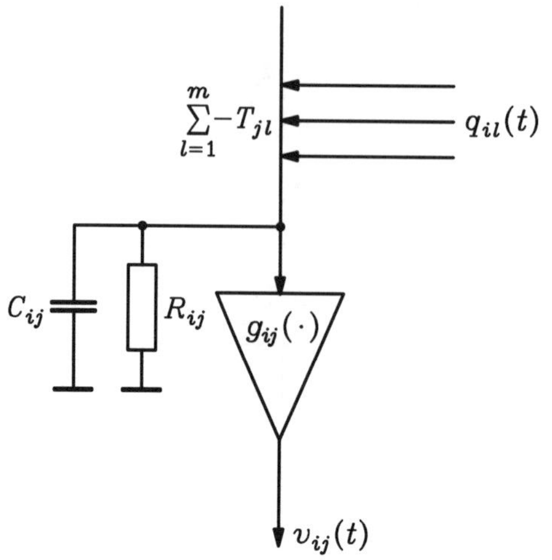

Figure 3.7. Neuron (i, j) in the main-network.

relationship curve $g_{ij}(u)$ and has an input resistor R_{ij} and an input capacitor C_{ij}. $v_{ij}(t)$ and $u_{ij}(t)$ stand for the output and input voltages of the neuron (i, j), respectively. The relationship function $g_{ij}(u)$ is selected to guarantee that

(1) $\int_0^{v_{ij}(t)} g_{ij}^{-1}(v)dv$ (for $i = 1, 2, \ldots, p$; $j = 1, 2, \ldots, n$) is bounded from below and

(2) $g_{ij}(u)$ (for $i = 1, 2, \ldots, p$; $j = 1, 2, \ldots, n$) is a monotonically increasing function.

The subnetwork has $p \times m$ neurons with the input–output relationship $f(z) = K_2 z$; that is, all neurons in the subnetwork are transresistance amplifiers. The output voltage of the neuron (i, l) is expressed as $q_{il}(t)$ (for $i = 1, 2, \ldots, p$; $l = 1, 2, \ldots, m$). b_{il} is the bias current of the neuron (i, l).

The output $q_{il}(t)$ of the neuron (i, l) in the subnetwork injects a current to the neuron (i, j) in the main-network by an amount proportional to $-T_{jl}$. The matrix $\boldsymbol{T} = \{T_{jl}\}$ (for $j = 1, 2, \ldots, n$; $l = 1, 2, \ldots, m$) is called the connection strength matrix. Obviously, for each neuron (i, j) in the main-network, we have

$$C_{ij}\frac{du_{ij}(t)}{dt} = -\frac{u_{ij}(t)}{R_{ij}} - \sum_{l=1}^{m} T_{jl} q_{il}(t)$$

$$(i = 1, 2, \ldots, p; \quad j = 1, 2, \ldots, n), \qquad (3.89)$$

which is shown in Figure 3.7.

The neuron (i, l) in the subnetwork receives its input from the neuron (i, j) of the main-network by an amount proportional to T_{jl} and is biased by the constant current b_{il}. As a result, one of the relationships between the main-network and the

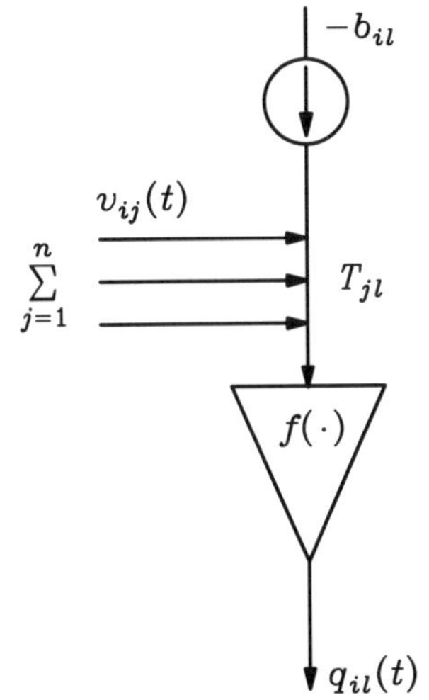

Figure 3.8. Neuron (i, l) in the subnetwork.

subnetwork is

$$q_{il}(t) = K_2\left(\sum_{j=1}^{n} T_{jl}v_{ij}(t) - b_{il}\right) \qquad (i = 1, 2, \ldots, p; \quad l = 1, 2, \ldots, m), \tag{3.90}$$

which is shown in Figure 3.8.

In order to guarantee that this 2-D neural network is stable, we define the energy function

$$E(t) = \frac{K_2}{2}tr\left\{(\boldsymbol{B} - \boldsymbol{V}(t))(\boldsymbol{B} - \boldsymbol{V}(t))^T\right\} + \sum_{i=1}^{p}\sum_{j=1}^{n}\frac{1}{R_{ij}}\int_0^{v_{ij}(t)} g_{ij}^{-1}(v)dv, \tag{3.91}$$

where $\boldsymbol{V}(t) = \{v_{ij}(t)\}$ is the output voltage matrix of the main-network and $\boldsymbol{B} = \{b_{il}\}$ is the bias current matrix of the subnetwork.

Because $tr\{(\boldsymbol{B} - \boldsymbol{V}(t))(\boldsymbol{B} - \boldsymbol{V}(t))^T\}$ is nonnegative and because $\frac{1}{R_{ij}}\int_0^{v_{ij}(t)} g_{ij}^{-1}(v)$ dv is bounded from below, $E(t)$ is bounded from below.

Moreover, from (3.91), we have the derivative of $E(t)$ with respect to time t as

$$\frac{dE(t)}{dt} = \sum_{i=1}^{p}\sum_{j=1}^{n}\left(\sum_{l=1}^{m} K_2\left[\sum_{k=1}^{n} T_{kl}v_{ik}(t) - b_{il}T_{jl} + \frac{u_{ij}(t)}{R_{ij}}\right]\frac{dv_{ij}(t)}{dt}\right). \tag{3.92}$$

Substituting for the bracketed expression of (3.92) from (3.89) and (3.90) gives

$$\frac{dE(t)}{dt} = -\sum_{i=1}^{p}\sum_{j=1}^{n} C_{ij}\frac{dv_{ij}(t)}{dt}\frac{du_{ij}(t)}{dt}$$
$$= -\sum_{i=1}^{p}\sum_{j=1}^{n} C_{ij}\left[g_{ij}^{-1}(v_{ij}(t))\right]'\left(\frac{dv_{ij}(t)}{dt}\right)^2. \tag{3.93}$$

Since C_{ij} is positive and $[g_{ij}^{-1}(v_{ij}(t))]'$ is a monotonically increasing function, each term in (3.93) is nonnegative and

$$\frac{dE(t)}{dt} \le 0, \qquad \frac{dE(t)}{dt} = 0 \rightarrow \frac{dv_{ij}(t)}{dt} = 0$$
$$(i = 1, 2, \ldots, p; \quad j = 1, 2, \ldots, n). \tag{3.94}$$

Together with the proof that $E(t)$ is bounded from below, (3.94) shows that the time evolution of this neural network is a motion in the state space that seeks out the minima of $E(t)$ and comes to a stop at such points.

Obviously, any optimization problem whose objective function is mapped to (3.91) can be solved by use of this 2-D neural network. One of the local minimum points of $E(t)$ is provided by $dv_{ij}(t)/dt = 0$ (for $i = 1, 2, \ldots, p$; $j = 1, 2, \ldots, n$).

Now we will show how to use this 2-D neural network to solve the problem (3.87) encountered in the multichannel LP spectral estimation methods. However, first we have to mention that this neural network also applies to the complex-valued case (details will be discussed in Chapter 7). In the complex-valued case, two neurons are used to represent a complex value. As a result, $p \times 2n$ and $p \times 2m$ neurons in the main-network and the subnetwork, respectively, are needed, and the dimensions of the connection strength matrix and the bias current matrix become $2n \times 2m$ and $p \times 2m$, respectively.

To use the proposed 2-D neural network to solve (3.87) or to compute (3.88), we choose:

(1) $\boldsymbol{T} = \boldsymbol{X}$ and $\boldsymbol{B} = \boldsymbol{Y}$. That is, the data matrices $\boldsymbol{X}$ and $\boldsymbol{Y}$ are directly taken as the connection strength matrix and the bias current matrix, respectively, of the network.

(2) $g_{ij}(u) = K_1 u$, which means that all the neurons in the main-network are modeled as linear amplifiers with voltage gains K_1. Obviously,

$$\frac{1}{R_{ij}}\int_0^{v_{ij}(t)} g_{ij}^{-1}(v)dv = \frac{1}{2R_{ij}K_1}(v_{ij}(t))^2 \ge 0;$$

this means that the above term is bounded from below.

(3) $C_{ij} = C$ and $R_{ij} = R$ (for $i = 1, 2, \ldots, p$; $j = 1, 2, \ldots, n$).

If the other parameters of the 2-D network and the neurons in the subnetwork are selected as proposed above, we know, according to the analyses just given, that this network used to solve (3.87) is stable and that the stationary state of the network can be obtained from $dv_{ij}(t)/dt = 0$ or $du_{ij}(t)/dt = 0$ (for $i = 1, 2, \ldots, p$; $j = 1, 2, \ldots, n$).

Combining (3.89) with (3.90) in the matrix form gives

$$C\frac{d\boldsymbol{U}(t)}{dt} = -\boldsymbol{U}(t)\left(\frac{1}{R}\boldsymbol{I} + K_1K_2\boldsymbol{X}\boldsymbol{X}^H\right) + K_2\boldsymbol{Y}\boldsymbol{X}^H, \tag{3.95}$$

where $\boldsymbol{U}(t) = \{u_{ij}(t)\}$ (for $i = 1, 2, \ldots, p$; $j = 1, 2, \ldots, n$) is the input voltage matrix of the main-network. Note that we have considered the complex-valued case.

The input matrix $\boldsymbol{U}_f$ and output matrix $\boldsymbol{V}_f$ in the stationary state can be obtained by letting $du_{ij}(t)/dt = 0$ for all i, j, that is

$$\boldsymbol{U}_f = \lim_{t\to\infty} \boldsymbol{U}(t) = K_2\boldsymbol{Y}\boldsymbol{X}^H\left(\frac{1}{R}\boldsymbol{I} + K_1K_2\boldsymbol{X}\boldsymbol{X}^H\right)^{-1} \tag{3.96}$$

and

$$\boldsymbol{V}_f = \lim_{t\to\infty} \boldsymbol{V}(t) = \boldsymbol{Y}\boldsymbol{X}^H\left(\frac{1}{RK_1K_2}\boldsymbol{I} + \boldsymbol{X}\boldsymbol{X}^H\right)^{-1}. \tag{3.97}$$

Moreover, according to the definition of the pseudoinverse [20], Equation (3.88) can be written as

$$\boldsymbol{D} = \boldsymbol{Y} \lim_{\alpha\to 0} \boldsymbol{X}^H(\alpha\boldsymbol{I} + \boldsymbol{X}\boldsymbol{X}^H)^{-1}. \tag{3.98}$$

If we compare (3.97) with (3.98), we see that $\boldsymbol{V}_f$ is an approximation of the coefficient matrix $\boldsymbol{D}$ of the multichannel LP spectral estimation methods. The error is

$$\begin{aligned} \|\boldsymbol{V}_f - \boldsymbol{D}\| &= \frac{1}{RK_1K_2}\left\|\boldsymbol{Y}\sum_{i=1}^{r}\frac{1}{\lambda_i + \frac{1}{RK_1K_2}}\frac{1}{\lambda_i}\boldsymbol{\Lambda}_i\boldsymbol{\Gamma}_i^H\right\| \\ &\le \frac{1}{RK_1K_2}\|\boldsymbol{Y}\|\left\|\sum_{i=1}^{r}\frac{1}{\lambda_i + \frac{1}{RK_1K_2}}\frac{1}{\lambda_i}\boldsymbol{\Lambda}_i\boldsymbol{\Gamma}_i^H\right\|, \end{aligned} \tag{3.99}$$

where we used

$$\boldsymbol{D} = \boldsymbol{Y}\sum_{i=1}^{r}\frac{1}{\lambda_i}\boldsymbol{\Lambda}_i\boldsymbol{\Gamma}_i^H$$

and

$$\boldsymbol{V}_f = \boldsymbol{Y}\sum_{i=1}^{r}\frac{1}{\lambda_i + \frac{1}{RK_1K_2}}\boldsymbol{\Lambda}_i\boldsymbol{\Gamma}_i^H.$$

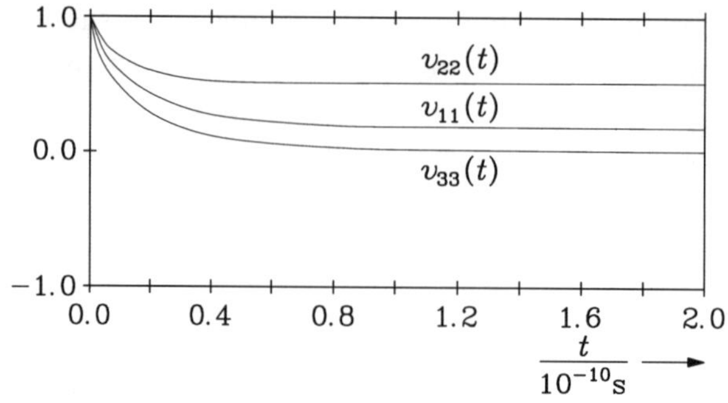

Figure 3.9. The dynamic curves of the diagonal elements of the output matrix $\boldsymbol{V}(t)$ in Example 3.3.

Here r is the rank of the matrix $\boldsymbol{X}$, the λ_i (for $i = 1, 2, \ldots, r$) are the nonzero eigenvalues of $\boldsymbol{X}^H\boldsymbol{X}$, and $\boldsymbol{\Lambda}_i$ and $\boldsymbol{\Gamma}_i$ are the right and left singular vectors corresponding to the eigenvalue λ_i.

Obviously, the difference between $\boldsymbol{V}_f$ and $\boldsymbol{D}$ can be made arbitrarily small by appropriately selecting R, K_1, and K_2.

From the uniqueness of $(\frac{1}{R}\boldsymbol{I} + K_1K_2\boldsymbol{X}\boldsymbol{X}^H)^{-1}$, we know that this neural network has only a stationary state, that is, we do not encounter the local minimum problem. Thus, the proposed 2-D neural network provides the best approximate solution of the coefficient matrix $\boldsymbol{D}$ in the sense of the Euclidean norm. Moreover, the convergence time can be quantitatively found out from the eigenvalues of $(\frac{1}{RC} + \frac{K_1K_2}{C}\boldsymbol{X}\boldsymbol{X}^H)$, which is, in general, on the order of hundreds of nanoseconds.

The above analysis shows that this neural network can provide on the order of hundreds of nanoseconds the best approximate solution to the desired coefficient matrix $\boldsymbol{D}$ of the multichannel spectral estimation in the sense of the Euclidean norm. As a result, this network is suitable for real-time applications of multichannel spectral estimation methods.

In effect, this two-dimensional neural network can be applied to solving any matrix equation

$$\boldsymbol{D}\boldsymbol{X} = \boldsymbol{Y} \tag{3.100}$$

in the sense of minimization of the Euclidean norm $tr\{(\boldsymbol{Y} - \boldsymbol{D}\boldsymbol{X})(\boldsymbol{Y} - \boldsymbol{D}\boldsymbol{X})^H\}$. This class of matrix equations is widely encountered in many other fields of signal processing besides the multichannel spectral estimation ((2.2) is a special case of (3.100)). As further illustrations of this 2-D neural network for solving matrix equations $\boldsymbol{D}\boldsymbol{X} = \boldsymbol{Y}$, we present two sets of simulation results (Examples 3.3 and 3.4).

In these two sets of simulation results, we used $K_1 = K_2 = 1$, $C = 100$ pF, and $R = 10^5\ \Omega$. $\boldsymbol{D} = \boldsymbol{Y}\boldsymbol{X}^+$ is the exact solution of the matrix equation (3.100);

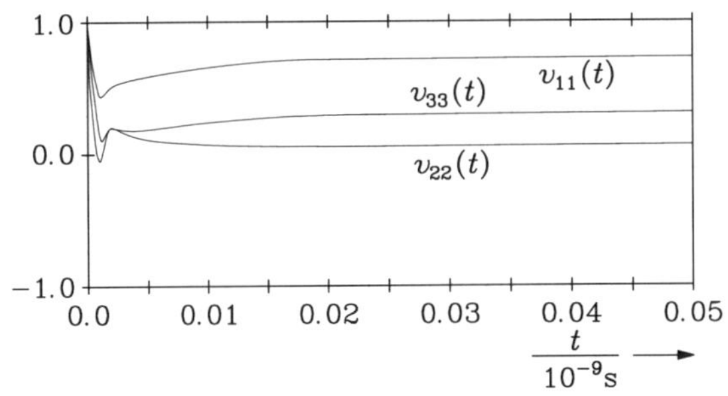

Figure 3.10. The dynamic curves of the diagonal elements of the output matrix $\boldsymbol{V}(t)$ in Example 3.4.

$\boldsymbol{V}_f$ is the solution provided by the neural network. The output matrix $\boldsymbol{V}(t)$ of the main-network from the initial state to the stationary state are also given. Figures 3.9 and 3.10 show the dynamic curves of the diagonal elements of the output matrix $\boldsymbol{V}(t)$ in these two examples.

Example 3.3

$$\boldsymbol{X} = \begin{pmatrix} 1 & -1 & -1 & 1 & 1 & 1 & -1 & -1 \\ -1 & -1 & 1 & 1 & 1 & -1 & -1 & 1 \\ 1 & 1 & 1 & -1 & 1 & -1 & 1 & 1 \end{pmatrix},$$

$$\boldsymbol{Y} = \begin{pmatrix} -1 & 1 & 1 & 1 & 1 & 1 & -1 & -1 \\ -1 & -1 & 1 & -1 & 1 & -1 & -1 & -1 \\ -1 & -1 & -1 & -1 & 1 & -1 & 1 & -1 \end{pmatrix},$$

$$\boldsymbol{D} = \begin{pmatrix} 0.167 & 0.250 & -0.167 \\ 0.000 & 0.500 & 0.000 \\ 0.000 & 0.000 & 0.000 \end{pmatrix},$$

$$\boldsymbol{V}_f = \begin{pmatrix} 0.167 & 0.250 & -0.167 \\ 0.000 & 0.500 & 0.000 \\ 0.000 & 0.000 & 0.000 \end{pmatrix},$$

$$\boldsymbol{V}(0) = \begin{pmatrix} 1.000 & 0.000 & 0.000 \\ 0.000 & 1.000 & 0.000 \\ 0.000 & 0.000 & 1.000 \end{pmatrix},$$

$$\boldsymbol{V}(5 \times 10^{-12}) = \begin{pmatrix} 0.750 & 0.085 & 0.065 \\ 0.000 & 0.830 & 0.000 \\ 0.144 & 0.000 & 0.672 \end{pmatrix},$$

$$\boldsymbol{V}(10^{-11}) = \begin{pmatrix} 0.592 & 0.141 & 0.073 \\ 0.000 & 0.717 & 0.000 \\ 0.193 & 0.000 & 0.472 \end{pmatrix},$$

$$V(1.5 \times 10^{-11}) = \begin{pmatrix} 0.487 & 0.178 & 0.055 \\ 0.000 & 0.643 & 0.000 \\ 0.200 & 0.000 & 0.345 \end{pmatrix},$$

$$\vdots$$

$$V(1.95 \times 10^{-10}) = \begin{pmatrix} 0.167 & 0.250 & -0.166 \\ 0.000 & 0.500 & 0.000 \\ 0.000 & 0.000 & 0.000 \end{pmatrix},$$

$$V(2 \times 10^{-10}) = \begin{pmatrix} 0.167 & 0.025 & -0.167 \\ 0.000 & 0.500 & 0.000 \\ 0.000 & 0.000 & 0.000 \end{pmatrix}$$

Example 3.4

$$X = \begin{pmatrix} -2.460 & 4.600 & 1.150 & -2.460 & 2.650 & 0.347 & -3.380 & 2.000 \\ 7.150 & -1.970 & -2.340 & 2.190 & 3.650 & 1.460 & 3.210 & -2.340 \\ 1.970 & 1.460 & 3.210 & 2.340 & 7.150 & 3.650 & -1.390 & 4.290 \end{pmatrix},$$

$$Y = \begin{pmatrix} 6.210 & -0.767 & -2.620 & -3.110 & 0.292 & 0.583 & -3.110 & 0.002 \\ 0.854 & 1.940 & 0.000 & 1.070 & -0.485 & 3.111 & -2.700 & -0.365 \\ 1.970 & -1.460 & 1.970 & 2.850 & -2.190 & -0.365 & 0.486 & 0.384 \end{pmatrix},$$

$$D = \begin{pmatrix} 0.730 & 0.790 & -0.381 \\ 0.219 & 0.066 & 0.076 \\ -0.798 & -0.303 & 0.311 \end{pmatrix},$$

$$V_f = \begin{pmatrix} 0.730 & 0.790 & -0.381 \\ 0.219 & 0.066 & 0.076 \\ -0.798 & -0.303 & 0.311 \end{pmatrix},$$

$$V(0) = \begin{pmatrix} 1.000 & 0.000 & 0.000 \\ 0.000 & 1.000 & 0.000 \\ 0.000 & 0.000 & 1.000 \end{pmatrix},$$

$$V(5 \times 10^{-12}) = \begin{pmatrix} 0.602 & 0.707 & -0.312 \\ 0.342 & 0.145 & 0.009 \\ -0.619 & -0.188 & 0.213 \end{pmatrix},$$

$$V(10^{-11}) = \begin{pmatrix} 0.667 & 0.749 & -0.347 \\ 0.279 & 0.105 & 0.043 \\ 0.710 & -0.246 & 0.264 \end{pmatrix},$$

$$V(1.5 \times 10^{-11}) = \begin{pmatrix} 0.699 & 0.770 & -0.365 \\ 0.249 & 0.085 & 0.060 \\ -0.755 & -0.275 & 0.288 \end{pmatrix},$$

$$\vdots$$

$$
V(1.1 \times 10^{-10}) = \begin{pmatrix} 0.730 & 0.790 & -0.381 \\ 0.219 & 0.066 & 0.076 \\ -0.798 & -0.303 & 0.311 \end{pmatrix},
$$

$$
V(1.15 \times 10^{-10}) = \begin{pmatrix} 0.730 & 0.790 & -0.381 \\ 0.219 & 0.066 & 0.076 \\ -0.798 & -0.303 & 0.311 \end{pmatrix}
$$

In addition, it is worth mentioning that this 2-D neural network can also be used to compute matrix inversion in real time. Given a matrix $\boldsymbol{X}$, we take the matrix $\boldsymbol{X}$ and the identity matrix $\boldsymbol{I}$ as the connection strength matrix $\boldsymbol{T}$ and the bias current matrix $\boldsymbol{B}$ of the network, respectively, and the other parameters of the network are selected as those for the coefficient matrix $\boldsymbol{D}$ of the multichannel LP spectral estimation methods. Then such a constructed neural network will provide the solution

$$
\boldsymbol{V}_f = \boldsymbol{I}\boldsymbol{X}^H \left(\frac{1}{RK_1K_2}\boldsymbol{I} + \boldsymbol{X}\boldsymbol{X}^H \right)^{-1} \approx \boldsymbol{X}^{-1}, \tag{3.101}
$$

that is, an approximation of the inverse of $\boldsymbol{X}$. The error is

$$
\begin{aligned}
\|\boldsymbol{V}_f - \boldsymbol{X}^{-1}\| &\leq \frac{\|\boldsymbol{X}^{-1}(\boldsymbol{X}^H)^{-1}\boldsymbol{X}^{-1}\|}{RK_1K_2} \\
&\leq \frac{\|\boldsymbol{X}^{-1}\|^2 \|(\boldsymbol{X}^H)^{-1}\|}{RK_1K_2},
\end{aligned} \tag{3.102}
$$

which can be made arbitrarily small by appropriately selecting R, K_1, and K_2.

We now give two simulation results of computing the matrix inversion by this 2-D neural network. In these results, $\boldsymbol{X}$ and $\boldsymbol{X}^{-1}$ are the given matrix and its exact inverse, respectively; $\boldsymbol{V}_f$ is the solution provided by the neural network; $\boldsymbol{V}(0)$ is the initial output matrix of the main-network; the values of $\boldsymbol{V}(t)$ describe the dynamics of the network; $\boldsymbol{I}_1 = \boldsymbol{X}\boldsymbol{V}_f$, which should approximate the identity matrix $\boldsymbol{I}$; and the error between $\boldsymbol{V}_f$ and $\boldsymbol{X}^{-1}$ is measured by the norm $\|\boldsymbol{V}_f - \boldsymbol{X}^{-1}\|$ and is denoted by e_2. Figures 3.11 and 3.12 give the dynamic curves of the diagonal elements of $\boldsymbol{V}(t)$ from the initial state to the stationary state of the network.

Example 3.5

$$
X = \begin{pmatrix} 1.5730 & 4.5559 & 1.1432 & -2.1145 \\ 1.5337 & -1.1922 & 5.3358 & 1.8928 \\ 2.1928 & 2.5991 & 3.2117 & 2.3583 \\ -7.2136 & -1.7270 & 2.6214 & -5.1220 \end{pmatrix},
$$

$$X^{-1} = \begin{pmatrix} 0.092150 & -0.127955 & -0.397306 & -0.266510 \\ 0.109640 & 0.108585 & 0.342846 & 0.151213 \\ 0.048378 & 0.167166 & 0.028199 & 0.054663 \\ -0.141987 & 0.229148 & 0.458381 & 0.157097 \end{pmatrix},$$

$$V_f = \begin{pmatrix} 0.092149 & -0.127953 & -0.397303 & -0.266510 \\ 0.109640 & 0.108584 & 0.342844 & 0.151212 \\ 0.048378 & 0.167166 & 0.028199 & 0.054663 \\ -0.141987 & 0.229147 & 0.458378 & 0.157096 \end{pmatrix},$$

$$I_1 = \begin{pmatrix} 0.999997 & 0.000002 & 0.000000 & 0.000003 \\ 0.000002 & 0.999998 & 0.000000 & -0.000002 \\ 0.000000 & 0.000000 & 1.000000 & -0.000001 \\ 0.000003 & -0.000002 & -0.000001 & 0.999997 \end{pmatrix},$$

$e_2 = 0.00000189$,

$$V(0) = \begin{pmatrix} 1.000000 & 0.0000000 & 0.000000 & 0.000000 \\ 0.000000 & 1.0000000 & 0.000000 & 0.000000 \\ 0.000000 & 0.0000000 & 1.000000 & 0.000000 \\ 0.000000 & 0.0000000 & 0.000000 & 1.000000 \end{pmatrix},$$

$$V(10^{-11}) = \begin{pmatrix} 0.107689 & -0.002823 & -0.118859 & -0.126053 \\ 0.111626 & 0.116380 & 0.324694 & 0.142012 \\ -0.026206 & 0.386316 & 0.561992 & 0.312960 \\ -0.117843 & 0.184898 & 0.359019 & 0.110232 \end{pmatrix},$$

$$V(2 \times 10^{-11}) = \begin{pmatrix} 0.063075 & -0.025424 & -0.157255 & -0.149954 \\ 0.111318 & 0.103786 & 0.330970 & 0.145467 \\ -0.013281 & 0.362531 & 0.488266 & 0.277425 \\ -0.129790 & 0.192238 & 0.371407 & 0.115044 \end{pmatrix},$$

$$V(3 \times 10^{-11}) = \begin{pmatrix} 0.064597 & -0.039674 & -0.189646 & -0.165935 \\ 0.111011 & 0.104247 & 0.332628 & 0.146266 \\ -0.004766 & 0.336297 & 0.426190 & 0.247386 \\ -0.131908 & 0.197170 & 0.383122 & 0.120657 \end{pmatrix},$$

$$\vdots$$

$$V(9.9 \times 10^{-10}) = \begin{pmatrix} 0.092149 & -0.127953 & -0.397303 & -0.266510 \\ 0.109640 & 0.108584 & 0.342844 & 0.151212 \\ 0.048378 & 0.167166 & 0.028199 & 0.054663 \\ -0.141987 & 0.229147 & 0.458378 & 0.157096 \end{pmatrix},$$

$$V(10^{-9}) = \begin{pmatrix} 0.092149 & -0.127953 & -0.397303 & -0.266510 \\ 0.109640 & 0.108584 & 0.342844 & 0.151212 \\ 0.048378 & 0.167166 & 0.028199 & 0.054663 \\ -0.141987 & 0.229147 & 0.458378 & 0.157096 \end{pmatrix}$$

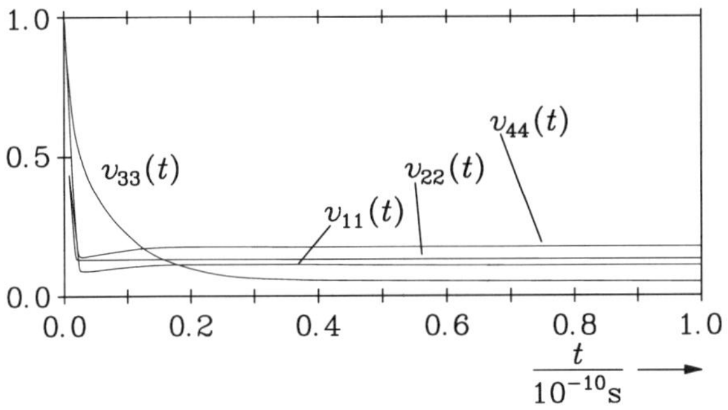

Figure 3.11. The dynamic curves of the diagonal elements of the output matrix $\boldsymbol{V}(t)$ in Example 3.5.

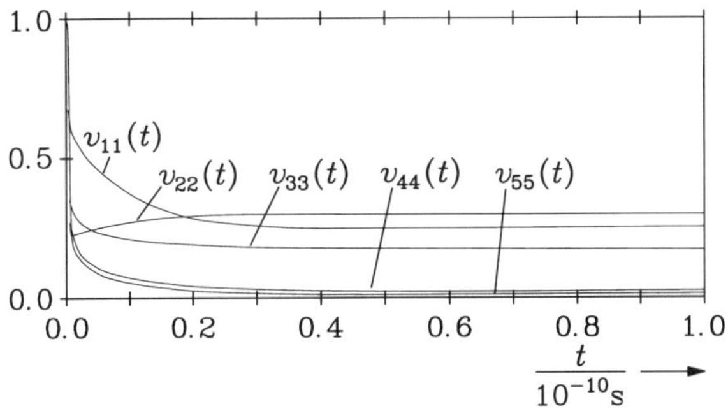

Figure 3.12. The dynamic curves of the diagonal elements of the output matrix $\boldsymbol{V}(t)$ in Example 3.6.

Example 3.6

$$\boldsymbol{X} = \begin{pmatrix} -1.2273 & 4.2256 & 1.2215 & 3.4573 & 0.8834 \\ 2.1225 & -1.9227 & 9.3358 & -3.1330 & -5.1575 \\ 9.7281 & 2.1459 & -0.2117 & 6.1633 & 1.3246 \\ 19.2186 & -3.7670 & -2.6222 & -0.1068 & 5.9124 \\ 1.5437 & -0.6227 & -8.5189 & 0.2522 & -0.1271 \end{pmatrix},$$

$$\boldsymbol{X}^{-1} = \begin{pmatrix} 0.110897 & 0.075195 & -0.026309 & 0.057603 & 0.081230 \\ 0.591626 & 0.171010 & -0.252524 & 0.127499 & 0.239269 \\ -0.034345 & -0.001200 & 0.023615 & -0.004139 & -0.122939 \\ -0.380994 & -0.148464 & 0.305273 & -0.148052 & -0.179345 \\ -0.005647 & -0.138684 & -0.059384 & 0.058618 & -0.169361 \end{pmatrix},$$

$$\boldsymbol{V}_f = \begin{pmatrix} 0.110896 & 0.075195 & -0.026309 & 0.057603 & 0.081230 \\ 0.591621 & 0.171008 & -0.252522 & 0.127498 & 0.239267 \\ -0.034345 & -0.001200 & 0.023614 & -0.004139 & -0.122938 \\ -0.380990 & -0.148463 & 0.305271 & -0.148051 & -0.179343 \\ -0.005646 & -0.138684 & -0.059384 & 0.058619 & -0.169360 \end{pmatrix},$$

$$I_1 = \begin{pmatrix} 1.000000 & -0.000001 & 0.000000 & 0.000000 & 0.000000 \\ -0.000001 & 0.999995 & 0.000000 & 0.000004 & 0.000000 \\ 0.000000 & 0.000000 & 1.000000 & 0.000000 & 0.000000 \\ 0.000000 & 0.000004 & 0.000000 & 0.999997 & 0.000000 \\ 0.000000 & 0.000000 & 0.000000 & 0.000000 & 0.999999 \end{pmatrix},$$

$e_2 = 0.00000219,$

$$V(0) = \begin{pmatrix} 1.000000 & 0.000000 & 0.000000 & 0.000000 & 0.000000 \\ 0.000000 & 1.000000 & 0.000000 & 0.000000 & 0.000000 \\ 0.000000 & 0.000000 & 1.000000 & 0.000000 & 0.000000 \\ 0.000000 & 0.000000 & 0.000000 & 1.000000 & 0.000000 \\ 0.000000 & 0.000000 & 0.000000 & 0.000000 & 1.000000 \end{pmatrix},$$

$$V(10^{-11}) = \begin{pmatrix} 0.534696 & 0.200412 & -0.266979 & 0.182999 & 0.247189 \\ 0.268973 & 0.103756 & -0.051351 & 0.018681 & 0.154887 \\ -0.275973 & -0.037689 & 0.183316 & -0.092314 & -0.165403 \\ 0.149837 & 0.002069 & -0.000495 & 0.012149 & 0.019116 \\ 0.252620 & 0.037379 & -0.142750 & 0.087823 & 0.080899 \end{pmatrix},$$

$$V(2 \times 10^{-11}) = \begin{pmatrix} 0.482808 & 0.199479 & -0.229496 & 0.160889 & 0.248401 \\ 0.315093 & 0.083752 & -0.098910 & 0.048258 & 0.122674 \\ -0.238233 & -0.062921 & 0.138538 & -0.063804 & -0.204997 \\ 0.083467 & 0.005577 & 0.052028 & -0.018508 & 0.027676 \\ 0.244676 & -0.036745 & -0.183833 & 0.119466 & -0.029508 \end{pmatrix},$$

$$\vdots$$

$$V(9.9 \times 10^{-10}) = \begin{pmatrix} 0.110900 & 0.075196 & -0.026311 & 0.057604 & 0.081232 \\ 0.591618 & 0.171007 & -0.252520 & 0.127497 & 0.239266 \\ -0.034347 & -0.001201 & 0.023616 & -0.004140 & -0.122940 \\ -0.380985 & -0.148461 & 0.305269 & -0.148050 & -0.179341 \\ -0.005644 & -0.138683 & -0.059386 & 0.058619 & -0.169359 \end{pmatrix},$$

$$V(10^{-9}) = \begin{pmatrix} 0.110900 & 0.075196 & -0.026311 & 0.057604 & 0.081232 \\ 0.591618 & 0.171007 & -0.252520 & 0.127497 & 0.239266 \\ -0.034347 & -0.001201 & 0.023615 & -0.004140 & -0.122939 \\ -0.380986 & -0.148461 & 0.305269 & -0.148050 & -0.179341 \\ -0.005644 & -0.138683 & -0.059386 & 0.058619 & -0.169359 \end{pmatrix}$$

3.4 Neural Networks for Two-Dimensional Spectral Estimation

In this section, we will show how to apply the neural network of Figure 2.2 to two-dimensional spectral estimation. Similar to one-dimensional spectral estimation, the two-dimensional spectral estimation problem may be stated briefly in the following way [21, 22, 23]. Given samples of a stationary and homogeneous zero-mean random field as

$$\begin{pmatrix} x(1,1) & x(1,2) & \cdots & x(1,N_2) \\ x(2,1) & x(2,2) & \cdots & x(2,N_2) \\ \vdots & \vdots & \ddots & \vdots \\ x(N_1,1) & x(N_1,2) & \cdots & x(N_1,N_2) \end{pmatrix}, \tag{3.103}$$

estimate its power spectrum, which is defined by the two-dimensional Fourier transform of the autocorrelation function,

$$s_x(f_1, f_2) = \sum_{m_1=-\infty}^{+\infty} \sum_{m_2=-\infty}^{+\infty} R(m_1, m_2) \exp(-j2\pi m_1 f_1 - j2\pi m_2 f_2), \tag{3.104}$$

where $R(m_1, m_2) = E[x^*(n_1+m_1, n_2+m_2)x(n_1, n_2)]$ is the autocorrelation function of $x(n_1, n_2)$.

As shown in Section 3.1, for the one-dimensional spectral estimation, the maximum entropy method is equivalent to the autoregressive signal modeling. That is, the spectrum obtained from the estimated AR coefficient vectors is identical to the maximum entropy spectral estimation. Although for two-dimensional signals this is no longer the case, it is widely appreciated that the ARMA (AR or MA) model [24, 25],

$$\begin{aligned} x(n_1, n_2) = & \sum_{k=0}^{q_1} \sum_{m=0}^{q_2} d_{km} \epsilon(n_1 - k, n_2 - m) \\ & - \sum_{k=0}^{p_1} \sum_{m=0}^{p_2} a_{km} x(n_1 - k, n_2 - m), \end{aligned} \tag{3.105}$$

is generally the most effective model and can provide a superior resolution compared with conventional methods such as the Fourier transform. Here $\{p_1, p_2, q_1, q_2\}$ is the set designating the model order, $\epsilon(n_1, n_2)$ is a white noise with zero mean and variance σ^2, and $\{d_{km}\}$ and $\{a_{km}\}$ are the coefficients of the model. Note that k and m in the second term on the right-hand side of (3.105) are not simultaneously zero.

With the above conditions, the power spectrum of the field $x(n_1, n_2)$ is given by

$$s_x(f_1, f_2) = \sigma^2 \frac{\left| \sum_{k=0}^{q_1} \sum_{m=0}^{q_2} d_{km} \exp(-j2\pi k f_1 - j2\pi m f_2) \right|^2}{\left| \sum_{k=0}^{p_1} \sum_{m=0}^{p_2} a_{km} \exp(-j2\pi k f_1 - j2\pi m f_2) \right|^2}, \tag{3.106}$$

where $a_{00} = 1$.

It has been shown [24] that the autoregressive coefficient vector $\boldsymbol{a} = [a_{01}, a_{02}, \ldots, a_{0p_2}, \ldots, a_{p_10}, a_{p_11}, \ldots, a_{p_1p_2}]^T$ (with $p_1p_2 - 1$ elements) satisfies the equations

$$\boldsymbol{Ra} = \boldsymbol{R}_1, \tag{3.107}$$

where $\boldsymbol{R}$ is the $(p_1p_2-1)\times(p_1p_2-1)$ matrix given by its elements

$$R_{jl}^{km}=\sum_{k_1=q_1+1}^{N_1-1}\sum_{k_2=q_2+1}^{N_2-1}\sum_{n_1=k_1+1}^{N_1}\sum_{n_2=k_2+1}^{N_2}\sum_{s_1=k_1+1}^{N_1}\sum_{s_2=k_2+1}^{N_2}W(k_1,k_2)$$
$$\times\, x(n_1-k,n_2-m)x(s_1-k_1,s_2-k_2)$$
$$\times\, x^*(s_1-j,s_2-l)x^*(n_1-k_1,n_2-k_2), \tag{3.108}$$

$\boldsymbol{R}_1$ is a $(p_1p_2-1)\times 1$ vector with elements R_{00}^{km}, and $W(k_1,k_2)$ is the weight, which is a nonnegative parameter generally selected as

$$W(k_1,k_2)=(N_1-k_1)(N_2-k_2)$$
$$(p_1+1\le k_1\le N_1-1;\, p_2+1\le k_2\le N_2-1). \tag{3.109}$$

Note that R_{jl}^{km} denotes that the elements are arranged consistently with those of the vector $\boldsymbol{a}$ from a_{km}.

Clearly, the neural network in Figure 2.2 can solve (3.107) in real time provided that we take the matrix $\boldsymbol{R}$ as the connection strength matrix $\boldsymbol{T}$ and the vector $\boldsymbol{R}_1$ as the bias current vector $\boldsymbol{B}$. The autoregressive coefficient vector provided by the network is

$$\boldsymbol{a}=\left(\frac{1}{RK_1K_2}\boldsymbol{I}+\boldsymbol{R}^H\boldsymbol{R}\right)^{-1}\boldsymbol{R}^H\boldsymbol{R}_1. \tag{3.110}$$

Once the autoregressive coefficient vector has been determined, we can immediately obtain the denominator term $A(f_1,f_2)$ of the power spectrum (3.106) of $x(n_1,n_2)$:

$$A(f_1,f_2)=\left|\sum_{k=0}^{p_1}\sum_{m=0}^{p_2}a_{km}\exp(-j2\pi kf_1-j2\pi mf_2)\right|^2. \tag{3.111}$$

The numerator term,

$$d(f_1,f_2)=\left|\sum_{k=0}^{q_1}\sum_{m=0}^{q_2}d_{km}\exp(-j2\pi kf_1-j2\pi mf_2)\right|^2, \tag{3.112}$$

can also be estimated from the autoregressive coefficient vector provided by the neural network and the sample matrix (3.103).

From (3.105), we can obtain a new sequence $r(n_1,n_2)$ as

$$r(n_1,n_2)=\sum_{k=0}^{q_1}\sum_{m=0}^{q_2}d_{km}\epsilon(n_1-k,n_2-m) \tag{3.113}$$

$$=\sum_{k=0}^{p_1}\sum_{m=0}^{p_2}a_{km}x(n_1-k,n_2-m). \tag{3.114}$$

It follows from (3.113) that

$$s_r(f_1, f_2) = d(f_1, f_2), \tag{3.115}$$

where $s_r(f_1, f_2)$ is the power spectrum of the sequence $r(n_1, n_2)$.

With this in mind, the estimation problem of $d(f_1, f_2)$ becomes that of $s_r(f_1, f_2)$. For $s_r(f_1, f_2)$, we can use Equation (3.114) to obtain the following matrix from the available autoregressive coefficient vector and the sample matrix (3.103):

$$\begin{pmatrix} r(p_1+1, p_2+1) & r(p_1+1, p_2+2) & \cdots & r(p_1+1, N_2) \\ r(p_1+2, p_2+1) & r(p_1+2, p_2+2) & \cdots & r(p_1+2, N_2) \\ \vdots & \vdots & \ddots & \vdots \\ r(N_1, p_2+1) & r(N_1, p_2+2) & \cdots & r(N_1, N_2) \end{pmatrix}. \tag{3.116}$$

Segmenting (3.116) into L_1L_2 partitions with the size $(q_1+1) \times (q_1+1)$ yields

$$\begin{gathered} r_{k_1k_2}(n_1, n_2) = w(n_1, n_2) r(n_1 + 1 + p_1 + k_1 i_1, n_2 + 1 + p_2 + k_2 i_2) \\ (0 \le n_1 \le q_1, \quad 0 \le n_2 \le q_2; \quad 0 \le k_1 \le L_1 - 1, \quad 0 \le k_2 \le L_2 - 1), \end{gathered} \tag{3.117}$$

where $w(n_1, n_2)$ is a data window and i_1, i_2 are positive integers specifying the spatial shift between adjacent portions. These individual partitions will overlap for a shift selection of $i_1 \le q_1, i_2 \le q_2$. Furthermore, the above parameters must be selected to satisfy $p_j + q_j + 1 + (L_j - 1)i_j < N_j$ $(j = 1, 2)$.

We can now estimate $d(f_1, f_2)$ (or say $s_r(f_1, f_2)$) as

$$\begin{aligned} d(f_1, f_2) = & \frac{1}{L_1L_2} \sum_{k_1=0}^{L_1-1} \sum_{k_2=0}^{L_2-1} \frac{1}{(q_1+1)(q_2+1)} \\ & \times \left| \sum_{n_1=0}^{q_1} \sum_{n_2=0}^{q_2} w(n_1, n_2) r(n_1 + p_1 + k_1 i_1, n_2 + 1 + p_2 + k_2 i_2) \right. \\ & \left. \times \exp(-j2\pi n_1 f_1 - j2\pi n_2 f_2) \right|^2 . \end{aligned} \tag{3.118}$$

The above approach for the two-dimensional spectral estimation is summarized as follows:

(1) Compute the matrix $\boldsymbol{R}$ and the vector $\boldsymbol{R}_1$ from the available sample matrix (3.103) by use of (3.108).

(2) Use the neural network in Figure 2.2 to compute the autoregressive coefficient vector $\boldsymbol{a}$.

(3) Compute the sequence $r(n_1, n_2)$ using the relationship (3.114).

(4) Compute the denominator term $A(f_1, f_2)$ and the numerator term $d(f_1, f_2)$ of the power spectrum using (3.111) and (3.118), respectively, and then obtain the desired spectral estimation (3.106).

It should be noted that, in this approach, the neural network is only used to provide the autoregressive coefficient vector $\boldsymbol{a}$; hence, additional computations are required so as to obtain the desired two-dimensional spectral estimation. Zhuang et al. [26] have proposed another neural network algorithm by use of which the power spectrum at uniformly distributed frequencies can be obtained directly from the given autocorrelation function (if not given, this can also be estimated from the available samples in a manner similar to (3.108)). One of the disadvantages of this neural network approach is that the resolution depends on how to discretize the frequency domain. For more detail, see Reference [26].

3.5 Neural Networks for Higher-Order Spectral Estimation

During the past decade higher-order spectra have found a wide use in signal processing. Higher-order spectra are defined in terms of higher-order cumulants. Particular cases of higher-order spectra are the third-order spectrum (also called bispectrum), which is the Fourier transform of the third cumulants, and the trispectrum (the fourth-order spectrum), which is the Fourier transform of the fourth cumulants. Compared with signal processing with the power spectrum, signal processing with higher-order spectra has several advantages [27, 28, 29, 30]:

(1) Additive colored Gaussian noise of an unknown power spectrum can be suppressed. This is based on the property that for Gaussian signals only, all cumulant spectra of order greater than two are identically zero.

(2) Nonminimum phase systems can be identified and nonminimum phase signals can be reconstructed. This is based on the fact that higher-order spectra preserve the true phase characteristics of signals.

(3) Nonlinear properties in signals can be detected and characterized and nonlinear systems can be identified.

In addition, using higher-order spectra we can obtain more information about the "real world" signals because most real world signals are non-Gaussian and thus have nonzero higher-order spectra. This can be very useful in signal detection and classification problems where distinct classification features can be extracted from higher-order spectral domains.

A large number of theories and applications of high-order spectral estimation have been published. Similar to power spectrum estimation techniques, higher-order spectral estimation techniques have split into two camps: nonparametric and parametric methods. Because the nonparametric higher-order spectral methods are subject to the same problems that plague nonparametric power spectrum methods, that is, high variances and low resolution; the current emphasis is on the parametric higher-order spectral methods. Parametric higher-order spectral methods first estimate the parameters of an underlying data-generating model from the sampled data and then compute the higher-order spectrum by use of the model and the estimated parameters.

There are two principal factors for the parametric higher-order spectral estimation methods: One is the estimation performance (estimation variances and resolution); the other is the computational complexity. Based on the available power spectral estimation techniques, a number of high-resolution higher-order spectral estimation techniques have been reported. The ARMA-model-based method and Pisarenko's harmonic method are two examples. However, these methods are in general computationally intensive. In this section, we will develop neural network approaches for higher-order spectral estimation. We will concentrate on the ARMA-model-based bispectral estimation method and on Pisarenko's harmonic method. First, we introduce some definitions and properties of higher-order spectra.

Consider a zero-mean real stationary signal $x(n)$. Its second-, third-, and fourth-order cumulants are given as follows [31, 32, 33].

(1) The second-order cumulants are

$$C_2(m) = E\left[x(n+m)x(n)\right]. \tag{3.119}$$

(2) The third-order cumulants are

$$C_3(m,k) = E\left[x(n+m)x(n+k)x(n)\right]. \tag{3.120}$$

(3) The fourth-order cumulants are

$$\begin{aligned} C_4(m,k,l) = E\left[x(n+m)x(n+k)x(n+l)x(n)\right] &- C_2(m)C_2(k-l) \\ - C_2(k)C_2(l-m) &- C_2(l)C_2(m-k). \end{aligned} \tag{3.121}$$

The bispectrum $s_{2,x}(f_1, f_2)$ and trispectrum $s_{3,x}(f_1, f_2, f_3)$ of $x(n)$ are defined as the Fourier transform of the third-order cumulants and fourth-order cumulants, respectively, that is,

$$s_{2,x}(f_1, f_2) = \sum_{m=-\infty}^{+\infty} \sum_{k=-\infty}^{+\infty} C_3(m,k) \exp(-j2\pi m f_1 - j2\pi k f_2) \tag{3.122}$$

and

$$\begin{aligned} s_{3,x}(f_1, f_2, f_3) = \sum_{m=-\infty}^{+\infty} \sum_{k=-\infty}^{+\infty} \sum_{l=-\infty}^{+\infty} C_4(m,k,l) \\ \times \exp(-j2\pi m f_1 - j2\pi k f_2 - j2\pi l f_3). \end{aligned} \tag{3.123}$$

Based on the ARMA model [34, 35], $x(n)$ can be written as

$$x(n) = -\sum_{i=1}^{p} a_i x(n-i) + \sum_{i=0}^{q} d_i \epsilon(n-i), \tag{3.124}$$

where $\epsilon(n)$ is a non-Gaussian, independent identically distributed (i.i.d.) sequence with zero mean, $E[\epsilon(n)\epsilon(n+k)] = \sigma^2\delta(k)$, and $E[\epsilon(n)\epsilon(n+k)\epsilon(n+m)] = \gamma\delta(m,k)$. In addition, we assume that the above ARMA model is causal, nonminimum phase, and free of pole-zero cancellations. Using this model, the bispectrum can be written in the form

$$s_{2,x}(f_1, f_2) = \gamma \frac{\sum_{m=-q}^{q}\sum_{k=-q}^{q} \beta(m,k)\exp(-j2\pi m f_1 - j2\pi k f_2)}{\sum_{m=-p}^{p}\sum_{k=-p}^{p} \alpha(m,k)\exp(-j2\pi m f_1 - j2\pi k f_2)}, \tag{3.125}$$

where $\beta(m,k)$ and $\alpha(m,k)$ are related to the ARMA parameters via

$$\beta(m,k) = \sum_{i=0}^{q} d_i d_{m+i} d_{k+i} \qquad (-q \le m, k \le q) \tag{3.126}$$

and

$$\alpha(m,k) = \sum_{i=0}^{p} a_i a_{m+i} a_{k+i} \qquad (-p \le m, k \le p). \tag{3.127}$$

According to [34], we have

$$\gamma\beta(m,k) = \sum_{i=-p}^{p}\sum_{l=-p}^{p} \alpha(i,l) C_3(m-i, k-l) \qquad (-q \le m, k \le q). \tag{3.128}$$

Substituting (3.128) into (3.125) yields

$$s_{2,x}(f_1, f_2) = \frac{\sum_{m=-q}^{q}\sum_{k=-q}^{q}\sum_{i=-p}^{p}\sum_{l=-p}^{p} \alpha(i,l) C_3(m-i, k-l)\exp(-j2\pi m f_1 - j2\pi k f_2)}{\sum_{m=-p}^{p}\sum_{k=-p}^{p} \alpha(m,k)\exp(-j2\pi m f_1 - j2\pi k f_2)}. \tag{3.129}$$

Moreover, it follows from (3.127) that

$$s_{2,x}(f_1, f_2) = \frac{\sum_{m=-q}^{q}\sum_{k=-q}^{q}\sum_{i=-p}^{p}\sum_{l=-p}^{p}\sum_{r=0}^{p} a_r a_{r+i} a_{r+l} C_3(m-i, k-l)\exp[-j2\pi(m f_1 + k f_2)]}{\sum_{m=-p}^{p}\sum_{k=-p}^{p}\sum_{i=0}^{p} a_i a_{m+i} a_{k+i} \exp(-j2\pi m f_1 - j2\pi k f_2)}. \tag{3.130}$$

Clearly, the bispectrum can be obtained by determining the coefficients a_i (for $i = 1, 2, \ldots, p$). It has been shown [36, 37] that the coefficients a_i satisfy the linear equations

$$\boldsymbol{C}\boldsymbol{a} = \boldsymbol{C}_1, \tag{3.131}$$

where

$$\boldsymbol{C} = \begin{pmatrix} C_3(q+1-p, q-p) & \cdots & C_3(q, q-p) \\ C_3(q+1-p, q-p+1) & \cdots & C_3(q, q-p+1) \\ \vdots & \ddots & \vdots \\ C_3(q+1-p, q) & \cdots & C_3(q, q) \\ C_3(q+1-p, q-1) & \cdots & C_3(q, q-1) \\ \vdots & \ddots & \vdots \\ C_3(q, q-p) & \cdots & C_3(q+p-1, q-p) \\ C_3(q, q-p+1) & \cdots & C_3(q+p-1, q-p+1) \\ \vdots & \ddots & \vdots \\ C_3(q, q) & \cdots & C_3(q+p-1, q) \end{pmatrix}, \tag{3.132}$$

$$\boldsymbol{a} = \begin{pmatrix} a_p \\ a_{p-1} \\ \vdots \\ a_1 \end{pmatrix}, \qquad \boldsymbol{C}_1 = \begin{pmatrix} C_3(q+1, q-p) \\ C_3(q+1, q-p+1) \\ \vdots \\ C_3(q+1, q) \\ C_3(q+1, q-1) \\ \vdots \\ C_3(q+p, q-p) \\ C_3(q+p, q-p+1) \\ \vdots \\ C_3(q+p, q) \end{pmatrix}. \tag{3.133}$$

Comparing (3.131) with (3.107), we can immediately use the neural network in Figure 2.2 to solve (3.131) by taking $\boldsymbol{C}$ as the connection strength matrix $\boldsymbol{T}$ and $\boldsymbol{C}_1$ as the bias current vector $\boldsymbol{B}$ of the network. If the other parameters of the network are selected as proposed in Section 2.1, such a constructed network is stable and will provide the solution

$$\boldsymbol{a}_n = \left(\frac{1}{RK_1K_2}\boldsymbol{I} + \boldsymbol{C}^H\boldsymbol{C}\right)^{-1} \boldsymbol{C}^H\boldsymbol{C}_1, \tag{3.134}$$

which is the best approximation to the exact solution of (3.131) in the sense of the Euclidean norm (that is, under the LS criterion). The time required by this neural network to provide the solution is within a few characteristic time constants of the network. The error between $\boldsymbol{a}_n$ and the exact coefficient vector $\boldsymbol{a}$ determined by (3.131) can be made arbitrarily small by appropriately selecting the other parameters such as R, K_1, and K_2 of the network.

In summary, the above bispectral estimation approach based on the ARMA model and neural networks includes mainly the following three steps:

(1) Estimation of the third-order cumulants from the available samples of $x(n)$. In terms of the definition (3.120), $C_3(m, k)$ is usually estimated by

$$C_3(m, k) = \frac{1}{L - N} \sum_{n=1}^{L-N} x(n)x(n + k)x(n + m), \tag{3.135}$$

where L is the number of samples and $N = \max\{|m|, |k|\}$.

(2) Using the neural network in Figure 2.2 to solve (3.131) so as to obtain the coefficients a_i (for $i = 1, 2, \ldots, p$).

(3) Computation of the bispectrum from the available coefficients a_i via (3.130).

Thus, in this approach the ARMA model and the neural networks are used, respectively, to provide the high-resolution and real-time performance.

For the harmonic retrieval problem dealt with in Section 3.2, the methods based on higher-order cumulants are much more effective than that based on the second-order cumulants in the additive colored noise case. In this case, the received signal can be represented as

$$x(n) = \sum_{i=1}^{P} \alpha_i \cos(\omega_i n + \theta_i) + \epsilon(n) \qquad (n = 1, 2, \ldots, L), \tag{3.136}$$

where the various variables are the same as those in (3.19) except that $\epsilon(n)$ is assumed to be colored Gaussian noise.

With this in mind, we can prove that all the third-order cumulants are identically zero and that the fourth-order cumulant is [30]

$$C_4(m, k, l) = -\frac{1}{8} \sum_{i=1}^{P} \alpha_i^4 \Big[\cos\big(\omega_i(m - k - l)\big) + \cos\big(\omega_i(k - m - l)\big) + \cos\big(\omega_i(l - k - m)\big) \Big]. \tag{3.137}$$

If we let $m = k = l$, then we have

$$C_4(k) = -\frac{3}{8} \sum_{i=1}^{P} \alpha_i^4 \cos(\omega_i k). \tag{3.138}$$

Comparing (3.138) with (3.21), we know that Pisarenko's method can immediately be applied just by replacing the autocorrelation function $R(k)$ with the fourth-order cumulants $C_4(k)$. Consequently, the neural network approaches proposed in

Section 3.2 can be used to provide the eigenvector corresponding to the smallest eigenvalue of the fourth-order cumulant matrix $\boldsymbol{C}_4$,

$$\boldsymbol{C}_4 = \begin{pmatrix} C_4(0) & C_4(1) & \cdots & C_4(N-1) \\ C_4(-1) & C_4(0) & \cdots & C_4(N-2) \\ \vdots & \vdots & \ddots & \vdots \\ C_4(-N+1) & C_4(-N+2) & \cdots & C_4(0) \end{pmatrix}. \tag{3.139}$$

Note that the fourth-order cumulant matrix $\boldsymbol{C}_4$ is positive semidefinite (for $N \geq 2P+1$), that is, the smallest eigenvalue is zero. As a result, the connection strength matrix of the neural networks in Figures 3.2 and 3.5 should be $\boldsymbol{C}_4 + K\boldsymbol{I}$ instead of $\boldsymbol{C}_4$, where K is a positive constant.

We can summarize the above approach as follows:

(1) Estimate the fourth-order cumulant matrix $\boldsymbol{C}_4$ (3.139) of the size $N \times N$ from the measured samples.

(2) Use the neural network in Figure 3.2 (or in Figure 3.5) to compute the eigenvector corresponding to the smallest eigenvalue of the estimated matrix $\boldsymbol{C}_4$. All the parameters of the network are selected as proposed in Section 3.2 except that the connection matrix is taken as $\boldsymbol{C}_4 + K\boldsymbol{I}$.

(3) Compute the roots of the polynomial formed by the elements of the above eigenvector. This polynomial will have $2P$ roots located at $\exp(\pm j\omega_i)$ (for $i = 1, 2, \ldots, P$).

Finally, we note that there is currently ongoing development of neural networks that can be used to provide in real time not only the desired eigenvector but also the roots of the related polynomial.

Bibliography

[1] Haykin, S., *Nonlinear Methods of Spectral Analysis*, Springer-Verlag, New York, 1983.

[2] Marple Jr., S. L., *Digital Spectral Analysis with Applications,* Prentice-Hall, Englewood Cliffs, NJ, 1987.

[3] Kay, S. M., *Modern Spectral Estimation*, Prentice-Hall, Englewood Cliffs, NJ, 1988.

[4] Burg, J. P., "Maximum Entropy Spectral Analysis," Stanford Univ. Thesis, 1975.

[5] Haykin, S. and Cadzow, J. A., *Spectral Estimation, Proc. IEEE*, Special Issue, Vol. 70, No. 9, 1982.

[6] Haykin, S., *Adaptive Filter Theory*, Prentice-Hall, Englewood Cliffs, NJ, 1991.

[7] Rahman, M. A. and Yu, K. B.,"Total Least Squares Approach for Frequency Estimation Using Linear Prediction," *IEEE. Trans. on Acoustics, Speech and Signal Processing*, Vol. 35, No. 10, 1987, pp. 1440–54.

[8] Fuhrmann, D. R. and Liu, B., "Rotational Search Methods for Adaptive Pisarenko's Harmonic Retrieval," *IEEE Trans. on Acoustics, Speech and Signal Processing*, Vol. 34, 1986, pp. 1550–65.

[9] Banjanian, Z., Cruz, J. R., and Zrnic, D. S., "Eigen-Decomposition Methods for Frequency Estimation: A Unified Approach," *Proc. IEEE Int. Conf. on Acoustics, Speech and Signal Processing*, 1990, pp. 2595–98.

[10] Pisarenko, V. F., "The Retrieval of Harmonics from a Covariance Function," *Geophys. J. Royal Astron. Soc.*, Vol. 33, 1973, pp. 347–66.

[11] Larimore, M. G., "Adaption Convergence of Spectral Estimation Based on Pisarenko's Harmonic Retrieval," *IEEE Trans. on Acoustics, Speech and Signal Processing*, Vol. 31, 1983, pp. 955–62.

[12] Reddy, V. U., Egardt, B., and Kailath, T., "Least Squares Type Algorithm for Adaptive Implementation of Pisarenko's Harmonic Retrieval Method," *IEEE Trans. on Acoustics, Speech and Signal Processing*, Vol. 30, 1982, pp. 399–405.

[13] Karhunen, J., "Recursive Estimation of Eigenvectors of Correlation Type Matrices for Signal Processing Applications," PhD Dissertation, Helsinki Univ., Finland, 1984.

[14] Oja, E. and Karhunen, J., "On Stochastic Approximation of the Eigenvectors and Eigenvalues of the Expectation of a Random Matrix," *J. Math. Analysis Appl.*, Vol. 106, 1985, pp. 69–84.

[15] Takeda, M. and Goodman, J. W., "Neural Networks for Computation: Numerical Representation and Programming Complexity," *Appl. Optics*, Vol. 25, 1986, pp. 3033–52.

[16] Mathew, G. and Reddy, V. U., "Development and Analysis of a Neural Network Approach to Pisarenko's Harmonic Retrieval Method," *IEEE Trans. on Signal Processing*, Vol. 42, No. 3, 1994, pp. 663–67.

[17] Luenberger, D. G., *Linear and Nonlinear Programming*, Addision-Wesley, Reading, MA, 1978, pp. 366–69.

[18] Marple Jr., S. L. and Nuttall, A. H., "Experimental Comparison of Three Multichannel Linear Prediction Spectral Estimators," *IEE Proc.*, Pt. F, Vol. 130, No. 3, 1983, pp. 218–29.

[19] Morf, M., Vieira, A., Lee, D. T. L., and Kailath, T., "Recursive Multichannel Maximum Entropy Spectral Estimation," *IEEE Trans. on GE*, 1978, pp. 85–94.

[20] Golub, G. H. and Van Loan, C. T., *Matrix Computation*, Johns Hopkins Univ. Press, Baltimore, MD, 1983.

[21] McClellan, J. H. and Lang, S. W., "Duality for Multidimensional MEM Spectral Analysis," *IEE Proc.*, Vol. 130, Pt. F, No. 3, 1983, pp. 231–35.

[22] McClellan, J. H., "Multidimensional Spectral Estimation," *Proc. IEEE*, Vol. 70, No. 9, 1982, pp. 1029–38.

[23] Lang, S. W. and McClellan, J. H., "Multidimensional MEM Spectral Estimation," *IEEE. Trans. on Acoustics, Speech and Signal Processing*, Vol. 30, 1982, pp. 880–86.

[24] Cadzow, J. A. and Ogino, K., "Two-Dimensional Spectral Estimation," *IEEE Trans. on Acoustics, Speech and Signal Processing*, Vol. 29, No. 3, 1981, pp. 396–401.

[25] Lim, J. S. and Malik, N. A., "A New Algorithm for Two-Dimensional Maximum Entropy Power Spectrum Estimation," *IEEE Trans. on Acoustics, Speech and Signal Processing*, Vol. 29, No. 3, 1981, pp. 401–3.

[26] Zhuang, X. H., Zhao, Y. X., and Huang, T. S., "A Neural Net Algorithm for Multidimensional Maximum Entropy Spectral Estimation," *Neural Networks*, Vol. 4, 1991, pp. 619–26.

[27] Lohmann, A. W. and Wirnitzer, B., "Triple Correlations," *Proc. IEEE*, Vol. 72, 1984, pp. 889–901.

[28] Klein, S. A. and Tyler, C. W., "Phase Discrimination of Compound Gratings: Generalized Autocorrelation Analysis," *J. Optical Soc. Am. A*, Vol. 3, 1986, pp. 868–79.

[29] Nikias, C. L. and Raghuveer, M. R., "Bispectrum Estimation: A Digital Signal Processing Framework," *Proc. IEEE*, Vol. 75, 1987, pp. 869–91.

[30] Mendel, J. M., "Tutorial on Higher-Order Statistics (Spectra) in Signal Processing and Systems Theory: Theoretical Results and Some Applications," *Proc. IEEE*, Vol. 79, 1991, pp. 278–305.

[31] Nikias, C. L. and Mendel, J. M., "Signal Processing with Higher-Order Spectra," *IEEE Signal Processing Magazine*, Vol. 10, July, 1993, pp. 10–37.

[32] Nikias, C. L. and Petropulu, A. P., *Higher-Order Spectral Analysis: A Nonlinear Processing Framework*, Prentice-Hall, Englewood Cliffs, NJ, 1993.

[33] Proakis, J. G., Rader, C. M., Ling, F., and Nikias, C.L., *Advanced Digital Signal Processing*, Macmillan, New York, 1992.

[34] Giannakis, G. B., "On the Identifiability of Non-Gaussian ARMA Models Using Cumulants," *IEEE Trans. on Automatic Control*, Vol. 35, No. 1, 1990, pp. 18–26.

[35] Giannakis, G. B. and Swami, A., "On Estimating Noncausal Nonminimum Phase ARMA Models of Non-Gaussian Processes," *IEEE Trans. on Acoustics, Speech and Signal Processing*, Vol. 38, No. 3, 1990, pp. 478–95.

[36] Swami, A. and Mendel, J. M., "ARMA Parameter Estimation Using Only Output Cumulants," *IEEE Trans. on Acoustics, Speech and Signal Processing*, Vol. 38, No. 7, 1990, pp. 1257–65.

[37] Zhang, X. D. and Zhou, Y. L., "A Novel Recursive Approach to Estimation MA Parameters of Causal ARMA Models from Cumulants," *IEEE Trans. on Signal Processing*, Vol. 40, No. 1, 1992, pp. 2870–73.

[38] Osowski, S., "Neural Networks for Estimation of Parameters of Sinewaves," *Electronics Lett.*, Vol. 26, 1990, pp. 689–90.

[39] Wang, C. J. and Wickert, M. A., "Three-Layer Neural Networks for Spectral Estimation," *Proc. IEEE Int. Conf. on Acoustics, Speech and Signal Processing*, 1990, pp. 881–84.

[40] Luo, F. L., Yang, W. H., and Shi, X. N., "The Detection of Target from Clutter by Use of the Third-Order Cumulant," *Proc. IEEE Int. Conf. on Information Theory (ISIT'91)*, 1991, p. 357, Budapest, Hungary.

[41] Shi, X. N. and Luo, F. L., "A Bearing Estimation Algorithm Using the Third-Order Cumulant," *J. Electronics, China*, Vol. 14, No. 1, 1992, pp. 37–42.

[42] Luo, F. L. and Unbehauen, R., "Neural Network Approach to Pisarenko's Frequency Estimation," *Proc. Int. Conf. on Artificial Neural Networks and Genetic Algorithms, France*, 1995, pp. 261–64.

[43] Luo, F. L. and Bao, Z., "Neural Network Approach to Computing Matrix Inversion," *Appl. Math. and Computation*, Vol. 47, No. 2, 1992, pp. 109–20.

[44] Luo, F. L. and Li, Y. D., "A 2-D Neural Network with Its Applications to Solving a Class of Matrix Equations," *Int. J. Circuit Theory Appl.*, Vol. 21, 1993, pp. 539–49.

[45] Luo, F. L. and Li, Y. D., "Real-Time Computation of the Eigenvector Corresponding to the Smallest Eigenvalue of a Positive Definite Matrix," *IEEE Trans. on Circuits and Systems*, I, Vol. 41, No. 8, 1994, pp. 550–53.

[46] Luo, F. L. and Li, Y. D.,"Real-Time Computation of the Eigenvector Corresponding to the Largest Eigenvalue of a Positive Definite Matrix," *Neurocomputing*, Vol. 7, No. 2, 1995, pp. 145–57.

[47] Luo, F. L., Li, Y. D., and He, C. X., "Neural Network Approach to Total Least Square Linear Prediction Frequency Estimation Problem," *Neurocomputing*, Vol. 11, 1996, pp. 31–42.

[48] Luo, F. L. and Unbehauen, R., "Recurrent Neural Network for Eigen-Decomposition of Positive Definite Matrix," *Proc. 1995 IEEE Workshop on Nonlinear Signal Processing and Image Processing*, 1995, pp. 46–49.

4

Neural Networks for Signal Detection

Applications of signal detection are found in many areas including radar, sonar, communications, and automatic control [1, 2]. Signal detection involves inferring from (imperfect) observational data whether or not a target signal is present [3, 4]. The design and realization of a detector under a certain criterion are our key tasks. In general, the available observational data are the input to the detector and the output from the detector; these can have one of two possible values, either 1 or 0. The value "1" signifies the presence of the target signal and "0" signifies the absence of the target (or say, the noise alone). There are two common optimum criteria for the design and realization of a signal detector: the Bayes criterion and the Neyman–Pearson criterion. These two criteria usually result in a so-called likelihood-ratio detector. The likelihood-ratio detector involves essentially two parts, as shown in Figure 4.1. The first part enables one to determine the likelihood ratio of the probability density function in the presence of the target signal to that in the absence of the target signal. The second part compares the likelihood ratio with a threshold and outputs "1" if the determined likelihood ratio is greater than the threshold and "0" if the ratio is less than or equal to the threshold. In many practical applications, the likelihood ratio can be replaced by a sufficient statistic, as shown in Figure 4.2. The likelihood ratio (sufficient statistic) depends mainly on the probability density function of the noise. If the noise is white and Gaussian, then a sufficient statistic that is a liner mapping of the input exists and it is not too difficult to find and realize this linear detector. However, if the noise is non-Gaussian, the likelihood ratio has a complicated nonlinear relationship with the input and in most cases an explicit relationship with the input does not exist, which makes it difficult to design and to realize a detector. In fact, a likelihood-ratio detector can be considered to be a special nonlinear mapping. Because an MLP neural network (see Section 1.4) can approximate any nonlinear mapping, applications of neural networks to signal detection have recently attracted much attention [5, 6, 7, 8, 9, 10]. In addition, for the purpose of improving the detection performance, researchers utilize some

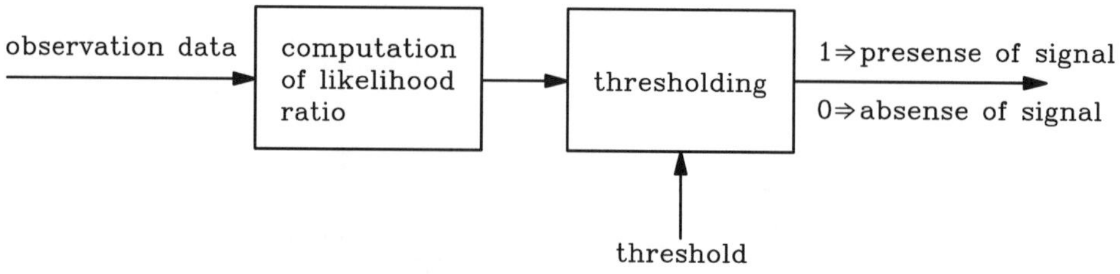

Figure 4.1. The likelihood-ratio signal detector.

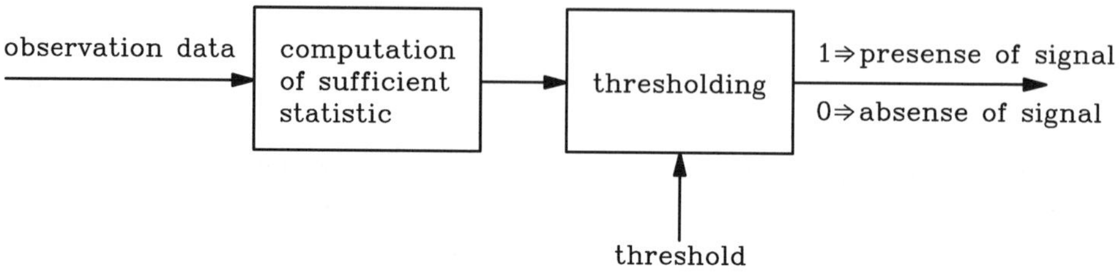

Figure 4.2. An alternative signal detector using a sufficient statistic.

preprocessing or postprocessing, such as pulse-compression and constant-false-alarm-rate processing involving the likelihood-ratio detector. The preprocessing and the postprocessing are usually computationally complex or highly nonlinear, and neural networks are also very powerful here. In this chapter, we will deal in detail with applications of neural networks to the design, realization, preprocessing, and postprocessing of signal detection.

4.1 A Likelihood-Ratio Neural Network Detector

In this section, we will first recall the fundamentals of the likelihood-ratio detector and then give the structure of the detector based on neural networks.

4.1.1 Fundamentals of the Likelihood-Ratio Detector

Let us consider a data vector $\boldsymbol{X}(t) = [x_1(t), x_2(t), \ldots, x_N(t)]^T$ as the input of the detector in Figure 4.1. Using the additive observation model, we have

$$\boldsymbol{X}(t) = \boldsymbol{S}(t) + \boldsymbol{C}(t) \tag{4.1}$$

for the hypothesis that the target signal is present (denoted by H_1) and

$$\boldsymbol{X}(t) = \boldsymbol{C}(t) \tag{4.2}$$

for the hypothesis that the target signal is absent (denoted by H_0), where $\boldsymbol{S}(t) = [s_1(t), s_2(t), \ldots, s_N(t)]^T$ and $\boldsymbol{C}(t) = [c_1(t), c_2(t), \ldots, c_N(t)]^T$ are the target signal vector (which is known a priori) and the noise vector, respectively. The likelihood ratio is defined by

$$\Lambda(\boldsymbol{X}(t)) = \frac{P(\boldsymbol{X}(t)/H_1)}{P(\boldsymbol{X}(t)/H_0)}, \tag{4.3}$$

where $P(\boldsymbol{X}(t)/H_1)$ and $P(\boldsymbol{X}(t)/H_0)$ are the jointly conditional probability density functions of $\boldsymbol{X}(t)$ under H_1 and H_0, respectively. Denoting the decision threshold by η, we choose H_1 (the output of the detector is 1) if $\Lambda(\boldsymbol{X}(t)) > \eta$; otherwise, we choose H_0 (the output of the detector is 0). The two commonly used measures to assess the performance of the detector are the probability of detection P_d and the probability of false alarm P_{fa}. These are defined by

$$P_d = P(H_1/H_1) \tag{4.4}$$

and

$$P_{fa} = P(H_1/H_0). \tag{4.5}$$

That is, P_d is defined as the probability of choosing H_1 given that H_1 is true, and P_{fa} is defined as the probability of choosing H_1 given that H_0 is true. Other forms of (4.4) and (4.5) are

$$P_d = P\Big(\Lambda(\boldsymbol{X}(t)) > \eta/H_1\Big) \tag{4.6}$$

and

$$P_{fa} = P\Big(\Lambda(\boldsymbol{X}(t)) > \eta/H_0\Big). \tag{4.7}$$

Assuming that the target $\boldsymbol{S}(t)$ is known and that $\boldsymbol{C}(t)$ is a zero-mean, white, Gaussian noise vector, the likelihood ratio $\Lambda(\boldsymbol{X}(t))$ can be replaced by a sufficient statistic $Z(t)$ that is a linear combination of each component $x_i(t)$ of the input $\boldsymbol{X}(t)$, that is,

$$Z(t) = \sum_{i=1}^{N} s_i(t) x_i(t). \tag{4.8}$$

Equation (4.8) indicates that the sufficient statistic $Z(t)$ is the output of a matched filter of the target signal $\boldsymbol{S}(t)$. As a result, this detector is also called the matched filter detector. In this case, the probability of detection P_d and the probability of false alarm P_{fa} can be computed using the following explicit expressions:

$$P_d = \frac{1}{\sqrt{2\pi}} \int_{A_1}^{\infty} e^{-\frac{z^2}{2}} dz \tag{4.9}$$

and

$$P_{fa} = \frac{1}{\sqrt{2\pi}} \int_{A_2}^{\infty} e^{-\frac{z^2}{2}} dz, \tag{4.10}$$

where

$$A_1 = \frac{\ln(\eta) - \beta}{(2\beta)^{\frac{1}{2}}}, \tag{4.11}$$

$$A_2 = \frac{\ln(\eta) + \beta}{(2\beta)^{\frac{1}{2}}}, \tag{4.12}$$

$$\beta = \frac{1}{2}\sum_{i=1}^{N} \frac{s_i^2(t)}{\sigma_i^2}, \tag{4.13}$$

and σ_i^2 is the variance of each component of the noise vector $\boldsymbol{C}(t)$. If the Bayes criterion is used, one can first determine the threshold η from the specified risks and then compute P_d and P_{fa} using (4.9) and (4.10). If the Neyman–Pearson criterion is used, one can first determine A_2 from the specified P_{fa} using (4.10), then find the threshold η from (4.12), third compute A_1 using (4.11), and finally compute P_d by use of (4.9).

In most cases, since the noise vector does not have a Gaussian probability density function, the likelihood ratio is a complicated nonlinear function of the input $\boldsymbol{X}(t)$, which makes it very difficult to design and to realize the detectors. Although some simpler detectors such as locally optimum detectors have been designed for a specific non-Gaussian noise [3, 4], their performance will greatly degrade when the related assumptions are violated. In the next subsection, we will present a new kind of signal detector based on the MLP network, which was shown in Section 1.4 to be very powerful in nonlinear mapping situations.

4.1.2 Structure of the Likelihood-Ratio Neural Network Detector

A neural network for likelihood-ratio detection is shown in Figure 4.3. It consists of three layers called the input layer, the hidden layer, and the output layer. The input layer has N neurons, which simply feed the input vector $\boldsymbol{X}(t)$ to the second layer without any modification. The hidden layer has M neurons with nonlinear transfer

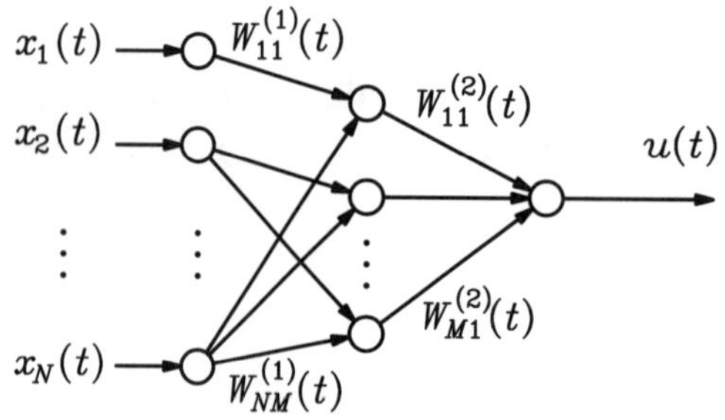

Figure 4.3. A neural network detector based on the MLP network.

functions (such as the sigmoid function). The output layer has only one neuron whose input–output relationship should be a sigmoid function with large steepness so that it approximates the two possible states "1" or "0" of the detector. The output neuron is also biased by a threshold that depends mainly on the detection threshold. If the weighted summation of the neuron outputs of the hidden layer is greater than the threshold, then the output neuron will give a value close to 1, indicating that the target signal is present; otherwise, the output neuron will give a value close to 0, which means that the target is absent.

Let $W_{ij}^{(1)}$ denote the connection weight between the i'th neuron in the input layer and the j'th neuron in the hidden layer (for $i = 1, 2, \ldots, N$; $j = 1, 2, \ldots, M$); let $y_j(t)$ and $f_j(\cdot)$ be the output and activation function of the j'th neuron in the hidden layer, respectively; let $W_{j1}^{(2)}$ denote the connection weight between the j'th neuron in the hidden layer and the neuron in the output layer; let $u(t)$, $v(t)$, $g(\cdot)$, and θ be the input, output, activation function, and threshold of the output neuron, respectively. Then we have

$$v(t) = g\big(u(t) - \theta\big) = \frac{1}{1 + e^{-\alpha(u(t)-\theta)}}, \tag{4.14}$$

$$u(t) = \sum_{j=1}^{M} W_{j1}^{(2)} y_j(t), \tag{4.15}$$

and

$$y_j(t) = f_j\left(\sum_{i=1}^{N} W_{ij}^{(1)} x_i(t)\right), \tag{4.16}$$

where α is a sufficiently large positive constant.

Substituting (4.16) into (4.15) gives

$$u(t) = \sum_{j=1}^{M} W_{j1}^{(2)} f_j\left(\sum_{i=1}^{N} W_{ij}^{(1)} x_i(t)\right). \tag{4.17}$$

Equation (4.17) shows that $u(t)$ has a nonlinear relationship with the input $x_i(t)$ and that this relationship is completely determined by the connection weights $W_{ij}^{(1)}$ and $W_{j1}^{(2)}$ (for $i = 1, 2, \ldots, N$; $j = 1, 2, \ldots, M$). This suggests that we can take $u(t)$ as a sufficient statistic to replace the likelihood ratio $\Lambda(\boldsymbol{X}(t))$ for the signal detection. In other words, for inputs with different probability density functions, we can use some algorithms to train the weights of the network so as to get different sufficient statistics. Furthermore, the threshold θ of the output neuron can represent the detection threshold η. As a result, the neural network in Figure 4.3 can function as a likelihood-ratio detector. The training and testing of this detector based on the MLP neural network will be the focus of the next section.

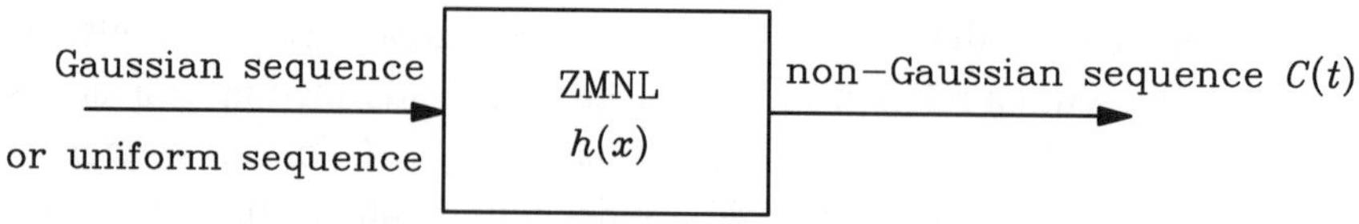

Figure 4.4. The scheme for generating non-Gaussian noise sequences.

4.2 Neural Networks for Signal Detection in Non-Gaussian Noise

From the last section, we know that the key problem of the neural network detector is to determine the connection weights $W_{ij}^{(1)}$ and $W_{j1}^{(2)}$ (for $i = 1, 2, \ldots, N$; $j = 1, 2, \ldots, M$) and the threshold θ according to prior information of the noise and target signal. This can be accomplished by use of the BP algorithm or other available algorithms [11, 12]. Details of the procedures are as follows:

(1) Generate L samples of the noise vector $\boldsymbol{C}(t)$ and the signal-plus-noise vector $\boldsymbol{S}(t) + \boldsymbol{C}(t)$ according to the probability density function of the noise and the a priori information of the target signal.

In general, a non-Gaussian noise sequence with the specified probability density function $p(x)$ is generated by driving a Gaussian sequence or uniform sequence through an appropriate zero-memory nonlinear (ZMNL) system, as shown in Figure 4.4. Let $F(x)$ denote the distribution function corresponding to $p(x)$; then the function $h(x)$ of the ZMNL system should be $h(x) = F^{-1}F_u(x)$ if the Gaussian sequence is the input of the ZMNL system and $h(x) = F^{-1}(x)$ if the uniform sequence is the input of the ZMNL system, where $F_u(x)$ is the Gaussian distribution function with zero mean and unit variance [13, 14].

There have also been some special methods to generate random sequences with a log-normal [15], a χ^2 [16], a spherically invariant random process [17], a Weibull [18, 19], or a Rayleigh probability density function [20]. For the random sequences mentioned above, the corresponding special methods have better performance than the general scheme in Figure 4.4. Recently, there has also been an increasing interest in using chaos to generate non-Gaussian sequences [21]. Once the samples of the noise vector $\boldsymbol{C}(t)$ are obtained, the signal-plus-noise sequence can be immediately obtained by use of (4.1).

(2) Initialize randomly all connection weights $W_{ij}^{(1)}$ and $W_{j1}^{(2)}$ (for $i = 1, 2, \ldots, N$; $j = 1, 2, \ldots, M$) and the threshold θ. In practical applications, we can let the threshold θ be a constant (for simplicity, in the following we let θ be zero).

(3) Define the noise-alone vector $\boldsymbol{C}(t)$ as the input vector $\boldsymbol{X}(t)$ of the network and set the desired output $\hat{v}(t)$ to 0.

(4) Compute (by proceeding forward) the output $y_j(t)$ (for $j = 1, 2, \ldots, M$) of the hidden layer and the output $v(t)$ using (4.14)–(4.17) and the available input vector $\boldsymbol{X}(t)$.

(5) Compute (by proceeding backward) the error propagation terms $\delta_1^{(2)}$ and $\delta_j^{(1)}$ (for $j = 1, 2, \ldots, M$):

$$\delta_1^{(2)} = -v(t)\Big[g\big(u(t)\big)\Big]', \tag{4.18}$$

$$\delta_j^{(1)} = \Big[f_j\big(u_j^{(1)}(t)\big)\Big]'\delta_1^{(2)}W_{j1}^{(2)}(t), \tag{4.19}$$

where

$$\Big[f_j\big(u_j^{(1)}(t)\big)\Big]' = \frac{\partial f_j(u)}{\partial u}\bigg|_{u=u_j^{(1)}(t)} \tag{4.20}$$

and

$$u_j^{(1)}(t) = \sum_{i=1}^{N} W_{ij}^{(1)}x_i(t). \tag{4.21}$$

(6) Adjust the connection weights according to

$$W_{j1}^{(2)}(t+1) = W_{j1}^{(2)}(t) + \gamma\delta_1^{(2)}y_j(t) \tag{4.22}$$

and

$$W_{ij}^{(1)}(t+1) = W_{ij}^{(1)}(t) + \gamma\delta_j^{(1)}x_i(t)$$
$$(i = 1, 2, \ldots, N; \quad j = 1, 2, \ldots, M), \tag{4.23}$$

where γ is the learning-rate parameter.

(7) Define the signal-plus-noise vector $\boldsymbol{C}(t) + \boldsymbol{X}(t)$ as the input vector $\boldsymbol{X}(t)$ of the network and set the desired output $\hat{v}(t)$ to 1.

(8) Repeat Steps (4)–(6) except that (4.18) becomes

$$\delta_1^{(2)} = \big(1 - v(t)\big)\Big[g\big(u(t)\big)\Big]' \tag{4.24}$$

(9) Use the next sample of the available noise-alone sequence and the signal-plus-noise sequence and iterate the computation by returning to Step (3) until convergence occurs.

The convergence is in the sense that for each sample of the noise-alone sequence, the output $v(t)$ must be close to 0 (at least less than 0.5), and for each sample of the signal-plus-noise sequence, the output must be close to 1 (at least greater than 0.5).

After convergence, the network will function as a likelihood-ratio detector. For each observation vector $\boldsymbol{X}(t)$, using the network after convergence, we will know that the target signal is present in the observation vector $\boldsymbol{X}(t)$ (the input of the network) if the corresponding output $v(t)$ is close to 1 ($u(t)$ is greater than 0) and that the noise is alone in the observation vector $\boldsymbol{X}(t)$ if the corresponding output $v(t)$ is close to 0 ($u(t)$ is less than 0). Since we have made no assumption about the probability density function of the noise, this neural network detector applies to any kind of noise background. Moreover, we can envisage, by applying a sufficiently long training sequence, that this neural network will have better performance than the locally optimum detectors in the presence of non-Gaussian noise. Extensive simulation results given in [22] have supported this point of view, although an analytical proof is not given.

We now make several additional comments concerning this neural network detector.

Comment 1 The performance of the detector depends greatly on the generated noise samples. In practical applications, there are usually two cases. One case is that the noise has one specific probability density function but some of its parameters such as the covariances are time-varying. In other words, the form of the probability density function of the noise corrupting the observation is known a priori but the related parameters are unknown. For example, the noise could have a double exponential probability density function $p(x) = [\exp(-|x|/\sigma)]/2\sigma$, but the parameter σ would be unknown. In order to guarantee that the detector has acceptable performance under all possible values of the related parameters, we should generate noise sequences using different parameter values and train the weights of the network. The other case is that the form of the probability density function of the noise is unknown. Clearly, in this case, noise sequences with different probability density functions should be generated to train the network. The performance of the network trained by these sequences will be independent of the probability density functions. A disadvantage in this case is that the convergence speed of the network will be very slow. One way to overcome this disadvantage is to increase the neuron number M of the hidden layer and to set the threshold for each neuron in the hidden layer so that we have more parameters to adjust.

Comment 2 To analyze quantitatively the probability of detection P_d and the probability of false alarm P_{fa} of this neural network detector is quite difficult. However, we can make the detection probability P_d as large as possible and the probability of false alarm P_{fa} as small as possible by adjusting all the parameters of the network during the training period.

Comment 3 This neural network detector can also serve as an alternative to the matched filter detector in Gaussian noise. Ramamurti et al. [22] and Watterson [23] have shown that both detectors have almost the same detection performance in Gaussian noise. However, because of its parallel-processing property, this neural network is more suitable than the matched filter detector in real-time signal detection. In addition, better detection performance will be delivered if the neural network detector replaces the matched filter detector in pulse-compression detection, which will be discussed in the next section. In the case of the Gaussian noise, the neural network detector will become much simpler. First, only one hidden neuron is sufficient (that is, $M = 1$); second, the weights $W_{i1}^{(1)}$ (for $i = 1, 2, \ldots, N$) can be obtained immediately by use of the following expression [23]:

$$W_{i1}^{(1)}(t) = k\frac{s_i(t)}{\sigma_i^2}, \tag{4.25}$$

where k is any chosen nonzero real number and the other variables are the same as those in (4.13).

Comment 4 In many applications, sequential detection is often needed because more information becomes available as time progresses. We thereby avoid having to infer whether the target signal is present before the next observation sample comes if the value $v(t)$ of the output neuron is close to 0.5 (in other words, if the output $v(t)$ is neither close to 1 nor to 0). However, because the network in Figure 4.3 employs the supervised learning and window-shift processing (the neuron number in the input layer, which determines the observation sample number involved in the detection, is fixed), more samples do not result in better detection performance. To take advantage of more information, one has to develop some unsupervised and recursive learning algorithms for the network in Figure 4.3, and thus the weights of the network are changed not only during the training phase but also during the testing phase.

Comment 5 This approach can be generalized for detecting multiple signals. In this case, more neurons are needed in the output layer. In the training of the network (Step (7)), there is at least one output neuron whose desired output is close to unity. This represents the presence of at least one target signal. In addition, since the detection of multiple signals can be viewed as a classification problem, the training and testing of the network can be performed in the same way as for classification problems [11].

Ramamurti et al. [22] have reported simulation results for this neural network detector in

(1) Gaussian noise, $p(x) = \frac{1}{\sqrt{2\pi\sigma^2}} e^{-\frac{x^2}{2\sigma^2}}$;

(2) double exponential noise, $p(x) = \frac{e^{-\frac{|x|}{\sigma}}}{2\sigma}$;

(3) contaminated Gaussian noise, $p(x) = \frac{1-\epsilon}{\sqrt{2\pi\sigma_0^2}} e^{-\frac{x^2}{2\sigma_0^2}} + \frac{\epsilon}{\sqrt{2\pi\sigma_1^2}} e^{-\frac{x^2}{2\sigma_1^2}}$; and in

(4) Cauchy noise, $p(x) = \frac{\sigma}{\pi(\sigma^2+x^2)}$.

For more detail, see Reference [22].

4.3 Neural Networks for Pulse Signal Detection

In radar and sonar systems, the target signal is usually a pulse signal. For the purpose of improving the detection performance, a pulse-compression technique is always employed. A pulse-compression technique involves the transmission of a long-duration wide-bandwidth signal code and the compression of the received signal to a narrow pulse. This compression can be achieved by the matched filter mentioned in Section 4.1, which is actually a correlator of the pulse signal. The output of the matched filter may serve as a sufficient statistic and can be used to detect whether or not the target signal is present. If the noise is Gaussian, the matched filter is optimum under all the Bayes criterion, the Neyman–Pearson criterion, and the criterion of maximizing the signal-to-noise ratio. If the noise is non-Gaussian, the matched filter is optimum only under the criterion of maximizing the signal-to-noise ratio. The key factor for measuring the performance of a pulse compressor is the signal-to-sidelobe ratio, which is defined as the ratio of the peak signal to the maximum sidelobe. A bound from above of the signal-to-sidelobe ratio of the matched filter has been given. For example, for the 13-element Barker code $(1, 1, 1, 1, 1, -1, -1, 1, 1, -1, 1, -1, 1)$ the signal-to-sidelobe ratio of the matched filter cannot be greater than 22.3 dB. Although some sidelobe reduction methods can be used to increase the signal-to-sidelobe ratio to some extent (for the 13-element Barker code, the signal-to-sidelobe ratio can be increased to 24 dB), it is difficult to deliver a higher signal-to-sidelobe ratio if the traditional pulse-compression techniques are used. As an alternative, this section presents a neural network approach for pulse compression that will be shown to have a much higher signal-to-sidelobe ratio than those of the traditional approaches.

The neural network for the pulse compression is exactly the same as shown in Figure 4.3. In this network, the number of neurons in the input layer should be equal to the number of code elements. The output layer has only one neuron that gives the desired sufficient statistic. The choice of the number of neurons of the hidden layer will be discussed later. All neurons in the hidden layer and the output layer take a sigmoid function as their activation function. For simplicity, we can let the thresholds of all neurons be zero.

Now we will take the 13-element Barker code as an example to show how to use the BP algorithm to train this neural network. We will use the same notation as that in Figure 4.3. For the 13-element Barker code, we have 26 possible sequences for training, namely,

(1) $1, 0, 0, 0, 0, 0, 0, 0, 0, 0, 0, 0, 0$;
(2) $-1, 1, 0, 0, 0, 0, 0, 0, 0, 0, 0, 0, 0$;
(3) $1, -1, 1, 0, 0, 0, 0, 0, 0, 0, 0, 0, 0$;
(4) $-1, 1, -1, 1, 0, 0, 0, 0, 0, 0, 0, 0, 0$;
(5) $1, -1, 1, -1, 1, 0, 0, 0, 0, 0, 0, 0, 0$;
(6) $1, 1, -1, 1, -1, 1, 0, 0, 0, 0, 0, 0, 0$;
(7) $-1, 1, 1, -1, 1, -1, 1, 0, 0, 0, 0, 0, 0$;
(8) $-1, -1, 1, 1, -1, 1, -1, 1, 0, 0, 0, 0, 0$;
(9) $1, -1, -1, 1, 1, -1, 1, -1, 1, 0, 0, 0, 0$;
(10) $1, 1, -1, -1, 1, 1, -1, 1, -1, 1, 0, 0, 0$;
(11) $1, 1, 1, -1, -1, 1, 1, -1, 1, -1, 1, 0, 0$;
(12) $1, 1, 1, 1, -1, -1, 1, 1, -1, 1, -1, 1, 0$;
(13) $1, 1, 1, 1, 1, -1, -1, 1, 1, -1, 1, -1, 1$;
(14) $0, 1, 1, 1, 1, 1, -1, -1, 1, 1, -1, 1, -1$;
(15) $0, 0, 1, 1, 1, 1, 1, -1, -1, 1, 1, -1, 1$;
(16) $0, 0, 0, 1, 1, 1, 1, 1, -1, -1, 1, 1, -1$;
(17) $0, 0, 0, 0, 1, 1, 1, 1, 1, -1, -1, 1, 1$;
(18) $0, 0, 0, 0, 0, 1, 1, 1, 1, 1, -1, -1, 1$;
(19) $0, 0, 0, 0, 0, 0, 1, 1, 1, 1, 1, -1, -1$;
(20) $0, 0, 0, 0, 0, 0, 0, 1, 1, 1, 1, 1, -1$;
(21) $0, 0, 0, 0, 0, 0, 0, 0, 1, 1, 1, 1, 1$;
(22) $0, 0, 0, 0, 0, 0, 0, 0, 0, 1, 1, 1, 1$;
(23) $0, 0, 0, 0, 0, 0, 0, 0, 0, 0, 1, 1, 1$;
(24) $0, 0, 0, 0, 0, 0, 0, 0, 0, 0, 0, 1, 1$;
(25) $0, 0, 0, 0, 0, 0, 0, 0, 0, 0, 0, 0, 1$; and
(26) $0, 0, 0, 0, 0, 0, 0, 0, 0, 0, 0, 0, 0$.

For the 13th sequence, the desired output $\hat{v}(t)$ of the network is 1, and for the remaining sequences the desired output $\hat{v}(t)$ is zero. If the above sequences are presented to the network sequentially, we get the desired output response shown in Figure 4.5. Details of the procedures for training the weights are as follows:

(1) Initialize randomly all connection weights $W_{ij}^{(1)}$ and $W_{j1}^{(2)}$ (for $i = 1, 2, \ldots, N$; $j = 1, 2, \ldots, M$).
(2) Input the above 26 sequences successively and compute (by proceeding forward) the output $y_j(t)$ of the hidden layer and the output $v(t)$ (for $j = 1, 2, \ldots, M$) using Equations (4.14)–(4.17) and the available input $X(t)$ (for $t = 1, 2, \ldots, 26$).
(3) Compute (by proceeding backward) the error propagation terms $\delta_1^{(2)}(t)$

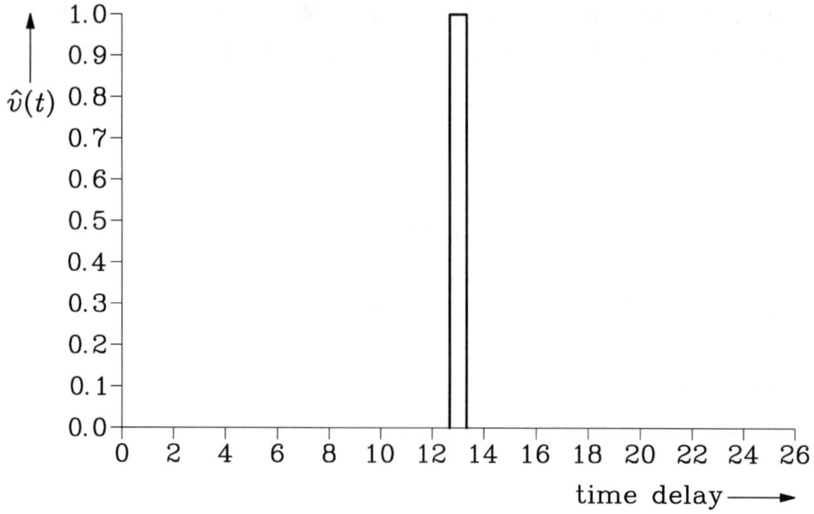

Figure 4.5. The desired output response of the network to the 13-element Barker code sequence.

and $\delta_j^{(1)}(t)$ (for $j = 1, 2, \ldots, M$):

$$\delta_1^{(2)}(t) = -v(t)\big[g\big(u(t)\big)\big]' \qquad (t = 1, 2, \ldots, 12, 14, \ldots, 26), \tag{4.26}$$

but for $t = 13$

$$\delta_1^{(2)}(t) = \big(1 - v(t)\big)\big[g\big(u(t)\big)\big]', \tag{4.27}$$

and

$$\delta_j^{(1)}(t) = \Big[f_j\big(u_j^{(1)}(t)\big)\Big]' \delta_1^{(2)}(t) W_{j1}^{(2)}(n), \tag{4.28}$$

where

$$\Big[f_j\big(u_j^{(1)}(t)\big)\Big]' = \left.\frac{\partial f_j(u)}{\partial u}\right|_{u=u_j^{(1)}(t)} \tag{4.29}$$

and

$$u_j^{(1)}(t) = \sum_{i=1}^{N} W_{ij}^{(1)}(n) x_i(t) \qquad (t = 1, 2, \ldots, 26). \tag{4.30}$$

(4) Adjust the connection weights according to

$$W_{j1}^{(2)}(n+1) = W_{j1}^{(2)}(n) + \gamma \sum_{t=1}^{26} \delta_1^{(2)}(t) y_j(t) \tag{4.31}$$

and

$$W_{ij}^{(1)}(n+1) = W_{ij}^{(1)}(n) + \gamma \sum_{t=1}^{26} \delta_j^{(1)}(t) x_i(t)$$
$$(i = 1, 2, \ldots, N; \quad j = 1, 2, \ldots, M), \qquad (4.32)$$

where γ is the learning-rate parameter.

(5) Compute the total error

$$e = \sum_{t=1}^{26} \left[\hat{v}(t) - v(t)\right]^2 \qquad (4.33)$$

and iterate the computation by returning to Step (2) until convergence occurs (i.e., the point at which error e is less than the specified one).

After training, the output of the network will have a high signal-to-sidelobe ratio, and this network can be used as a detector for the pulse signal. Kwan and Lee [24] have shown with the aid of simulation results that the signal-to-sidelobe ratio of this network can reach 42.73 dB for the 13-element Barker code and 49.71 dB for the 63-bit m-sequence, which is much higher than those (22.3 dB and 17 dB) obtained by the corresponding matched filters. Comments 6–8 relate specifically to this scheme.

Comment 6 In the above training, no noise has been added. For more practical applications, the network should be trained by the above (26) code sequences plus noise sequences with different kinds of probability density functions. Clearly, in this case, the signal-to-sidelobe ratio of the trained network will degrade but will still be much higher than that obtained by traditional methods [24].

Comment 7 A higher signal-to-sidelobe ratio can be achieved by increasing the number of neurons in the hidden layer and adding an adjustable threshold to each neuron in the hidden layer and the output layer. The algorithm to adjust the thresholds is the same as that to adjust the weights. In effect, the threshold of each neuron functions like an additional neuron. The input of the additional neuron is fixed and equal to unity. The output of the additional neuron is connected to the corresponding neuron with a weight whose value is the same as that of the threshold. Kwan and Lee [24] point out that although using more hidden neurons could give higher signal-to-sidelobe ratios, using three hidden neurons proves to be a good choice. This is because more hidden neurons result in a larger number of interconnections and make the network more complicated. Secondly, if one or two hidden neurons are used, the system may not be sufficiently robust.

Comment 8 To speed up the convergence of the training, we can use many other learning algorithms, such as the extended Kalman-filtering-based learning algorithm [25] and the least-squares learning algorithm [12], instead of the BP algorithm used in the above. More interesting is that not only faster convergence but also higher signal-to-sidelobe ratio can be achieved by these fast algorithms than by the BP algorithm. However, the computation invested in these fast algorithms is much more extensive than that in the BP algorithm. As a result, one has to trade off between the two in specific applications.

Together with the fact that the hardware implementation of the above compressor based on the MLP network is not difficult, the above analysis shows that this neural network approach is suitable for pulse signal detection.

Finally, we should mention that since nonlinear activation functions are employed in this neural network approach, the waveform of the output $v(t)$, which contains much other information about the target, will be distorted. This may result in a degradation of the resolution and new difficulty in extracting further information after the detection. Further investigation of this approach is desirable.

4.4 Neural Networks for Weak Signal Detection in High-Noise Environments

In many applications of signal detection, it is desirable for the detector to maintain a reasonable false alarm probability. To achieve this purpose, we usually set a high threshold for the detector, particularly, in a high-noise environment. However, by selecting this threshold setting, detection of a weak target signal can be missed. One method for solving this problem is to employ the temporal and spatial contacts (correlations) of the current sample with several past samples of the target signal and to make detections on target trajectories formed from the stored contacts. This method is called postdetection processing. The major disadvantage of this method is that it involves extensive computations. In this section, we will show how to use neural networks to implement effectively this kind of postdetection processing [26]. However, we first introduce briefly the principles of postdetection processing.

Let us divide the space to be considered into N resolution cells [26, 27], and let x_i be a binary variable (equal to 1 or 0) that denotes the output after thresholding a sufficient statistic in the i'th cell. $x_i = 1$ signifies that the sufficient statistic is greater than the threshold, and $x_i = 0$ signifies that the sufficient statistic is less than or equal to the threshold. The probabilities in the case of the noise alone are given by

$$P_c(x_i) = \begin{cases} \alpha, & x_i = 1, \\ 1 - \alpha, & x_i = 0 \end{cases} \tag{4.34}$$

for the i'th resolution cell and

$$P_c(x_1, x_2, \ldots, x_N) = \prod_{i=1}^{N} P_c(x_i) \tag{4.35}$$

for the whole space of the N cells, where the parameter α is clearly the false alarm probability. Equation (4.35) can be rewritten as

$$P_c(x_i) = \alpha^{x_i}(1-\alpha)^{1-x_i}. \tag{4.36}$$

Without loss of generality, we assume that each cell is independent and has the same parameter α. Equation (4.36) then becomes

$$\begin{aligned} P_c(x_1, x_2, \ldots, x_N) &= \alpha^{\Sigma_{i=1}^N x_i}(1-\alpha)^{N-\Sigma_{i=1}^N x_i} \\ &= e^{a+b\Sigma_{i=1}^N x_i}, \end{aligned} \tag{4.37}$$

where

$$a = N\ln(1-\alpha) \tag{4.38}$$

and

$$b = \ln(\alpha) - \ln(1-\alpha). \tag{4.39}$$

If we assume that the detection probability corresponding to the threshold of the false alarm probability is β, then the probability of x_i in the case of the target plus noise is

$$P_s(x_i) = \beta^{x_i}(1-\beta)^{1-x_i}. \tag{4.40}$$

If the target makes a track given by $S = \{i/i \in \text{target track}\}$ then the probability that the target makes the track is

$$P_t(x_i/x_i \in S) = e^{c+d\Sigma_{i\in S}x_i}, \tag{4.41}$$

where

$$c = M\ln(1-\beta), \tag{4.42}$$

$$d = \ln(\beta) - \ln(1-\beta), \tag{4.43}$$

and M is the number of the cells in the track S. Note that all the cells outside the target track contain just noise. Consequently, the total probability for all the cells given a track S is

$$P_t(x_1, x_2, \ldots, x_N/S) = e^{c+h+d\Sigma_{i\in S}x_i+b\Sigma_{i\notin S}x_i}, \tag{4.44}$$

where $h = (N - M)\ln(1 - \alpha)$. With these in mind, we have the likelihood ratio from (4.3) as

$$\Lambda(x_1, x_2, \ldots, x_N) = \frac{P_t(x_1, x_2, \ldots, x_N/S)}{P_c(x_1, x_2, \ldots, x_N)}. \tag{4.45}$$

For simplicity, we use the log-likelihood ratio

$$\begin{aligned} L(x_1, x_2, \ldots, x_N) &= \ln\big(\Lambda(x_1, x_2, \ldots, x_N)\big) \\ &= \ln\left[\frac{P_t(x_1, x_2, \ldots, x_N/S)}{P_c(x_1, x_2, \ldots, x_N)}\right] \\ &= A + B \sum_{i \in S} x_i \end{aligned} \tag{4.46}$$

as a sufficient statistic to replace the likelihood ratio $\Lambda(x_1, x_2, \ldots, x_N)$, where $A = c + h - a$ and $B = d - b$. Therefore, we can set a threshold and declare the target signal to be present if $L(x_1, x_2, \ldots, x_N)$ is greater than this threshold; otherwise, the noise-alone hypothesis H_0 is chosen. Moreover, since A and B are constants, the term $Z = \sum_{i \in S} x_i$ can be directly taken as a sufficient statistic for target detection. In order to use a neural network to implement the above processing, we rewrite $Z = \sum_{i \in S} x_i$ as

$$Z = \sum_{i=1}^{N} x_i^S x_i, \tag{4.47}$$

where

$$x_i^S = \begin{cases} 1, & i \in S, \\ 0, & i \notin S. \end{cases} \tag{4.48}$$

Clearly, the target detection using Z as a sufficient statistic can easily be implemented by a two-layer feedforward neural network, as shown in Figure 4.6. The input layer has N neurons that simply feed the inputs x_i to the second layer without any modification. The output layer has only one neuron, which has the heavy step function

$$g(u) = \begin{cases} 1, & u > 0, \\ 0, & u \leq 0 \end{cases} \tag{4.49}$$

and is biased by a threshold θ. The threshold θ of the output neuron should be equal to that in the detection with the variable Z given by (4.47) as a sufficient statistic. The connection weights W_i (for $i = 1, 2, \ldots, N$) between the i'th input neuron

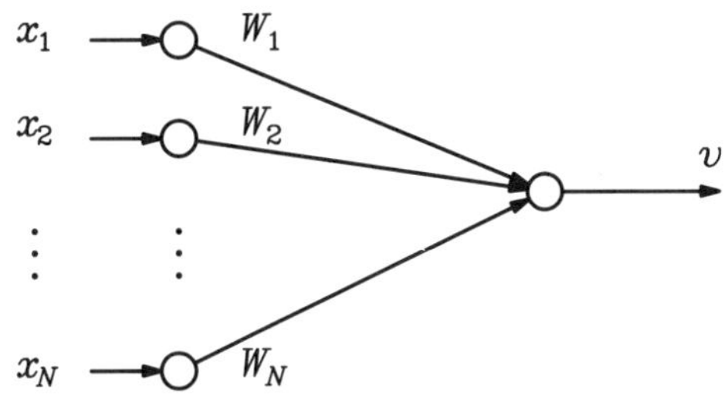

Figure 4.6. A two-layer feedforward neuron network for postdetection processing.

and output neuron are selected as $W_i = x_i^S$. As a result, the output neuron provides

$$v = g(u - \theta) = g\left(\sum_{i=1}^{N} W_i x_i - \theta\right), \tag{4.50}$$

and we choose H_1 if $v = 1$ and H_0 if $v = 0$.

Note that although the sufficient statistic Z in (4.47) has the same form as that in the output (4.8) of the matched filter detector employed in Gaussian noise, their inputs are different. The inputs of the matched filter detector have continuous values, but the inputs of Z in (4.47) have binary values. Consequently, the weights of the network in Figure 4.6 have binary values as well. It has been shown that it is easier to implement a feedforward neural network with binary-valued weights than one with continuous-valued weights [28].

Since the sufficient statistic Z in (4.47) has a binomial probability distribution in either the presence or absence of the target signal, the probability of detection P_d and the probability of false alarm P_{fa} of the detector in Figure 4.6 can easily be calculated by

$$P_d = \sum_{l=\theta}^{M} \binom{M}{l} \beta^l (1-\beta)^{M-l} \tag{4.51}$$

and

$$P_{fa} = \sum_{l=\theta}^{M} \binom{M}{l} \alpha^l (1-\alpha)^{M-l}. \tag{4.52}$$

In addition, this approach can be generalized for the case in which there exist multiple target tracks. In this case, the number of neuron in the output layer must be equal to the number of target tracks.

Roth [26] has also shown that the Hopfield neural network can be used to perform the above postdetection processing. For the single track case, the feedforward network and the Hopfield network can achieve the same detection performance.

However, for the multitrack case, the Hopfield network needs the assumption that the tracks are orthogonal. How to eliminate this assumption remains a problem.

4.5 Neural Networks for Moving-Target Detection

In Sections 4.1 and 4.2, we dealt with general principles of signal detection and related neural network approaches. In different environments and applications, as given in Sections 4.3 and 4.4, different preprocessing and postprocessing techniques are employed to improve the detection performance. In the case of moving-target detection, we usually first use the difference of the power spectra of the moving target and the noise to design a noise-rejection filter and then take the output of the noise-rejection filter as the observation vector (input) of the likelihood-ratio detector to make a detection. Since the output of the noise-rejection filter has a much higher signal-to-noise ratio than the input, the detection performance with the noise-rejection filter is much better than that without the filter. Furthermore, according to the central limit theorem, the probability density function of the output of the noise-rejection filter can be regarded as being Gaussian. As a result, the corresponding likelihood-ratio detector approximates a matched filter detector. This scheme is shown in Figure 4.7.

In general, the detection performance depends greatly on the design of the noise-rejection filter. It is desirable that the transfer function of the filter traces the power spectrum of the noise as closely as possible so as to reject the noise as much as possible. If we use the criterion to minimize the power of the output noise, the coefficients a_i (for $i = 1, 2, \ldots, N$) of the noise-rejection filter should satisfy the Wiener–Hopf equations

$$\boldsymbol{Ra} = \boldsymbol{R}_1, \tag{4.53}$$

where N is the order of the filter, $\boldsymbol{a} = [a_1, a_2, \ldots, a_N]^T$ is the coefficient vector, and $\boldsymbol{R}$ and $\boldsymbol{R}_1$ are the autocorrelation matrix and the autocorrelation vector of the noise, respectively. In most practical applications, $\boldsymbol{R}$ and $\boldsymbol{R}_1$ are not known a priori and have to be estimated from the observation samples. In this case, (4.53) becomes

$$\hat{\boldsymbol{R}}\boldsymbol{a} = \hat{\boldsymbol{R}}_1, \tag{4.54}$$

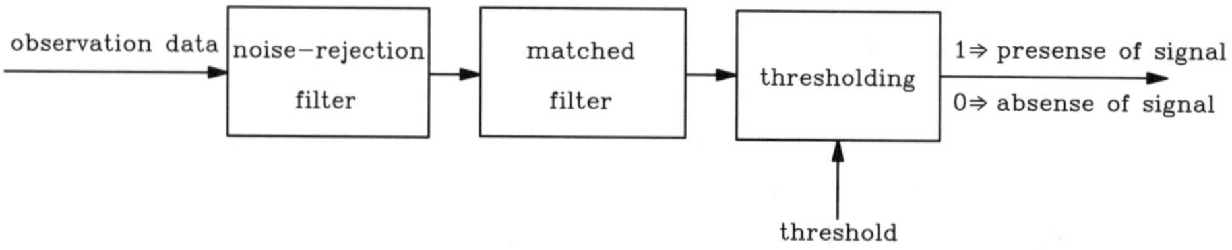

Figure 4.7. The general principle of the moving-target detection.

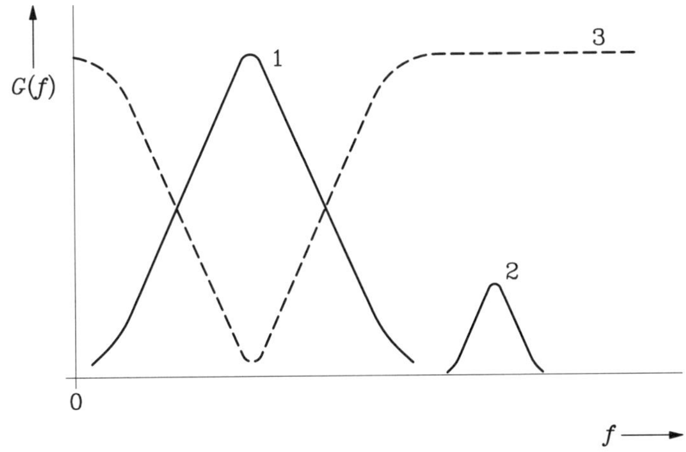

Figure 4.8. The spectral relationship without the signal-cancellation problem. Curves 1 and 2 represent the spectra of the noise and signal, respectively. Curve 3 is the transfer function of the noise-rejection filter obtained by (4.53).

where $\hat{\boldsymbol{R}}$ and $\hat{\boldsymbol{R}}_1$ are the estimated autocorrelation matrix and the autocorrelation vector, respectively. If the target signal is present in the observation samples, then $\hat{\boldsymbol{R}}$ and $\hat{\boldsymbol{R}}_1$ each consists of two parts:

$$\hat{\boldsymbol{R}} = \hat{\boldsymbol{R}}_c + \hat{\boldsymbol{R}}_s \tag{4.55}$$

and

$$\hat{\boldsymbol{R}}_1 = \hat{\boldsymbol{R}}_{1c} + \hat{\boldsymbol{R}}_{1s}, \tag{4.56}$$

where $\hat{\boldsymbol{R}}_c$, $\hat{\boldsymbol{R}}_s$ and $\hat{\boldsymbol{R}}_{1c}$, $\hat{\boldsymbol{R}}_{1s}$ are the estimated autocorrelation matrix and the estimated autocorrelation vector corresponding to the noise and the target signal, respectively. In addition, it is assumed that the noise and the target signal are independent of each other.

The filter with the coefficients obtained by (4.54)–(4.56) will minimize not only the noise power but also the signal power. This is referred to as the signal-cancellation problem. For further illustration, Figures 4.8 and 4.9 show the spectral relationship of the filter, noise, and signal with and without the signal-cancellation problem.

Overcoming the signal-cancellation problem is the current emphasis in the investigation of moving-target detection. In [29], an effective detection method based on the spectral estimation is proposed. This method includes essentially three steps: (1) estimation of the spectrum (or parameters) of the observation samples, (2) calculation of the frequency values of the peaks of the estimated spectrum, and (3) determination of whether the target signal is present or not according to these frequency values. However, because this method based on spectral estimation

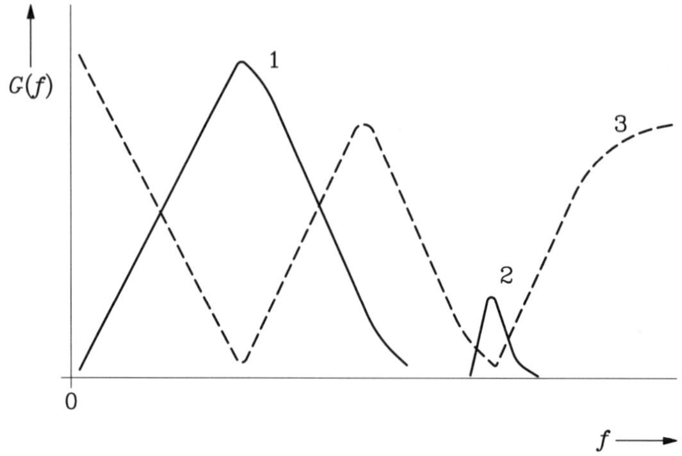

Figure 4.9. The spectral relationship with the signal-cancellation problem. Curves 1 and 2 represent the spectra of the noise and signal, respectively. Curve 3 is the transfer function of the noise-rejection filter obtained by (4.54)–(4.56).

involves very extensive computation, it was impractical in the past. Neural networks for spectral estimation discussed in Chapter 3 offer new opportunities to implement this method. In this section, we will first present this detection method and then show how to use neural networks to implement it.

Let us consider a time sequence $x(n)$ described by

$$\begin{cases} H_1: & x(n) = s(n) + c(n) \\ H_0: & x(n) = c(n), \end{cases} \tag{4.57}$$

where $s(n)$ and $c(n)$ are the signal sequence and noise sequence, respectively, and H_1 and H_0 represent the presence and absence of the target signal, respectively. Equation (4.57) can be mapped to

$$\begin{cases} H_1: & R_x(m) = R_s(m) + R_c(m) \\ H_0: & R_x(m) = R_c(m) \end{cases} \tag{4.58}$$

and

$$\begin{cases} H_1: & G_x(f) = G_s(f) + G_c(f) \\ H_0: & G_x(f) = G_c(f), \end{cases} \tag{4.59}$$

where $R_x(m)$, $R_s(m)$, and $R_c(m)$ and $G_x(f)$, $G_s(f)$, and $G_c(f)$ are the autocorrelation functions and corresponding power spectra of the observation sequence $x(n)$, the signal sequence $s(n)$, and the noise sequence $c(n)$, respectively. Since the motion speeds of the moving target and the noise environments are different, the power spectra will be different. Using this difference, we can detect the presence of the target signal. For convenience of description, we first assume that the spectral

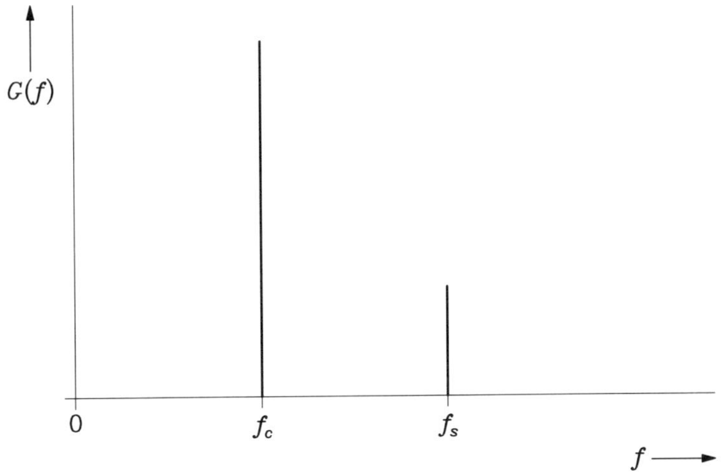

Figure 4.10. The spectra of the noise and signal with zero width.

bandwidths of the target and noise are zero and are located at f_s and f_c, respectively, as shown in Figure 4.10. With this in mind, $x(n)$ can be modeled by a special AR model, namely,

$$\begin{cases} H_1: & x(n) = a_1 x(n-1) + a_2 x(n-2) \\ H_0: & x(n) = b_1 x(n-1) + b_2 x(n-2), \end{cases} \tag{4.60}$$

where the coefficients a_1, a_2, b_1, and b_2 satisfy the Yule–Walker equations and can be computed under ideal conditions by using the following expressions:

$$b_1 = \frac{R_{x/H_0}(0)}{R_{x/H_0}(1)}, \tag{4.61}$$

$$b_2 = 0, \tag{4.62}$$

$$a_1 = \frac{R_{x/H_1}(-1)R_{x/H_1}(2) - R_{x/H_1}(0)R_{x/H_1}(1)}{|R_{x/H_1}(1)|^2 - |R_{x/H_1}(0)|^2}, \tag{4.63}$$

and

$$a_2 = \frac{(R_{x/H_1}(1))^2 - R_{x/H_1}(0)R_{x/H_1}(2)}{|R_{x/H_1}(1)|^2 - |R_{x/H_1}(0)|^2}, \tag{4.64}$$

where $R_{x/H_i}(m)$ (for $i = 0, 1$; $m = -1, 0, 1, 2$) is the autocorrelation value of the sequence $x(n)$ under the related hypothesis. Note that b_2 is no longer zero in any practical application because $R_{x/H_i}(m)$ is estimated from finite samples.

Using the coefficients a_1, a_2 and b_1, b_2 to construct the polynomials, we obtain

$$\begin{cases} H_1: & 1 - a_1 z^{-1} - a_2 z^{-2} \\ H_0: & 1 - b_1 z^{-1} - b_2 z^{-2}. \end{cases} \tag{4.65}$$

Their roots are

$$\begin{cases} H_1: & z_1 = e^{j2\pi f_s T_r}, \quad z_2 = e^{j2\pi f_c T_r} \\ H_0: & z_1 = e^{j2\pi f_c T_r}, \quad z_2 = e^{j2\pi f_c T_r}, \end{cases} \tag{4.66}$$

where T_r is the sampling period. Equations (4.66) show that there are two different roots if the signal is present and there are two repeated roots if the signal is absent. This means that we can use the following procedures to determine whether the signal is present or not:

(1) Estimate the autocorrelation functions $R_x(m)$ (for $m = -1, 0, 1, 2$) using the available samples.

(2) Calculate the coefficients a_1 and a_2 by using

$$a_1 = \frac{R_x(-1)R_x(2) - R_x(0)R_x(1)}{|R_x(1)|^2 - |R_x(0)|^2} \tag{4.67}$$

and

$$a_2 = \frac{(R_x(1))^2 - R_x(0)R_x(2)}{|R_x(1)|^2 - |R_x(0)|^2}. \tag{4.68}$$

(3) Calculate the roots of the polynomial

$$1 - a_1 z^{-1} - a_2 z^{-2}. \tag{4.69}$$

(4) Declare the signal to be present if there are two different roots; otherwise, the hypothesis of the absence of the signal is chosen.

In practical applications, although the widths of the noise spectrum and signal spectrum are no longer zero, $x(n)$ can still be approximately modeled by an AR model, namely,

$$\begin{cases} H_1: & x(n) = a_1 x(n-1) + a_2 x(n-2) + \sigma_1^2 \delta(n) \\ H_0: & x(n) = b_1 x(n-1) + b_2 x(n-2) + \sigma_0^2 \delta(n), \end{cases} \tag{4.70}$$

where σ_1^2 and σ_0^2 are the variances of the related white noises under H_1 and H_0, respectively, and $\delta(n)$ is the white noise whose variance is unity. Note that in (4.70), the order of the AR model was assumed to be 2. The case that the order is greater than 2 will be discussed later. The parameters a_1 and a_2 in (4.70) are still computed by (4.63) and (4.64), but the expressions for computing b_1 and b_2 become

$$b_1 = \frac{R_{x/H_0}(-1)R_{x/H_0}(2) - R_{x/H_0}(0)R_{x/H_0}(1)}{|R_{x/H_0}(1)|^2 - |R_{x/H_0}(0)|^2} \tag{4.71}$$

and

$$b_2 = \frac{(R_{x/H_0}(1))^2 - R_{x/H_0}(0)R_{x/H_0}(2)}{|R_{x/H_0}(1)|^2 - |R_{x/H_0}(0)|^2}. \tag{4.72}$$

In this case, the two roots of the polynomial $1-b_1z^{-1}-b_2z^{-2}$ are no longer repeated, but both are close to $= e^{j2\pi f_c T_r}$. The two roots of the polynomial $1-a_1z^{-1}-a_2z^{-2}$ are close to $= e^{j2\pi f_c T_r}$ and $= e^{j2\pi f_s T_r}$, respectively.

Since the roots correspond to the peaks of the related power spectra,

$$\begin{cases} H_1: \quad G(f) = \dfrac{\sigma_1^2}{|1-\sum_{k=1}^{2} a_k e^{-j2\pi k f T_r}|^2} \\ H_0: \quad G(f) = \dfrac{\sigma_0^2}{|1-\sum_{k=1}^{2} a_k e^{-j2\pi k f T_r}|^2}, \end{cases} \tag{4.73}$$

the frequency values of the two peaks will have a large difference in the case of signal plus noise and a small difference in the case of noise alone. Therefore, we can take $|f_1 - f_2|$ as a sufficient statistic to infer whether the target signal is present or not, that is,

$$\begin{cases} |f_1 - f_2| > \eta_f \Rightarrow H_1 \\ |f_1 - f_2| \le \eta_f \Rightarrow H_0, \end{cases} \tag{4.74}$$

where

$$f_1 = \frac{1}{2\pi T_r} \arctan\left[\frac{\text{Im}(z_1)}{\text{Re}(z_1)}\right] \tag{4.75}$$

and

$$f_2 = \frac{1}{2\pi T_r} \arctan\left[\frac{\text{Im}(z_2)}{\text{Re}(z_2)}\right], \tag{4.76}$$

$\text{Im}(\cdot)$ and $\text{Re}(\cdot)$ denote the imaginary part and real part of the two roots z_1 and z_2, respectively, and η_f is the threshold, which depends mainly on the bandwidths and the signal-to-noise ratio. In any case, η_f should be greater than the bandwidth of the noise spectrum.

If the bandwidths of the noise and signal are relatively large, then the order P of the AR model should be greater than 2. With this observation, (4.70) becomes

$$\begin{cases} H_1: \quad x(n) = \sum_{k=1}^{P} a_k x(n-k) + \sigma_1^2 \delta(n) \\ H_0: \quad x(n) = \sum_{k=1}^{P} b_k x(n-k) + \sigma_0^2 \delta(n). \end{cases} \tag{4.77}$$

The coefficients a_k and b_k (for $k = 1, 2, \ldots, P$) satisfy

$$\begin{pmatrix} R_{x/H_1}(0) & R_{x/H_1}(-1) & \cdots & R_{x/H_1}(1-P) \\ R_{x/H_1}(1) & R_{x/H_1}(0) & \cdots & R_{x/H_1}(2-P) \\ \vdots & \vdots & \ddots & \vdots \\ R_{x/H_1}(P-1) & R_{x/H_1}(P-2) & \cdots & R_{x/H_1}(0) \end{pmatrix} \begin{pmatrix} a_1 \\ a_2 \\ \vdots \\ a_P \end{pmatrix}$$

$$= - \begin{pmatrix} R_{x/H_1}(1) \\ R_{x/H_1}(2) \\ \vdots \\ R_{x/H_1}(P) \end{pmatrix} \tag{4.78}$$

for H_1 and

$$\begin{pmatrix} R_{x/H_0}(0) & R_{x/H_0}(-1) & \cdots & R_{x/H_0}(1-P) \\ R_{x/H_0}(1) & R_{x/H_0}(0) & \cdots & R_{x/H_0}(2-P) \\ \vdots & \vdots & \ddots & \vdots \\ R_{x/H_0}(P-1) & R_{x/H_0}(P-2) & \cdots & R_{x/H_0}(0) \end{pmatrix} \begin{pmatrix} b_1 \\ b_2 \\ \vdots \\ b_P \end{pmatrix}$$

$$= - \begin{pmatrix} R_{x/H_0}(1) \\ R_{x/H_0}(2) \\ \vdots \\ R_{x/H_0}(P) \end{pmatrix} \tag{4.79}$$

for H_0.

The polynomials corresponding to (4.77) are

$$\begin{cases} H_1: & 1 - \sum_{k=1}^{P} a_k z^{-k} \\ H_0: & 1 - \sum_{k=1}^{P} b_k z^{-k}, \end{cases} \tag{4.80}$$

which have P roots (denoted by $z_1, z_2, \ldots, z_P$). The frequency values of the peaks corresponding to these roots are

$$f_i = \frac{1}{2\pi T_r} \arctan\left[\frac{\mathrm{Im}(z_i)}{\mathrm{Re}(z_i)}\right] \qquad (i = 1, 2, \ldots, P). \tag{4.81}$$

If the signal is absent, all roots will be located closely to $z_c = e^{j2\pi f_c T_r}$. If the signal is present and the signal-to-noise ratio is not too small, then there is at least one root that is closer to $z_s = e^{j2\pi f_s T_r}$ than to $z_c = e^{j2\pi f_c T_r}$. From these facts, we can formulate a sufficient statistic v as

$$v = \max_{i,j} |f_i - f_j| \qquad (i, j = 1, 2, \ldots, P) \tag{4.82}$$

and

$$\begin{cases} v > \eta_f \Rightarrow H_1 \\ v \le \eta_f \Rightarrow H_0. \end{cases} \tag{4.83}$$

The value of the threshold η_f should be different from that in (4.74).

In summary, we can write the procedures of this detection method as:

(1) Estimate the $P + 1$ autocorrelation values from the available samples $x(0)$, $x(1), \ldots, x(N-1)$,

$$R_x(k) = \frac{1}{N} \sum_{n=0}^{N-k-1} x(n+k)x(n) \qquad (k = 0, 1, \ldots, P). \tag{4.84}$$

(2) Construct the Yule–Walker equations as shown in (4.78) and (4.79):

$$\begin{pmatrix} R_x(0) & R_x(-1) & \cdots & R_x(1-P) \\ R_x(1) & R_x(0) & \cdots & R_x(2-P) \\ \vdots & \vdots & \ddots & \vdots \\ R_x(P-1) & R_x(P-2) & \cdots & R_x(0) \end{pmatrix} \begin{pmatrix} a_1 \\ a_2 \\ \vdots \\ a_P \end{pmatrix} = - \begin{pmatrix} R_x(1) \\ R_x(2) \\ \vdots \\ R_x(P) \end{pmatrix}. \tag{4.85}$$

(3) Find a_k (for $k = 1, 2, \ldots, P$) by solving (4.85).

(4) Calculate the roots of the polynomial $1 - \sum_{k=1}^{P} a_k z^{-k}$ and the corresponding frequency values f_i (for $i = 1, 2, \ldots, P$) by use of (4.81).

(5) Calculate the values of the variable v using (4.82).

(6) Compare v with the threshold η_f. If $v > \eta_f$, then H_1 is chosen; otherwise, H_0 is chosen.

Note that the computational complexity in the above steps are very intensive (e.g., the multiplication complexity required from Step (1) to Step (3) is $O(P^3)$). As a result, this method was considered to be impractical in the past.

In the following, we will show how to use the neural network in Figure 2.2 to perform in real time all the computations required from Step (1) to Step (3).

Substituting (4.84) into (4.85) gives

$$\boldsymbol{X}^T \boldsymbol{X} \boldsymbol{a} = \boldsymbol{X}^T \boldsymbol{X}_1, \tag{4.86}$$

where

$$
\boldsymbol{X} = \begin{pmatrix}
x(0) & 0 & \cdots & 0 \\
x(1) & x(0) & \cdots & 0 \\
\vdots & \vdots & \ddots & \vdots \\
x(P-1) & x(P-2) & \cdots & x(0) \\
\vdots & \vdots & \ddots & \vdots \\
x(N-1) & x(N-2) & \cdots & x(N-P) \\
\vdots & \vdots & \ddots & \vdots \\
0 & \cdots & x(0) & x(1) \\
0 & \cdots & 0 & x(0)
\end{pmatrix}_{(N+P)\times P}, \tag{4.87}
$$

$$
\boldsymbol{a} = \begin{pmatrix} a_1 \\ a_2 \\ \vdots \\ a_P \end{pmatrix}_{P\times 1}, \quad \text{and} \quad \boldsymbol{X}_1 = - \begin{pmatrix} x(1) \\ x(2) \\ \vdots \\ x(N-1) \\ 0 \\ \vdots \\ 0 \end{pmatrix}_{(N+P)\times 1}. \tag{4.88}
$$

$\boldsymbol{X}$ and $\boldsymbol{X}_1$ are referred to as the data matrix and data vector, respectively, and are in effect exactly the same as those given in (3.12) and (3.13) except for the sign; $\boldsymbol{a}$ is the coefficient vector. The solution of (4.86) is

$$
\boldsymbol{a} = (\boldsymbol{X}^T\boldsymbol{X})^{-1}\boldsymbol{X}^T\boldsymbol{X}_1. \tag{4.89}
$$

Comparing (4.89) with (2.3) and (2.14), we know that if we let the data matrix $\boldsymbol{X}$ and vector $\boldsymbol{X}_1$ be taken as the connection strength matrix $\boldsymbol{T}$ and bias current vector $\boldsymbol{B}$ of the neural network of Figure 2.2, respectively, then such a constructed network is stable and provides the solution

$$
\boldsymbol{V}_f = \left(\frac{1}{RK_1K_2}\boldsymbol{I} + \boldsymbol{X}^T\boldsymbol{X}\right)^{-1}\boldsymbol{X}^T\boldsymbol{X}_1, \tag{4.90}
$$

which is an approximation of (4.89). The error can be made arbitrarily small by appropriately selecting the parameters K_1, K_2, and R. Together with the high-speed computational capability of the neural network shown in Section 2.1, the above results means that the neural network can perform the computation required from Step (1) to Step (3) of the signal detection method based on the spectral estimation during an elapsed time of only a few characteristic time constants of the network with an arbitrarily small error. Consequently, this neural network approach is suitable for real-time detection of the moving target.

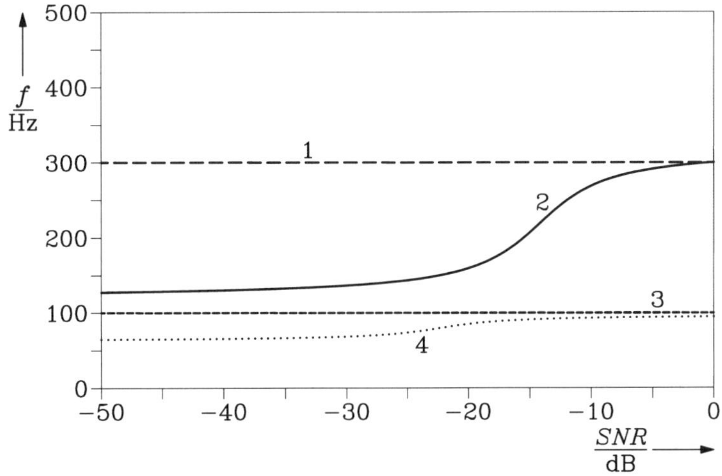

Figure 4.11. Relationship between estimated frequency values and the SNR. $B_c = 30$ Hz, $B_s = 5$ Hz, $f_c = 100$ Hz, and $f_s = 300$ Hz.

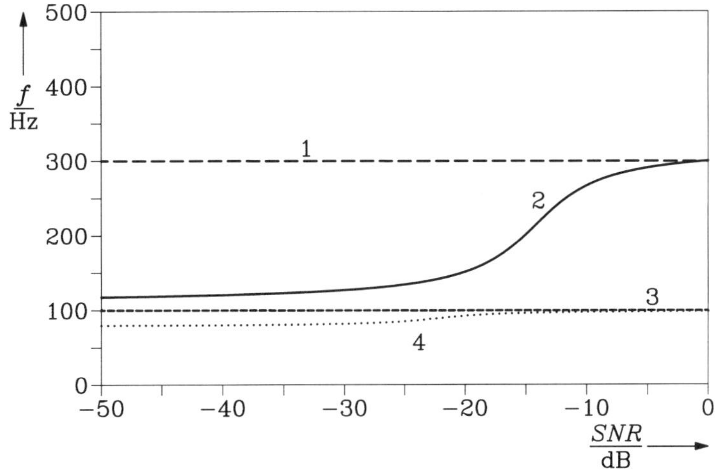

Figure 4.12. f vs. SNR for $B_c = 20$ Hz, $B_s = 5$ Hz, $f_c = 100$ Hz, and $f_s = 300$ Hz.

We next present several simulation results accompanied by further discussion. In these simulations, we selected the order of the AR model to be 2. We denote B_c and B_s as the bandwidths of the noise spectrum and the target signal spectrum, respectively. SNR is the signal-to-noise ratio. Figures 4.11–4.14 show the relationships between the estimated frequency values of the roots and the SNR. Figure 4.15 shows the relationship between the estimated frequency values of the roots and the bandwidth B_c of the noise spectrum. In these figures, Curves 1 and 3 represent the central frequency values of the noise and signal, respectively. Curves 2 and 4 represent the estimated frequency values of the two roots.

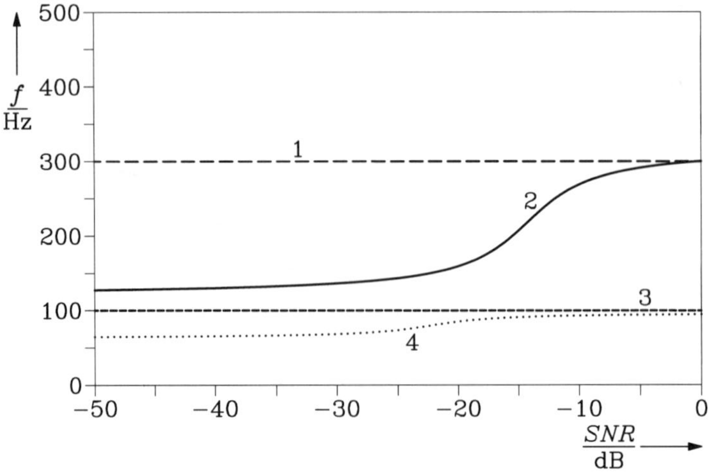

Figure 4.13. f vs. SNR for $B_c = 30$ Hz, $B_s = 3$ Hz, $f_c = 100$ Hz, and $f_s = 300$ Hz.

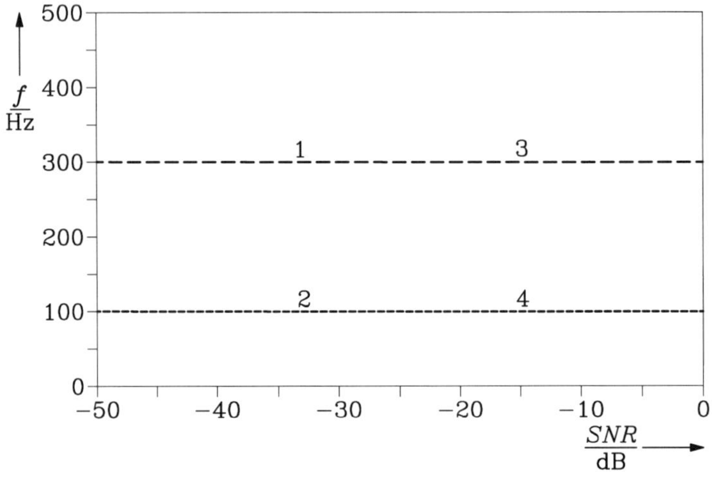

Figure 4.14. f vs. SNR for $B_c = 0$ Hz, $B_s = 0$ Hz, $f_c = 100$ Hz, and $f_s = 300$ Hz.

From the simulation results, we know that the performance of detection depends on many factors, such as the signal-to-noise ratio, the bandwidths and central frequencies of the noise spectrum, and the signal spectrum. In the case that the bandwidths of the noise spectrum and the signal spectrum are zeros (Figure 4.14), the signal can be found even if the SNR $= -60$ dB. However, in general, any signal whose SNR is less than -25 dB will be missed mainly because the signal has a very small effect on the spectral estimation of the noise plus signal. In other words, the spectrum of the noise plus signal is approximately equal to that of the noise alone. This allows us to neglect the signal-cancellation problem. In this case, we can still use the detection scheme shown in Figure 4.7. In contrast, any signal whose

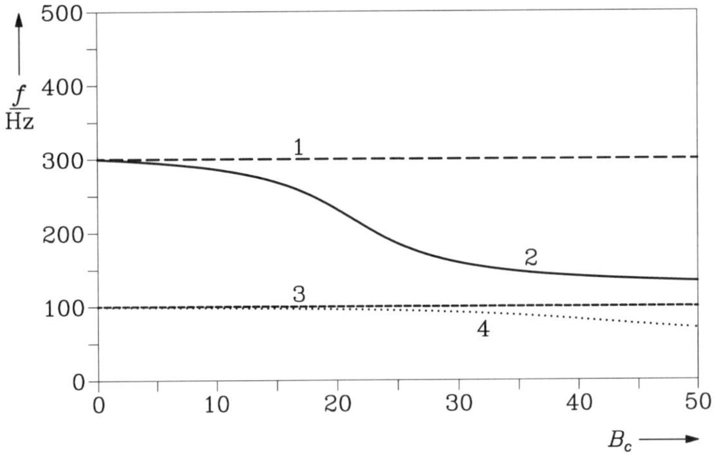

Figure 4.15. f vs. SNR for $B_s = 5$ Hz, $f_c = 100$ Hz, $f_s = 300$ Hz, and SNR$= -20$ dB.

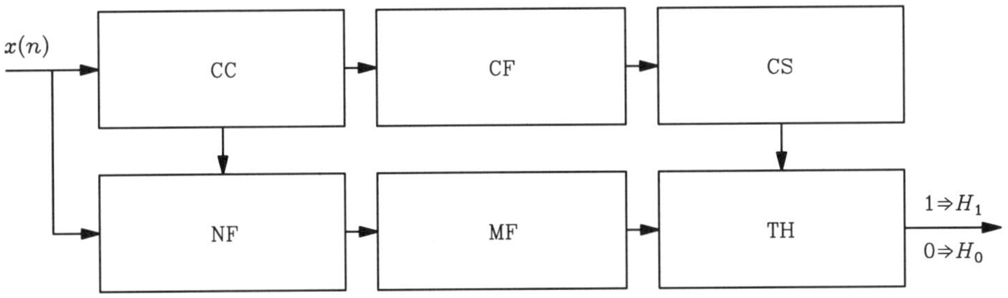

Figure 4.16. A mixed scheme for moving target detection. CC: computation of the coefficients by neural networks; CF: computation of the frequency values of the roots; CS: computation of the sufficient statistic by (4.82); NF: the noise-rejection filter; MF: the matched filter; TH: thresholding.

SNR is large enough to give rise to the signal-cancellation problem can be found by this method based on the spectral estimation. This is because the spectrum of the noise plus signal with large SNR must be different from that of the noise alone. These facts suggest that we should mix these two methods so as to deliver better performance for moving-target detection.

It is worth noting that the coefficients of the noise-rejection filter are exactly the same as those used in the spectral estimation (computed from (4.89)). That is, the output $u(n)$ of the noise-rejection filter can be written as

$$u(n) = x(n) - \sum_{k=1}^{P} a_k x(n-k). \tag{4.91}$$

This means that we can also use the neural network of Figure 2.2 to compute in real time the coefficients of the noise-rejection filter. Based on this argument, we can devise a mixed scheme for moving-target detection, as shown in Figure 4.16. Together with the high-speed computational capability of the neural network, the above analysis and simulations demonstrate the suitability of this mixed scheme for moving-target detection.

Finally, we should note that any signal whose central frequency value f_s satisfies $|f_s - f_c| < B_c$ will be missed if the above scheme is used. Developing techniques for overcoming this problem are desired.

Bibliography

[1] Gibson, J. D. and Melsa, J. L., *Introduction to Nonparametric Detection with Applications*, Academic Press, New York, 1975.

[2] Srinath, M. D. and Rajasekaran, P. K., *An Introduction to Statistical Signal Processing with Applications*, Wiley, New York, 1979.

[3] Poor, H. V., *An Introduction to Signal Detection and Estimation*, Springer-Verlag, New York, 1988.

[4] Kassam, S. A., *Signal Detection in Non-Gaussian Noise*, Springer-Verlag, New York, 1988.

[5] Lippmann, R. P. and Bekman, P., "Adaptive Neural Net Preprocessing for Signal Detection in Non-Gaussian Noise," *Adv. Neural Information Processing System*, Vol. 1, 1989, pp. 124–32.

[6] Michapoulou, Z., Notle, L., and Alexanderou, D., "ROC Performance Evolution of Multilayer Perceptrons in the Detection of One M Orthogonal Signals," *Proc. IEEE Int. Conf. on Acoustics, Speech and Signal Processing*, Vol. 2, 1992, pp. 309–12.

[7] Malkoff, D., "A Neural Network Approach to the Detection Problem Using Joint-Time Frequency Distribution," *Proc. IEEE Int. Conf. on Acoustics, Speech and Signal Processing*, 1990, pp. 2739–42.

[8] Wilson, E., Umesh, S., and Tufts, D. W., "Multistage Neural Network Structure for Transient Detection and Feature Extraction," *Proc. IEEE Int. Conf. on Acoustics, Speech and Signal Processing*, Vol. 1, 1993, pp. 489–92.

[9] Liou, R. J., Azimi-Sadjadi, M. R., and Dent, R., "Detection of Dim Targets in High Cluttered Background Using High Order Correlation Neural Networks," *Proc. Int. Joint Conf. on Neural Networks*, 1991, pp. 701–6.

[10] Liou, R. J. and Azimi-Sadjadi, M. R., "Dim Target Detection Using High Order Correlation Method," *IEEE Trans. on Aerospace and Electronics*, Vol. 29, No. 3, 1995, pp. 841–57.

[11] Haykin, S., *Neural Networks, A Comprehensive Foundation*, IEEE Press, New York, 1994.

[12] Kollias, S. and Anastassion, D., "An Adaptive Least Squares Algorithm for the Efficient Training of Artificial Neural Networks," *IEEE Trans. on Circuits and Systems*, Vol. 36, No. 8, 1989, pp. 1092–101.

[13] Liu, B. and Munson Jr., D. C., "Generation of a Random Sequence Having Jointly Specified Marginal Distribution and Autocovariance," *IEEE Trans. on Acoustics, Speech and Signal Processing*, Vol. 30, 1982, pp. 973–83.

[14] Luo, F. L. and Yang, W. H., "The Simulation of the Stochastic Processes with the Cross Correlation Considered," *Proc. 12th Symp. of Information Theory and Its Applications*, 1989, pp. 199–203.

[15] Luo, F. L. and Yang, W. H., "Two Techniques for Simulating Correlated Log-Normal Sequences," *Proc. Int. Conf. of Modelling and Simulation*, 1988, pp. 291–99.

[16] Luo, F. L. and Yang, W. H., "The Simulation of the Correlated χ^2 Sequences," *AMSE Review*, Vol. 15, No. 1, 1990, pp. 27–33.

[17] Luo, F. L. and Yang, W. H., "One Scheme for Simulating Spherically Invariant Random Process," *AMSE Review*, Vol. 15, No. 3, 1991, pp. 1–10.

[18] Luo, F. L. and Li, Y. D., "The Simulation of the Radar Clutter with the Spatial Correlation Considered," *Adv. Model. Analysis*, Vol. 33, No. 2, 1995, pp. 55–63.

[19] Li, G. and Yu, K. B., "Modelling and Simulation of Coherent Weibull Clutter," *IEE Proc.*, Pt. F, Vol. 136, No. 1, 1989, pp. 1–9.

[20] Szajowski, W. J., "The Generation of Correlated Weibull Clutter for Signal Detection Problem," *IEEE Trans. on Aerospace and Electronic Systems*, Vol. 13, No. 5, 1977, pp. 536–40.

[21] Davis, T., "Circuits and Systems for Generation of Random and Pseudorandom Signals," *Tutorial of Int. Symp. on Circuits and Systems*, 1995.

[22] Ramamurti, V., Rao, S. S., and Gandhi, P. P., "Neural Detectors for Signals in Non-Gaussian Noise," *Proc. IEEE Int. Conf. on Acoustics, Speech and Signal Processing*, Vol. 1, 1993, pp. 481–94.

[23] Watterson, J. W., "An Optimum Multilayer Perceptron Neural Receiver for Signal Detection," *IEEE Trans. on Neural Networks*, Vol. 1, No. 4, 1990, pp. 298–300.

[24] Kwan, H. K. and Lee, C. K., "A Neural Network Approach to Pulse Radar Detection," *IEEE Trans. on Aerospace and Electronics*, Vol. 29, No. 1, 1993, pp. 9–21.

[25] Rao, K. D. and Sridhar, G., "Improving Performance in Pulse Radar Detection Using Neural Networks," *IEEE Trans. on Aerospace and Electronics*, Vol. 31, No. 3, 1995, pp. 1193–98.

[26] Roth, M. W., "Neural Networks for Extraction of Weak Targets in High Clutter Environments," *IEEE Trans. on Systems, Man, and Cybernetics*, Vol. 19, No. 5, 1989, pp. 1211–17.

[27] Duda, R. O. and Hart, P. E., *Pattern Classification and Scene Analysis*, Wiley, New York, 1973.

[28] Piret, P.,"Analysis of a Modified Hebbian Rule," *IEEE Trans. on Information Theory*, Vol. 36, No. 6, 1990, pp. 1391–97.

[29] Luo, F. L. and Li, J. J., "A Method for Detecting the Moving Target," *Radar*, No. 1, 1991, pp. 12–18.

5

Neural Networks for Signal Reconstruction

Signal reconstruction is regarded as an important part of signal processing, arising mainly in communications, image, and seismic signal processing [1, 2, 3]. The task of signal reconstruction is to find the unknown signal $x(n)$ from a corrupted and imperfect observation $y(n)$. Current topics include the implementation of optimal reconstruction algorithms and the development of nonlinear and blind reconstruction algorithms. There are many available algorithms for signal reconstruction under various optimization criteria (such as the maximum entropy criteria and the minimum L_p norm criteria). However, these optimal algorithms suffer from intrinsic and intensive computation complexity and require a considerable amount of computer time, especially in the case of large image reconstruction. This drawback makes it difficult or even impossible to apply these optimal algorithms to real-time problems. In addition, most of these algorithms deal only with the case of linear operation, that is, the observation $y(n)$ is assumed to be the result of passing $x(n)$ through a linear system. The reconstruction performance will be greatly degraded if this assumption is violated. As a result, algorithms that deal with nonlinear operations are desired. Blind reconstruction is another hot topic in this field [4]. Blind reconstruction algorithms are very useful in cases where the transfer function of the related system through which the desired signal passes is unknown a priori. Recently, neural networks have been successfully applied in signal reconstruction. We can take advantage of the high-speed computational capability of neural networks to implement many optimal reconstruction algorithms. Based on the nonlinear mapping functions of MLP networks and RBF networks, we can develop some effective nonlinear reconstruction algorithms so as to deliver better performance than linear ones can give. Blind reconstruction may be viewed as a self-organized learning process, self-organized in the sense that the reconstruction is performed in the absence of prior knowledge. In this respect, much can be gained from neural networks, in particular, that part of the subject that deals with self-organization and unsupervised learning algorithms [5]. These three aspects of neural networks for signal reconstruction will be detailed in this chapter. Section 5.1 will present neural networks to implement optimal reconstruction algorithms in real time with emphasis on the maximum entropy algorithm. Reconstruction of binary signals based on nonlinear mapping neural networks will be dealt with in Sections 5.2–5.4. In the final section of this chapter, we will give a blind reconstruction algorithm based on higher-order statistics and neural networks.

5.1 Maximum Entropy Signal Reconstruction by Neural Networks

In this section, we will show how to use neural networks to implement maximum entropy signal reconstruction algorithms. Before proceeding, we first present briefly the formulation of the problem of signal reconstruction.

Let us consider an unknown signal $x(n)$ that passes through a linear system $h(n)$ (the impulse response). The available measured data ($y(n)$) are then the convolution of the unknown signal $x(n)$ with the impulse response $h(n)$ plus an additive noise $\epsilon(n)$, that is,

$$y(n) = \sum_{i=-\infty}^{\infty} x(i)h(n-i) + \epsilon(n) = x(n) * h(n) + \epsilon(n). \tag{5.1}$$

In general, only a finite number of samples of $y(n)$ are available and the impulse response is known a priori. Our task is to obtain $x(n)$ (for $n = 1, 2, \ldots, N$) from the available data $y(1), y(2), \ldots, y(M)$ and the known $h(n)$, where N is the number of samples to be considered.

Equation (5.1) can be written in a matrix form as

$$\boldsymbol{Y} = \boldsymbol{H}\boldsymbol{X} + \boldsymbol{q}, \tag{5.2}$$

where $\boldsymbol{Y} = [y(1), y(2), \ldots, y(M)]^T$ and $\boldsymbol{X} = [x(1), x(2), \ldots, x(N)]^T$ are the M-dimensional measured data vector and the N-dimensional unknown signal vector, respectively, and $\boldsymbol{q} = [\epsilon(1), \epsilon(2), \ldots, \epsilon(M)]^T$ is the noise vector, which is not known, either. $\boldsymbol{H}$ is the $M \times N$ dimensional matrix consisting of the known impulse response $h(n)$.

The simplest way to find $\boldsymbol{X}$ is to solve the following least-squares (corresponding to the minimum L_2 criterion) problem:

$$\min_{\boldsymbol{X}} \|\boldsymbol{Y} - \boldsymbol{H}\boldsymbol{X}\|^2. \tag{5.3}$$

Its solution is

$$\boldsymbol{X} = \boldsymbol{H}^+\boldsymbol{Y} = \lim_{\alpha\to 0}(\boldsymbol{H}^T\boldsymbol{H} + \alpha\boldsymbol{I})^{-1}\boldsymbol{H}^T\boldsymbol{Y}, \tag{5.4}$$

where H^+ is the pseudoinverse of H. In practical applications, (5.4) can be approximated by

$$X = H^+Y \approx (H^T H + kI)^{-1} H^T Y, \tag{5.5}$$

where k is a small positive constant. If $H^T H$ or HH^T is not of full rank (this is the most common case), then k can decrease the effect of this singularity to some extent.

It is very difficult to solve (5.4) in real time because of its intensive computational complexity, which involves a matrix inversion and matrix multiplications [6].

Comparing (5.4) with (2.3) and (2.14), we know that if we let the known matrix H and the vector Y be taken as the connection strength matrix T and the bias current vector B of the neural network of Figure 2.2, respectively, while the other parameters are the same as those for (2.3), then such a constructed network is stable and provides the solution

$$V_f = \left(\frac{1}{RK_1K_2} I + H^T H\right)^{-1} H^T Y \tag{5.6}$$

during an elapsed time of only a few characteristic time constants of the network. Clearly, Equation (5.6) is an approximation of Equation (5.4). The error can be made arbitrarily small by appropriately selecting the parameters K_1, K_2, and R. In addition, the effects of some uncertainties such as the singularities of the matrix H can be decreased by the term $1/(RK_1K_2)$ as by k in (5.5) [7].

In practical applications, for the purpose of achieving a better performance, other optimal criteria such as the maximum entropy criteria are used to reconstruct the desired signal. In this case, the reconstruction problem can be formulated as the problem

$$\begin{cases} \max_X G(X) \\ s.t. \quad Y = HX, \end{cases} \tag{5.7}$$

where the objective function may take different forms. In general, we use some entropy measure such as those given in [8], [9], [10], and [11]. These include:

(1) Shannon's entropy:

$$G(X) = -\sum_{n=1}^{N} x(n) \ln\big(x(n)\big) \qquad x(n) > 0; \tag{5.8}$$

(2) Burg's entropy:

$$G(X) = -\sum_{n=1}^{N} \ln\big(x(n)\big) \qquad x(n) > 0; \tag{5.9}$$

(3) Frieden's and Zoltani's entropy (bounded entropy)

$$G(\boldsymbol{X}) = -\sum_{n=1}^{N}\left[x(n)\ln\big(x(n)\big) + \big(r - x(n)\big)\ln\big(r - x(n)\big)\right]$$

$$0 < x(n) < r, \tag{5.10}$$

where r is a bound from above on the value of $x(n)$; and

(4) Skilling's entropy:

$$G(\boldsymbol{X}) = \sum_{n=1}^{N}\left[x(n) - \hat{x}(n) - x(n)\ln\big(x(n)\big) + x(n)\ln\big(\hat{x}(n)\big)\right], \tag{5.11}$$

where $\hat{x}(n)$ (for $n = 1, 2, \ldots, N$) is a prior estimation of $x(n)$.

In actuality, Equations (5.9)–(5.11) are different modifications of Shannon's entropy (5.8).

It is difficult to find explicitly the solution of the constrained optimization problem (5.7). On the basis of the penalty method, we can change (5.7) to

$$\max_{\boldsymbol{X}}\left[G(\boldsymbol{X}) + K\sum_{i=1}^{M}P\big(e_i(\boldsymbol{X})\big)\right] \tag{5.12}$$

or

$$\min_{\boldsymbol{X}}\left[-G(\boldsymbol{X}) + K\sum_{i=1}^{M}P\big(e_i(\boldsymbol{X})\big)\right], \tag{5.13}$$

where K is a sufficiently large positive constant and

$$e_i(\boldsymbol{X}) = \sum_{j=1}^{N}H_{ij}x(j) - y(i), \tag{5.14}$$

which is the i'th element of the vector $\boldsymbol{HX} - \boldsymbol{Y}$. $P(\cdot)$ is the penalty function, which usually takes forms such as

(1) $P(e) = \frac{1}{2}e^2$,

(2) $P(e) = \frac{1}{\rho}|e|^{\rho}$, $\rho \geq 1$,

(3) $P(e) = \begin{cases} \frac{1}{2}e^2, & |e| \leq \beta, \\ \beta|e| - \frac{1}{2}\beta^2, & |e| > \beta, \end{cases}$

(4) $P(e) = \beta^2\ln\cosh(\frac{e}{\beta})$, $\beta > 0$, or

(5) $P(e) = \frac{1}{\rho}|e|^{\rho} + \frac{1}{2}e^2$, $\rho > 1$.

From Reference [12], we know that (5.13) and (5.7) have the same solution if K tends to infinity. However, if K is large enough, the solution of (5.13) can approximate that of (5.7).

With this in mind, we now will develop the neural network approach for solving (5.13). First, we construct an energy function as follows:

$$E(t) = -G\big(\boldsymbol{V}(t)\big) + K \sum_{i=1}^{M} P\Big[e_i\big(\boldsymbol{V}(t)\big)\Big], \tag{5.15}$$

where $\boldsymbol{V}(t) = [v_1(t), v_2(t), \ldots, v_N(t)]^T$ is an N-dimensional column vector. The derivative of (5.15) with respect to time t is

$$\begin{aligned}\frac{dE(t)}{dt} &= -\sum_{i=1}^{N} \frac{\partial G(\boldsymbol{V}(t))}{\partial v_i(t)} \frac{dv_i(t)}{dt} + K \sum_{j=1}^{M} \sum_{i=1}^{N} \frac{\partial P[e_j(\boldsymbol{V}(t))]}{\partial e_j(t)} \frac{\partial e_j(t)}{\partial v_i(t)} \frac{dv_i(t)}{dt} \\ &= -\sum_{i=1}^{N} \frac{dv_i(t)}{dt} \left(\frac{\partial G(\boldsymbol{V}(t))}{\partial v_i(t)} - K \sum_{j=1}^{M} \frac{\partial P[e_j(\boldsymbol{V}(t))]}{\partial e_j(t)} \frac{\partial e_j(t)}{\partial v_i(t)} \right).\end{aligned} \tag{5.16}$$

If we let

$$\frac{dv_i(t)}{dt} = \frac{\partial G(\boldsymbol{V}(t))}{\partial v_i(t)} - K \sum_{j=1}^{M} \frac{\partial P[e_j(\boldsymbol{V}(t))]}{\partial e_j(t)} \frac{\partial e_j(t)}{\partial v_i(t)} \tag{5.17}$$

and substitute (5.17) into (5.16), then we obtain

$$\frac{dE(t)}{dt} = -\sum_{i=1}^{N} \left(\frac{dv_i(t)}{dt} \right)^2. \tag{5.18}$$

Since the entropy function $-G(\cdot)$ and penalty function $P(\cdot)$ are bounded from below and since each term on the right-hand side of (5.18) is nonnegative, we have

$$\frac{dE(t)}{dt} \leq 0, \qquad \frac{dE(t)}{dt} = 0 \quad \rightarrow \quad \frac{dv_i(t)}{dt} = 0 \qquad (i = 1, 2, \ldots, N). \tag{5.19}$$

From (5.19) we see that the time evolution of the energy function $E(t)$ is a motion in state space that seeks out the minimum of $E(t)$ and comes to a stop at such a point.

Accordingly, we can use (5.17) to construct a neural network as shown in Figure 5.1. This network consists of integrators, adders (summing amplifiers) with corresponding connection weights, and nonlinear neurons whose activation functions are

$$f(e_j) = \frac{\partial P[e_j(\boldsymbol{V}(t))]}{\partial e_j(t)} \qquad (\text{for } j = 1, 2, \ldots, M)$$

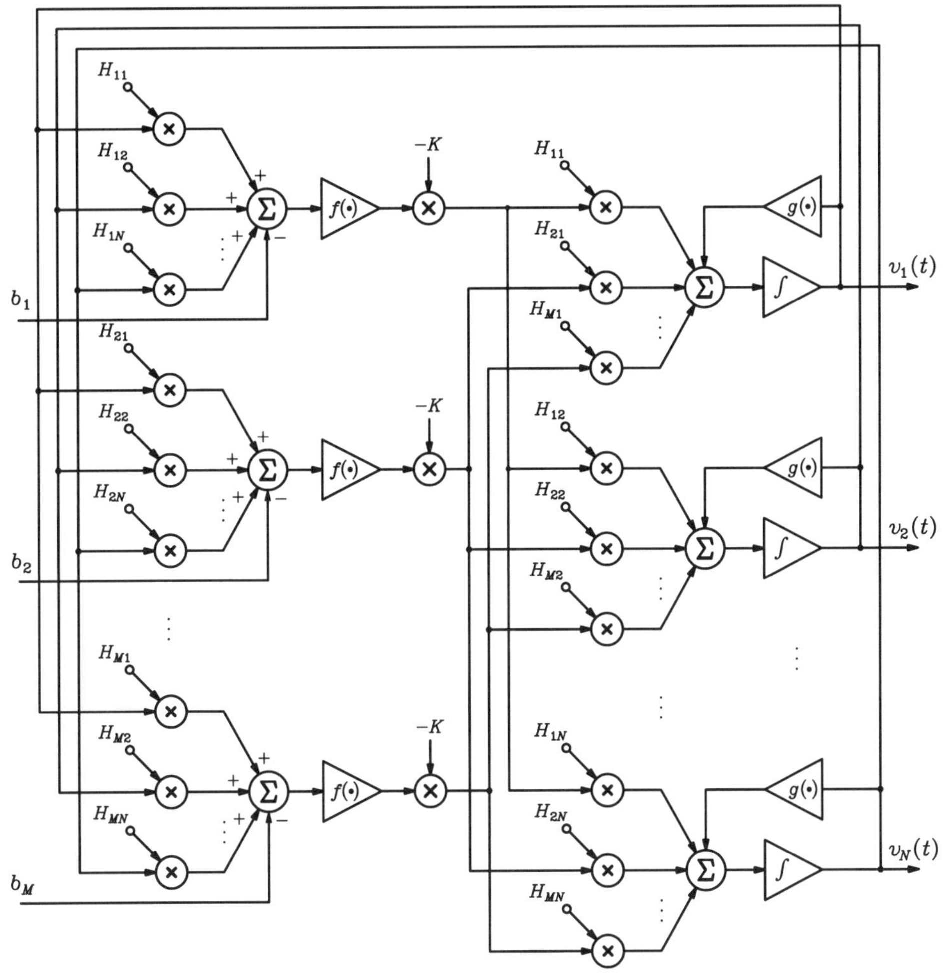

Figure 5.1. A neural network for solving the signal reconstruction problem (5.13).

and

$$g(v_i) = \frac{\partial G(\boldsymbol{V}(t))}{\partial v_i(t)} \qquad (\text{for } i = 1, 2, \ldots, N).$$

(Exemplary plots of the activation function $f(e_j)$ are shown in Figure 5.2.) In this network, the known element H_{ij} (for $i = 1, 2, \ldots, M$; $j = 1, 2, \ldots, N$) is taken as the connection strength of the network, which can be fixed or time variable depending on the field of application of the signal reconstruction. In practice, they can be realized by using VLSI analog multipliers, tunable (voltage-controlled) transconductors, programmable switched-capacitors, or high-resistivity polysilicon thin-film resistors. The observation data vector $\boldsymbol{Y}$ is directly taken as the bias current

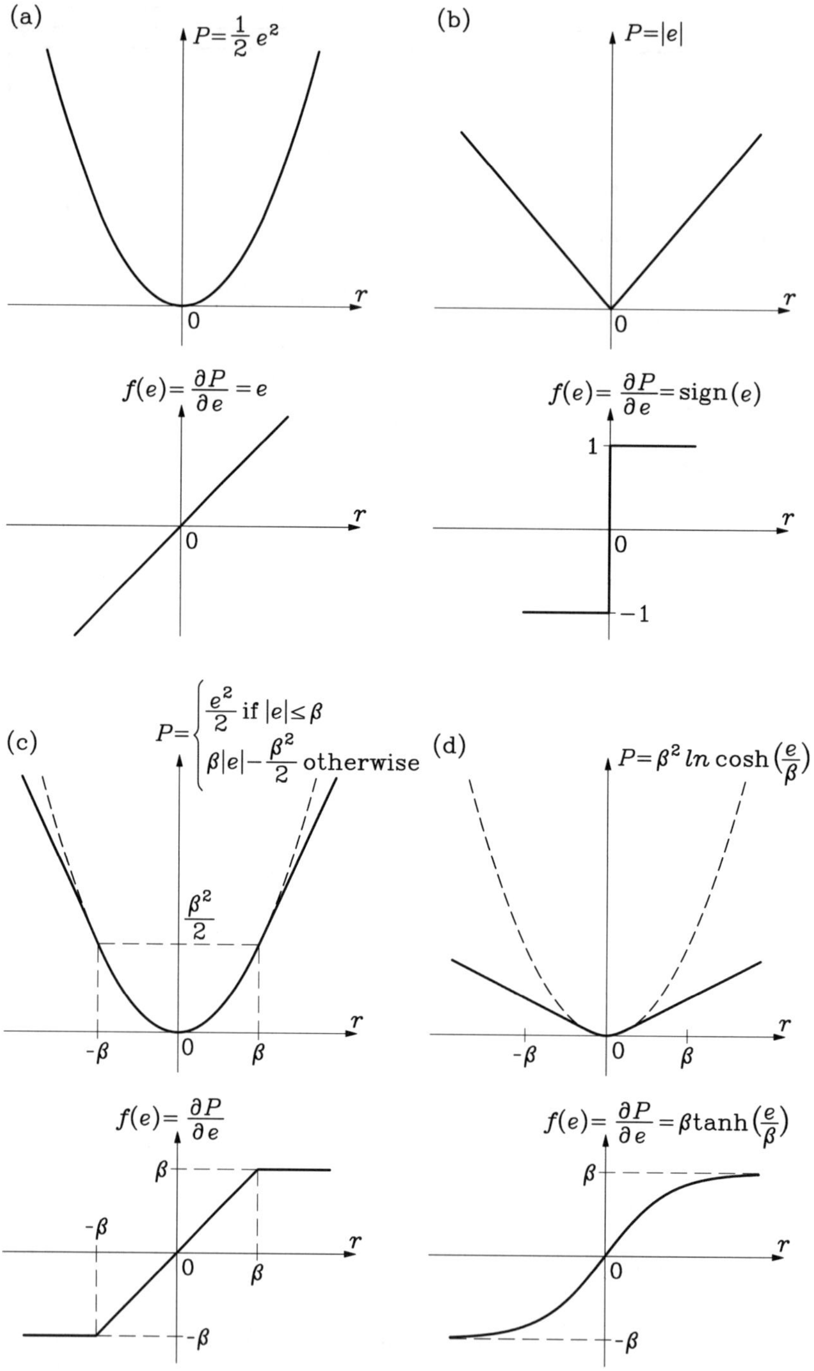

Figure 5.2. Exemplary plots of the activation function $f(e_j)$ for the signal reconstruction problem (5.13).

vector of the network $\boldsymbol{B} = [b_1, b_2, \ldots, b_M]^T$ without any computation. The vector $\boldsymbol{V}(t)$ serves as the output vector of the network.

In terms of Kirchhoff's laws, the differential equations describing this network are exactly the same as (5.17). As a result, according to (5.15) and (5.19), we know that this network is stable and the stationary state $\boldsymbol{V}_f$ of the network is a minimum point of the maximum entropy problem (5.13), that is, the signal vector $\boldsymbol{X}$ to be reconstructed. The time of the network to provide the solution (the time to approach the stationary state from the initial state) is an elapsed time of only a few characteristic time constants of the network (in general, on the order of hundreds of nanoseconds). In other words, no matter what the initial value is (provided it is nonzero), this neural network can provide the desired signal vector under maximum entropy criteria during an elapsed time on the order of hundreds of nanoseconds. As we know, it is impossible for any available digital and sequential scheme to provide the desired signal reconstruction during hundreds of nanoseconds. Therefore, for maximum entropy signal reconstruction, this neural network is much more powerful than any available digital and sequential scheme.

We can make some further comments on this neural network approach for the signal reconstruction problem:

(1) Although the connection strengths H_{ij} may be time varying, they must be kept unchanged during one time evolution of the network from the initial state to the approach to the stationary state.

(2) This neural network approach can easily be generalized to signal reconstruction under the criterion of minimizing the L_p norm as long as we replace the function $-G(\boldsymbol{X})$ in (5.13) by $\frac{1}{p}\sum_{n=1}^{N}|x(n)|^p$, where $1 \leq p < \infty$ [13]. The activation functions

$$g(v_i) = \frac{\partial G(\boldsymbol{V}(t))}{\partial v_i(t)} \qquad (\text{for } i = 1, 2, \ldots, N)$$

with different values of the parameter p are shown in Figure 5.3. Note that in the case of $p = 2$, the related problem is equivalent to the least-squares problem (5.3).

(3) Although the solution provided by this network approach can approximate the exact solution of (5.13) with arbitrarily small error, the difference between the network solution and the real maximum entropy solution (by (5.7)) greatly depends on the constant K. The larger K is, the smaller the difference will be. However, a large constant K will give rise to difficulties in hardware implementation.

(4) The local minimum problem still exists in this neural network approach.

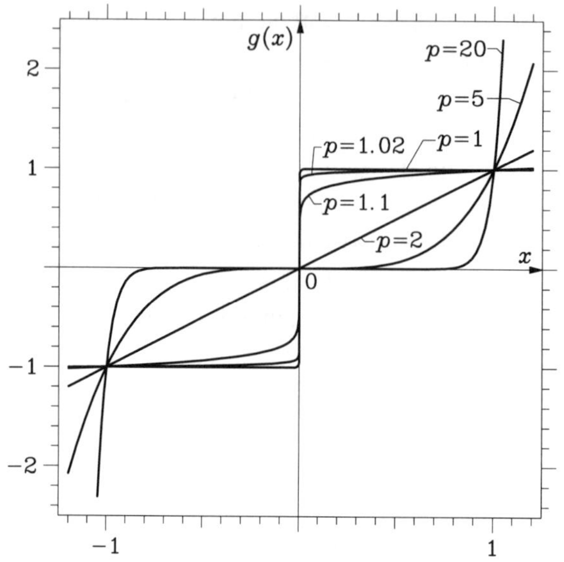

Figure 5.3. Exemplary plots of the activation function $g(v_i)$ for the signal reconstruction problem under the criterion of minimizing the L_p norm.

However, if we use the L_2 norm and the quadratic penalty function $P(e) = \frac{1}{2}e^2$, then the local minimum problem will disappear.

On the basis of the above four entropy forms, an alternative network for solving (5.13) can be constructed. This network is shown in Figure 5.4. The input–output relationships of the neurons in the left and right parts of this network are denoted by $g_i(u)$ (for $i = 1, 2, \ldots, N$) and $f_j(e)$ (for $j = 1, 2, \ldots, M$), respectively. The left part has N neurons and the right part has M neurons. R_i and C_i (for $i = 1, 2, \ldots, N$) are the input resistance and capacitance of the i'th neuron in the left part, respectively (here, for mathematical convenience, we let $R_i = R$ and $C_i = C$). $\boldsymbol{T} = \{T_{ji}\}$ (for $j = 1, 2, \ldots, M; i = 1, 2, \ldots, N$) is the connection strength matrix. $\boldsymbol{A} = [a_1, a_2, \ldots, a_N]^T$ and $\boldsymbol{B} = [b_1, b_2, \ldots, b_M]^T$ are the bias current vectors of the left and right parts, respectively. $v_i(t)$ and $q_j(t)$ (for $i = 1, 2, \ldots, N; j = 1, 2, \ldots, M$) are the neuron outputs of the left and right parts of the network, respectively. $u_i(t)$ is the neuron input of the left part (for $i = 1, 2, \ldots, N$). Clearly, this network resembles the one shown in Figure 2.2. The only difference is that each neuron of the left part has a bias current a_i (for $i = 1, 2, \ldots, N$). The activation function $g_i(\cdot)$ of the left part is selected such that

(1) $\int_0^{v_i(t)} g_i^{-1}(v)dv$ (for $i = 1, 2, \ldots, N$) is bounded from below and

(2) $g_i(u)$ (for $i = 1, 2, \ldots, N$) is a monotonically increasing function.

The activation function $f_j(e)$ (for $j = 1, 2, \ldots, M$) of the right part will be discussed later.

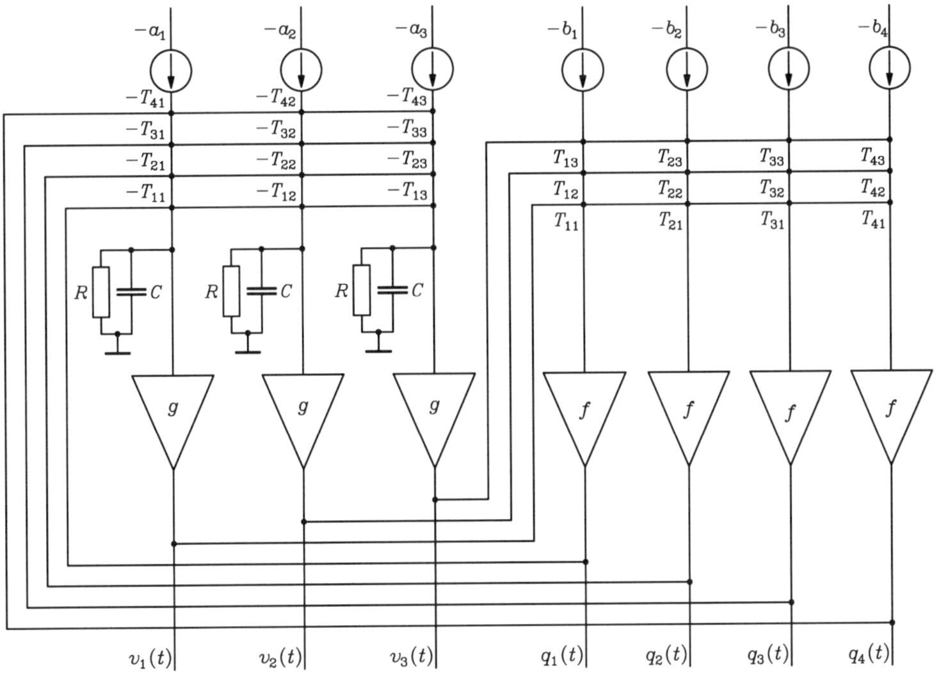

Figure 5.4. An alternative neural network for solving the signal reconstruction problem (5.13). $N = 3$, $M = 4$.

For this network, we can define an energy function similar to (2.5) as

$$E(t) = \sum_{j=1}^{M} F\left(\sum_{i=1}^{N} T_{ji} v_i(t) - b_j\right) + \sum_{i=1}^{N}\left(\frac{1}{R}\int_0^{v_i(t)} g_i^{-1}(v)dv + a_i v_i(t)\right) \tag{5.20}$$

and obtain the dynamic equations of this network as

$$C\frac{du_i(t)}{dt} = -\sum_{j=1}^{M} T_{ji} f\left(\sum_{k=1}^{N} T_{jk} v_k(t) - b_j\right) - \frac{u_i(t)}{R} - a_i \qquad (i = 1, 2, \ldots, N). \tag{5.21}$$

It is easy to prove that $E(t)$ is nonincreasing during the time evolution (that is, $\frac{dE(t)}{dt} \leq 0$, for $t > 0$). If we choose appropriate $f(\cdot)$ and $g(\cdot)$ so that $E(t)$ is bounded from below, then the network described by (5.21) will be stable and provide one of the minima of the energy function (5.20).

We can make Equation (5.20) exactly the same as (5.13) by choosing

(1) $T_{ji} = H_{ij}$ and $b_j = y(j)$ (for $i = 1, 2, \ldots, N$; $j = 1, 2, \ldots, M$), that is, the known element H_{ij} and $y(j)$ are directly taken as the interconnection strengths and bias currents of the network, respectively,

(2) $f(e) = K\frac{dP(e)}{de}$, which means that the derivative of the penalty function is taken as the activation function of the neurons in the right part of the network, and

(3) $\sum_{i=1}^{N}(\frac{1}{R}\int_0^{v_i(t)} g_i^{-1}(v)dv + a_i v_i(t)) = -G(v_i(t))$.

Because $P(\cdot)$ and $G(\cdot)$ are bounded from below, such a constructed network can solve the maximum entropy reconstruction problem (5.13).

Moreover, we can obtain the activation function $g(\cdot)$ and the bias current a_i for Shannon's entropy (5.8) as

$$g(u) = e^{\frac{u}{R}}, \qquad a_i = 1$$

and for Skilling's entropy (5.11) [14] as

$$g(u) = e^{\frac{u}{R}}, \qquad a_i = -\ln(\hat{x}(i)) \qquad (i = 1, 2, \ldots, N).$$

Unfortunately, we are unable to give an explicit form of $g(u)$ for either Burg's entropy (5.9) or for Frieden's and Zoltani's entropy (5.10). This remains an open problem.

It is interesting to note that if we replace the entropy by the L_2 norm and use the quadratic penalty function, then the network in Figure 5.4 will be exactly the same as the network in Figure 2.2.

5.2 Reconstruction of Binary Signals Using MLP Networks

The reconstruction of binary signals constitutes the major problem in signal reconstruction, arising mainly in digital communications. A digital communication system is shown in Figure 5.5. The transmitted sequence $x(n)$, which takes binary values $\{-1, 1\}$ with equal probability that are assumed to be independent of one another, is passed through a dispersive channel. The channel output $\hat{y}(n)$ is corrupted by an additive noise $\epsilon(n)$, which is independent of the transmitted sequence $x(n)$. As a result, the observation data $y(n)$ are equal to $\hat{y}(n)$ plus $\epsilon(n)$, that is, $y(n) = \hat{y}(n) + \epsilon(n)$. Our task is to use the information represented by the observed data $y(n), y(n-1), \ldots, y(n-m+1)$ to produce an estimate of the values of $x(n-d)$ (that is, the symbol of $x(n-d)$ because the absolute value of $x(n-d)$ is always unity). A device to perform this function is known as an equalizer. Consequently,

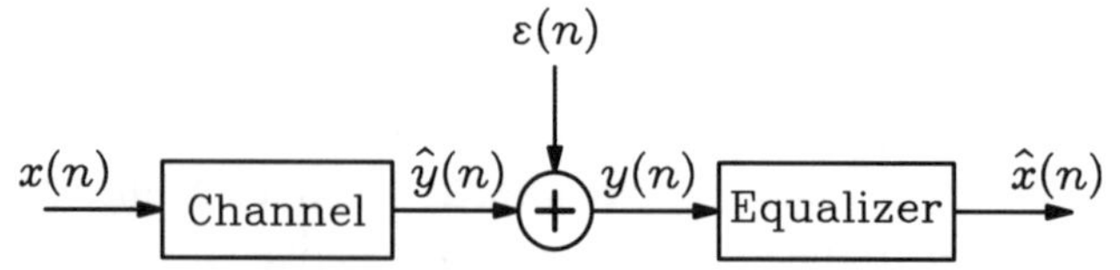

Figure 5.5. Schematic of a digital communication system.

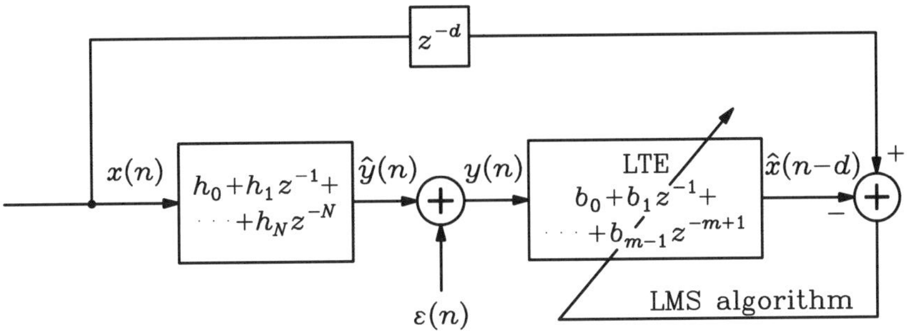

Figure 5.6. Schematic of a linear transversal equalizer (LTE).

the reconstruction of binary-value signals is referred to as equalization, with the integers m and d known as the order and the delay of the equalizer, respectively.

Traditionally, the channel is modeled by an FIR filter with the transfer function

$$H(z) = \sum_{i=0}^{N} h_i z^{-i}, \tag{5.22}$$

and the common equalizer employed for this kind of channel is the linear transversal equalizer (LTE) as shown in Figure 5.6. The transfer function of the LTE is

$$H_e(z) = \sum_{i=0}^{m-1} b_i z^{-i}, \tag{5.23}$$

and the estimate $\hat{x}(n-d)$ of the transmitted signal $x(n-d)$ is given by

$$\hat{x}(n-d) = \operatorname{sgn}\left[\sum_{j=0}^{m-1} b_j y(n-j)\right], \tag{5.24}$$

where the coefficients b_j (for $j = 0, 1, \ldots, m-1$) of the LTE can be determined by an algorithm such as the least-mean-squares (LMS) algorithm.

Clearly, an LTE estimates $x(n-d)$ to be 1 or -1 according to which side of the hyperplane ($\sum_{j=0}^{m-1} b_j y(n-j) = 0$) the observed data $y(n), y(n-1), \ldots, y(n-m+1)$ are situated. Gibson et al. [15] have shown that the performance of the LTE depends primarily on the channel transfer function and the signal-to-noise ratio. When the signal-to-noise ratio is high and the channel transfer function has minimum phase, then the LTE can reconstruct the transmitted signal $x(n)$ with an acceptable performance. However, if the signal-to-noise ratio is not high enough or the channel transfer function has nonminimum phase, the LTE may exhibit a considerable bit-error rate (BER), mainly because of the nonlinearly separated boundaries. In addition, Chen et al. [16] have shown that if the channel system is nonlinear and cannot be well approximated by a linear system, then the equalization performance of the LTE will also be greatly degraded. These problems motivate the investigation and development of nonlinear equalizers.

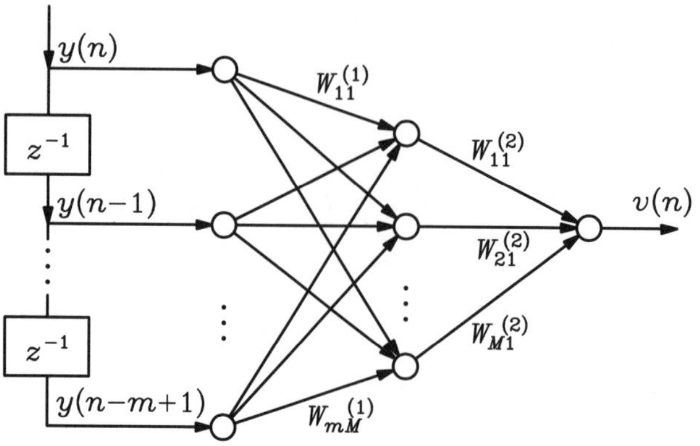

Figure 5.7. A nonlinear equalizer based on MLP networks.

Recently, nonlinear equalizers based on nonlinear mapping neural networks such as MLP networks, RBF networks, and high-order networks have been proposed and extensively analyzed [16, 17, 18]. The nonlinear mapping capability of these neural networks lay a foundation for using them to construct nonlinear equalizers. In this section and the next two sections, we will present these nonlinear equalizers based on MLP networks, RBF networks, and high-order networks, respectively.

The nonlinear equalizer based on MLP networks is shown in Figure 5.7. In general, the network has three layers. The input layer has m neurons; these simply feed the observation vector $\boldsymbol{Y}(n) = [y(n), y(n-1), \ldots, y(n-m+1)]^T$ to the second layer without any modification. The hidden layer has M neurons with nonlinear transfer functions (such as the sigmoid function). The output layer has only one neuron whose input–output relationship should be

$$\frac{1-e^{-\alpha x}}{1+e^{-\alpha x}},$$

where α is a sufficiently large positive constant.

Let $W_{ij}^{(1)}$ denote the connection weight between the i'th neuron in the input layer and the j'th neuron in the hidden layer (for $i = 1, 2, \ldots, m$; $j = 1, 2, \ldots, M$); $q_j(n)$ and $f_j(\cdot)$ (for $j = 1, 2, \ldots, M$) be the output and activation function of the j'th neuron in the hidden layer, respectively; $W_{j1}^{(2)}$ denote the connection weight between the j'th neuron in the hidden layer and the neuron in the output layer; and $u(n)$, $v(n)$, $g(\cdot)$, and θ be the input, output, activation function, and threshold of the output neuron, respectively. Then we have

$$v(n) = g\big(u(n) - \theta\big) = \frac{1-e^{-\alpha(u(n)-\theta)}}{1+e^{-\alpha(u(n)-\theta)}}, \tag{5.25}$$

$$u(n) = \sum_{j=1}^{M} W_{j1}^{(2)} q_j(n), \tag{5.26}$$

and

$$q_j(n) = f_j\left(\sum_{i=1}^{m} W_{ij}^{(1)} y(n-i+1)\right) \qquad (j = 1, 2, \ldots, M). \tag{5.27}$$

For simplicity, in the following we choose the threshold θ to be zero.

Substituting (5.27) into (5.26) gives

$$u(n) = \sum_{j=1}^{M} W_{j1}^{(2)} f_j\left(\sum_{i=1}^{m} W_{ij}^{(1)} y(n-i+1)\right). \tag{5.28}$$

Equations (5.25) and (5.28) both show that $v(n)$ has a nonlinear relationship with the inputs $y(n), y(n-1), \ldots, y(n-m+1)$ and that this relationship is completely determined by the connection weights $W_{ij}^{(1)}$ and $W_{j1}^{(2)}$ (for $i = 1, 2, \ldots, m; j = 1, 2, \ldots, M$). The functions of these weights are similar to those of the coefficients b_j in the LTE. For different channels, the connection weights will be different. Before the nonlinear equalizer begins to estimate the values of the transmitted signal from the observation data, one has to determine all the connection weights. This can be accomplished by use of the BP algorithm or other available algorithms [19] and by use of the known samples of the transmitted signals (for $n = m, m + 1, \ldots, N + m - 1$). Note that, in the communication equalization, N samples of the transmitted signals are known a priori by the receiver and can be used as the desired output $\hat{v}(n)$ (for $n = m, m + 1, \ldots, N + m - 1$) of the above nonlinear equalizer. If these samples are not available, then blind equalizers will be needed. Blind equalization will be discussed in Section 5.5.

The details of the procedures to determine the connection weights by use of the BP algorithm are:

(1) Initialize randomly all connection weights $W_{ij}^{(1)}$ and $W_{j1}^{(2)}$ (for $i = 1, 2, \ldots, m; j = 1, 2, \ldots, M$).

(2) Pass N observation vectors $\boldsymbol{Y}(n) = [y(n), y(n-1), \ldots, y(n-m+1)]^T$ successively to the input layer of the nonlinear equalizer.

(3) Compute by proceeding forward the output $q_j(n)$ (for $j = 1, 2, \ldots, M$) of the hidden layer and the output $v(n)$ using (5.25)–(5.27) and the available input vectors $\boldsymbol{Y}(n)$.

(4) Compute by proceeding backward the error propagation terms $\delta_1^{(2)}(n)$ and $\delta_j^{(1)}(n)$ (for $j = 1, 2, \ldots, M$) by using

$$\delta_1^{(2)}(n) = (\hat{v}(n) - v(n))\left[g(u(n))\right]' \tag{5.29}$$

and

$$\delta_j^{(1)}(n) = \left[f_j(u_j^{(1)}(n))\right]' \delta_1^{(2)}(n) W_{j1}^{(2)}(t), \tag{5.30}$$

where

$$\left[f_j\left(u_j^{(1)}(n)\right)\right]' = \left.\frac{\partial f_j(u)}{\partial u}\right|_{u=u_j^{(1)}(n)} \tag{5.31}$$

and

$$u_j^{(1)}(n) = \sum_{i=1}^{m} W_{ij}^{(1)}(t) y(n-i+1)$$

$$(n = m, m+1, \ldots, N+m-1). \tag{5.32}$$

Note that in (5.30)–(5.32), $W_{ij}^{(1)}(t)$ and $W_{j1}^{(2)}(t)$ denote the related connection weights at the t'th iteration. In addition, the values of $\hat{v}(n)$ are always either -1 or 1.

(5) Adjust the connection weights according to

$$W_{j1}^{(2)}(t+1) = W_{j1}^{(2)}(t) + \gamma \sum_{n=m}^{N+m-1} \delta_1^{(2)}(n) q_j(n) \tag{5.33}$$

and

$$W_{ij}^{(1)}(t+1) = W_{ij}^{(1)}(t) + \gamma \sum_{n=m}^{N+m-1} \delta_j^{(1)}(n) y(n-i+1)$$

$$(i = 1, 2, \ldots, m; \quad j = 1, 2, \ldots, M), \tag{5.34}$$

where γ is the learning-rate parameter.

(6) Compute the total error

$$e = \sum_{n=m}^{N+m-1} \left|\hat{v}(n) - v(n)\right|^2 \tag{5.35}$$

and iterate the computation by returning to Step (2) until this error is less than a specified one.

This is a kind of batch-processing, where in all N observation vectors $\boldsymbol{Y}(m)$, $\boldsymbol{Y}(m+1), \ldots, \boldsymbol{Y}(N+m-1)$ are available when the training of the connection weights is initiated. If each new observation vector becomes available during the training, then the following procedures can be used:

(1) Initialize randomly all connection weights $W_{ij}^{(1)}(n)$ and $W_{j1}^{(2)}(n)$ (for $i = 1, 2, \ldots, m$; $j = 1, 2, \ldots, M$) with $n = m$. This means that the training starts only after the m'th observation $y(m)$ becomes available.

(2) Pass one observation vector $\boldsymbol{Y}(n) = [y(n), y(n-1), \ldots, y(n-m+1)]^T$ to the input layer of the nonlinear equalizer.

(3) Compute by proceeding forward the output $q_j(n)$ (for $j = 1, 2, \ldots, M$) of the hidden layer and the output $v(t)$ using (5.25)–(5.27) and the available input vectors $\boldsymbol{Y}(n)$.

(4) Compute by proceeding backward the error propagation terms $\delta_1^{(2)}(n)$ and $\delta_j^{(1)}(n)$ (for $j = 1, 2, \ldots, M$) by using

$$\delta_1^{(2)}(n) = (\hat{v}(n) - v(n))\left[g\left(u(n)\right)\right]' \tag{5.36}$$

and

$$\delta_j^{(1)}(n) = \left[f_j\left(u_j^{(1)}(n)\right)\right]' \delta_1^{(2)}(n) W_{j1}^{(2)}(n), \tag{5.37}$$

where

$$\left[f_j\left(u_j^{(1)}(n)\right)\right]' = \left.\frac{\partial f_j(u)}{\partial u}\right|_{u=u_j^{(1)}(n)} \tag{5.38}$$

and

$$u_j^{(1)}(n) = \sum_{i=1}^{m} W_{ij}^{(1)}(n) y(n-i+1). \tag{5.39}$$

(5) Adjust the connection weights according to

$$W_{j1}^{(2)}(n+1) = W_{j1}^{(2)}(n) + \gamma \delta_1^{(2)}(n) q_j(n) \tag{5.40}$$

and

$$W_{ij}^{(1)}(n+1) = W_{ij}^{(1)}(n) + \gamma \delta_j^{(1)}(n) y(n-i+1)$$
$$(i = 1, 2, \ldots, m; \quad j = 1, 2, \ldots, M), \tag{5.41}$$

where γ is the learning-rate parameter whose value may be different from that in (5.33) and (5.34). Note the difference between (5.30) and (5.37), (5.32) and (5.39), (5.33) and (5.40), and between (5.34) and (5.41).

(6) Increase the iteration number to $n+1$ and return to Step (2) by taking the next observation vector as the input vector of the nonlinear equalizer if the next observation sample is available. Otherwise, iterate the computation by returning to Step (2) and by still using the available observation vector until the next observation sample is available.

(7) Stop the training if N observation vectors have all become available and if the total error

$$e = \sum_{n=m}^{N+m-1} \left|\hat{v}(n) - v(n)\right|^2 \tag{5.42}$$

is less than the specified one. Note that in (5.42), $v(n)$ should be computed by using the weights at time $N+m-1$ instead of those at time n, that is,

$$v(n) = g\big(u(n)\big) = \frac{1 - e^{-\alpha u(n)}}{1 + e^{-\alpha u(n)}}, \tag{5.43}$$

$$u(n) = \sum_{j=1}^{M} W_{j1}^{(2)}(N + m - 1) f_j \left(\sum_{i=1}^{m} W_{ij}^{(1)}(N + m - 1) y(n - i + 1) \right). \tag{5.44}$$

In addition, in many high-speed communication systems, even though the error e is not less than the specified one, the training must be stopped once the unknown signal has been transmitted. To attack this problem, we need to speed up the convergence of the above training, and thus some fast training algorithms instead of the BP algorithm should be used in these high-speed applications. Unfortunately, these fast algorithms, with their intensive computational complexity, make the hardware structure of this nonlinear equalizer more complicated. Another possible way to speed up the convergence and to increase the degree of the nonlinearity is to use more neurons in the second layer (the hidden layer) and add an adjustable threshold to each neuron in the second layer and in the output neuron. As pointed out in Section 4.3, the function of the threshold of each neuron is equivalent to that of an additional neuron.

After the training, the nonlinear equalizer will reflect the channel characteristics and can be used to provide the estimates of the values of the transmitted signal. Because of the high dimensionality and degrees of the nonlinearity it is difficult to analyze theoretically the equalization performance (such as BER) of this MLP nonlinear equalizer. However, many simulation results have shown that this nonlinear equalizer based on the MLP network can achieve better performance than the LTE under adverse conditions such as low signal-to-noise ratio (white and colored noises), nonminimum phase of the linear channel [15], and finite nonlinear channels represented by

$$y(n) = h\big(x(n), x(n-1), \ldots, x(n-m+1); \Theta\big) + \epsilon(n), \tag{5.45}$$

where $h(\cdot)$ is some nonlinear function and Θ denotes the parameter set of the nonlinear channel [16].

In summary, this nonlinear equalizer based on the MLP network is considered to be very promising and more investigations would be desirable.

5.3 Reconstruction of Binary Signals Using RBF Networks

The nonlinear equalizer based on RBF networks may serve as an alternative for the reconstruction of binary signals in communication systems. As pointed out in

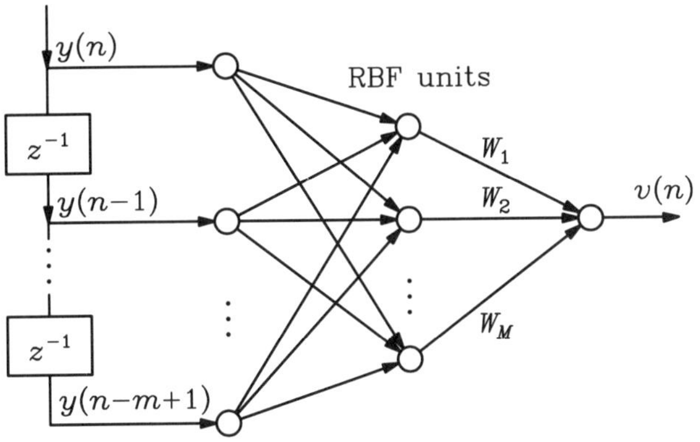

Figure 5.8. Schematic of the nonlinear equalizer based on an RBF network.

Section 1.6, the RBF network possesses nonlinear mapping ability and yet has a linear-in-the-parameter structure. The former property is essential to realize the nonlinear equalization solution and the latter is beneficial in practical hardware implementations. In this section, we will present the details of the nonlinear equalizer based on RBF networks.

The structure of the nonlinear equalizer based on an RBF network is shown in Figure 5.8. Like the MLP-based equalizer, this equalizer consists of three layers. The input layer has neurons with a linear function that simply feeds the input signals to the hidden layer. Moreover, the connections between the input layer and the hidden layer are not weighted – each hidden neuron receives each corresponding input value unaltered. The neuron numbers in the hidden layer and input layer are denoted by M and m, respectively. The hidden neurons are processing units that perform the radial basis function. The transfer function of the hidden neurons in this RBF network can be, for example, the

(1) Gaussian function: $f(r) = e^{-\frac{r^2}{\sigma^2}}$,
(2) multiquadratic function: $f(r) = (r^2 + \sigma^2)^{\frac{1}{2}}$,
(3) inverse multiquadratic function: $f(r) = (r^2 + \sigma^2)^{-\frac{1}{2}}$,
(4) thin-plate-spline function: $f(r) = r^2 \log(r)$,
(5) piecewise-linear function: $f(r) = \frac{1}{2}(|r+1| - |r-1|)$, or the
(6) cubic approximation function: $f(r) = \frac{1}{2}(|r^3+1| - |r^3-1|)$,

where σ is a real parameter (called a scaling parameter) and r is the distance between the input vector $\boldsymbol{Y}(n) = [y(n), y(n-1), \ldots, y(n-m+1)]^T$ and the center vector $\boldsymbol{C}_j$ (for $j = 1, 2, \ldots, M$). The distance is usually measured by the Euclidean norm, that is,

$$r_j = \|\boldsymbol{Y}(n) - \boldsymbol{C}_j\| \qquad (j = 1, 2, \ldots, M). \tag{5.46}$$

Because the choice of the nonlinearity of the RBF is not crucial for the approximation performance of the network, we will employ the Gaussian function in the following for simplicity. The output of each neuron in the hidden layer is

$$h_i(n) = f_i\left(\|\boldsymbol{Y}(n) - \boldsymbol{C}_i\|\right) \qquad (i = 1, 2, \ldots, M). \tag{5.47}$$

The output layer has only one neuron with the activation function

$$g(x) = \frac{1 - e^{-\alpha x}}{1 + e^{-\alpha x}}$$

with large steepness. Note that the output neuron of the RBF network given in Section 1.6 performs only simple summations, that is, the activation function is a linear function instead of the nonlinear function

$$g(x) = \frac{1 - e^{-\alpha x}}{1 + e^{-\alpha x}}.$$

The output neuron in this nonlinear equalizer is also biased by a threshold θ. The connections between each neuron in the hidden layer and the output neuron are weighted. If W_j denotes the connection weight between the j'th neuron in the hidden layer and the output neuron, then the input $u(n)$ and the output $v(n)$ of the neuron in the output layer at time n are

$$u(n) = \sum_{j=1}^{M} W_j h_j(n) - \theta = \sum_{j=1}^{M} W_j f_j\left(\|\boldsymbol{Y}(n) - \boldsymbol{C}_j\|\right) - \theta \tag{5.48}$$

and

$$v(n) = g\left(\sum_{j=1}^{M} W_j h_j(n) - \theta\right) = g\left(\sum_{j=1}^{M} W_j f_j\left(\|\boldsymbol{Y}(n) - \boldsymbol{C}_j\|\right) - \theta\right). \tag{5.49}$$

If the threshold θ is considered as a special hidden neuron whose output is unity and the value of the connection weight of which with the output neuron is θ, then Equations (5.48) and (5.49) can be written as

$$u(n) = \sum_{j=0}^{M} W_j f_j\left(\|\boldsymbol{Y}(n) - \boldsymbol{C}_j\|\right) \tag{5.50}$$

and

$$v(n) = g\left(\sum_{j=0}^{M} W_j f_j\left(\|\boldsymbol{Y}(n) - \boldsymbol{C}_j\|\right)\right), \tag{5.51}$$

with $W_0 = \theta$ and $\boldsymbol{C}_0 = \boldsymbol{Y}(n)$, that is, $h_0 = 1$.

These three sets of unknown parameters – the scaling parameter σ, the center vectors $\boldsymbol{C}_j$, and the weights W_j (for $j = 1, 2, \ldots, M$) – must be determined before the nonlinear equalizer can begin to estimate the values of the transmitted signal. They depend on the communication channel and can in principle be determined by use of the available observation data and some known samples of the transmitted signal. The procedures to determine the parameters of the nonlinear equalizer based on the RBF network can be described similarly to those presented in Section 1.6. There are three steps:

(1) Determination of the Center Vectors $\boldsymbol{C}_j$ (for $j = 1, 2, \ldots, M$)
There are several ways to choose the center vectors for this RBF nonlinear equalizer. The simplest technique is to choose these vectors randomly from a subset of the observation data. Another way is to use the known transmitted samples to construct the center vectors. If N samples $(x(m), x(m+1), \ldots, x(N+m-1))$ of the transmitted signal are known, then we can take

$$\boldsymbol{C}_j = \left[x(j+2m-2), x(j+2m-3), \ldots, x(j+m-1)\right]^T \qquad (j = 1, 2, \ldots, M). \tag{5.52}$$

In this case, each element of the center vectors has binary values (either -1 or 1). Obviously, we can select $M = N - m + 1$ (i.e., the number of the hidden neurons must not be greater than $N - m + 1$). Because the number N is usually not great, the equalizer with the $N - m + 1$ hidden neurons may not be enough to approximate the desired nonlinear mapping required in the nonlinear equalization. Based on the above, Chen et al. [20] suggest using the "k-means clustering algorithm" as shown in Reference [21] and Section 1.6. The main disadvantage of this cluster algorithm is its computational complexity. This algorithm involves m multiplications, m divisions, and $m + 1$ additions for finding a center vector. Determining the center vectors more accurately and more effectively would be a worthwhile subject of further investigation.

(2) Determination of the Scaling Parameters σ_j (for $j = 1, 2, \ldots, M$)
If the hidden neurons take the Gaussian function as their activation functions, that is,

$$f_j(r) = e^{-\frac{r^2}{\sigma_j^2}} \qquad (j = 1, 2, \ldots, M), \tag{5.53}$$

then it is necessary to determine the scaling parameters σ_j. An appropriate method to determine the scaling parameters is based on the P-nearest neighbor heuristic,

$$\sigma_j = \frac{1}{P}\sum_{i=1}^{P} \|\boldsymbol{C}_j - \boldsymbol{C}_i\|^2 \qquad (j = 1, 2, \ldots, M), \tag{5.54}$$

where $\boldsymbol{C}_i$ (for $i = 1, 2, \ldots, P$) are the P-nearest neighbors of $\boldsymbol{C}_j$. However, for simplicity, we can set all scaling parameters σ_j to be identical and on the order of the noise power. It has been shown that the scaling parameters are less influential in comparison with the center vectors and weights [20].

(3) Determination of the Connection Weights W_j (for $j = 0, 1, \ldots, M$)
Because the output neuron in this RBF equalizer possesses the nonlinear activation function

$$g(x) = \frac{1 - e^{-\alpha x}}{1 + e^{-\alpha x}}$$

instead of performing simple weighted summation, the method given in Section 1.6 cannot be directly used to determine the weights W_j (for $j = 0, 1, \ldots, M$). However, the fact that the output layer has only one neuron makes it simple and easy to determine the weights. Once the center vectors and the scaling parameters have been determined, the weights can be obtained by solving the optimization problem

$$\min_{\boldsymbol{W}} \sum_{n=m}^{N+m-1} |v(n) - \hat{v}(n)|^2, \tag{5.55}$$

where $\boldsymbol{W} = [W_0, W_1, \ldots, W_M]^T$ is the $(M+1)$-dimensional vector of the connection weights and $\hat{v}(n)$ (for $n = m, m+1, \ldots, N+m-1$) are the same as those in the MLP equalizer (i.e., the known samples of the transmitted signal). Using the gradient descent procedure, we can obtain a recursive algorithm to determine the weights as follows:

(1) Initialize randomly the weight vector $\boldsymbol{W}$.
(2) Pass N observation vectors $\boldsymbol{Y}(n) = [y(n), y(n-1), \ldots, y(n-m+1)]^T$ successively to the input layer of the nonlinear equalizer.
(3) Compute the output $h_j(n)$ (for $j = 1, 2, \ldots, M$) of the hidden layer, the input $u(n)$, and the output $v(n)$ of the output neuron by using (5.47)–(5.51) and the available input vectors $\boldsymbol{Y}(n)$.
(4) Compute the error term $\delta(n)$ by

$$\delta(n) = (\hat{v}(n) - v(n)) \left[g(u(n))\right]' \tag{5.56}$$

$$\left[g(u(n))\right]' = \left.\frac{dg(u)}{du}\right|_{u=u(n)} \qquad (n = m, m+1, \ldots, N+m-1). \tag{5.57}$$

(5) Adjust the connection weights according to

$$W_j(t+1) = W_j(t) + \gamma \sum_{n=m}^{N+m-1} \delta(n) h_j(n) \qquad (j = 0, 1, \ldots, M), \tag{5.58}$$

where γ is the learning-rate parameter. Note that the related weights at the current iteration are denoted by $W_j(t)$ (for $j = 0, 1, \ldots, M$).

(6) Compute the total error

$$e = \sum_{n=m}^{N+m-1} \left|\hat{v}(n) - v(n)\right|^2 \tag{5.59}$$

and iterate the computation by returning to Step (2) until this error is less than the specified one.

This is a kind of batch-processing. If each new observation vector becomes available during training, then the computations in Step (5) and Step (6) become

$$W_j(n+1) = W_j(n) + \gamma\delta(n)h_j(n) \qquad (j = 0, 1, \ldots, M) \tag{5.60}$$

and

$$e = \sum_{k=m}^{n} \left|\hat{v}(k) - v(k)\right|^2, \tag{5.61}$$

where the related weights at the current iteration are denoted by $W_j(n)$ (for $j = 0, 1, \ldots, M$). Note that the variable $v(k)$ in (5.61) should be computed by use of the weights at the iteration n, that is,

$$v(k) = g\left(\sum_{j=0}^{M} W_j(n) f_j\left(\left\|\boldsymbol{Y}(k) - \boldsymbol{C}_j\right\|\right)\right), \tag{5.62}$$

and in Step (2) only one observation vector is passed to the RBF equalizer for each iteration. If all N samples have become available but the error e is still greater than the specified one, then we can take the weights at time $N + m - 1$ as the new initialized values and use the above batch-processing algorithm without any modifications to continue to train the weights. However, once the unknown signal has been transmitted the training must be stopped, even though the error e may not be less than the specified one.

To reduce the complexity of the above algorithms, we can take the known samples of the transmitted signal as the desired input $\hat{u}(n)$ of the output neuron instead of the desired output $\hat{v}(n)$. Under this condition, Equation (5.55) becomes

$$\min_{\mathbf{W}} \sum_{n=m}^{N+m-1} \left|u(n) - \hat{u}(n)\right|^2. \tag{5.63}$$

Substituting (5.50) into (5.63), we obtain

$$\min_{\mathbf{W}} \sum_{n=m}^{N+m-1} \left\|\sum_{j=0}^{M} W_j h_j(n) - \hat{u}(n)\right\|^2 \tag{5.64}$$

and

$$\min_{\boldsymbol{W}} \sum_{n=m}^{N+m-1} \left\|\boldsymbol{W}^T \boldsymbol{H}(n) - \hat{u}(n)\right\|^2, \tag{5.65}$$

where $\boldsymbol{H}(n) = [1, h_1(n), h_2(n), \ldots, h_M(n)]^T$. Moreover, it follows that

$$\min_{\boldsymbol{W}} \|\boldsymbol{H}\boldsymbol{W} - \hat{\boldsymbol{U}}\|^2, \tag{5.66}$$

where $\boldsymbol{H} = [\boldsymbol{H}(m), \boldsymbol{H}(m+1), \ldots, \boldsymbol{H}(N+m-1)]^T$ and $\hat{\boldsymbol{U}} = [\hat{u}(m), \hat{u}(m+1), \ldots, \hat{u}(N+m-1)]^T$ are an $N \times (M+1)$-dimensional matrix and $N \times 1$-dimensional vector, respectively. Clearly, (5.66) is a least-squares problem similar to (1.72). Its solution is

$$\boldsymbol{W} = \boldsymbol{H}^+\hat{\boldsymbol{U}} = \lim_{\alpha \to 0}(\boldsymbol{H}^T\boldsymbol{H} + \alpha\boldsymbol{I})^{-1}\boldsymbol{H}^T\hat{\boldsymbol{U}}. \tag{5.67}$$

Using the gradient descent procedure, we can also solve (5.66) by the following recursive algorithm:

(1) Initialize randomly the weight vector $\boldsymbol{W}$.

(2) Pass N observation vectors $\boldsymbol{Y}(n) = [y(n), y(n-1), \ldots, y(n-m+1)]^T$ successively to the input layer of the nonlinear equalizer.

(3) Compute the output $h_j(n)$ (for $j = 1, 2, \ldots, M$) of the hidden layer and the input $u(n)$ of the output neuron using Equations (5.47)–(5.50) and using the available input vector $\boldsymbol{Y}(n)$.

(4) Compute the error term $\delta(n)$ by

$$\delta(n) = \hat{u}(n) - u(n)$$

$$(n = m, m+1, \ldots, N+m-1). \tag{5.68}$$

(5) Adjust the connection weights according to

$$W_j(t+1) = W_j(t) + \gamma \sum_{n=m}^{N+m-1} \delta(n)h_j(n) \qquad (j = 0, 1, \ldots, M). \tag{5.69}$$

Note that the related weights at the current iteration are denoted by $W_j(t)$ (for $j = 0, 1, \ldots, M$).

(6) Compute the total error

$$e = \sum_{n=m}^{N+m-1} \left|\hat{u}(n) - u(n)\right|^2 \tag{5.70}$$

and iterate the computation by returning to Step (2) until this error is less than the specified one.

For the case in which each new sample set becomes available recursively, the above algorithm becomes the following:

(1) Initialize randomly all connection weights.
(2) Pass one observation vector $\boldsymbol{Y}(n) = [y(n), y(n-1), \ldots, y(n-m+1)]^T$ to the input layer of the nonlinear equalizer.
(3) Compute the output $h_j(n)$ (for $j = 1, 2, \ldots, M$) of the hidden layer and the input $u(n)$ of the output neuron by using Equations (5.47)–(5.50) and the available input vectors $\boldsymbol{Y}(n)$.
(4) Compute the error term $\delta(n)$ by

$$\delta(n) = \hat{u}(n) - u(n). \tag{5.71}$$

(5) Adjust the connection weights according to

$$\begin{aligned} W_j(n+1) &= W_j(n) + \gamma\delta(n)h_j(n) \\ &= W_j(n) + \gamma\big(\hat{u}(n) - u(n)\big)h_j(n) \qquad (j = 0, 1, \ldots, M). \end{aligned} \tag{5.72}$$

(6) Increase the iteration number to $n+1$ and return to Step (2) by taking the next observation vector as the input of the nonlinear equalizer if the next observation sample is available. Otherwise, iterate the computation by returning to Step (2) and by still using the available observation vector until the next observation sample arrives.
(7) Stop the training if N observation vectors have all become available and if the total error

$$e = \sum_{n=m}^{N+m-1} \left|\hat{u}(n) - u(n)\right|^2 \tag{5.73}$$

is less than the specified one. Note that in (5.73), $u(n)$ should be computed by using the weights at time $N+m-1$ instead of those at time n, that is,

$$u(n) = \sum_{j=0}^{M} W_j(N+m-1) f_j\Big(\big\|\boldsymbol{Y}(n) - \boldsymbol{C}_j\big\|\Big). \tag{5.74}$$

Obviously, Equation (5.72) is in effect the LMS algorithm for determining the weights of the nonlinear RBF equalizer. Chen et al. [22] proposed an orthogonal least-squares learning algorithm that can deliver faster convergence and hence may serve as an alternative to the above two recursive algorithms.

It is worth mentioning that in the case of very high speed communication equalization, the analog neural network in Figure 2.2 can be employed to perform (5.66) in real time as long as we let the known matrix $\boldsymbol{H}$ and vector $\hat{\boldsymbol{U}}$ be taken as the connection strength matrix $\boldsymbol{T}$ and bias current vector $\boldsymbol{B}$ of the neural network of Figure 2.2, respectively, with the other parameters of the network taken as the same as those for (2.3). The detailed analysis of the neural network in Figure 2.2 for performing (5.66) can be taken to be the same as that for (2.3).

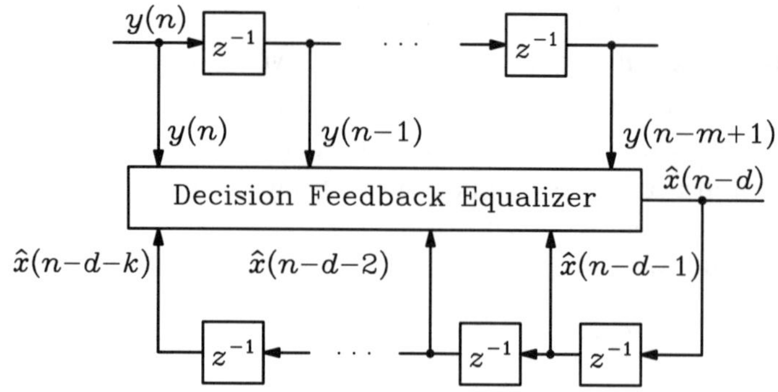

Figure 5.9. Schematic of a decision feedback equalization (DFE).

Chen et al. [17, 20, 23] have shown that the nonlinear RBF equalizer can provide the optimal Bayesian solution of the symbol-decision equalizer, which explains clearly why this neural network equalizer outperforms the traditional linear equalizer.

The nonlinear equalizers based on neural networks can also be extended to decision feedback equalization [24, 25]. Decision feedback equalization (DFE) is another widely used technique that makes decisions on a symbol-by-symbol basis. The architecture of a decision feedback equalizer is shown in Figure 5.9. The advantage of the DFE is that intersymbol interference is eliminated without enhancement of noise by using the past decisions, but a disadvantage is the error propagation caused by decision errors. The operation of the DFE at each sample instant is based on the m most recent channel observation data and the k past decisions. The conventional DFE employs two linear finite filters called the feedforward and feedback filters. It makes the decisions according to

$$\hat{x}(n-d) = \operatorname{sgn}\left(\sum_{j=0}^{m-1} b_j y(n-j) + \sum_{j=1}^{k} a_j \hat{x}(n-d-j)\right), \tag{5.75}$$

where b_j and a_j are the coefficients of the feedforward and feedback filters, respectively. In fact, the feedforward filter is exactly the same as the FIR filter used in the LTE. It can easily be shown that the optimal solution of the DFE is a nonlinear problem (a nonlinear decision boundary) and that the traditional linear DFE cannot realize this nonlinear decision boundary. However, if the linear feedforward and feedback filters are replaced by an RBF network, then this problem can effectively be overcome. Moreover, we can also use the MLP network to replace the linear feedforward and feedback filters of the DFE. In References [17] and [25] it is shown by simulations and analytically that the DFE based on the MLP and RBF network can provide superior performance over the traditional linear DFE.

In some communication systems, the signals and related channels are represented in complex-valued forms [26, 27]. The above nonlinear RBF equalizer can also be generalized to the complex-valued case. In this generalization, the hidden neurons have complex centers and the weights between the hidden neurons and the output neuron have complex values, but the nonlinearity of the hidden neurons remains a real-valued function. The activation function of the output neuron is linear, that is, $v(n) = ku(n)$, where k is a positive constant. For simplicity, we let $k = 1$. With this in mind, the output of the nonlinear RBF equalizer in the complex-valued case is

$$v(n) = \sum_{j=0}^{M} W_j h_j(n) = \sum_{j=0}^{M} W_j f_j\left((\boldsymbol{Y}(n) - \boldsymbol{C}_j)^H (\boldsymbol{Y}(n) - \boldsymbol{C}_j)\right). \tag{5.76}$$

Consequently, the algorithms listed above can directly be generalized for determining the related centers and weights. Chen et al. [27] have shown both analytically and by simulation results that the optimal Bayesian equalizer is structurally equivalent to an RBF network with complex values. For more detail, see Reference [27].

5.4 Reconstruction of Binary Signals Using High-Order Neural Networks

Based on the principles of the nonlinear MLP equalizers and RBF equalizers, we can also use the high-order neural network presented in Section 1.7 to construct a nonlinear equalizer. Such an equalizer is shown in Figure 5.10. This nonlinear equalizer consists of three layers: the input layer, the high-order layer, and the output layer. Similar to the input layer of the MLP and RBF equalizers, the input layer of the high-order neural network simply feeds the observation vector $\boldsymbol{Y}(n) = [y(n), y(n-1), \ldots, y(n-m+1)]^T$ to the second layer (the high-order

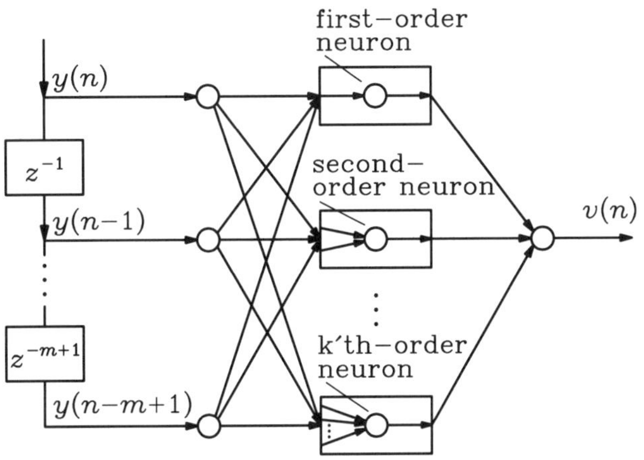

Figure 5.10. Schematic of the nonlinear equalizer based on a high-order neural network.

layer) without modification. This means that there are m neurons in the input layer. For convenience, we renotate $\boldsymbol{Y}(n)$ as $[y_1(n), y_2(n), \ldots, y_m(n)]^T$, that is, $y_i(n) = y(n-i+1)$. The units of the high-order layer are in effect multipliers whose outputs are computed as follows:

(1) The outputs of the first-order neurons are

$$q_i^{(1)}(n) = y_i(n) \qquad (i = 1, 2, \ldots, m). \tag{5.77}$$

(2) The outputs of the second-order neurons are determined by

$$q_{ij}^{(2)}(n) = y_i(n) y_j(n) \qquad (i, j = 1, 2, \ldots, m). \tag{5.78}$$

(3) The outputs of the k'th-order neurons are determined by

$$q_{i_1 i_2 \ldots i_k}^{(k)}(n) = y_{i_1}(n) y_{i_2}(n) \ldots y_{i_k}(n)$$
$$(i_j = 1, 2, \ldots, m; \quad j = 1, 2, \ldots, k). \tag{5.79}$$

The output layer has only one neuron with an activation function

$$g(x) = \frac{1 - e^{-\alpha x}}{1 + e^{-\alpha x}}.$$

with large steepness. The connections between the neurons in the high-order layer and the output neuron are weighted. The output $v(n)$ of the output neuron is

$$v(n) = g\big(u(n)\big) = g\Bigg[\sum_{i=1}^{m} W(i) q_i^{(1)}(n) + \sum_{i=1}^{m}\sum_{j=1}^{m} W(ij) q_{ij}^{(2)}(n) + \cdots + \sum_{i_1=1}^{m}\sum_{i_2=1}^{m} \cdots \sum_{i_k=1}^{m} W(i_1 i_2 \cdots i_k) q_{i_1 i_2 \ldots i_k}^{(k)}(n)\Bigg], \tag{5.80}$$

where $u(n)$ is the input of the output neuron and $W(\cdot)$ is the corresponding connection weight between the high-order neurons and the output neuron. Note that in (5.80), we let the threshold of the output neuron be zero. If more parameters are required, an adjustable nonzero threshold can be added to the output neuron.

Before this equalizer based on a high-order network estimates the values of the transmitted signals in different channels, it is necessary to train the weights of the network by using the N known samples (training data) of the transmitted signals.

The connection weights can be obtained from the training data by solving the optimization problem

$$\min_{\boldsymbol{W}} \sum_{n=m}^{N+m-1} \left| v(n) - \hat{v}(n) \right|^2, \tag{5.81}$$

where $\boldsymbol{W}$ denotes the vector consisting of all the connection weights.

Clearly, the simplest way to solve (5.81) is with the gradient descent procedure. There are five steps:

(1) Initialize randomly all connection weights.

(2) Compute the output $v(n)$ by use of (5.77)–(5.80) and the available observation data $\boldsymbol{Y}(n)$ (for $n = m, m+1, \ldots, N+m-1$).

(3) Compute the error term $\delta(n)$ of the output neuron as

$$\delta(n) = \left(\hat{v}(n) - v(n)\right)\left[g\left(u(n)\right)\right]', \tag{5.82}$$

where

$$\left[g\left(u(n)\right)\right]' = \left.\frac{\partial g(u)}{\partial u}\right|_{u=u(n)} \tag{5.83}$$

and

$$u(n) = \sum_{i=1}^{m} W(i) q_i^{(1)}(n) + \sum_{i=1}^{m}\sum_{j=1}^{m} W(ij) q_{ij}^{(2)}(n) + \cdots + \sum_{i_1=1}^{m}\sum_{i_2=1}^{m}\cdots\sum_{i_k=1}^{m} W(i_1 i_2 \ldots i_k) q_{i_1 i_2 \ldots i_k}^{(k)}(n) \quad (n = m, m+1, \ldots, N+m-1). \tag{5.84}$$

(4) Update the connection weights by the correction term

$$\Delta W(i) = \gamma_1 \sum_{n=m}^{N+m-1} \delta(n) q_i^{(1)}(n) \tag{5.85}$$

for the weights related to the first-order neurons, by

$$\Delta W(ij) = \gamma_2 \sum_{n=m}^{N+m-1} \delta(n) q_{ij}^{(2)}(n) \tag{5.86}$$

for the weights related to the second-order neurons, and by

$$\Delta W(i_1 i_2 \ldots i_k) = \gamma_k \sum_{n=m}^{N+m-1} \delta(n) q_{i_1 i_2 \ldots i_k}^{(k)}(n) \tag{5.87}$$

for the weights related to the k'th-order neurons, where γ_1, γ_2, and γ_k are the learning-rate parameters. They can all take the same value.

(5) Compute the total error

$$e = \sum_{n=m}^{N+m-1} \left|v(n) - \hat{v}(n)\right|^2 \tag{5.88}$$

and iterate the computation by returning to Step (2) until this error is less than the specified one.

Similarly to (5.40) and (5.60), for the case in which each new sample set becomes recursively available during the above training, Equations (5.85)–(5.87) will become

$$\Delta W(i) = \gamma_1 \delta(n) q_i^{(1)}(n), \tag{5.89}$$

$$\Delta W(ij) = \gamma_2 \delta(n) q_{ij}^{(2)}(n), \tag{5.90}$$

and

$$\Delta W(i_1 i_2 \ldots i_k) = \gamma_k \delta(n) q_{i_1 i_2 \ldots i_k}^{(k)}(n). \tag{5.91}$$

Moreover, the error should be computed by using

$$e = \sum_{l=m}^{n} \left|\hat{v}(l) - v(l)\right|^2, \tag{5.92}$$

and the variable $v(l)$ in (5.92) should be computed by use of the weights at the iteration n instead of those at the iteration l.

In this recursive algorithm, if all N samples have become available but the error e is still greater than the specified one, then we can take the weights at time $N+m-1$ as the new initialized values and use the above batch-processing algorithm without any modifications to continue to train the weights. However, once the unknown signal has been transmitted the training must be stopped, even though the error e may not be less than the specified one.

In addition, as pointed out in Section 5.3, if we take the known samples of the transmitted signal as the desired input $\hat{u}(n)$ of the output neuron instead of the desired output $\hat{v}(n)$, then (5.81) will become a linear least-squares problem. As a result, the connection weights can be given in an explicit form and can also be obtained by a simple algorithm similar to the LMS algorithm.

We see that the number of the connection weights increases exponentially as k increases. In practical applications, restricting k to equal 3 or 5 is often adequate, and this is also supported by the simulation results given by Chen et al. [18]. The selection of the neuron number m of the input layer is to a large extent influenced by the noise power. Even under high-noise conditions, one prefers to employ a low m because the increase of the number m will also increase the total noise power, which tends to diminish any advantages gained by increasing the number m. On the contrary, it could be argued that increasing m may only lead to an increase in complexity, training time, and misadjustment, leading ultimately to a decrease in efficiency.

Usually, increasing the computational complexity and dimensionality is the price one pays for employing the nonlinear equalizer based on high-order neural networks. However, the structure and operation of the equalizer based on high-order neural networks are simpler than those of the nonlinear MLP equalizers.

We should also mention that using alternative nonlinear mapping neural networks such as functional link neuron networks [28] is another effective way to realize nonlinear equalization [29].

5.5 Blind Equalization Using Neural Networks

The equalizers discussed in Sections 5.2–5.4 require the transmission of a training sequence. In other words, some samples of the transmitted signals are known at the receiver, and these samples provide the desired response for training the connection weights and the other parameters related to equalizers. Once the training process is completed, the equalizer is switched into the model for regular data transmission over the unknown channel. There are, however, practical situations where it is not feasible to use a training sequence. For example, in a digital radio system, the received signal suffers from the multipath problem, which arises from the fact that the transmitted signal reaches the receiver via a multiplicity of paths. The presence of multipaths can produce severe channel fading and therefore system outage, characterized by a significant reduction in the received signal power. If the outage occurs during the training process, the equalizer in the receiver is deprived of its desired response and the equalization process is thereby seriously impaired. In such a situation, we are compelled to use some form of blind equalization, which does not require the use of a training sequence for the adjustment of the parameters such as the connection weights.

Many blind equalization algorithms have been proposed and extensively analyzed [4]. In these available blind equalization algorithms, the algorithms based on the higher-order spectra are most appealing, mainly because these blind algorithms can identify nonminimum-phase communication channels and would not be affected by white additive Gaussian noise [30, 31]. As has been mentioned in Section 5.2, the transmitted sequence $x(n)$, which takes binary values $\{-1, 1\}$ with equal probability that are assumed to be independent of one another, possesses a non-Gaussian probability density function. The transmitted signal is passed through a dispersive channel, which can be described as an FIR filter with the transfer function described by Equation (5.22). The channel output $\hat{y}(n)$ is corrupted by a zero-mean, additive Gaussian noise $\epsilon(n)$, which is independent of the transmitted sequence $x(n)$. As a result, the observation data (the input of the blind equalizer) $y(n)$ is equal to $\hat{y}(n)$ plus $\epsilon(n)$, that is,

$$y(n) = \hat{y}(n) + \epsilon(n) = \sum_{i=0}^{L} h_i x(n-i) + \epsilon(n). \tag{5.93}$$

According to the definitions of the higher-order cumulants of a random sequence, we can obtain the fourth-cumulants $C_{4y}(m,k,l)$ of $y(n)$ as

$$\begin{aligned} C_{4y}(m,k,l) &= E\left[y(n+m)y(n+k)y(n+l)y(n)\right] - C_{2y}(m)C_{2y}(k-l) \\ &\quad - C_{2y}(k)C_{2y}(l-m) - C_{2y}(l)C_{2y}(m-k) \\ &= C_{4\hat{y}}(m,k,l) + C_{4\epsilon}(m,k,l), \end{aligned} \tag{5.94}$$

where $C_{2y}(\cdot)$ is the related second-order cumulant and $C_{4\hat{y}}(m,k,l)$ and $C_{4\epsilon}(m,k,l)$ are the fourth-order cumulants of $\hat{y}(n)$ and $\epsilon(n)$, respectively. Since $\epsilon(n)$ is Gaussian, its fourth-order cumulants are zero, which means

$$C_{4y}(m,k,l) = C_{4\hat{y}}(m,k,l). \tag{5.95}$$

With this in mind, it is easy to show that [31]

$$C_{4y}(m,k,l) = \gamma_x \sum_{t=0}^{L-\max(m,k,l)} h_t h_{t+m} h_{t+k} h_{t+l}, \tag{5.96}$$

where $\gamma_x = E[x^4(n)] - 3(E[x^2(n)])^2$. In addition, in (5.96), we assumed $E[x(n)] = E[x^3(n)] = 0$. If the fourth-order cumulants are known, we can obtain the coefficients h_j (for $j = 0, 1, 2, \ldots, L$) of the channel by solving (5.96) and then use the inverse filter of the channel to estimate the values of the transmitted signal $x(n)$. In practical applications, although the fourth-order cumulants are not known exactly they can be estimated from the available observation data. We can summarize the processing mentioned above as the following procedures:

(1) Estimate the fourth-order cumulants using the available observation data.
(2) Determine the coefficients of the channel by solving Equation (5.96).
(3) Construct an inverse filter of the channel by using the determined coefficients.
(4) Estimate the values of the transmitted signal by taking the observation data as the input of the inverse filter. That is, use the output of the inverse filter to provide the desired estimates of the transmitted signal.

Since we require neither the training sequence nor the transfer function of the channel, the above processing is a blind equalization. There are, however, existing problems involving the determination of the coefficients and the accuracy of the constructed inverse filter of the channel. Traditionally, we use a linear FIR filter (that is, an LTE) to approximate the inverse of the channel. However, the approximation performance of the linear FIR filter will be greatly degraded under the conditions of short observation data and low signal-to-noise ratio [31]. As a result, blind equalization based on nonlinear structures is desirable. In this section, we will present a nonlinear blind equalizer based on MLP networks that will be shown to offer advantages over the traditional linear equalizers. In addition, using the same

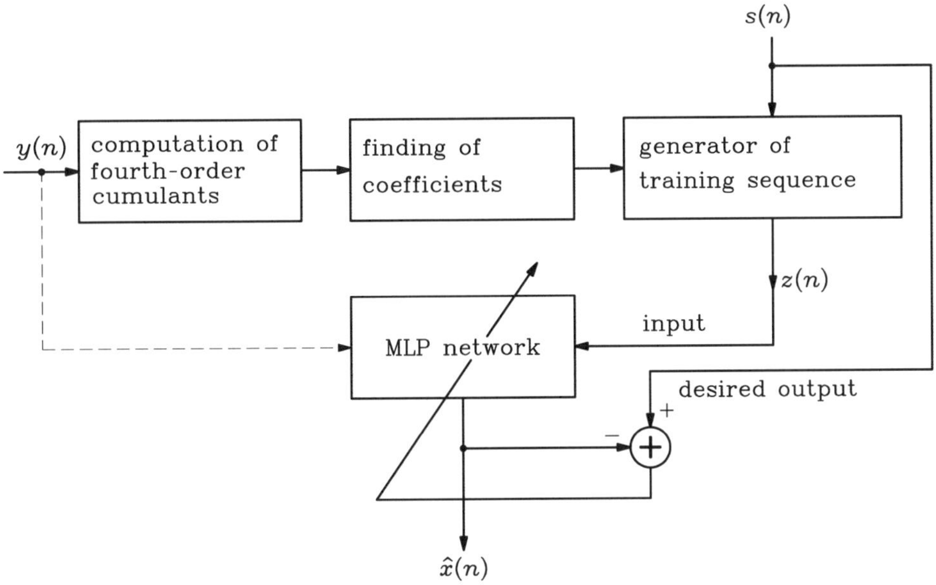

Figure 5.11. Schematic of the nonlinear blind equalizer based on an MLP network.

principles, we can easily develop blind equalizers based on other nonlinear mapping networks such as RBF networks and high-order neural networks.

A schematic structure of the nonlinear blind equalizer based on MLP networks is shown in Figure 5.11. This scheme was proposed by Mo and Shafai in 1994 [31]. The procedures concerning this nonlinear blind equalizer include mainly three parts, to be presented next.

Determination of the Coefficients of the Channel, or say, the Reconstruction of the Channel

There are many available methods to determine the coefficients from the estimated fourth-order cumulants. In effect, from (5.96), we know that the coefficients satisfy the equations [31]

$$\boldsymbol{C}\boldsymbol{H} = \boldsymbol{c}, \tag{5.97}$$

where $\boldsymbol{C}$ and $\boldsymbol{c}$ are the matrix and vector consisting of the estimated fourth-order cumulants and $\boldsymbol{H} = [h_1, h_2, \ldots, h_L]^T$ is the unknown coefficient vector. The solution of (5.97) can be given in an explicit form as

$$\boldsymbol{H} = \boldsymbol{C}^+\boldsymbol{c} = \lim_{\alpha\to 0}(\boldsymbol{C}^T\boldsymbol{C} + \alpha\boldsymbol{I})^{-1}\boldsymbol{C}^T\boldsymbol{c}. \tag{5.98}$$

Once the vector $\boldsymbol{H}$ has been determined, h_0 can immediately be obtained by solving (5.96). Note that before determining h_0, we should find γ_x, which is equal to the estimated value of $C_{4y}(0, 0, 0)$.

It is easily seen that (5.98) can also be performed in real time by the analog neural network in Figure 2.2, provided that we let the known matrix $\boldsymbol{C}$ and vector $\boldsymbol{c}$ be taken as the connection strength matrix $\boldsymbol{T}$ and the bias current vector $\boldsymbol{B}$ of the neural network of Figure 2.2, respectively, with the other parameters of the network being the same as those for (2.3).

For the case that each new sample set becomes available recursively, we can use the available recursive algorithms such as the RLS algorithm [32] to obtain the coefficients. If a faster solution is desired, the analog neural network approach proposed in Section 2.2 may be employed [33]. The difference is that in solving (5.97) the connection strengths and bias currents of the network must be computed by the estimated fourth-order cumulants instead of the observation data themselves. Mo and Shafai [31] have also proposed another alternative way to speed up the convergence of the recursive algorithms. For more detail concerning the procedures of this alternative method, one can refer to their paper [31].

Training of the MLP Network

The MLP network employed in this nonlinear equalizer is exactly the same as that in Figure 5.7. The main function of the MLP network is to approximate the inverse of the channel. The nonlinear mapping performed by the MLP network depends mainly on the connection weights. To determine the connection weights, we require training sequences. However, as pointed out in the beginning of this section, the desired training sequences are not available in the blind equalization. This problem can be overcome in the following way. Based on the available coefficients, we can generate some training sequences by using

$$z(n) = \sum_{i=0}^{L} h_i s(n-i), \tag{5.99}$$

where $s(n)$ is a random binary-valued sequence and acts as the input of the generator. Usually, $s(n)$ should have the same probability density function as that of the actual transmitted sequences. This PDF depends on the communication system and coding algorithms. Once some samples of $z(n)$ and $s(n)$ are available, we can train the network in the same way as that for the equalizer of Figure 5.7, provided that we take $z(n)$ as the observation data (the input of the network) and $s(n)$ as the desired output $\hat{v}(n)$.

Estimation (Reconstruction) of the Transmitted Signals

After the training of the network is completed, the received signal (observation data) $y(n)$ is fed into the input layer of the MLP network to get an output sequence $v(n)$ (computed by (5.25)), which is the estimate of the actual transmitted signal $x(n)$.

For a time-varying or nonstationary channel, the training should continue throughout the whole working operation so that the MLP network can track the change of the channel. In addition, the estimated transmitted signal may serve as the input ($s(n)$) of the training sequence generator. This is to some extent similar to the function of the DFE as mentioned in Section 5.3.

Mo and Shafai [31] have reported extensive simulation results and have shown that this nonlinear blind equalizer works well. The main disadvantages of this scheme are that the structure is more complex and that more computation is required than in conventional linear blind equalization systems.

Finally, it is interesting to note that the above principles also apply to the blind reconstruction of continuous-valued signals as long as the continuous-valued signals are non-Gaussian and the additive noise is Gaussian (as assumed in the beginning of this section). In the case of continuous values, one difference is that the activation function of the output neuron may no longer be

$$\frac{1 - e^{-\alpha x}}{1 + e^{-\alpha x}}$$

so that the output neuron can provide values greater than unity (of course, one can first perform normalization processing and then use

$$\frac{1 - e^{-\alpha x}}{1 + e^{-\alpha x}}$$

as the activation function). The other difference is that the input sequence $s(n)$ of the training sequence generator should take continuous values.

Bibliography

[1] Gull, S. F. and Daniell, B. J., "Image Reconstruction from Incomplete and Noisy Data," *Nature*, Vol. 272, 1978, p. 686.

[2] Proakis, J. G., *Digital Communications*, McGraw-Hill, New York, 1989.

[3] Claerbout, J., *Fundamentals of Geophysical Data Processing*, Blackwell Scientific, Duesseldorf, 1985.

[4] Haykin, S., *Blind Deconvolution*, Prentice Hall, Englewood Cliffs, NJ, 1994.

[5] Haykin, S., "Blind Equalization Formulated as a Self-Organized Learning Process," *Proc. Twenty-Sixth Asilomar Conf. on Signals, Systems and Computers*, Pacific Grove, CA, 1992, pp. 346–50.

[6] Zhou, Y. T., Chellappa, R., and Jenkiens, B. K., "Image Restoration Using a Neural Network, " *IEEE Trans. on Acoustics, Speech and Signal Processing*, Vol. 36, 1988, pp. 1141–52.

[7] Luo, F. L. and Bao, Z., "Neural Network Approach to Adaptive FIR Filtering and Deconvolution Problems," *Proc. IEEE Int. Conf. on Industrial Electronics (IECON'91)*, Kobe, Japan, 1991, pp. 1449–53.

[8] Ingman, D. and Merlis, Y., "Maximum Entropy Signal Reconstruction with Neural Networks," *IEEE Trans. on Neural Networks*, Vol. 3, No. 2, 1992, pp. 195–201.

[9] Frieden, B. R. and Zoltani, C. R., "Maximum Bounded Entropy: Applications to Tomographic Reconstruction," *Appl. Optics*, Vol. 24, No. 2, 1985, pp. 201–7.

[10] Skilling, J., "The Axiom of Maximum Entropy," in *Maximum Entropy and Bayesian Methods in Science and Engineering*, Vol. 1, Kluwer, Dordrecht, The Netherlands, 1988, pp. 173–88.

[11] Papoulis, A., "Maximum Entropy and Spectral Estimation: A Review," *IEEE Trans. on Acoustics, Speech and Signal Processing*, Vol. 29, 1981, pp. 1176–86.

[12] Luenberger, D. G., *Linear and Nonlinear Programming*, Addision-Wesley, Reading, MA, 1978, pp. 366–69.

[13] Cichocki, A., Unbehauen, R., Lendl, M., and Weinzierl, K., "Neural Networks for Linear Inverse Problems with Incomplete Data Especially in Applications to Signal and Image Reconstruction," *Neurocomputing*, Vol. 8, 1995, pp. 7–41.

[14] Marrian, C. R. K. and Peckerar, M. C., "Electronic Neural Net Algorithm for Maximum Entropy Solutions of Ill-Posed Problems," *IEEE Trans. on Circuits and Systems*, Vol. 36, No. 2, 1989, pp. 288–95.

[15] Gibson, G. J., Siu, S., and Cowan, C. F. N., "The Applications of Nonlinear Structures to the Reconstruction of Binary Signals," *IEEE Trans. on Signal Processing*, Vol. 39, No. 8, 1991, pp. 1877–84.

[16] Chen, S., Gibson, G. J., Cowan, C. F. N., and Grant, P. M., "Adaptive Equalization of Finite Nonlinear Channels Using Multilayer Perceptrons," *Signal Processing*, Vol. 20, 1990, pp. 107–19.

[17] Chen, S., Gibson, G. J., Cowan, C. F. N., and Grant, P. M., "Reconstruction of Binary Signals Using an Adaptive Radial-Basis-Function Equalizer," *Signal Processing*, Vol. 22, 1991, pp. 77–93.

[18] Chen, S., Gibson, G. J., and Cowan, C. F. N., "Adaptive Channel Equalization Using a Polynomial Perceptron Structure," *IEE Proc.*, Pt. I, Vol. 137, 1990, pp. 257–64.

[19] Haykin, S., *Neural Networks, A Comprehensive Foundation*, IEEE Press, New York, 1994.

[20] Chen, S., Mulgrew, B., and Grant, P. M., "A Clustering Techniques for Digital Communications Channel Equalization Using Radial Basis Function Networks," *IEEE Trans. on Neural Networks*, Vol. 4, No. 4, 1993, pp. 570–79.

[21] Moody, J. E. and Darken, C. J., "Fast Learning in Networks of Locally Tuned Processing Units," *Neural Computation*, Vol. 1, 1989, pp. 281–94.

[22] Chen, S., Cowan, C. F. N., and Grant, P. M., "Orthogonal Least Squares Learning Algorithm for Radial Basis Function Networks," *IEEE Trans. on Neural Networks*, Vol. 2, 1991, pp. 302–9.

[23] Chen, S. and Mulgrew, B., "Overcoming Co-Channel Interferences Using an Adaptive Radial Basis Function Equalizer," *Signal Processing*, Vol. 28, 1992, pp. 91–107.

[24] Chen, S., Mulgrew, B., and McLaughlin, S., "Adaptive Bayesian Equalizer with Decision Feedback," *IEEE Trans. on Signal Processing*, Vol. 41, No. 9, 1993, pp. 2918–27.

[25] Siu, S., Gibson, G. J., and Cowan, C. F. N., "Decision Feedback Equalization Using Neural Network Structure and Performance Comparison with Standard Architecture," *IEE Proc.*, Pt. I, Vol. 137, No. 4, 1990, pp. 221–25.

[26] Chen, S., McLaughlin, S., and Mulgrew, B., "Complex-Valued Radial Basis Function Network, Part I: Network Architecture and Learning Algorithms," *Signal Processing*, Vol. 35, 1994, pp. 19–31.

[27] Chen, S., McLaughlin, S., and Mulgrew, B., "Complex-Valued Radial Basis Function Network, Part II: Application to Digital Communication Channel Equalization," *Signal Processing*, Vol. 36, 1994, pp. 175–88.

[28] Arcens, S., Sueiro, J. C., and Vidal, A. R. F., "Pao Networks for Data Transmission Equalization," *Proc. Int. Joint Conf. on Neural Networks*, Baltimore, Vol. 2, 1992, pp. 963–68.

[29] Patra, J. C. and Pal, N., "A Functional Link Artificial Neural Network for Adaptive Equalization," *Signal Processing*, Vol. 43, 1995, pp. 181–95.

[30] Li, Y. and Ding, Z., "Convergence Analysis of Finite Length Blind Adaptive Equalizer," *IEEE Trans. on Signal Processing*, Vol. 43, No. 9, 1995, pp. 2120–29.

[31] Mo, S. and Shafai, B., "Blind Equalization Using Higher Order Cumulants and Neural Networks," *IEEE Trans. on Signal Processing*, Vol. 42, No. 11, 1994, pp. 3211–17.

[32] Gioffi, J. M. and Kailath, T., "Fast RLS Transversal Filters for Adaptive Filtering," *IEEE Trans. on Acoustics, Speech and Signal Processing*, Vol. 32, No. 2, 1984, pp. 304–37.

[33] Luo, F. L. and Li, Y. D., "Neural Networks for the Exact Adaptive RLS Algorithm," *Appl. Math. Computation*, Vol. 60, No. 2, 1994, pp. 103–12.

6

Neural Networks for Adaptive Extraction of Principal and Minor Components

The eigenvectors corresponding to the largest and smallest eigenvalues of the autocorrelation matrix of input signals are referred to as the principal components and minor components, respectively [1, 2, 3, 4]. Consequently, the subspaces spanned by these eigenvectors are called the principal subspace and minor subspace, respectively. In general, the principal components (subspace) contain the desired information and main features of the input signals, whereas the minor components (subspace) represent the statistics of the additive noise corrupting the input signals. Adaptively extracting these parameters is a primary requirement in many fields of signal processing, including data compression and coding [5, 6, 7], feature extraction and pattern classification [8, 9, 10, 11, 12, 13, 14], vector quantization [7, 15], spectral estimation [16, 17], total-least-squares processing [18, 19], eigen-based bearing estimation [20], digital beamforming [21], moving-target indication, and clutter cancellation [22, 23].

The main purpose of this chapter is to present neural networks and adaptive unsupervised learning algorithms for extracting the principal components and minor components and to analyze their performances.

6.1 Adaptive Extraction of the First Principal Component

Let us consider a bounded continuous-valued stationary ergodic data vector $\boldsymbol{X}(t) = [x_1(t), x_2(t), \ldots, x_N(t)]^T$ with finite second-order moments, whose autocorrelation matrix $\boldsymbol{R}$ is defined as

$$\boldsymbol{R} = E\left[\boldsymbol{X}(t)\boldsymbol{X}^T(t)\right]. \tag{6.1}$$

If $\lambda_1 \geq \lambda_2 \geq \cdots \geq \lambda_N \geq 0$ denote the eigenvalues and $\boldsymbol{S}_1, \boldsymbol{S}_2, \ldots, \boldsymbol{S}_N$ denote the corresponding orthonormal eigenvectors of the matrix $\boldsymbol{R}$, then, according to the definitions mentioned above, we have that

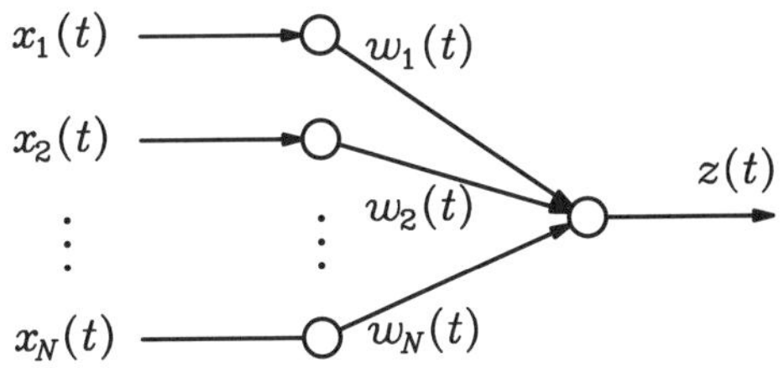

Figure 6.1. The neural network for extracting the first principal component.

(1) $\boldsymbol{S}_1$ is the first principal component,
(2) $\boldsymbol{S}_N$ is the first minor component,
(3) $\boldsymbol{S}_1, \boldsymbol{S}_2, \ldots, \boldsymbol{S}_M$ ($M < N$, and in general, $\lambda_M \gg \lambda_{M+1}$) are M principal components,
(4) $\boldsymbol{S}_{N-L+1}, \boldsymbol{S}_{N-L+2}, \ldots, \boldsymbol{S}_N$ ($L < N$, and in general, $\lambda_{N-L} \gg \lambda_{N-L+1}$) are L minor components,
(5) $\boldsymbol{Q}_M = span\{\boldsymbol{S}_1, \boldsymbol{S}_2, \ldots, \boldsymbol{S}_M\}$ ($M < N$, and in general, $\lambda_M \gg \lambda_{M+1}$) is an M-dimensional principal subspace, and
(6) $\boldsymbol{Q}_L = span\{\boldsymbol{S}_{N-L+1}, \boldsymbol{S}_{N-L+2}, \ldots, \boldsymbol{S}_N\}$ ($L < N$, and in general, $\lambda_{N-L} \gg \lambda_{N-L+1}$) is an L-dimensional minor subspace.

Our task is to develop neural networks and related algorithms that can adaptively extract these different features for different application fields of signal processing.

The neural network for extracting the first principal component is shown in Figure 6.1 and consists of two layers. The first layer is the input layer and has N neurons that simply feed the input vector $\boldsymbol{X}(t)$ to the second layer without any modification. The second layer is the output layer and has only one neuron whose input–output relationship is a linear function so that it performs simple summations. The connections between the input layer and output layer are weighted. The connection weight vector is denoted as $\boldsymbol{W}(t) = [w_1(t), w_2(t), \ldots, w_N(t)]^T$, where $w_i(t)$ is the connection strength between the i'th input neuron and the output neuron. With these in mind, the output of the network is

$$z(t) = \sum_{i=1}^{N} w_i(t)x_i(t) = \boldsymbol{W}^T(t)\boldsymbol{X}(t). \tag{6.2}$$

In 1982, Oja [24] proposed a very simple unsupervised learning algorithm to update the weight vector $\boldsymbol{W}(t)$ of the network as

$$w_i(t+1) = w_i(t) + \gamma(t)\big(z(t)x_i(t) - z^2(t)w_i(t)\big) \qquad (i = 1, 2, \ldots, N). \tag{6.3}$$

We can write this algorithm in the vector form

$$\boldsymbol{W}(t+1) = \boldsymbol{W}(t) + \gamma(t)z(t)\big[\boldsymbol{X}(t) - z(t)\boldsymbol{W}(t)\big]. \tag{6.4}$$

According to the stochastic approximation theory in References [25] and [26] and further explanations in References [6] and [27], if certain conditions are satisfied, then the asymptotic limits of the above discrete learning algorithm can be solved by applying the corresponding continuous-time differential equations

$$\begin{aligned}\frac{d\boldsymbol{W}(t)}{dt} &= z(t)\boldsymbol{X}(t) - z^2(t)\boldsymbol{W}(t)\\ &= \boldsymbol{X}(t)\boldsymbol{X}^T(t)\boldsymbol{W}(t) - \boldsymbol{W}^T(t)\boldsymbol{X}(t)\boldsymbol{X}^T(t)\boldsymbol{W}(t)\boldsymbol{W}(t)\end{aligned} \tag{6.5}$$

and the averaging differential equations

$$\frac{d\boldsymbol{W}(t)}{dt} = \boldsymbol{R}\boldsymbol{W}(t) - \boldsymbol{W}^T(t)\boldsymbol{R}\boldsymbol{W}(t)\boldsymbol{W}(t). \tag{6.6}$$

These conditions are:

(1) $\gamma(t)$ is a sequence of positive real numbers such that $\gamma(t) \to 0$, $\sum_t \gamma^p(t) < \infty$ for some p, and $\sum_t \gamma(t) = \infty$;

(2) $\boldsymbol{X}(t)$ is zero mean, stationary, and bounded with probability 1; and

(3) the right-hand side term of (6.5) is continuously differentiable in $\boldsymbol{W}(t)$ and $\boldsymbol{X}(t)$ and its derivative is bounded in time.

For details, one can read References [25], [26], and [27].

The solutions $\boldsymbol{W}(t)$ of (6.4) tend to the asymptotically stable points of (6.6) with probability one. It should be noted that this equivalence property between the discrete learning algorithms and the corresponding continuous-time differential equations will be used in all the learning algorithms dealt with in this chapter unless we mention an exception.

For the asymptotic stability of (6.6), we have the following theorem:

Theorem 6.1 *If the initial weight vector* $\boldsymbol{W}(0)$ *satisfies* $\boldsymbol{W}^T(0)\boldsymbol{S}_1 \neq 0$, *then*

$$\lim_{t\to\infty} \boldsymbol{W}(t) = \pm\boldsymbol{S}_1 \tag{6.7}$$

holds.

Proof: Setting

$$\boldsymbol{W}(t) = \sum_{i=1}^{N} y_i(t)\boldsymbol{S}_i \tag{6.8}$$

and substituting (6.8) into (6.6) give

$$\begin{aligned}\sum_{i=1}^{N} \frac{dy_i(t)}{dt}\boldsymbol{S}_i &= \sum_{i=1}^{N} \boldsymbol{R}y_i(t)\boldsymbol{S}_i - \boldsymbol{W}^T(t)\boldsymbol{R}\boldsymbol{W}(t)\sum_{i=1}^{N} y_i(t)\boldsymbol{S}_i\\ &= \sum_{i=1}^{N} \lambda_i y_i(t)\boldsymbol{S}_i - \boldsymbol{W}^T(t)\boldsymbol{R}\boldsymbol{W}(t)\sum_{i=1}^{N} y_i(t)\boldsymbol{S}_i.\end{aligned} \tag{6.9}$$

Moreover,

$$\frac{dy_i(t)}{dt} = \lambda_i y_i(t) - \boldsymbol{W}^T(t)\boldsymbol{R}\boldsymbol{W}(t)y_i(t) \qquad (i = 1, 2, \ldots, N). \tag{6.10}$$

According to the assumption of the initial value $\boldsymbol{W}(0)$, we may define [6, 27]

$$c_i(t) = \frac{y_i(t)}{y_1(t)} \qquad (i = 2, 3, \ldots, N) \tag{6.11}$$

and

$$\frac{dc_i(t)}{dt} = \frac{y_1(t)\frac{dy_i(t)}{dt} - y_i(t)\frac{dy_1(t)}{dt}}{(y_1(t))^2}. \tag{6.12}$$

Combining (6.10) and (6.12) gives

$$\begin{aligned}\frac{dc_i(t)}{dt} &= \frac{1}{(y_1(t))^2}\Big[\big(\lambda_i y_i(t) - \boldsymbol{W}^T(t)\boldsymbol{R}\boldsymbol{W}(t)\big)y_i(t)\Big]y_1(t) \\ &\quad - \Big[\big(\lambda_1 y_1(t) - \boldsymbol{W}^T(t)\boldsymbol{R}\boldsymbol{W}(t)y_1(t)\big)y_i(t)\Big] \\ &= (\lambda_i - \lambda_1)c_i(t) \qquad (i = 2, 3, \ldots, N)\end{aligned} \tag{6.13}$$

and

$$c_i(t) = K_i \exp\big((\lambda_i - \lambda_1)t\big) \qquad (i = 2, 3, \ldots, N), \tag{6.14}$$

where K_i is a constant depending on the initial values and the eigenvalues of the matrix $\boldsymbol{R}$.

Using (6.11) yields

$$y_i(t) = K_{i1} y_1(t)\exp\big((\lambda_i - \lambda_1)t\big) \qquad (i = 2, 3, \ldots, N). \tag{6.15}$$

If the largest eigenvalue λ_1 is single but not multiple, that is, $\lambda_i < \lambda_1$ (for $i = 2, 3, \ldots, N$), then we have from (6.15),

$$\lim_{t\to\infty} y_i(t) = 0 \qquad (i = 2, 3, \ldots, N). \tag{6.16}$$

Together with (6.8), (6.16) shows that

$$\lim_{t\to\infty} \boldsymbol{W}(t) = \lim_{t\to\infty} y_1(t)\boldsymbol{S}_1. \tag{6.17}$$

Multiplying (6.6) by $\boldsymbol{W}^T(t)$ on the left yields

$$\begin{aligned}\frac{d\|\boldsymbol{W}(t)\|^2}{dt} &= 2\big[\boldsymbol{W}^T(t)\boldsymbol{R}\boldsymbol{W}(t) - \boldsymbol{W}^T(t)\boldsymbol{R}\boldsymbol{W}(t)\boldsymbol{W}^T(t)\boldsymbol{W}(t)\big] \\ &= 2\boldsymbol{W}^T(t)\boldsymbol{R}\boldsymbol{W}(t)\big(1 - \|\boldsymbol{W}(t)\|^2\big).\end{aligned} \tag{6.18}$$

Since $\boldsymbol{W}^T(t)\boldsymbol{R}\boldsymbol{W}(t)$ is nonnegative, we know that

(1) if $\|\boldsymbol{W}(0)\|^2 = 0$, then $\|\boldsymbol{W}(t)\|^2 = 0$ for $t > 0$;

(2) if $\|\boldsymbol{W}(0)\|^2 = 1$, then $\|\boldsymbol{W}(t)\|^2 = 1$ for $t > 0$;

(3) if $\|\boldsymbol{W}(0)\|^2 > 1$, then $\|\boldsymbol{W}(t)\|^2$ will decrease during the time evolution and will approach 1, that is, $\lim_{t\to\infty} \|\boldsymbol{W}(t)\|^2 = 1$;

(4) if $\|\boldsymbol{W}(0)\|^2 < 1$, then $\|\boldsymbol{W}(t)\|^2$ will increase during the time evolution and will approach 1, that is, $\lim_{t\to\infty} \|\boldsymbol{W}(t)\|^2 = 1$;

which means that there are only two possible stationary points in (6.18): 0 and 1. With this fact and the assumption about the initial value of the weight vector ($y_1(0) \neq 0$), we have

$$\lim_{t\to\infty} \|\boldsymbol{W}(t)\|^2 = 1. \tag{6.19}$$

Substituting (6.17) into (6.19), we obtain

$$\lim_{t\to\infty} y_1(t) = \pm 1 \tag{6.20}$$

and

$$\lim_{t\to\infty} \boldsymbol{W}(t) = \pm \boldsymbol{S}_1, \tag{6.21}$$

which is Equation (6.7).

In the case that the largest eigenvalue λ_1 is multiple but not single, that is, $\lambda_1 = \lambda_2 = \cdots = \lambda_K = \lambda$, then (6.15) still holds but (6.17) becomes

$$\lim_{t\to\infty} \boldsymbol{W}(t) = \sum_{i=1}^{K} \lim_{t\to\infty} y_i(t)\boldsymbol{S}_i. \tag{6.22}$$

Multiplying the above equation by $\boldsymbol{R}$ on the left yields

$$\begin{aligned}\lim_{t\to\infty} \boldsymbol{R}\boldsymbol{W}(t) &= \sum_{i=1}^{K} \lim_{t\to\infty} y_i(t)\boldsymbol{R}\boldsymbol{S}_i = \sum_{i=1}^{K} \lim_{t\to\infty} \lambda_i y_i(t)\boldsymbol{S}_i \\ &= \lambda \sum_{i=1}^{K} \lim_{t\to\infty} y_i(t)\boldsymbol{S}_i = \lambda \lim_{t\to\infty} \boldsymbol{W}(t),\end{aligned} \tag{6.23}$$

which means that $\lim_{t\to\infty} \boldsymbol{W}(t)$ is still the eigenvector corresponding to the multiple eigenvalue λ of the matrix $\boldsymbol{R}$. This fact and (6.21) both guarantee that (6.7) holds, and this concludes the proof. QED

Theorem 6.1 shows that the network in Figure 6.1 and its learning algorithm (6.4) can adaptively extract the desired first principal component $\boldsymbol{S}_1$.

Concerning the learning algorithm (6.4) and the corresponding differential equations (6.6), we make the following comments:

Comment 1 Compared with the well-known Hebbian learning rule, a constraint term $-\gamma(t)z^2(t)w_i(t)$ is added in Oja's learning algorithm (6.4). This term is the key to guaranteeing the convergence of the algorithm. As a result, the algorithm (6.4) is usually referred to as the constrained Hebbian learning rule.

Comment 2 The initial weight vector $\boldsymbol{W}(0)$ of the network can be selected almost arbitrarily because we only have the assumption that $\boldsymbol{W}(0)$ is not orthogonal to the first principal component $\boldsymbol{S}_1$ (i.e., $y_1(0) = \boldsymbol{W}^T(0)\boldsymbol{S}_1 \neq 0$). From (6.10)–(6.17), we know that the weight vector of the network will converge to the second principal component $\boldsymbol{S}_2$ if $y_1(0) = \boldsymbol{W}^T(0)\boldsymbol{S}_1 = 0$ and $y_2(0) = \boldsymbol{W}^T(0)\boldsymbol{S}_2 \neq 0$. We can also conclude that the weight vector $\boldsymbol{W}(t)$ of the network will converge to the eigenvector $\boldsymbol{S}_{i+1}$ corresponding to the eigenvalue λ_{i+1} if $y_1(0) = y_2(0) = \cdots = y_i(0) = 0$ and $y_{i+1}(0) \neq 0$.

Comment 3 This learning algorithm can easily be generalized to the case in which the input signal vector $\boldsymbol{X}(t)$ takes complex values. In effect, the eigendecomposition problem $\boldsymbol{R}\boldsymbol{S}_1 = \lambda_1\boldsymbol{S}_1$ of the autocorrelation matrix $\boldsymbol{R}$ ($\boldsymbol{R} = E[\boldsymbol{X}(t)\boldsymbol{X}^H(t)]$, where "$H$" denotes the conjugate transpose) with complex values can be changed to

$$[\boldsymbol{R}_r + j\boldsymbol{R}_i][\boldsymbol{S}_{1r} + j\boldsymbol{S}_{1i}] = \lambda_1[\boldsymbol{S}_{1r} + j\boldsymbol{S}_{1i}]$$
$$= \boldsymbol{R}_r\boldsymbol{S}_{1r} - \boldsymbol{R}_i\boldsymbol{S}_{1i} + j[\boldsymbol{R}_r\boldsymbol{S}_{1i} + \boldsymbol{R}_i\boldsymbol{S}_{1r}] = \lambda_1\boldsymbol{S}_{1r} + j\lambda_1\boldsymbol{S}_{1i}.$$

Moreover,

$$\boldsymbol{R}_c\boldsymbol{S}_{1c} = \lambda_1\boldsymbol{S}_{1c}, \tag{6.24}$$

where

$$\boldsymbol{R}_c = \begin{pmatrix} \boldsymbol{R}_r & -\boldsymbol{R}_i \\ \boldsymbol{R}_i & \boldsymbol{R}_r \end{pmatrix} \tag{6.25}$$

and

$$\boldsymbol{S}_{1c} = \begin{pmatrix} \boldsymbol{S}_{1r} \\ \boldsymbol{S}_{1i} \end{pmatrix}, \tag{6.26}$$

which shows that the first principal component $\boldsymbol{S}_1$ of the input signal $\boldsymbol{X}(t)$ can be obtained by computing the vector $\boldsymbol{S}_{1c}$. For the purpose of constructing a learning algorithm to extract adaptively the eigenvector $\boldsymbol{S}_{1c}$ from the input $\boldsymbol{X}(t)$, we write the input–output relationship ($z(t) = \boldsymbol{X}^H(t)\boldsymbol{W}(t)$) of the neural network in real-valued form.

$$z_r(t) = \boldsymbol{W}_r^T(t)\boldsymbol{X}_r(t) + \boldsymbol{W}_i^T(t)\boldsymbol{X}_i(t) \tag{6.27}$$

and

$$z_i(t) = -\boldsymbol{W}_r^T(t)\boldsymbol{X}_i(t) + \boldsymbol{W}_i^T(t)\boldsymbol{X}_r(t), \tag{6.28}$$

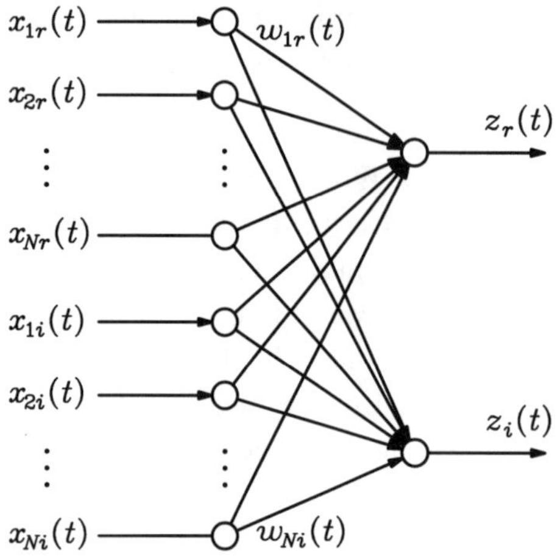

Figure 6.2. The neural network for extracting the first principal component in the complex-valued case.

where $\boldsymbol{X}(t) = \boldsymbol{X}_r(t) + j\boldsymbol{X}_i(t)$, $\boldsymbol{W}(t) = \boldsymbol{W}_r(t) + j\boldsymbol{W}_i(t)$, and $z(t) = z_r(t) + jz_i(t)$. Moreover, Equations (6.27) and (6.28) can be written as

$$z_c(t) = \boldsymbol{X}_c^T(t)\boldsymbol{W}_c(t), \tag{6.29}$$

where

$$z_c(t) = \begin{pmatrix} z_r(t) \\ z_i(t) \end{pmatrix}, \quad \boldsymbol{X}_c(t) = \begin{pmatrix} \boldsymbol{X}_r(t) & -\boldsymbol{X}_i(t) \\ \boldsymbol{X}_i(t) & \boldsymbol{X}_r(t) \end{pmatrix}, \quad \boldsymbol{W}_c(t) = \begin{pmatrix} \boldsymbol{W}_r(t) \\ \boldsymbol{W}_i(t) \end{pmatrix}.$$

It is obvious that two units are needed for each complex value output, as shown in Figure 6.2.

For $\boldsymbol{W}_c(t)$, we have the following algorithm:

$$\boldsymbol{W}_c(t+1) = \boldsymbol{W}_c(t) + \gamma(t)\{\boldsymbol{X}_c(t)z_c(t) - z_c^T(t)z_c(t)\boldsymbol{W}_c(t)\}. \tag{6.30}$$

The corresponding continuous-time differential equations are

$$\frac{d\boldsymbol{W}_c(t)}{dt} = \boldsymbol{R}_c\boldsymbol{W}_c(t) - \boldsymbol{W}_c^T(t)\boldsymbol{R}_c\boldsymbol{W}_c(t)\boldsymbol{W}_c(t), \tag{6.31}$$

where we used $\boldsymbol{R}_c = E[\boldsymbol{X}_c(t)\boldsymbol{X}_c^T(t)]$. This equation can be proved as follows:

$$\begin{aligned}
\boldsymbol{R} &= E\left[\boldsymbol{X}(t)\boldsymbol{X}^H(t)\right] \\
&= E\Big[(\boldsymbol{X}_r(t) + j\boldsymbol{X}_i(t))\left(\boldsymbol{X}_r^T(t) - j\boldsymbol{X}_i^T(t)\right)\Big] \\
&= E\left[\boldsymbol{X}_r(t)\boldsymbol{X}_r^T(t)\right] + E\left[(\boldsymbol{X}_i(t)\boldsymbol{X}_i^T(t)\right] \\
&\quad + j\Big\{E\left[\boldsymbol{X}_i(t)\boldsymbol{X}_r^T(t)\right] - E\left[\boldsymbol{X}_r(t)\boldsymbol{X}_i^T(t)\right]\Big\}.
\end{aligned} \tag{6.32}$$

As a result,

$$\boldsymbol{R}_r = E\left[\boldsymbol{X}_r(t)\boldsymbol{X}_r^T(t)\right] + E\left[\boldsymbol{X}_i(t)\boldsymbol{X}_i^T(t)\right] \tag{6.33}$$

and

$$\boldsymbol{R}_i = E\left[\boldsymbol{X}_i(t)\boldsymbol{X}_r^T(t)\right] - E\left[\boldsymbol{X}_r(t)\boldsymbol{X}_i^T(t)\right]. \tag{6.34}$$

Substituting (6.33) and (6.34) into (6.25) gives

$$\boldsymbol{R}_c = \begin{pmatrix} E[\boldsymbol{X}_r(t)\boldsymbol{X}_r^T(t) + \boldsymbol{X}_i(t)\boldsymbol{X}_i^T(t)] & -E[\boldsymbol{X}_i(t)\boldsymbol{X}_r^T(t) - \boldsymbol{X}_r(t)\boldsymbol{X}_i^T(t)] \\ E[\boldsymbol{X}_i(t)\boldsymbol{X}_r^T(t) - \boldsymbol{X}_r(t)\boldsymbol{X}_i^T(t)] & E[\boldsymbol{X}_r(t)\boldsymbol{X}_r^T(t) + \boldsymbol{X}_i(t)\boldsymbol{X}_i^T(t)] \end{pmatrix}. \tag{6.35}$$

In contrast, using $\boldsymbol{X}_c(t)$ in (6.29), we get

$$\begin{aligned} E[\boldsymbol{X}_c(t)\boldsymbol{X}_c^T(t)] &= E\left[\begin{pmatrix} \boldsymbol{X}_r(t) & -\boldsymbol{X}_i(t) \\ \boldsymbol{X}_i(t) & \boldsymbol{X}_r(t) \end{pmatrix} \begin{pmatrix} \boldsymbol{X}_r^T(t) & \boldsymbol{X}_i^T(t) \\ -\boldsymbol{X}_i^T(t) & \boldsymbol{X}_r^T(t) \end{pmatrix}\right] \\ &= \begin{pmatrix} E[\boldsymbol{X}_r(t)\boldsymbol{X}_r^T(t) + \boldsymbol{X}_i(t)\boldsymbol{X}_i^T(t)] & -E[\boldsymbol{X}_i(t)\boldsymbol{X}_r^T(t) - \boldsymbol{X}_r(t)\boldsymbol{X}_i^T(t)] \\ E[\boldsymbol{X}_i(t)\boldsymbol{X}_r^T(t) - \boldsymbol{X}_r(t)\boldsymbol{X}_i^T(t)] & E[\boldsymbol{X}_r(t)\boldsymbol{X}_r^T(t) + \boldsymbol{X}_i(t)\boldsymbol{X}_i^T(t)] \end{pmatrix}, \end{aligned} \tag{6.36}$$

which is equal to (6.35), that is, $\boldsymbol{R}_c = E[\boldsymbol{X}_c(t)\boldsymbol{X}_c^T(t)]$. This equation also shows that $\boldsymbol{R}_c$ is nonnegative definite.

Since (6.31) has the same form as (6.6), we know from Theorem 6.1 that if the initial weight vector $\boldsymbol{W}_c(0)$ satisfies $\boldsymbol{W}_c^T(0)\boldsymbol{S}_{1c} \neq 0$, then $\boldsymbol{W}_c(t)$ will converge to the eigenvector $\boldsymbol{S}_{1c}$ corresponding to the largest eigenvalue of the matrix $\boldsymbol{R}_c$. Consequently, $\boldsymbol{W}(t)$ will converge to the first principal component $\boldsymbol{S}_1$.

Comment 4 There are two modified learning algorithms based on (6.4). One is

$$\boldsymbol{W}(t+1) = \boldsymbol{W}(t) + \gamma(t)z(t)\left(\boldsymbol{X}(t) - \frac{z(t)}{\boldsymbol{W}^T(t)\boldsymbol{W}(t)}\boldsymbol{W}(t)\right). \tag{6.37}$$

The other is

$$\boldsymbol{W}(t+1) = \boldsymbol{W}(t) + \gamma(t)z(t)\left[\boldsymbol{W}^T(t)\boldsymbol{W}(t)\boldsymbol{X}(t) - z(t)\boldsymbol{W}(t)\right]. \tag{6.38}$$

The corresponding continuous-time differential equations of (6.37) and (6.38) are

$$\begin{aligned} \frac{d\boldsymbol{W}(t)}{dt} &= z(t)\left(\boldsymbol{X}(t) - \frac{z(t)}{\boldsymbol{W}^T(t)\boldsymbol{W}(t)}\boldsymbol{W}(t)\right) \\ &= \boldsymbol{X}(t)\boldsymbol{X}^T(t)\boldsymbol{W}(t) - \frac{\boldsymbol{W}^T(t)\boldsymbol{X}(t)\boldsymbol{X}^T(t)\boldsymbol{W}(t)}{\boldsymbol{W}^T(t)\boldsymbol{W}(t)}\boldsymbol{W}(t) \end{aligned} \tag{6.39}$$

and

$$\frac{dW(t)}{dt} = z(t)\left[W^T(t)W(t)X(t) - z(t)W(t)\right]$$
$$= W^T(t)W(t)X(t)X^T(t)W(t) - W^T(t)X(t)X^T(t)W(t)W(t), \tag{6.40}$$

and the averaging differential equations are

$$\frac{dW(t)}{dt} = RW(t) - \frac{W^T(t)RW(t)}{W^T(t)W(t)}W(t) \tag{6.41}$$

and

$$\frac{dW(t)}{dt} = W^T(t)W(t)RW(t) - W^T(t)RW(t)W(t). \tag{6.42}$$

For (6.41) and (6.42), if $W^T(0)S_1 \neq 0$, then we have

$$\|W(t)\|^2 = \|W(0)\|^2 \qquad t \geq 0 \tag{6.43}$$

and

$$\lim_{t\to\infty} W(t) = \pm\|W(0)\|S_1. \tag{6.44}$$

To prove (6.43), we multiply (6.41) and (6.42) by $W^T(t)$ on the left. This yields

$$W^T(t)\frac{dW(t)}{dt} = W^T(t)RW(t) - W^T(t)RW(t) = 0 \tag{6.45}$$

and

$$W^T(t)\frac{dW(t)}{dt} = W^T(t)RW(t)W^T(t)W(t)$$
$$- W^T(t)RW(t)W^T(t)W(t) = 0. \tag{6.46}$$

Both (6.45) and (6.46) show that

$$\frac{d\|W(t)\|^2}{dt} = 2W^T(t)\frac{dW(t)}{dt} = 0, \tag{6.47}$$

that is, (6.43) holds.

The proof of (6.44) is exactly the same as that of (6.7).

Equation (6.43) shows that the norm of the weight vector $W(t)$ of the differential equations (6.41) and (6.42) is invariant and equal to the norm of the initial vector $W(0)$ during the time evolution. According to the stochastic approximation of (6.41) to (6.37) and (6.42) to (6.38), the norm of the weight $W(t)$ provided by the learning algorithms (6.37) and (6.38) is close to the norm of the initial vector with large probability. This property is useful in designing the hardware implementation of the corresponding network. Equation (6.43) also implies that we can write (6.37) as

$$\boldsymbol{W}(t+1) = \boldsymbol{W}(t) + \gamma(t)z(t)\left(\boldsymbol{X}(t) - \frac{z(t)}{\boldsymbol{W}^T(0)\boldsymbol{W}(0)}\boldsymbol{W}(t)\right) \tag{6.48}$$

and (6.38) as

$$\boldsymbol{W}(t+1) = \boldsymbol{W}(t) + \gamma(t)z(t)\left[\boldsymbol{W}^T(0)\boldsymbol{W}(0)\boldsymbol{X}(t) - z(t)\boldsymbol{W}(t)\right]. \tag{6.49}$$

Moreover,

$$\boldsymbol{W}(t+1) = \boldsymbol{W}(t) + \gamma(t)z(t)\left[\boldsymbol{X}(t) - a_2 z(t)\boldsymbol{W}(t)\right] \tag{6.50}$$

and

$$\boldsymbol{W}(t+1) = \boldsymbol{W}(t) + \gamma(t)z(t)\left[a_1\boldsymbol{X}(t) - z(t)\boldsymbol{W}(t)\right], \tag{6.51}$$

where a_1 and a_2 are positive constants. From (6.50) and (6.51), we can obtain a general learning algorithm as

$$\boldsymbol{W}(t+1) = \boldsymbol{W}(t) + a_1\gamma(t)z(t)\boldsymbol{X}(t) - a_2\gamma(t)z^2(t)\boldsymbol{W}(t) \tag{6.52}$$

whose corresponding differential equations and statistically averaging differential equations are

$$\begin{aligned}\frac{d\boldsymbol{W}(t)}{dt} &= a_1 z(t)\boldsymbol{X}(t) - a_2 z^2(t)\boldsymbol{W}(t)\\ &= a_1\boldsymbol{X}(t)\boldsymbol{X}^T(t)\boldsymbol{W}(t) - a_2\boldsymbol{W}^T(t)\boldsymbol{X}(t)\boldsymbol{X}^T(t)\boldsymbol{W}(t)\boldsymbol{W}(t)\end{aligned} \tag{6.53}$$

and

$$\frac{d\boldsymbol{W}(t)}{dt} = a_1\boldsymbol{R}\boldsymbol{W}(t) - a_2\boldsymbol{W}^T(t)\boldsymbol{R}\boldsymbol{W}(t)\boldsymbol{W}(t). \tag{6.54}$$

Concerning the asymptotic stability of (6.54), we have

$$\lim_{t\to\infty}\boldsymbol{W}(t) = \pm\sqrt{\frac{a_1}{a_2}}\boldsymbol{S}_1, \tag{6.55}$$

which is valid on the condition that the initial weight vector $\boldsymbol{W}(0)$ satisfies $\boldsymbol{W}^T(0)\boldsymbol{S}_1 \neq 0$.

The proof of (6.55) is as follows: Multiplying (6.54) by $\boldsymbol{S}_i^T$ on the left yields

$$\begin{aligned}\boldsymbol{S}_i^T\frac{d\boldsymbol{W}(t)}{dt} &= a_1\boldsymbol{S}_i^T\boldsymbol{R}\boldsymbol{W}(t) - a_2\boldsymbol{W}^T(t)\boldsymbol{R}\boldsymbol{W}(t)\boldsymbol{S}_i^T\boldsymbol{W}(t)\\ &= a_1\lambda_i\boldsymbol{S}_i^T\boldsymbol{W}(t) - a_2\boldsymbol{W}^T(t)\boldsymbol{R}\boldsymbol{W}(t)\boldsymbol{S}_i^T\boldsymbol{W}(t) \quad (i = 1, 2, \ldots, N).\end{aligned} \tag{6.56}$$

Since $W^T(0)S_1 \neq 0$, we can define

$$c_i(t) = \frac{S_i^T W(t)}{S_1^T W(t)} \qquad (i = 2, 3, \ldots, N) \tag{6.57}$$

and obtain

$$\frac{dc_i(t)}{dt} = a_1(\lambda_i - \lambda_1)c_i(t) \qquad (i = 2, 3, \ldots, N). \tag{6.58}$$

Moreover,

$$c_i(t) = K_i \exp\big(a_1(\lambda_i - \lambda_1)t\big) \qquad (i = 2, 3, \ldots, N), \tag{6.59}$$

which is the same as Equation (6.14) except for the constant term a_1. As a result, we have

$$\lim_{t\to\infty} W(t) = \pm \lim_{t\to\infty} \|W(t)\| S_1. \tag{6.60}$$

Multiplying (6.54) by $W^T(t)$ on the left yields

$$\begin{aligned}\frac{d\|W(t)\|^2}{dt} &= 2\big[a_1 W^T(t)RW(t) - a_2 W^T(t)RW(t)W^T(t)W(t)\big]\\ &= 2a_1 W^T(t)RW(t)\left(1 - \frac{a_2}{a_1}\|W(t)\|^2\right).\end{aligned} \tag{6.61}$$

Using the same analysis as for (6.18) (based on the four cases: $\|W(0)\|^2 = 0$, $\|W(0)\|^2 = \frac{a_1}{a_2}$, $\|W(0)\|^2 > \frac{a_1}{a_2}$, and $\|W(0)\|^2 < \frac{a_1}{a_2}$), we know that (6.61) has two possible stationary points: 0 and $\frac{a_1}{a_2}$. With this fact and the assumption about the initial values of the weight vector, we get

$$\lim_{t\to\infty} \|W(t)\|^2 = \frac{a_1}{a_2}. \tag{6.62}$$

Substituting this into (6.60) gives

$$\lim_{t\to\infty} W(t) = \pm\sqrt{\frac{a_1}{a_2}} S_1,$$

which is (6.55) and completes the proof of this equation.

Clearly, algorithms (6.4), (6.37), and (6.38) are special cases of the general form (6.52). In (6.4), $a_1 = a_2 = 1$; in (6.37), $a_1 = 1, a_2 = \frac{1}{W^T(0)W(0)}$ (which specifies the initial values of the weight vector if a_2 is given a priori); in (6.38), $a_1 = W^T(0)W(0)$ (which specifies the initial values of the weight vector too, if a_1 is given a priori), $a_2 = 1$. Moreover, we also know that (6.4) is the special case of (6.37) and (6.38) with $W^T(0)W(0) = 1$.

One of the advantages of Equation (6.52) over (6.4), (6.37), and (6.38) is that we have more parameters to adjust so as to speed up the convergence. The convergence

Table 6.1. *The dynamic values of the weight vector* $\boldsymbol{W}(t)$

Iteration	$w_1(t)$	$w_2(t)$	$w_3(t)$
0	1.000000	1.000000	1.000000
1	0.982499	1.028549	0.998880
2	0.961780	1.061152	0.975556
⋮	⋮	⋮	⋮
43	−0.824183	1.659876	−0.410696
44	−0.824189	1.659367	−0.411359

of $\boldsymbol{W}(t)$ can be divided into two parts: the direction convergence and norm convergence. The direction convergence is the convergence of the direction of $\boldsymbol{W}(t)$ to the direction of $\boldsymbol{S}_1$. The norm convergence deals with the convergence of the norm of $\boldsymbol{W}(t)$ to $\sqrt{\frac{a_1}{a_2}}$. Since (6.59) is independent of a_2, the direction convergence depends mainly on a_1. We can speed up the direction convergence by increasing a_1, but this slows down the norm convergence. This negative effect can be overcome by appropriately selecting a_2.

In addition, if we select $\boldsymbol{W}^T(0)\boldsymbol{W}(0) = \frac{a_1}{a_2}$, then the norm of $\boldsymbol{W}(t)$ will be invariant and be equal to $\sqrt{(a_1/a_2)}$, which means that under this condition only the direction convergence is involved.

Exactly how a_1 and a_2 affect the stochastic approximation between the learning algorithm (6.52) and the corresponding equations (6.54) remains unclear and is a worthy subject of more detailed investigation.

With appropriate initial values and parameters a_1 and a_2, the above algorithms can also be used to find the first minor component just by reversing the sign of the corresponding term. More details will be discussed in Section 6.4.

For further illustration, we will present two simulation examples [28] (Examples 6.1 and 6.2) concerning the dynamics of (6.54). In these simulation results, we selected $a_1 = \boldsymbol{W}^T(0)\boldsymbol{W}(0)$ and $a_2 = 1$. $\boldsymbol{R}$ is the autocorrelation matrix. $\boldsymbol{W}(0)$ and $\boldsymbol{W}(f)$ are the weight vector in the initial state and in the convergent state, respectively. $\lambda(f)$ is the eigenvalue computed by the weight vector $\boldsymbol{W}(f)$ $[\lambda(f) = (\boldsymbol{W}^T(f)\boldsymbol{R}\boldsymbol{W}(f))/(\boldsymbol{W}^T(f)\boldsymbol{W}(f))]$. $\lambda(a)$ is the exact largest eigenvalue of the matrix $\boldsymbol{R}$. Obviously, $\lambda(f) = \lambda(a)$. Tables 6.1 and 6.2 and Figures 6.3 and 6.4 describe the dynamics of the weight vector $\boldsymbol{W}(t)$.

Example 6.1

$$\boldsymbol{R} = \begin{pmatrix} 1.5000 & -2.0000 & 0.5000 \\ -2.0000 & 4.5000 & -1.0000 \\ 0.5000 & -1.0000 & 0.7500 \end{pmatrix}$$

Table 6.2. *The dynamic values of the weight vector* $\mathbf{W}(t)$

Iteration	$w_1(t)$	$w_2(t)$	$w_3(t)$	$w_4(t)$
0	1.000000	1.000000	1.000000	1.000000
1	0.998174	1.000800	0.985327	1.015828
2	0.996199	1.001495	0.970292	1.031799
⋮	⋮	⋮	⋮	⋮
170	0.311728	0.530007	−0.755498	1.767993
171	0.311235	0.529574	−0.755980	1.767875

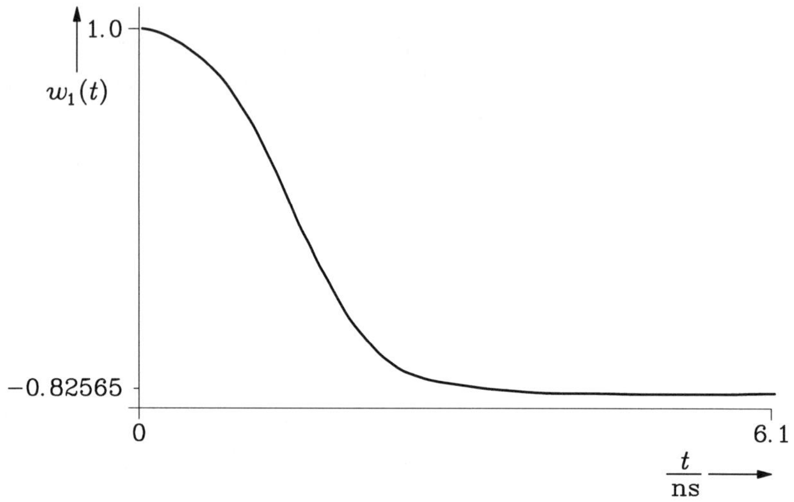

Figure 6.3. The dynamic curve of the variable $w_1(t)$ in Example 6.1.

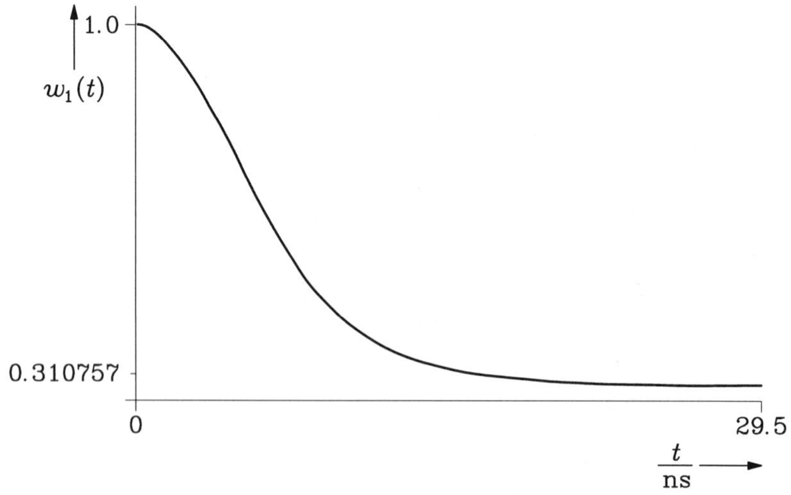

Figure 6.4. The dynamic curve of the variable $w_1(t)$ in Example 6.2.

$$W(0) = \begin{pmatrix} 1.000000 \\ 1.000000 \\ 1.000000 \end{pmatrix}, \qquad W(f) = \begin{pmatrix} -0.825645 \\ 1.658971 \\ -0.411899 \end{pmatrix}$$

$\lambda(f) = 5.749962, \qquad \lambda(a) = 5.749999$

Example 6.2

$$R = \begin{pmatrix} 0.5144 & 0.0252 & -0.0384 & 0.0864 \\ 0.0252 & 0.5441 & -0.0672 & 0.1512 \\ -0.0384 & -0.0672 & 0.6024 & -0.2304 \\ 0.0864 & 0.1512 & -0.2304 & 1.0184 \end{pmatrix}$$

$$W(0) = \begin{pmatrix} 1.000000 \\ 1.000000 \\ 1.000000 \\ 1.000000 \end{pmatrix}, \qquad W(f) = \begin{pmatrix} 0.310757 \\ 0.529154 \\ -0.757040 \\ 1.767760 \end{pmatrix}$$

$\lambda(f) = 1.179082, \qquad \lambda(a) = 1.179300$

6.2 Adaptive Extraction of the Principal Subspace

In this section, we will present neural networks and unsupervised learning algorithms that can be used to extract adaptively the principal subspace (i.e., the subspace spanned by the eigenvectors corresponding to the principal eigenvalues of the autocorrelation matrix $\boldsymbol{R}$ of the input signal $\boldsymbol{X}(t)$).

A neural network for extracting the principal subspace is shown in Figure 6.5. Like the neural network in Figure 6.1, this network consists of two layers. The first layer is the input layer and has N neurons that simply feed the input vector $\boldsymbol{X}(t)$ to the second layer without any modification. The second layer is the output layer and has M neurons. The input–output relationship of each neuron in the output layer is a linear function so that they perform simple summations. The connections between the input layer and output layer are weighted. The connection weight vector is denoted by $\boldsymbol{W}_j(t) = [w_{j1}(t), w_{j2}(t), \ldots, w_{jN}(t)]^T$ (for $j = 1, 2, \ldots, M$), where $w_{ji}(t)$ is the connection strength between the i'th input neuron and the j'th output neuron, and M is the dimension of the principal subspace. With these in mind, the output of the j'th neuron in the output layer is

$$z_j(t) = \sum_{i=1}^{N} w_{ji}(t)x_i(t) = \boldsymbol{W}_j^T(t)\boldsymbol{X}(t) \qquad (j = 1, 2, \ldots, M). \tag{6.63}$$

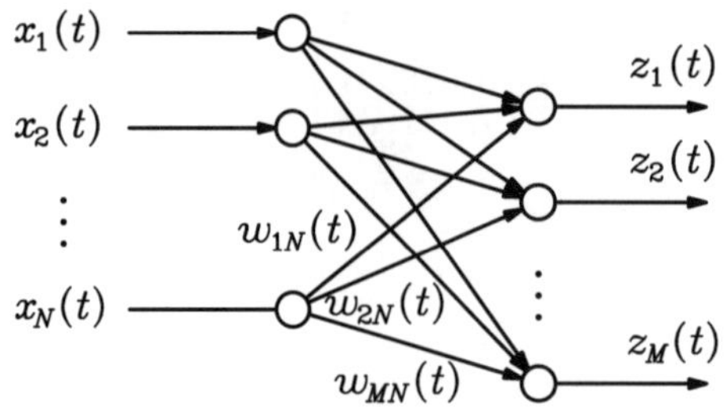

Figure 6.5. The neural network for extracting the principal subspace.

Oja [29] proposed an unsupervised learning algorithm for updating the weight vectors of the neural network as

$$\boldsymbol{W}_j(t+1) = \boldsymbol{W}_j(t) + \gamma(t) z_j(t) \left(\boldsymbol{X}(t) - \sum_{i=1}^{M} z_i(t) \boldsymbol{W}_i(t) \right) \quad (j = 1, 2, \ldots, M). \tag{6.64}$$

Equation (6.64) is rewritten in the matrix form

$$\boldsymbol{W}(t+1) = \boldsymbol{W}(t) + \gamma(t)\left[\boldsymbol{X}(t)\boldsymbol{Z}^T(t) - \boldsymbol{W}(t)\boldsymbol{Z}(t)\boldsymbol{Z}^T(t)\right], \tag{6.65}$$

where $\boldsymbol{W}(t) = [\boldsymbol{W}_1(t), \boldsymbol{W}_2(t), \ldots, \boldsymbol{W}_M(t)]$ and $\boldsymbol{Z}(t) = [z_1(t), z_1(t), \ldots, z_M(t)]^T$.

This algorithm can be obtained by minimizing the mean-square representation error [30, 31]. Consider the linear approximation $\hat{\boldsymbol{X}}(t)$ of the input vector $\boldsymbol{X}(t)$ in terms of M basis vectors $\boldsymbol{W}_1(t), \boldsymbol{W}_2(t), \ldots, \boldsymbol{W}_M(t)$:

$$\hat{\boldsymbol{X}}(t) = \sum_{j=1}^{M} \boldsymbol{W}_j^T(t)\boldsymbol{X}(t)\boldsymbol{W}_j(t) = \sum_{j=1}^{M} z_j(t)\boldsymbol{W}_j(t). \tag{6.66}$$

Clearly, the output of each neuron in the output layer is the coefficient of the above approximation. The mean-square representation error is

$$\begin{aligned} E(t) &= E\left[\|\boldsymbol{e}(t)\|^2\right] = E\left[\|\boldsymbol{X}(t) - \hat{\boldsymbol{X}}(t)\|^2\right] \\ &= E\left[\left\| \boldsymbol{X}(t) - \sum_{j=1}^{M} \boldsymbol{W}_j^T(t)\boldsymbol{X}(t)\boldsymbol{W}_j(t) \right\|^2\right], \end{aligned} \tag{6.67}$$

where

$$\begin{aligned} \boldsymbol{e}(t) &= \boldsymbol{X}(t) - \sum_{j=1}^{M} \boldsymbol{W}_j^T(t)\boldsymbol{X}(t)\boldsymbol{W}_j(t) \\ &= \boldsymbol{X}(t) - \sum_{j=1}^{M} z_j(t)\boldsymbol{W}_j(t) \end{aligned} \tag{6.68}$$

is the error vector.

Using the gradient descent method and the standard instantaneous gradient estimates [31], we can get

$$\boldsymbol{W}_j(t+1) = \boldsymbol{W}_j(t) + \gamma(t)\big[\boldsymbol{W}_j^T(t)\boldsymbol{e}(t)\boldsymbol{X}(t) + \boldsymbol{X}^T(t)\boldsymbol{W}_j(t)\boldsymbol{e}(t)\big]. \tag{6.69}$$

Furthermore, we can assume that the average value of $\boldsymbol{W}_j^T(t)\boldsymbol{e}(t)$ is close to zero because the error vector $\boldsymbol{e}(t)$ should be relatively small after the initial convergence and its sign can be either positive or negative. Hence (6.69) becomes

$$\begin{aligned}\boldsymbol{W}_j(t+1) &= \boldsymbol{W}_j(t) + \gamma(t)\boldsymbol{X}^T(t)\boldsymbol{W}_j(t)\boldsymbol{e}(t) = \boldsymbol{W}_j(t) + \gamma(t)z_j(t)\boldsymbol{e}(t)\\ &= \boldsymbol{W}_j(t) + \gamma(t)z_j(t)\left(\boldsymbol{X}(t) - \sum_{i=1}^{M} z_i(t)\boldsymbol{W}_i(t)\right),\end{aligned} \tag{6.70}$$

which is (6.64).

The corresponding differential equations of (6.64) are

$$\frac{d\boldsymbol{W}(t)}{dt} = \boldsymbol{R}\boldsymbol{W}(t) - \boldsymbol{W}(t)\boldsymbol{W}^T(t)\boldsymbol{R}\boldsymbol{W}(t) \tag{6.71}$$

or

$$\frac{d\boldsymbol{W}_j(t)}{dt} = \boldsymbol{R}\boldsymbol{W}_j(t) - \sum_{i=1}^{M} \boldsymbol{W}_j^T(t)\boldsymbol{R}\boldsymbol{W}_i(t)\boldsymbol{W}_i(t) \qquad (j = 1, 2, \ldots, M). \tag{6.72}$$

It has been shown [32] that if $\boldsymbol{W}(0)$ is of full column rank M, then the solution of (6.71) converges to the principal subspace $\boldsymbol{Q}_M$, that is,

$$Span\{\boldsymbol{W}_1(f), \boldsymbol{W}_2(f), \ldots, \boldsymbol{W}_M(f)\} = \boldsymbol{Q}_M, \tag{6.73}$$

where

$$\boldsymbol{W}_j(f) = \lim_{t\to\infty} \boldsymbol{W}_j(t) \qquad (j = 1, 2, \ldots, M). \tag{6.74}$$

Note that in the above we used the assumption that $\lambda_M > \lambda_{M+1}$.

The principal subspace analysis (PSA) algorithm (6.64) can be modified in a more general form as

$$\boldsymbol{W}_j(t+1) = \boldsymbol{W}_j(t) + \gamma(t)z_j(t)\left(\boldsymbol{X}(t) - \theta_j\sum_{i=1}^{M} z_i(t)\boldsymbol{W}_i(t)\right) \qquad (j = 1, 2, \ldots, M), \tag{6.75}$$

where θ_j (for $j = 1, 2, \ldots, M$) are positive parameters. This general algorithm is called the weighted subspace algorithm [33, 34]. The dynamics and convergence of (6.75) can be changed using different values of this set of parameters. For example, as we will see in the next section, if $0 < \theta_1 < \theta_2 < \cdots < \theta_M$, then $\boldsymbol{W}_j(t)$ will converge to the true j'th eigenvector $\boldsymbol{S}_j$ (for $j = 1, 2, \ldots, M$).

For a wider application, it is necessary to generalize the PSA algorithm (6.64) for the case that $\boldsymbol{X}(t)$ takes complex values. This can be handled in a manner similar to that given in Section 6.1 for the extraction of the first principal component.

Let us use the notation

$$z_{jc}(t) = \begin{pmatrix} z_{jr}(t) \\ z_{ji}(t) \end{pmatrix}, \qquad \boldsymbol{W}_{jc}(t) = \begin{pmatrix} \boldsymbol{W}_{jr}(t) \\ \boldsymbol{W}_{ji}(t) \end{pmatrix},$$

and

$$z_{jr}(t) = \boldsymbol{W}_{jr}^T(t)\boldsymbol{X}_r(t) + \boldsymbol{W}_{ji}^T(t)\boldsymbol{X}_i(t), \tag{6.76}$$

$$z_{ji}(t) = -\boldsymbol{W}_{jr}^T(t)\boldsymbol{X}_i(t) + \boldsymbol{W}_{ji}^T(t)\boldsymbol{X}_r(t) \qquad (j = 1, 2, \ldots, M), \tag{6.77}$$

where $\boldsymbol{X}_r(t)$, $\boldsymbol{W}_{jr}(t)$, and $z_{jr}(t)$ and $\boldsymbol{X}_i(t)$, $\boldsymbol{W}_{ji}(t)$, and $z_{ji}(t)$ are the real and imaginary parts of the variables $\boldsymbol{X}(t)$, $\boldsymbol{W}_j(t)$, and $z_j(t)$, respectively. Then from (6.64) we have the algorithm to update the weights as

$$\boldsymbol{W}_{jc}(t+1) = \boldsymbol{W}_{jc}(t) + \gamma(t)\left(\boldsymbol{X}_c(t)\boldsymbol{z}_{jc}(t) - \sum_{i=1}^{M} \boldsymbol{z}^T{}_{jc}(t)\boldsymbol{z}_{ic}(t)\boldsymbol{W}_{ic}(t)\right)$$
$$(j = 1, 2, \ldots, M), \tag{6.78}$$

where

$$\boldsymbol{X}_c(t) = \begin{pmatrix} \boldsymbol{X}_r(t) & -\boldsymbol{X}_i(t) \\ \boldsymbol{X}_i(t) & \boldsymbol{X}_r(t) \end{pmatrix}. \tag{6.79}$$

The corresponding continuous-time differential equations are

$$\frac{d\boldsymbol{W}_{jc}(t)}{dt} = \boldsymbol{R}_c\boldsymbol{W}_{jc}(t) - \sum_{i=1}^{M} \boldsymbol{W}_{jc}^T(t)\boldsymbol{R}_c\boldsymbol{W}_{ic}(t)\boldsymbol{W}_{ic}(t)$$
$$(j = 1, 2, \ldots, M), \tag{6.80}$$

where we used $\boldsymbol{z}_{jc}(t) = \boldsymbol{X}_c^T(t)\boldsymbol{W}_{jc}(t)$ and $\boldsymbol{R}_c = E[\boldsymbol{X}_c(t)\boldsymbol{X}_c^T(t)]$. Since $\boldsymbol{R}_c$ has exactly the same property as that of the autocorrelation matrix $\boldsymbol{R}$ in the real-valued case, the solution of (6.80) will converge to the subspace spanned by the eigenvectors corresponding to the M largest eigenvalues of the matrix $\boldsymbol{R}_c$. Consequently, $\boldsymbol{W}(t) = [\boldsymbol{W}_{1r}(t) + j\boldsymbol{W}_{1i}(t), \boldsymbol{W}_{2r}(t) + j\boldsymbol{W}_{2i}(t), \ldots, \boldsymbol{W}_{Mr}(t) + j\boldsymbol{W}_{Mi}(t)]$ will converge to the desired principal subspace of the complex-valued autocorrelation matrix $E[\boldsymbol{X}(t)\boldsymbol{X}^H(t)]$.

The above algorithms for PSA are suitable for applications in which finding the eigen-subspace is sufficient, such as in bearing estimation and spectral estimation. However, in many other applications, it is necessary to extract the true eigenvectors. This problem will be dealt with in the next section.

6.3 Adaptive Extraction of the Principal Components

Many algorithms to extract adaptively the principal components have been proposed and extensively analyzed. In this section, we will give an overview of these algorithms.

The Stochastic Gradient Ascent (SGA) Algorithm

The SGA algorithm proposed by Oja and Karhunen can be written as [27, 35]

$$\boldsymbol{W}_j(t+1) = \boldsymbol{W}_j(t) + \gamma(t) z_j(t) \left(\boldsymbol{X}(t) - z_j(t)\boldsymbol{W}_j(t) - 2\sum_{i=1}^{j-1} z_i(t)\boldsymbol{W}_i(t) \right) \quad (j = 1, 2, \ldots, M), \tag{6.81}$$

whose corresponding differential equations are

$$\frac{d\boldsymbol{W}_j(t)}{dt} = \boldsymbol{R}\boldsymbol{W}_j(t) - \boldsymbol{W}_j^T(t)\boldsymbol{R}\boldsymbol{W}_j(t)\boldsymbol{W}_j(t) - 2\sum_{i=1}^{j-1} \boldsymbol{W}_j^T(t)\boldsymbol{R}\boldsymbol{W}_i(t)\boldsymbol{W}_i(t) \quad (j = 1, 2, \ldots, M), \tag{6.82}$$

where the various quantities are the same as those in the PSA algorithm (6.64) and Figure 6.5.

In comparison with the algorithm (6.4) for extracting the first principal component, the SGA algorithm has an additional term, $-2\sum_{i=1}^{j-1} z_i(t)\boldsymbol{W}_i(t)$, which is a consequence of using the Gram–Schmidt orthonormalization on each weight vector $\boldsymbol{W}_j(t)$ (for $j = 2, 3, \ldots, M$). The Gram–Schmidt orthonormalization can also be performed by replacing the coefficient 2 with unity. Thus, we have the following alternative algorithm.

The Generalized Hebbian Algorithm (GHA)

The GHA can be written as [36]

$$\boldsymbol{W}_j(t+1) = \boldsymbol{W}_j(t) + \gamma(t) z_j(t) \left[\boldsymbol{X}(t) - z_j(t)\boldsymbol{W}_j(t) - \sum_{i=1}^{j-1} z_i(t)\boldsymbol{W}_i(t) \right]. \tag{6.83}$$

Combining the second and third term in the brackets, we have

$$\boldsymbol{W}_j(t+1) = \boldsymbol{W}_j(t) + \gamma(t) z_j(t) \left[\boldsymbol{X}(t) - \sum_{i=1}^{j} z_i(t) \boldsymbol{W}_i(t) \right] \quad (j = 1, 2, \ldots, M), \tag{6.84}$$

which has corresponding differential equations of

$$\frac{d\boldsymbol{W}_j(t)}{dt} = \boldsymbol{R}\boldsymbol{W}_j(t) - \sum_{i=1}^{j} \boldsymbol{W}_j^T(t)\boldsymbol{R}\boldsymbol{W}_i(t)\boldsymbol{W}_i(t) \qquad (j = 1, 2, \ldots, M). \tag{6.85}$$

The GHA is also similar to the PSA algorithm (6.64) and can be obtained by minimizing the mean-square representation error [31]. The only difference between (6.64) and (6.84) is that in the GHA for the j'th weight vector the summation is up to j instead of M. Using the same notation as that in (6.65), we can write (6.85) in another matrix form,

$$\boldsymbol{W}(t+1) = \boldsymbol{W}(t) + \gamma(t)\left\{ \boldsymbol{X}(t)\boldsymbol{Z}^T(t) - \boldsymbol{W}(t)\ upper\ \left[\boldsymbol{Z}(t)\boldsymbol{Z}^T(t) \right] \right\}, \tag{6.86}$$

where the operator *upper* makes all subdiagonal elements of a matrix zero. Clearly, the GHA is in effect the result of replacing the matrix $\boldsymbol{Z}(t)\boldsymbol{Z}^T(t)$ of (6.65) by just its diagonal and superdiagonal.

Based on the above comparison and the weighted subspace algorithm (6.75) mentioned in Section 6.2, we have a more general algorithm

$$\boldsymbol{W}_j(t+1) = \boldsymbol{W}_j(t) + \gamma(t) z_j(t) \left(\boldsymbol{X}(t) - \theta_j \sum_{i=1}^{I(j)} z_i(t) \boldsymbol{W}_i(t) \right) \quad (j = 1, 2, \ldots, M), \tag{6.87}$$

where $I(j)$ is a positive integer and θ_j are positive weight coefficients. Clearly,

(1) in the GHA (6.84), $I(j) = j$ and $\theta_j = 1$ (for $j = 1, 2, \ldots, M$),
(2) in the PSA algorithm (6.64), $I(j) = M$ and $\theta_j = 1$ (for $j = 1, 2, \ldots, M$), and
(3) in the weighted PSA algorithm (6.75), $I(j) = M$ and $0 < \theta_1 < \theta_2 < \cdots \theta_M$.

The convergence of (6.87) will depend greatly on the integer $I(j)$ and weight coefficients θ_j (for $j = 1, 2, \ldots, M$).

The LEAP Algorithm

Chen and Liu [37, 38] proposed an unsupervised learning algorithm called LEAP as

$$\begin{aligned} \boldsymbol{W}_j(t+1) = \boldsymbol{W}_j(t) + \gamma(t)\big\{\boldsymbol{B}_j(t)\big[z_j(t)\boldsymbol{X}(t) - z_j^2(t)\boldsymbol{W}_j(t)\big] \\ - \boldsymbol{A}_j(t)\boldsymbol{W}_j(t)\big\} \qquad (j = 1, 2, \ldots, M), \end{aligned} \tag{6.88}$$

where

$$\boldsymbol{B}_j(t) = \boldsymbol{I} - \sum_{i=1}^{j-1} \boldsymbol{W}_i(t)\boldsymbol{W}_i^T(t), \tag{6.89}$$

$$\boldsymbol{A}_j(t) = \sum_{i=1}^{j-1} \boldsymbol{W}_i(t)\boldsymbol{W}_i^T(t) \qquad (j = 2, 3, \ldots, M), \tag{6.90}$$

and $\boldsymbol{B}_1(t) = \boldsymbol{I}$ and $\boldsymbol{A}_1(t) = \boldsymbol{O}$.

The corresponding differential equations of (6.88) are

$$\begin{aligned} \frac{d\boldsymbol{W}_j(t)}{dt} = \boldsymbol{B}_j(t)\big[\boldsymbol{R}\boldsymbol{W}_j(t) - \boldsymbol{W}_j(t)\boldsymbol{W}_j^T(t)\boldsymbol{R}\boldsymbol{W}_j(t)\big] - \boldsymbol{A}_j(t)\boldsymbol{W}_j(t) \\ (j = 1, 2, \ldots, M). \end{aligned} \tag{6.91}$$

This algorithm is in effect another version of combining the constraint Hebbian learning rule (the term $z_j(t)\boldsymbol{X}(t) - z_j^2(t)\boldsymbol{W}_j(t)$) shown in (6.4) with the Gram–Schmidt orthonormalization principle (performed mainly by the matrices $\boldsymbol{A}_j(t)$ and $\boldsymbol{B}_j(t)$). Clearly, if $j = 1$, (6.88) becomes the algorithm (6.4).

The Invariant-Norm Principal Component Analysis (PCA) Algorithm

A PCA algorithm with invariant-norm weight vectors is presented as [39]

$$\begin{aligned} \boldsymbol{W}_j(t+1) = \boldsymbol{W}_j(t) + \gamma(t)z_j(t)\left\{ \boldsymbol{W}_j^T(t)\boldsymbol{W}_j(t)\left[\boldsymbol{X}(t) - \sum_{i=1}^{j-1} z_i(t)\boldsymbol{W}_i(t)\right]\right. \\ \left. - z_j(t)\boldsymbol{W}_j(t) + \sum_{i=1}^{j-1} z_i(t)\boldsymbol{W}_j^T(t)\boldsymbol{W}_i(t)\boldsymbol{W}_j(t)\right\} \\ (j = 1, 2, \ldots, M). \end{aligned} \tag{6.92}$$

The corresponding continuous-time differential equations are

$$\begin{aligned} \frac{d\boldsymbol{W}_j(t)}{dt} = z_j(t)\left\{ \boldsymbol{W}_j^T(t)\boldsymbol{W}_j(t)\left[\boldsymbol{X}(t) - \sum_{i=1}^{j-1} z_i(t)\boldsymbol{W}_i(t)\right]\right. \\ \left. - z_j(t)\boldsymbol{W}_j(t) + \sum_{i=1}^{j-1} z_i(t)\boldsymbol{W}_j^T(t)\boldsymbol{W}_i(t)\boldsymbol{W}_j(t)\right\} \\ (j = 1, 2, \ldots, M) \end{aligned} \tag{6.93}$$

and the averaging differential equations are

$$\frac{d\mathbf{W}_j(t)}{dt} = \mathbf{W}_j^T(t)\mathbf{W}_j(t)\left[\mathbf{I} - \sum_{i=1}^{j-1}\mathbf{W}_i(t)\mathbf{W}_i^T(t)\right]\mathbf{R}\mathbf{W}_j(t)$$
$$- \mathbf{W}_j^T(t)\left[\mathbf{I} - \sum_{i=1}^{j-1}\mathbf{W}_i(t)\mathbf{W}_i^T(t)\right]\mathbf{R}\mathbf{W}_j(t)\mathbf{W}_j(t)$$
$$(j = 1, 2, \ldots, M). \tag{6.94}$$

If we notate

$$\mathbf{R}_j = \mathbf{R} - \sum_{i=1}^{j-1}\mathbf{W}_i(t)\mathbf{W}_i^T(t)\mathbf{R}, \tag{6.95}$$

$$\psi_j(t) = \mathbf{W}_j^T(t)\mathbf{W}_j(t), \tag{6.96}$$

and

$$\phi_j(t) = \mathbf{W}_j^T(t)\mathbf{R}_j\mathbf{W}_j(t) \qquad (j = 1, 2, \ldots, M), \tag{6.97}$$

then (6.94) can be written as

$$\frac{d\mathbf{W}_j(t)}{dt} = \psi_j(t)\mathbf{R}_j\mathbf{W}_j(t) - \phi_j(t)\mathbf{W}_j(t) \qquad (j = 1, 2, \ldots, M). \tag{6.98}$$

Multiplying (6.98) by $\mathbf{W}_j^T(t)$ on the left yields

$$\frac{d\|\mathbf{W}_j(t)\|^2}{dt} = 2\{\psi_j(t)\mathbf{W}_j^T(t)\mathbf{R}_j\mathbf{W}_j(t) - \phi_j(t)\mathbf{W}_j^T(t)\mathbf{W}_j(t)\}$$
$$(j = 1, 2, \ldots, M). \tag{6.99}$$

Substituting (6.96) and (6.97) into (6.99) gives

$$\frac{d\|\mathbf{W}_j(t)\|^2}{dt} = 2\{\psi_j(t)\phi_j(t) - \phi_j(t)\psi_j(t)\} = 0. \tag{6.100}$$

Moreover,

$$\|\mathbf{W}_j(t)\|^2 = \|\mathbf{W}_j(0)\|^2, \qquad t \geq 0 \qquad (j = 1, 2, \ldots, M), \tag{6.101}$$

which means that the norm of each weight vector $\mathbf{W}_j(t)$ is invariant during the time evolution and is equal to the norm of the initial vector $\mathbf{W}_j(0)$. This property is very useful in designing the hardware implementation of the proposed algorithm. Furthermore, with this property, we can replace the term $\mathbf{W}_j^T(t)\mathbf{W}_j(t)$ in (6.92) by a positive constant a_j and obtain a modified PCA algorithm,

$$\boldsymbol{W}_j(t+1) = \boldsymbol{W}_j(t) + \gamma(t) z_j(t) \left\{ a_j \left[\boldsymbol{X}(t) - \sum_{i=1}^{j-1} z_i(t) \boldsymbol{W}_i(t) \right] - z_j(t) \boldsymbol{W}_j(t) \right.$$
$$\left. + \sum_{i=1}^{j-1} z_i(t) \boldsymbol{W}_j^T(t) \boldsymbol{W}_i(t) \boldsymbol{W}_j(t) \right\}$$
$$(j = 1, 2, \ldots, M), \tag{6.102}$$

with corresponding differential equations of

$$\frac{d\boldsymbol{W}_j(t)}{dt} = a_j \boldsymbol{R}_j \boldsymbol{W}_j(t) - \phi_j(t) \boldsymbol{W}_j(t) \qquad (j = 1, 2, \ldots, M). \tag{6.103}$$

The norm of the vector $\boldsymbol{W}_j(t)$ in (6.103) is also invariant if the initial weight vector $\boldsymbol{W}_j(0)$ satisfies $\|\boldsymbol{W}_j(0)\| = \pm\sqrt{a_j}$. This can simply be proved as follows: Multiplying (6.103) by $\boldsymbol{W}_j^T(t)$ on the left yields

$$\frac{d\|\boldsymbol{W}_j(t)\|^2}{dt} = 2\{a_j \boldsymbol{W}_j^T(t) \boldsymbol{R}_j \boldsymbol{W}_j(t) - \phi_j(t) \boldsymbol{W}_j^T(t) \boldsymbol{W}_j(t)\}$$
$$(j = 1, 2, \ldots, M). \tag{6.104}$$

Substituting (6.97) into (6.104) gives

$$\frac{d\|\boldsymbol{W}_j(t)\|^2}{dt} = 2\left\{\phi_j(t)\left(a_j - \|\boldsymbol{W}_j(t)\|^2\right)\right\} = 0. \tag{6.105}$$

Obviously, if $\|\boldsymbol{W}_j(0)\|^2 = a_j$ then $d\|\boldsymbol{W}_j(t)\|^2/dt = 0$ for $t > 0$, that is, $\|\boldsymbol{W}_j(t)\|^2 = \|\boldsymbol{W}_j(0)\|^2$ for $t \geq 0$ (for $j = 1, 2, \ldots, M$). The computation of the multiplication required in (6.102) is less than that in (6.92) by $O(N)$. The disadvantage incurred by this reduction of computational complexity is that we have to select the initial weight vector $\boldsymbol{W}_j(0)$ to satisfy $\|\boldsymbol{W}_j(0)\| = \pm\sqrt{a_j}$.

APEX Algorithm

Kung, Diammantaras, and Taur have developed a new adaptive principal component extraction (APEX) network and learning algorithm [40, 41, 42].

The APEX network, as depicted in Figure 6.6, has two layers called the input layer and the output layer. In addition to the feedforward connection weights w_{ij} between the i'th input neuron and j'th output neuron, there exist lateral weights that connect all the first $j-1$ output neurons with the j'th output neuron. If we denote the $(j-1)$-dimensional lateral weight vector for the j'th output neuron as $\boldsymbol{C}_j(t)$, then the output of the j'th output neuron is

$$z_j(t) = \boldsymbol{W}_j^T(t) \boldsymbol{X}(t) - \boldsymbol{C}_j^T(t) \boldsymbol{A}_j(t) \boldsymbol{X}(t) \qquad (j = 1, 2, \ldots, M), \tag{6.106}$$

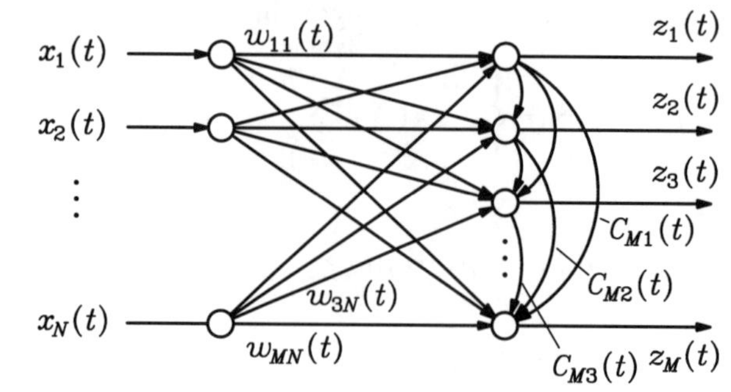

Figure 6.6. The model of the APEX network.

where $A_j(t) = [W_1(t), W_2(t), \ldots, W_{j-1}(t)]^T$ is a $(j-1) \times N$ dimensional matrix.

One of the learning algorithms for updating the weights $W_j(t)$ and $C_j(t)$ is

$$W_j(t+1) = W_j(t) + \gamma(t)z_j(t)\left[X(t) - z_j(t)W_j(t)\right], \tag{6.107}$$

$$C_j(t+1) = C_j(t) + \gamma(t)z_j(t)\left[A_j(t)X(t) - z_j(t)C_j(t)\right] \quad (j = 1, 2, \ldots, M). \tag{6.108}$$

Clearly, (6.107) is the same as (6.4) and hence the weights in (6.107), intuitively, work toward finding the most principal direction. Equation (6.107) is called the Hebbian part of the above algorithm. The weight vector $C_j(t)$, intuitively, subtracts the first $j-1$ components from the j'th neuron so that the j'th output neuron tends to become orthogonal to all the previous components. In other words, the main function of (6.108) is to perform the orthogonalization.

The associated differential equations corresponding to (6.107) and (6.108) are

$$\frac{dW_j(t)}{dt} = RW_j(t) - RA_j^T(t)C_j(t) - B_j(t)W_j(t) \tag{6.109}$$

and

$$\frac{dC_j(t)}{dt} = A_j(t)RW_j(t) - A_j(t)RA_j^T(t)C_j(t) - B_j(t)C_j(t) \quad (j = 1, 2, \ldots, M), \tag{6.110}$$

where

$$B_j(t) = \left[W_j(t) - A_j^T(t)C_j(t)\right]^T R\left[W_j(t) - A_j^T(t)C_j(t)\right]. \tag{6.111}$$

General Conclusion of Convergence

We conclude that under certain assumptions about the initial values of the weight vector, each weight vector of all the above PCA algorithms will converge to the corresponding principal component vector, that is,

$$\lim_{t\to\infty} \boldsymbol{W}_j(t) = \pm b_j \boldsymbol{S}_j, \tag{6.112}$$

where b_j is a positive constant determined by the initial values of the weight vector. For the invariant-norm PCA algorithm (6.92), $b_j = \sqrt{\boldsymbol{W}_j^T(0)\boldsymbol{W}_j(0)}$. For the other algorithms, $b_j = 1$. Assumptions about the initial values of the weight vector are

(1) $\boldsymbol{W}_j^T(0)\boldsymbol{S}_j \neq 0$ for the SGA, generalized Hebbian, and APEX algorithms and the weighted PSA algorithm,

(2) $\boldsymbol{W}_j^T(0)\boldsymbol{S}_j \neq 0$ and $\boldsymbol{W}_j^T(0)\boldsymbol{W}_j(0) \leq 1$ for the LEAP algorithm, and

(3) $\boldsymbol{W}_j^T(0)\boldsymbol{S}_j \neq 0$ and $\boldsymbol{W}_j^T(0)\boldsymbol{W}_j(0) \geq \lambda_{j+1}/\lambda_j$ (for simplicity, $\boldsymbol{W}_j^T(0)\boldsymbol{W}_j(0) \geq 1$) for the invariant-norm PCA algorithm (6.92).

In addition, the parameters θ_j of the weighted PSA algorithm should satisfy $\theta_1 < \theta_2 < \cdots < \theta_M$.

The proof of the above conclusion can be seen in the related references. It is worth noting that

(1) although it is assumed in the proof that the autocorrelation matrix $\boldsymbol{R}$ has distinct eigenvalues (that is, $\lambda_1 > \lambda_2 > \cdots > \lambda_M$), it is expected that (6.112) holds even for the case that $\boldsymbol{R}$ has nondistinct eigenvalues;

(2) although in practical applications M is less than N, (6.112) still holds even if $M = N$;

(3) any analog system represented by the continuous-time differential equations of the above PCA algorithms can be used to find in real time the eigenvectors corresponding to several or all eigenvalues of an arbitrary positive definite matrix.

Using the same notation as given in (6.76)–(6.80), the PCA algorithms and corresponding differential equations can be extended to the case that the input vector $\boldsymbol{X}(t)$ takes complex values as follows:

(1) For the SGA algorithm:

$$\begin{aligned}\boldsymbol{W}_{jc}(t+1) = \boldsymbol{W}_{jc}(t) + \gamma(t)\Bigg(&\boldsymbol{X}_c(t)\boldsymbol{z}_{jc}(t) - \boldsymbol{z}_{jc}^T(t)\boldsymbol{z}_{jc}(t)\boldsymbol{W}_{jc}(t)\\ &- 2\sum_{i=1}^{j-1}\boldsymbol{z}_{jc}^T(t)\boldsymbol{z}_{ic}(t)\boldsymbol{W}_{ic}(t)\Bigg),\end{aligned} \tag{6.113}$$

$$\begin{aligned}\frac{d\boldsymbol{W}_{jc}(t)}{dt} = {}& \boldsymbol{R}_c\boldsymbol{W}_{jc}(t) - \boldsymbol{W}_{jc}^T(t)\boldsymbol{R}_c\boldsymbol{W}_{jc}(t)\boldsymbol{W}_{jc}(t)\\ &- 2\sum_{i=1}^{j-1}\boldsymbol{W}_{jc}^T(t)\boldsymbol{R}_c\boldsymbol{W}_{ic}(t)\boldsymbol{W}_{ic}(t) \qquad (j = 1, 2, \ldots, M).\end{aligned} \tag{6.114}$$

(2) For the GHA:

$$W_{jc}(t+1) = W_{jc}(t) + \gamma(t)\left(X_c(t)z_{jc}(t) - \sum_{i=1}^{j} z_{jc}^T(t)z_{ic}(t)W_{ic}(t)\right), \tag{6.115}$$

$$\frac{dW_{jc}(t)}{dt} = R_c W_{jc}(t) - \sum_{i=1}^{j} W_{jc}^T(t)R_c W_{ic}(t)W_{ic}(t) \quad (j = 1, 2, \ldots, M). \tag{6.116}$$

(3) For the LEAP algorithm:

$$W_{jc}(t+1) = W_{jc}(t) + \gamma(t)\Big\{B_{jc}(t)\big[X_c(t)z_{jc}(t) - z_{jc}^T(t)z_{jc}(t)W_{jc}(t)\big] - A_{jc}(t)W_{jc}(t)\Big\}, \tag{6.117}$$

$$\frac{dW_{jc}(t)}{dt} = B_{jc}(t)\big[R_c W_{jc}(t) - W_{jc}^T(t)R_c W_{jc}(t)W_{jc}(t)\big] - A_{jc}(t)W_{jc}(t) \quad (j = 1, 2, \ldots, M), \tag{6.118}$$

where

$$B_{jc}(t) = I - \sum_{i=1}^{j-1} W_{ic}(t)W_{ic}^T(t), \tag{6.119}$$

$$A_{jc}(t) = \sum_{i=1}^{j-1} W_{ic}(t)W_{ic}^T(t) \qquad (j = 2, 3, \ldots, M), \tag{6.120}$$

and $B_{1c}(t) = I$ and $A_{1c}(t) = O$.

(4) For the invariant-norm PCA algorithm:

$$W_{jc}(t+1) = W_{jc}(t) + \gamma(t)\left\{ W_{jc}^T(t)W_{jc}(t)\left[X_c(t)z_{jc}(t) - \sum_{i=1}^{j-1} z_{jc}^T(t)z_{ic}(t)W_{ic}(t)\right] - z_{jc}^T(t)z_{jc}(t)W_{jc}(t) + \sum_{i=1}^{j-1} z_{jc}^T(t)z_{ic}(t)W_{jc}^T(t)W_{ic}(t)W_{jc}(t)\right\}, \tag{6.121}$$

$$\frac{d\boldsymbol{W}_{jc}(t)}{dt} = \boldsymbol{W}_{jc}^T(t)\boldsymbol{W}_{jc}(t)\left[\boldsymbol{I} - \sum_{i=1}^{j-1} \boldsymbol{W}_{ic}(t)\boldsymbol{W}_{ic}^T(t)\right]\boldsymbol{R}_c\boldsymbol{W}_{jc}(t)$$
$$- \boldsymbol{W}_{jc}^T(t)\left[\boldsymbol{I} - \sum_{i=1}^{j-1} \boldsymbol{W}_{ic}(t)\boldsymbol{W}_{ic}^T(t)\right]\boldsymbol{R}_c\boldsymbol{W}_{jc}(t)\boldsymbol{W}_{jc}(t)$$
$$(j = 1, 2, \ldots, M). \quad (6.122)$$

(5) For the APEX algorithm:

$$\boldsymbol{W}_{jc}(t+1) = \boldsymbol{W}_{jc}(t) + \gamma(t)\left[\boldsymbol{X}_c(t)\boldsymbol{z}_{jc}(t) - \boldsymbol{z}_{jc}^T(t)\boldsymbol{z}_{jc}(t)\boldsymbol{W}_{jc}(t)\right], \quad (6.123)$$

$$\boldsymbol{C}_{jc}(t+1) = \boldsymbol{C}_{jc}(t) + \gamma(t)\left[\boldsymbol{A}_{jx}(t)\boldsymbol{z}_{jc}(t) - \boldsymbol{z}_{jc}^T(t)\boldsymbol{z}_{jc}(t)\boldsymbol{C}_{jc}(t)\right], \quad (6.124)$$

$$\frac{d\boldsymbol{W}_{jc}(t)}{dt} = \boldsymbol{R}_c\boldsymbol{W}_{jc}(t) - \boldsymbol{R}_c\boldsymbol{A}_{jc}^T(t)\boldsymbol{C}_{jc}(t) - B_{jc}(t)\boldsymbol{W}_{jc}(t), \quad (6.125)$$

$$\frac{d\boldsymbol{C}_{jc}(t)}{dt} = \boldsymbol{A}_{jc}(t)\boldsymbol{R}_c\boldsymbol{W}_{jc}(t) - \boldsymbol{A}_{jc}(t)\boldsymbol{R}_c\boldsymbol{A}_{jc}^T(t)\boldsymbol{C}_{jc}(t) - B_{jc}(t)\boldsymbol{C}_{jc}(t)$$
$$(j = 1, 2, \ldots, M). \quad (6.126)$$

Note that in (6.123)–(6.126) we used

$$\boldsymbol{C}_{jc}(t) = [\boldsymbol{C}_{jr}^T(t), \boldsymbol{C}_{ji}^T(t)]^T,$$

$$\boldsymbol{A}_{jc}(t) = [\boldsymbol{W}_{1c}(t), \boldsymbol{W}_{2c}(t), \ldots, \boldsymbol{W}_{(j-1)c}(t)]^T,$$

$$\boldsymbol{z}_{jc}(t) = \begin{pmatrix} \boldsymbol{z}_{jr}(t) \\ \boldsymbol{z}_{ji}(t) \end{pmatrix} = \boldsymbol{X}_c^T(t)\boldsymbol{W}_{jc}(t) - \boldsymbol{A}_{jx}(t)^T \begin{pmatrix} \boldsymbol{C}_{ji}(t) \\ \boldsymbol{C}_{jr}(t) \end{pmatrix},$$

and

$$B_{jc}(t) = \left[\boldsymbol{W}_{jc}(t) - \boldsymbol{A}_{jc}^T(t)\boldsymbol{C}_{jc}(t)\right]^T \boldsymbol{R}_c\left[\boldsymbol{W}_{jc}(t) - \boldsymbol{A}_{jc}^T(t)\boldsymbol{C}_{jc}(t)\right]$$
$$(j = 1, 2, \ldots, M),$$

where

$$\boldsymbol{A}_{jx}(t) = \begin{pmatrix} \boldsymbol{A}_{jr}(t)\boldsymbol{X}_r(t) - \boldsymbol{A}_{ji}(t)\boldsymbol{X}_i(t) & -\boldsymbol{A}_{ji}(t)\boldsymbol{X}_r(t) - \boldsymbol{A}_{jr}(t)\boldsymbol{X}_i(t) \\ \boldsymbol{A}_{ji}(t)\boldsymbol{X}_r(t) + \boldsymbol{A}_{jr}(t)\boldsymbol{X}_i(t) & \boldsymbol{A}_{jr}(t)\boldsymbol{X}_r(t) - \boldsymbol{A}_{ji}(t)\boldsymbol{X}_i(t) \end{pmatrix}$$

and $\boldsymbol{C}_{jr}(t)$, $\boldsymbol{A}_{jr}(t)$ and $\boldsymbol{C}_{ji}(t)$, $\boldsymbol{A}_{ji}(t)$ are the real and imaginary parts of the variables $\boldsymbol{C}_j(t)$ and $\boldsymbol{A}_j(t)$, respectively.

In the final part of this section, we present two sets of simulation results concerning the invariant-norm PCA algorithm and its corresponding differential equations

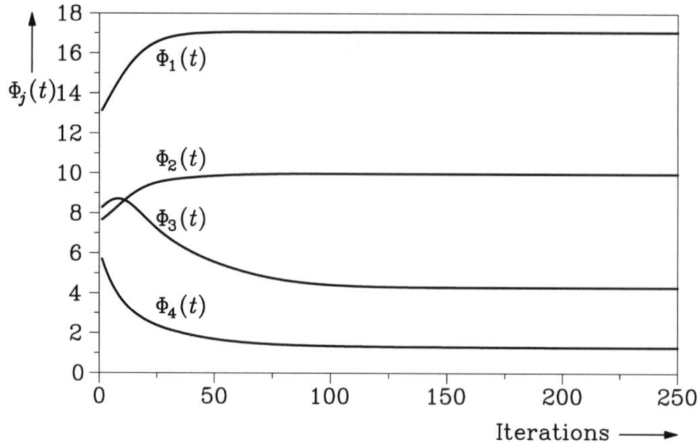

Figure 6.7. The dynamic curves of the variables $\phi_j(t)$ (for $j = 1, 2, 3, 4$) in Example 6.3.

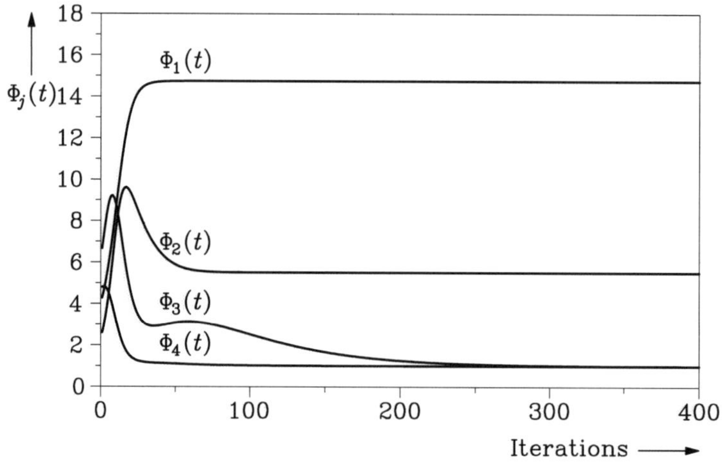

Figure 6.8. The dynamic curves of the variables $\phi_j(t)$ (for $j = 1, 2, 3, 4$) in Example 6.4.

[43]. The first example (Example 6.3) has distinct eigenvalues; the second example (Example 6.4) has nondistinct eigenvalues. In these simulations, $\boldsymbol{R}$ is the given autocorrelation matrix with $N \times N$ elements, $\boldsymbol{W}_j(f)$ and $\boldsymbol{W}_j(0)$ (for $j = 1, 2, \ldots, N$) are the weight vector of the equations (6.98) in the steady state and the initial state ($\|\boldsymbol{W}_j(0)\| = 1$), respectively, $\phi_j(f)$ is the eigenvalue provided by the variable $\phi_j(t)$ in the steady state, and $\lambda_j(a)$ (for $j = 1, 2, \ldots, N$) are the exact eigenvalues of the matrix $\boldsymbol{R}$. Obviously, $\phi_j(f) = \lambda_j(a)$ (for $j = 1, 2, \ldots, N$). Note that we selected $M = N$. For further illustrations, Figures 6.7 and 6.8 describe the time evolution of the variables $\phi_j(t)$. Finally, $\boldsymbol{W}^T(f)\boldsymbol{R}\boldsymbol{W}(f)$ and $\boldsymbol{W}^T(f)\boldsymbol{W}(f)$ ($\boldsymbol{W}(f) = [\boldsymbol{W}_1(f), \boldsymbol{W}_2(f), \ldots, \boldsymbol{W}_N(f)]$) are also given.

Example 6.3

$$\boldsymbol{R} = \begin{pmatrix} 7.5635 & 3.9025 & 3.5674 & -0.6843 \\ 3.9025 & 8.1487 & 5.7788 & -1.7451 \\ 3.5674 & 5.7788 & 6.1811 & -1.3944 \\ -0.6843 & -1.7451 & -1.3944 & 10.7305 \end{pmatrix}$$

$$\boldsymbol{W}_1(f) = \begin{pmatrix} 0.4762 \\ 0.6170 \\ 0.5274 \\ -0.3380 \end{pmatrix}, \qquad \boldsymbol{W}_2(f) = \begin{pmatrix} -0.2579 \\ -0.1740 \\ -0.1635 \\ -0.9362 \end{pmatrix}$$

$$\boldsymbol{W}_3(f) = \begin{pmatrix} -0.8390 \\ 0.4434 \\ 0.3004 \\ 0.0962 \end{pmatrix}, \qquad \boldsymbol{W}_4(f) = \begin{pmatrix} -0.0532 \\ -0.6263 \\ 0.7777 \\ -0.0048 \end{pmatrix}$$

$$\boldsymbol{W}_1(0) = \begin{pmatrix} 0.5000 \\ 0.5000 \\ 0.5000 \\ 0.5000 \end{pmatrix}, \qquad \boldsymbol{W}_2(0) = \begin{pmatrix} 1.0000 \\ 0.0000 \\ 0.0000 \\ 0.0000 \end{pmatrix}$$

$$\boldsymbol{W}_3(0) = \begin{pmatrix} 0.0000 \\ 1.0000 \\ 0.0000 \\ 0.0000 \end{pmatrix}, \qquad \boldsymbol{W}_4(0) = \begin{pmatrix} 0.0000 \\ 0.0000 \\ 1.0000 \\ 0.0000 \end{pmatrix}$$

$$\phi_1(f) = 17.0563, \quad \phi_2(f) = 9.9742, \quad \phi_3(f) = 4.3019, \quad \phi_4(f) = 1.2916$$

$$\lambda_1(a) = 17.0563, \quad \lambda_2(a) = 9.9742, \quad \lambda_3(a) = 4.3019, \quad \lambda_3(f) = 1.2916$$

$$\boldsymbol{W}^T(f)\boldsymbol{R}\boldsymbol{W}(f) = \begin{pmatrix} 17.0563 & 0.0000 & 0.0000 & -0.0001 \\ 0.0000 & 9.9742 & 0.0000 & 0.0002 \\ 0.0000 & 0.0000 & 4.3019 & 0.0003 \\ -0.0001 & 0.0002 & 0.0003 & 1.2916 \end{pmatrix}$$

$$\boldsymbol{W}^T(f)\boldsymbol{W}(f) = \begin{pmatrix} 1.0000 & 0.0000 & 0.0000 & 0.0000 \\ 0.0000 & 1.0000 & 0.0000 & 0.0000 \\ 0.0000 & 0.0000 & 1.0000 & 0.0001 \\ 0.0000 & 0.0000 & 0.0001 & 1.0000 \end{pmatrix}.$$

Example 6.4

$$\boldsymbol{R} = \begin{pmatrix} 2.3279 & 1.0825 & 0.8574 & -3.2503 \\ 1.0825 & 6.1087 & 4.3588 & -3.4051 \\ 0.8574 & 4.3588 & 4.7211 & -2.7944 \\ -3.2503 & -3.4051 & -2.7944 & 9.0905 \end{pmatrix}$$

$$W_1(f) = \begin{pmatrix} 0.2556 \\ 0.5243 \\ 0.4410 \\ -0.6821 \end{pmatrix}, \quad W_2(f) = \begin{pmatrix} 0.3088 \\ -0.5434 \\ -0.4822 \\ -0.6138 \end{pmatrix}$$

$$W_3(f) = \begin{pmatrix} -0.8451 \\ 0.2187 \\ -0.3283 \\ -0.3609 \end{pmatrix}, \quad W_4(f) = \begin{pmatrix} -0.3536 \\ -0.6180 \\ 0.6821 \\ -0.1666 \end{pmatrix}$$

$$W_1(0) = \begin{pmatrix} 0.5000 \\ 0.5000 \\ 0.5000 \\ 0.5000 \end{pmatrix}, \quad W_2(0) = \begin{pmatrix} 1.0000 \\ 0.0000 \\ 0.0000 \\ 0.0000 \end{pmatrix}$$

$$W_3(0) = \begin{pmatrix} 0.0000 \\ 1.0000 \\ 0.0000 \\ 0.0000 \end{pmatrix}, \quad W_4(0) = \begin{pmatrix} 0.0000 \\ 0.0000 \\ 1.0000 \\ 0.0000 \end{pmatrix}$$

$$\phi_1(f) = 14.7325, \quad \phi_2(f) = 5.5157, \quad \phi_3(f) = 1.0000, \quad \phi_4(f) = 1.0000$$

$$\lambda_1(a) = 14.7325, \quad \lambda_2(a) = 5.5157, \quad \lambda_3(a) = 1.0000, \quad \lambda_3(f) = 1.0000$$

$$W^T(f)RW(f) = \begin{pmatrix} 14.7325 & 0.0000 & 0.0005 & 0.0002 \\ 0.0000 & 5.5157 & -0.0003 & -0.0001 \\ 0.0005 & -0.0003 & 1.0000 & -0.0001 \\ 0.0002 & -0.0001 & -0.0001 & 1.0000 \end{pmatrix}$$

$$W^T(f)W(f) = \begin{pmatrix} 1.0000 & 0.0000 & 0.0000 & 0.0000 \\ 0.0000 & 1.0000 & -0.0001 & 0.0000 \\ 0.0000 & -0.0001 & 1.0000 & -0.0001 \\ 0.0000 & 0.0000 & -0.0001 & 1.0000 \end{pmatrix}.$$

6.4 Adaptive Extraction of the Minor Components

The minor components of an input signal usually represent the statistics of the additive noise in the input signal and play a very important role in noise cancellation. Since the minor components are the counterpart of the principal components, the available PCA algorithms provide a starting point for developing the algorithms of the minor component analysis (MCA) [44]. Moreover, some of the above PCA algorithms can directly be generalized to the extraction of the minor components by simply reversing the sign, which we will see in the next subsections.

6.4.1 Adaptive Extraction of the First Minor Component

With the same network and the same notation as that for the first principal component, we will present and analyze three algorithms for extracting the first minor component $\boldsymbol{S}_N$ of the input vector $\boldsymbol{X}(t)$.

The Constrained Anti-Hebbian Algorithm

The constrained anti-Hebbian algorithm can be obtained from the constrained Hebbian algorithm (6.4) by simply changing the sign of the Hebbian term and the constraint term, that is,

$$\boldsymbol{W}(t+1) = \boldsymbol{W}(t) - \gamma(t)z(t)\left[\boldsymbol{X}(t) - z(t)\boldsymbol{W}(t)\right]. \tag{6.127}$$

The corresponding differential equations are

$$\frac{d\boldsymbol{W}(t)}{dt} = -\boldsymbol{R}\boldsymbol{W}(t) + \boldsymbol{W}^T(t)\boldsymbol{R}\boldsymbol{W}(t)\boldsymbol{W}(t). \tag{6.128}$$

It has been shown [19, 44] that if $\boldsymbol{W}^T(0)\boldsymbol{S}_N \neq 0$ and $\|\boldsymbol{W}(0)\| = 1$, then the weight vector $\boldsymbol{W}(t)$ will converge to the minor component $\boldsymbol{S}_N$, that is,

$$\lim_{t\to\infty} \|\boldsymbol{W}(t)\|^2 = 1 \tag{6.129}$$

and

$$\lim_{t\to\infty} \boldsymbol{W}(t) = \pm\boldsymbol{S}_N. \tag{6.130}$$

Now we will show that if $\|\boldsymbol{W}(0)\| > 1$, then $\lim_{t\to\infty} \|\boldsymbol{W}(t)\|^2 = \infty$; hence, the convergence can not be guaranteed.

Multiplying (6.128) by $\boldsymbol{W}^T(t)$ on the left yields

$$\begin{aligned}\frac{d\|\boldsymbol{W}(t)\|^2}{dt} &= -2\left[\boldsymbol{W}^T(t)\boldsymbol{R}\boldsymbol{W}(t) - \boldsymbol{W}^T(t)\boldsymbol{R}\boldsymbol{W}(t)\boldsymbol{W}^T(t)\boldsymbol{W}(t)\right] \\ &= 2\boldsymbol{W}^T(t)\boldsymbol{R}\boldsymbol{W}(t)\left(\|\boldsymbol{W}(t)\|^2 - 1\right).\end{aligned} \tag{6.131}$$

Since $\boldsymbol{W}^T(t)\boldsymbol{R}\boldsymbol{W}(t) > 0$ and $\|\boldsymbol{W}(0)\| > 1$, we have $d\|\boldsymbol{W}(t)\|^2/dt > 0$ for $t > 0$. Consequently, $\lim_{t\to\infty} \|\boldsymbol{W}(t)\|^2 = \infty$. Equation (6.131) also shows that if $\|\boldsymbol{W}(0)\| < 1$, then $\lim_{t\to\infty} \|\boldsymbol{W}(t)\|^2 = 0$. These two facts show that $\|\boldsymbol{W}(t)\|^2 = 1$ is an unstable fixed point.

In order to overcome this problem, Xu, Oja, and Suen [19] proposed the normalized constrained anti-Hebbian algorithm.

The Normalized Constrained Anti-Hebbian Algorithm

This algorithm is given by [19, 45]

$$\boldsymbol{W}(t+1) = \boldsymbol{W}(t) - \gamma(t)z(t)\left(\boldsymbol{X}(t) - \frac{z(t)}{\boldsymbol{W}^T(t)\boldsymbol{W}(t)}\boldsymbol{W}(t)\right) \tag{6.132}$$

and its differential equations are

$$\frac{d\boldsymbol{W}(t)}{dt} = -\boldsymbol{R}\boldsymbol{W}(t) + \frac{\boldsymbol{W}^T(t)\boldsymbol{R}\boldsymbol{W}(t)}{\boldsymbol{W}^T(t)\boldsymbol{W}(t)}\boldsymbol{W}(t). \tag{6.133}$$

It is easy to see that this normalized constrained anti-Hebbian algorithm can be obtained by simply changing the sign of the second and third terms of the PCA algorithm (6.37). Consequently, for (6.133) it holds from (6.45) that

$$\|\boldsymbol{W}(t)\|^2 = \|\boldsymbol{W}(0)\|^2 \qquad (t \geq 0). \tag{6.134}$$

Moreover, if $\boldsymbol{W}^T(0)\boldsymbol{S}_N \neq 0$, we have

$$\lim_{t\to\infty} \boldsymbol{W}(t) = \pm\|\boldsymbol{W}(0)\|\boldsymbol{S}_N. \tag{6.135}$$

Equations (6.134) and (6.135) show that the weight vector provided by the normalized constrained anti-Hebbian algorithm will converge to the first minor component $\boldsymbol{S}_N$ with the norm of the initial vector.

Equation (6.135) can be proved in a fashion similar to that used for (6.7). Using (6.8) and (6.133), we get

$$\frac{dy_i(t)}{dt} = -\lambda_i y_i(t) + \frac{\boldsymbol{W}^T(t)\boldsymbol{R}\boldsymbol{W}(t)}{\boldsymbol{W}^T(t)\boldsymbol{W}(t)} y_i(t) \qquad (i = 1, 2, \ldots, N). \tag{6.136}$$

According to the assumption of the initial value $\boldsymbol{W}(0)$, we may define

$$c_i(t) = \frac{y_i(t)}{y_N(t)} \qquad (i = 1, 2, \ldots, N-1) \tag{6.137}$$

and obtain

$$\frac{dc_i(t)}{dt} = \frac{y_N(t)\frac{dy_i(t)}{dt} - y_i(t)\frac{dy_N(t)}{dt}}{(y_N(t))^2}. \tag{6.138}$$

Combining (6.136) and (6.138) gives

$$\begin{aligned}\frac{dc_i(t)}{dt} &= \frac{1}{y_N^2(t)}\Bigg[\Bigg(-\lambda_i y_i(t) + \frac{\boldsymbol{W}^T(t)\boldsymbol{R}\boldsymbol{W}(t)}{\boldsymbol{W}^T(t)\boldsymbol{W}(t)} y_i(t)\Bigg) y_N(t) \\ &\quad - \Bigg(-\lambda_N y_N(t) + \frac{\boldsymbol{W}^T(t)\boldsymbol{R}\boldsymbol{W}(t)}{\boldsymbol{W}^T(t)\boldsymbol{W}(t)} y_N(t)\Bigg) y_i(t)\Bigg] \\ &= (\lambda_N - \lambda_i)c_i(t) \qquad (i = 1, 2, \ldots, N-1)\end{aligned} \tag{6.139}$$

and

$$c_i(t) = K_{iN}\exp\big((\lambda_N - \lambda_i)t\big) \qquad (i = 1, 2, \ldots, N-1), \tag{6.140}$$

where K_{iN} is a constant depending on the initial values and the eigenvalues of the matrix $\boldsymbol{R}$.

Using (6.137) yields

$$y_i(t) = K_{iN} y_N(t) \exp\big((\lambda_N - \lambda_i)t\big) \qquad (i = 1, 2, \dots, N-1). \tag{6.141}$$

If the smallest eigenvalue λ_N is single but not multiple, that is, $\lambda_N < \lambda_i$ (for $i = 1, 2, \dots, N-1$), then we have, from (6.141),

$$\lim_{t\to\infty} y_i(t) = 0 \qquad (i = 1, 2, \dots, N-1) \tag{6.142}$$

and

$$\lim_{t\to\infty} \boldsymbol{W}(t) = \lim_{t\to\infty} y_N(t)\boldsymbol{S}_N. \tag{6.143}$$

Together with (6.134), Equations (6.143) shows that

$$\lim_{t\to\infty} y_N(t) = \pm\|\boldsymbol{W}(0)\| \tag{6.144}$$

and

$$\lim_{t\to\infty} \boldsymbol{W}(t) = \pm\|\boldsymbol{W}(0)\|\boldsymbol{S}_N, \tag{6.145}$$

which is Equation (6.135).

In the case that the smallest eigenvalue λ_N is multiple but not single (i.e., $\lambda_N = \lambda_{N-1} = \cdots = \lambda_K = \lambda$), (6.141) still holds but (6.143) becomes

$$\lim_{t\to\infty} \boldsymbol{W}(t) = \sum_{i=K}^{N} \lim_{t\to\infty} y_i(t)\boldsymbol{S}_i. \tag{6.146}$$

Multiplying the above equation by $\boldsymbol{R}$ on the left yields

$$\begin{aligned}\lim_{t\to\infty} \boldsymbol{R}\boldsymbol{W}(t) &= \sum_{i=K}^{N} \lim_{t\to\infty} y_i(t)\boldsymbol{R}\boldsymbol{S}_i = \sum_{i=K}^{N} \lim_{t\to\infty} \lambda_i y_i(t)\boldsymbol{S}_i \\ &= \lambda \sum_{i=K}^{N} \lim_{t\to\infty} y_i(t)\boldsymbol{S}_i = \lambda \lim_{t\to\infty} \boldsymbol{W}(t).\end{aligned} \tag{6.147}$$

Therefore, the $\lim_{t\to\infty} \boldsymbol{W}(t)$ is still the eigenvector corresponding to the multiple eigenvalue λ of the matrix $\boldsymbol{R}$. This fact and (6.145) both guarantee that (6.135) holds.

One of the shortcomings of the normalized constrained anti-Hebbian algorithm (6.132) is that it needs division computation in each iteration. To reduce the computational complexity incurred by the division computation, we propose an alternative algorithm known as the invariant-norm MCA algorithm.

Invariant-Norm MCA Algorithm

If we change the sign of the second and third terms of the PCA algorithm (6.38), we get

$$W(t+1) = W(t) - \gamma(t)z(t)\left[W^T(t)W(t)X(t) - z(t)W(t)\right], \tag{6.148}$$

which has the corresponding differential equations

$$\frac{dW(t)}{dt} = -W^T(t)W(t)RW(t) + W^T(t)RW(t)W(t). \tag{6.149}$$

For (6.149), we have

$$\|W(t)\|^2 = \|W(0)\|^2, \qquad t \geq 0, \tag{6.150}$$

and if $W^T(0)S_N \neq 0$, then

$$\lim_{t\to\infty} W(t) = \pm\|W(0)\|S_N. \tag{6.151}$$

In effect, Equation (6.149) is equivalent to Equation (3.30). Therefore, (6.150) and (6.151) can be obtained immediately from Theorem 3.1 and Theorem 3.2.

Compared with (6.132), Equation (6.148) has avoided the division computation. As a result, (6.148) is more suitable than (6.132) in practical applications.

In addition, the weight vector of algorithms (6.132) and (6.148) will converge to the second minor component S_{N-1} if $y_N(0) = W^T(0)S_N = 0$ and $y_{N-1}(0) = W^T(0)S_{N-1} \neq 0$. We can also conclude that the weight vector $W(t)$ of the network will converge to the eigenvector S_{i-1} corresponding to the eigenvalue λ_{i-1} if $y_N(0) = y_{N-1}(0) = \cdots = y_i(0) = 0$ and $y_{i-1}(0) \neq 0$.

Complex-Valued MCA Algorithms

For the case that $X(t)$ takes a complex value, these three MCA algorithms and their related differential equations will become:

(1) for the constrained anti-Hebbian algorithm:

$$W_c(t+1) = W_c(t) - \gamma(t)\left[X_c(t)z_c(t) - z_c^T(t)z_c(t)W_c(t)\right] \tag{6.152}$$

and

$$\frac{dW_c(t)}{dt} = -R_cW_c(t) + W_c^T(t)R_cW_c(t)W_c(t), \tag{6.153}$$

(2) for the normalized constrained anti-Hebbian algorithm:

$$W_c(t+1) = W_c(t) - \gamma(t)\left(X_c(t)z_c(t) - \frac{z_c^T(t)z_c(t)}{W_c^T(t)W_c(t)}W_c(t)\right) \tag{6.154}$$

Table 6.3. *The values of the weight vector in learning phase*

Iteration	$w_1(t)$	$w_2(t)$	$w_3(t)$
0	−0.5773	0.5773	0.5773
50	−0.6085	0.5698	0.5516
100	−0.6252	0.5814	0.5202
1000	−0.7640	0.6482	0.0421
4900	−0.4613	0.7774	−0.4510
5000	−0.4553	0.7841	−0.4459

and

$$\frac{d\boldsymbol{W}_c(t)}{dt} = -\boldsymbol{R}_c\boldsymbol{W}_c(t) + \frac{\boldsymbol{W}_c^T(t)\boldsymbol{R}_c\boldsymbol{W}_c(t)}{\boldsymbol{W}_c^T(t)\boldsymbol{W}_c(t)}\boldsymbol{W}_c(t), \tag{6.155}$$

and

(3) for the invariant-norm MCA algorithm:

$$\boldsymbol{W}_c(t+1) = \boldsymbol{W}_c(t) - \gamma(t)\left[\boldsymbol{W}_c^T(t)\boldsymbol{W}_c(t)\boldsymbol{X}_c(t)z_c(t) - z_c^T(t)z_c(t)\boldsymbol{W}_c(t)\right] \tag{6.156}$$

and

$$\frac{\boldsymbol{W}_c(t)}{dt} = -\boldsymbol{W}_c^T(t)\boldsymbol{W}_c(t)\boldsymbol{R}_c\boldsymbol{W}_c(t) + \boldsymbol{W}_c^T(t)\boldsymbol{R}_c\boldsymbol{W}_c(t)\boldsymbol{W}_c(t), \tag{6.157}$$

where we used the same notation as given in (6.24)–(6.30).

In the following, we will present two simulation examples (Examples 6.5 and Example 6.6) using the algorithm (6.148) for the adaptive extraction of the first minor component. In these simulation examples, we generate a random vector $\boldsymbol{X}(t)$ with the autocorrelation matrix $\boldsymbol{R}$ and take it as the input signal. For simplicity, in the simulations we let $\gamma(t)$ be 0.005 and leave it unchanged during the learning phase. A more sophisticated and better selection of $\gamma(t)$ could be made according to the Robbins–Monro stochastic approximation procedures [46] and the methods proposed in Reference [45]. $\boldsymbol{W}(0)$ and $\boldsymbol{W}(f)$ are the weight vector in the initial state and in the convergent state, respectively. $\lambda(f)$ is the eigenvalue computed by the weight vector $\boldsymbol{W}(f)[\lambda(f) = \boldsymbol{W}^T(f)\boldsymbol{R}\boldsymbol{W}(f)/\boldsymbol{W}^T(f)\boldsymbol{W}(f)]$. $\lambda(a)$ is the exact smallest eigenvalue of the autocorrelation matrix $\boldsymbol{R}$ of the input signals. Obviously, $\lambda(f) = \lambda(a)$. For further illustration, Tables 6.3 and 6.4 and Figures 6.9 and 6.10 describe the learning process of the weight vector $\boldsymbol{W}(t)$.

Table 6.4. *The values of the weight vector in learning phase*

Iteration	$w_1(t)$	$w_2(t)$	$w_3(t)$	$w_4(t)$
0	−0.5000	0.5000	0.5000	−0.5000
50	−0.5036	0.4993	0.4911	−0.5059
100	−0.5135	0.4917	0.4790	−0.5152
1000	−0.5643	0.4131	0.4461	−0.5608
10000	−0.3164	−0.0187	0.7240	−0.6321
30000	0.2220	−0.6104	0.7062	−0.3744

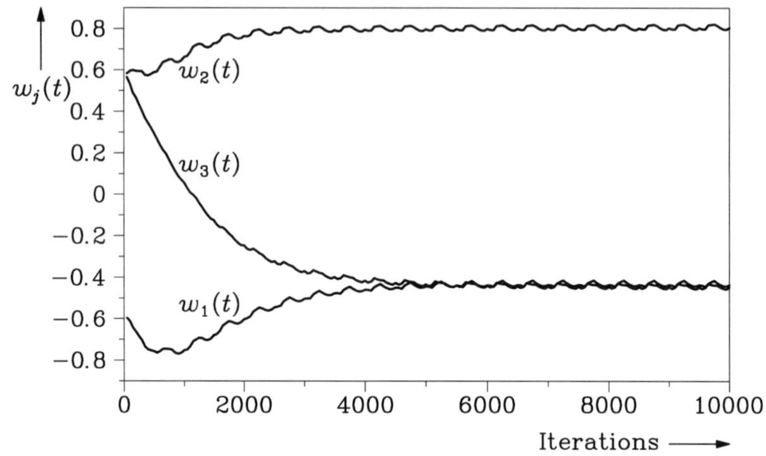

Figure 6.9. The learning process of the weight vector $\boldsymbol{W}(t)$ in Example 6.5.

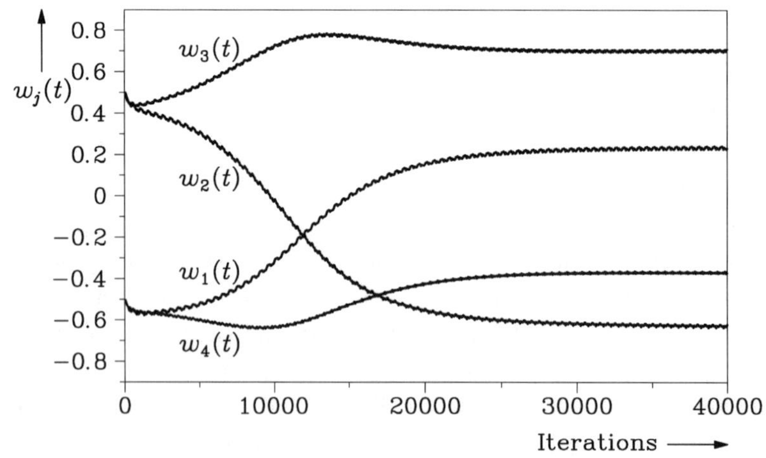

Figure 6.10. The learning process of the weight vector $\boldsymbol{W}(t)$ in Example 6.6.

Example 6.5

$$R = \begin{pmatrix} 0.4035 & 0.2125 & 0.0954 \\ 0.2125 & 0.3703 & 0.2216 \\ 0.0954 & 0.2216 & 0.4159 \end{pmatrix}$$

$$W(0) = \begin{pmatrix} -0.5773 \\ 0.5773 \\ 0.5773 \end{pmatrix}, \qquad W(f) = \begin{pmatrix} -0.4553 \\ 0.7841 \\ -0.4459 \end{pmatrix}$$

$$\lambda(f) = 0.1235, \qquad \lambda(a) = 0.1235.$$

Example 6.6

$$R = \begin{pmatrix} 0.2947 & 0.1982 & 0.1167 & 0.0377 \\ 0.1982 & 0.2989 & 0.2015 & 0.0953 \\ 0.1167 & 0.2015 & 0.2943 & 0.1869 \\ 0.0377 & 0.0953 & 0.1869 & 0.2734 \end{pmatrix}$$

$$W(0) = \begin{pmatrix} -0.5000 \\ 0.5000 \\ 0.5000 \\ -0.5000 \end{pmatrix}, \qquad W(f) = \begin{pmatrix} 0.2220 \\ -0.6104 \\ 0.7062 \\ -0.3744 \end{pmatrix}$$

$$\lambda(f) = 0.0552, \qquad \lambda(a) = 0.0552.$$

6.4.2 Adaptive Extraction of the Multiple Minor Components

Several algorithms for multiple minor components have been proposed. These are presented next.

Oja's MCA Algorithm

Oja [44] extended the constrained anti-Hebbian rule (6.127) for the multiple minor components as

$$\begin{aligned} W_j(t+1) = W_j(t) + \gamma(t)\Big\{ &- X(t)z_j(t) + \big[z_j^2(t) + 1 \\ &- W_j^T(t)W_j(t)\big] W_j(t) - \alpha \sum_{i>j} z_i(t)z_j(t)W_i(t)\Big\} \end{aligned}$$

$$(j = N, N-1, \ldots, N-M+1), \tag{6.158}$$

which has the differential equations

$$\frac{d\boldsymbol{W}_j(t)}{dt} = -\boldsymbol{R}\boldsymbol{W}_j(t) + \boldsymbol{W}_j^T(t)\boldsymbol{R}\boldsymbol{W}_j(t)\boldsymbol{W}_j(t) + \boldsymbol{W}_j(t) - \boldsymbol{W}_j^T(t)\boldsymbol{W}_j(t)\boldsymbol{W}_j(t) - \alpha\sum_{i>j}\boldsymbol{W}_i^T(t)\boldsymbol{R}\boldsymbol{W}_j(t)\boldsymbol{W}_i(t) \quad (j = N, N-1, \ldots, N-M+1), \tag{6.159}$$

where the network structure and all variables are the same as shown in Figure 6.5 except that the output neurons are indexed by the order $N, N-1, \ldots, N-M+1$.

Clearly, the first term $-\boldsymbol{X}(t)z_j(t)$ between the curly braces in (6.158) is the anti-Hebbian term in (6.127); $[z_j^2(t) + 1 - \boldsymbol{W}_j^T(t)\boldsymbol{W}_j(t)]\boldsymbol{W}_j(t)$ is the constraint term (also called the forgetting term), which is more complicated than that in the SGA algorithm or GHA. The third term $\alpha\sum_{i>j} z_i(t)z_j(t)\boldsymbol{W}_i(t)$ plays the key role in performing the Gram–Schmidt orthonormalization.

Oja [44] showed that if $\boldsymbol{W}_j^T(0)\boldsymbol{S}_j \neq 0$, $\|\boldsymbol{W}_j(0)\| = 1$, $\lambda_{N-M+1} < 1$, and $\alpha > \frac{\lambda_{N-M+1}}{\lambda_N} - 1$, then the weight vector in (6.159) will converge to the minor components $\boldsymbol{S}_N, \boldsymbol{S}_{N-1}, \ldots, \boldsymbol{S}_{N-M+1}$, that is,

$$\lim_{t\to\infty} \boldsymbol{W}_j(t) = \pm\boldsymbol{S}_j \qquad (j = N, N-1, \ldots, N-M+1). \tag{6.160}$$

Note that the condition that $\alpha > \frac{\lambda_{N-M+1}}{\lambda_N} - 1$ is essential in the sense that if it does not hold, then $\boldsymbol{W}_{N-M+1}(t)$ will not converge to $\boldsymbol{S}_{N-M+1}$. If α is larger than but close to $\frac{\lambda_{N-M+1}}{\lambda_N} - 1$, then the convergence will be slow. Fast convergence is obtained when α is large enough.

The Generalization of the Invariant-Norm MCA Algorithm

There are two forms for generalizing the invariant-norm MCA algorithm (6.148) to extract the multiple minor components. One is

$$\boldsymbol{W}_j(t+1) = \boldsymbol{W}_j(t) - \gamma(t)z_j(t)\Big(\boldsymbol{W}_j^T(t)\boldsymbol{W}_j(t)\boldsymbol{X}(t) - z_j(t)\boldsymbol{W}_j(t) - \beta\boldsymbol{W}_j^T(t)\boldsymbol{W}_j(t)\sum_{i>j} z_i(t)\boldsymbol{W}_i(t)\Big) \quad (j = N, N-1, \ldots, N-M+1). \tag{6.161}$$

Its corresponding differential equations are

$$\frac{d\boldsymbol{W}_j(t)}{dt} = -\boldsymbol{W}_j^T(t)\boldsymbol{W}_j(t)\Big[\boldsymbol{I} - \beta\sum_{i>j}\boldsymbol{W}_i(t)\boldsymbol{W}_i^T(t)\Big]\boldsymbol{R}\boldsymbol{W}_j(t) + \boldsymbol{W}_j^T(t)\boldsymbol{R}\boldsymbol{W}_j(t)\boldsymbol{W}_j(t) \quad (j = N, N-1, \ldots, N-M+1), \tag{6.162}$$

where β is a positive constant that affects the convergence speed of the algorithm (6.161). Unfortunately, (6.162) does not possess the invariant-norm property of (6.149) for $j < N$.

The other form for generalizing the invariant-norm MCA algorithm (6.148) is

$$\boldsymbol{W}_j(t+1) = \boldsymbol{W}_j(t) - \gamma(t) z_j(t) \left\{ \boldsymbol{W}_j^T(t)\boldsymbol{W}_j(t)\left[\boldsymbol{X}(t) - \sum_{i>j} z_i(t)\boldsymbol{W}_i(t)\right] - z_j(t)\boldsymbol{W}_j(t) + \sum_{i>j} z_i(t)\boldsymbol{W}_j^T(t)\boldsymbol{W}_i(t)\boldsymbol{W}_j(t) \right\}$$
$$(j = N, N-1, \ldots, N-M+1). \qquad (6.163)$$

The corresponding continuous-time differential equations of (6.163) are

$$\frac{d\boldsymbol{W}_j(t)}{dt} = -\boldsymbol{W}_j^T(t)\boldsymbol{W}_j(t)\left[\boldsymbol{I} - \sum_{i>j} \boldsymbol{W}_i(t)\boldsymbol{W}_i^T(t)\right]\boldsymbol{R}\boldsymbol{W}_j(t) + \boldsymbol{W}_j^T(t)\left[\boldsymbol{I} - \sum_{i>j} \boldsymbol{W}_i(t)\boldsymbol{W}_i^T(t)\right]\boldsymbol{R}\boldsymbol{W}_j(t)\boldsymbol{W}_j(t)$$
$$(j = N, N-1, \ldots, N-M+1). \qquad (6.164)$$

Using

$$\boldsymbol{R}_j = \boldsymbol{R} - \sum_{i>j} \boldsymbol{W}_i(t)\boldsymbol{W}_i^T(t)\boldsymbol{R} \qquad (j = N, N-1, \ldots, N-M-1) \qquad (6.165)$$

and Equations (6.96) and (6.97), we can write (6.164) as

$$\frac{d\boldsymbol{W}_j(t)}{dt} = -\psi_j(t)\boldsymbol{R}_j\boldsymbol{W}_j(t) + \phi_j(t)\boldsymbol{W}_j(t)$$
$$(j = N, N-1, \ldots, N-M+1). \qquad (6.166)$$

Clearly, (6.163) can be obtained by simply changing the sign of all the terms between the curly braces of the invariant-norm PCA algorithm (6.92). This shows that Equation (6.166) has the invariant-norm property, that is, the norm of each weight vector of (6.166) is invariant and equal to the norm of the initial vector during the time evolution.

Using the same method as that for the proof of the convergence of the PCA algorithms, we can also prove that if the initial vector satisfies $\boldsymbol{W}_j^T(0)\boldsymbol{S}_j \neq 0$ (for $j = N, N-1, \ldots, N-M+1$), then the weight vectors provided by (6.161) and (6.163) will converge to the corresponding minor components $\boldsymbol{S}_N, \boldsymbol{S}_{N-1}, \ldots, \boldsymbol{S}_{N-M+1}$, that is,

$$\lim_{t\to\infty} \boldsymbol{W}_j(t) = \pm b_j \boldsymbol{S}_j,$$

where b_j is a positive constant.

Note that for (6.163), $b_j = \sqrt{\boldsymbol{W}_j^T(0)\boldsymbol{W}_j(0)}$ but for (6.161), b_j is determined by the initial values of the weight vector and the eigenvalues.

Complex-Valued Multiple MCA Algorithms

For a complex-valued input vector $\boldsymbol{X}(t)$, we can generalize the algorithms (6.158), (6.161), and (6.163) and their associated differential equations as follows:

(1) For the algorithm (6.158):

$$\boldsymbol{W}_{jc}(t+1) = \boldsymbol{W}_{jc}(t) + \gamma(t)\Big\{ - \boldsymbol{X}_c(t) z_{jc}(t) + \big[z_{jc}^T(t) z_{jc}(t) + 1 - \boldsymbol{W}_{jc}^T(t)\boldsymbol{W}_{jc}(t)\big]\boldsymbol{W}_{jc}(t) - \alpha \sum_{i>j} z_{ic}^T(t) z_{jc}(t)\boldsymbol{W}_{ic}(t)\Big\}, \tag{6.167}$$

$$\frac{d\boldsymbol{W}_{jc}(t)}{dt} = -\boldsymbol{R}_c\boldsymbol{W}_{jc}(t) + \boldsymbol{W}_{jc}^T(t)\boldsymbol{R}_c\boldsymbol{W}_{jc}(t)\boldsymbol{W}_{jc}(t) + \boldsymbol{W}_{jc}(t) - \boldsymbol{W}_{jc}^T(t)\boldsymbol{W}_{jc}(t)\boldsymbol{W}_{jc}(t) - \alpha\sum_{i>j}\boldsymbol{W}_{ic}^T(t)\boldsymbol{R}_c\boldsymbol{W}_{jc}(t)\boldsymbol{W}_{ic}(t)$$
$$(j = N, N-1, \ldots, N-M+1). \tag{6.168}$$

(2) For the algorithm (6.161):

$$\boldsymbol{W}_{jc}(t+1) = \boldsymbol{W}_{jc}(t) - \gamma(t)\Big[\boldsymbol{W}_{jc}^T(t)\boldsymbol{W}_{jc}(t)\boldsymbol{X}_c(t) z_{jc}(t) - z_{jc}^T(t) z_{jc}(t)\boldsymbol{W}_{jc}(t) - \beta\boldsymbol{W}_{jc}^T(t)\boldsymbol{W}_{jc}(t)\sum_{i>j} z_{jc}^T(t) z_{ic}(t)\boldsymbol{W}_{ic}(t)\Big], \tag{6.169}$$

$$\frac{d\boldsymbol{W}_{jc}(t)}{dt} = -\boldsymbol{W}_{jc}^T(t)\boldsymbol{W}_{jc}(t)\Big[\boldsymbol{I} - \beta\sum_{i>j}\boldsymbol{W}_{ic}(t)\boldsymbol{W}_{ic}^T(t)\Big]\boldsymbol{R}_c\boldsymbol{W}_{jc}(t) + \boldsymbol{W}_{jc}^T(t)\boldsymbol{R}_c\boldsymbol{W}_{jc}(t)\boldsymbol{W}_{jc}(t)$$
$$(j = N, N-1, \ldots, N-M+1). \tag{6.170}$$

(3) For the algorithm (6.163):

$$\begin{aligned} W_{jc}(t+1) = W_{jc}(t) - \gamma(t)\Big\{ & W_{jc}^T(t)W_{jc}(t)\Big[X_c(t)z_{jc}(t) \\ & - \sum_{i>j} z_{jc}^T(t)z_{ic}(t)W_{ic}(t)\Big] - z_{jc}^T(t)z_{jc}(t)W_{jc}(t) \\ & + \sum_{i>j} z_{jc}^T(t)z_{ic}(t)W_{jc}^T(t)W_{ic}(t)W_{jc}(t)\Big\}, \end{aligned} \tag{6.171}$$

$$\begin{aligned} \frac{dW_{jc}(t)}{dt} = & -W_{jc}^T(t)W_{jc}(t)\Big[I - \sum_{i>j} W_{ic}(t)W_{ic}^T(t)\Big]R_c W_{jc}(t) \\ & + W_{jc}^T(t)\Big[I - \sum_{i>j} W_{ic}(t)W_{ic}^T(t)\Big]RW_{jc}(t)W_{jc}(t) \\ & (j = N, N-1, \ldots, N-M+1). \end{aligned} \tag{6.172}$$

In some practical applications, the smallest eigenvalue may be zero or very small; this could give rise to the divergence or very slow convergence of these MCA algorithms. This problem can be overcome by adding a random signal vector $\epsilon(t) = [\epsilon_1(t), \epsilon_2(t), \ldots, \epsilon_N(t)]^T$ to the input vector $X(t)$. The random signal vector $\epsilon(t)$ is generated by one of several available techniques [47] so as to be a stationary and ergodic process with zero mean and variance matrix $\sigma^2 I$, where σ^2 is a positive constant. It is easy to find that the smallest eigenvalue of the autocorrelation matrix of the signal vector $X(t) + \epsilon(t)$ is $\sigma^2 + \lambda_N$ (which can be made sufficiently large), but the eigenvector corresponding to the eigenvalue $\sigma^2 + \lambda_N$ is still S_N (i.e., the minor component of the input vector $X(t)$). As a result, all of these MCA algorithms can be used merely by replacing the input vector $X(t)$ with $X(t) + \epsilon(t)$.

Now we will present two simulation examples (Example 6.7 and Example 6.8) utilizing the algorithm (6.163) and its differential equations (6.164). The first example has distinct eigenvalues; the second example has nondistinct eigenvalues. In these simulations, we let $\gamma(t)$ be 0.001 and leave it unchanged during the learning phase. R is the given autocorrelation matrix with $N \times N$ elements; $W_j(f)$ and $W_j(0)$ (for $j = N, N-1, \ldots, N-M+1$) are the weight vector of the equations (6.164) in the steady state and the initial state ($\|W_j(0)\| = 1$), respectively. $\phi_j(f)$ is the eigenvalue provided by the variable $\phi_j(t)$ in the steady state. $\lambda_j(a)$ (for $j = N, N-1, \ldots, N-M+1$) are the exact eigenvalues of the matrix R. Obviously, $\phi_j(f) = \lambda_j(a)$ (for $j = N, N-1, \ldots, N-M+1$). Note that we selected $M = 2$. For further illustrations, Figures 6.11 and 6.12 describe the time evolution of the variables $\phi_j(t)$.

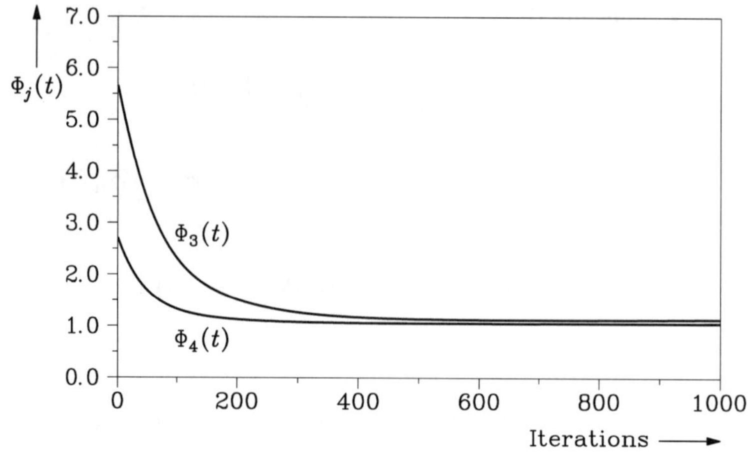

Figure 6.11. The dynamic curves of the variables $\phi_j(t)$ (for $j = 3, 4$) in Example 6.7.

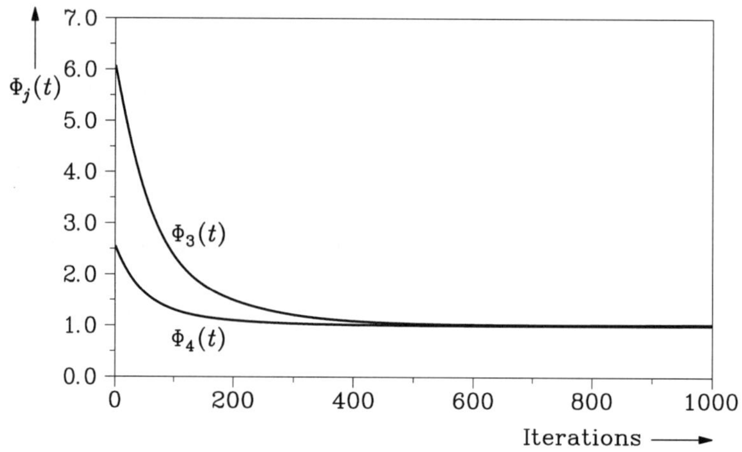

Figure 6.12. The dynamic curves of the variables $\phi_j(t)$ (for $j = 3, 4$) in Example 6.8.

Example 6.7

$$\boldsymbol{R} = \begin{pmatrix} 2.7262 & 1.1921 & 1.0135 & -3.7443 \\ 1.1921 & 5.7048 & 4.3103 & -3.0969 \\ 1.0135 & 4.3103 & 5.1518 & -2.6711 \\ -3.7443 & -3.0969 & -2.6711 & 9.4544 \end{pmatrix}$$

$$\boldsymbol{W}_4(f) = \begin{pmatrix} 0.9104 \\ 0.0183 \\ 0.0287 \\ 0.4206 \end{pmatrix}, \qquad \boldsymbol{W}_3(f) = \begin{pmatrix} 0.0093 \\ 0.7058 \\ -0.7271 \\ 0.0351 \end{pmatrix}$$

$$W_4(0) = \begin{pmatrix} 1.0000 \\ 0.0000 \\ 0.0000 \\ 0.0000 \end{pmatrix}, \qquad W_3(0) = \begin{pmatrix} 0.0000 \\ 1.0000 \\ 0.0000 \\ 0.0000 \end{pmatrix}$$

$$\phi_4(f) = 1.0486, \qquad \phi_3(f) = 1.1047$$

$$\lambda_4(a) = 1.0484, \qquad \lambda_3(a) = 1.1048.$$

Example 6.8

$$R = \begin{pmatrix} 2.5678 & 1.1626 & 0.9302 & -3.5269 \\ 1.1626 & 6.1180 & 4.3627 & -3.4317 \\ 0.9302 & 4.3627 & 4.7218 & -2.7970 \\ -3.5269 & -3.4317 & -2.7970 & 9.0905 \end{pmatrix}$$

$$W_4(f) = \begin{pmatrix} 0.9059 \\ 0.0403 \\ 0.0444 \\ 0.4261 \end{pmatrix}, \qquad W_3(f) = \begin{pmatrix} 0.0471 \\ 0.6701 \\ -0.7577 \\ 0.0457 \end{pmatrix}$$

$$W_4(0) = \begin{pmatrix} 1.0000 \\ 0.0000 \\ 0.0000 \\ 0.0000 \end{pmatrix}, \qquad W_3(0) = \begin{pmatrix} 0.0000 \\ 1.0000 \\ 0.0000 \\ 0.0000 \end{pmatrix}$$

$$\lambda_4(f) = 1.0001, \qquad \lambda_3(f) = 1.0003$$

$$\lambda_4(a) = 1.0000, \qquad \lambda_3(a) = 1.0000.$$

6.5 Robust and Nonlinear PCA Algorithms and Networks

The previous sections of this chapter were restricted to linear neural networks. There are, however, several limitations of linear neural networks and their learning algorithms:

(1) They can only provide and extract the second-order statistics (correlation) of the input signal vector $\boldsymbol{X}(t)$. Thus, they characterize completely only Gaussian data and stationary, linear processing operations. However, in applications of signal processing with higher-order statistics, it is necessary to extract adaptively the higher-order statistics of the input signal.

(2) They can realize only linear input–output mappings. Additional hidden layers with linear processing do not contribute anything new.

(3) Their outputs are mutually uncorrelated but usually not independent. Even though the input signal consists of a mixture of statistically independent subsignals, the uncorrelated outputs are generally some linear combinations

of the subsignals only. In many applications, such as signal separation, the subsignals themselves are desired, but linear networks cannot provide the desired subsignals directly.

(4) The learning algorithms discussed here are mainly based on the gradient-type methods and hence are of relatively slow convergence.

For these reasons, it is meaningful to add nonlinearities to the linear networks and related learning algorithms [30, 31, 48, 49, 50]. With nonlinearities in the network and algorithms, the situation will greatly change. First, the input–output mapping becomes generally nonlinear, which is in effect a major argument for using neural networks. Nonlinear processing of data is often more efficient. Second, the higher-order statistics of the input signal are introduced and can be extracted. Third, the independence of the outputs is generally increased so that the subsignals can directly be separated from the outputs of the network. Fourth, the convergence of the algorithms can be improved.

The direct way to add nonlinearities is to replace the linear input–output relationship of the output neurons in Figure 6.5 by a nonlinear function $f(\cdot)$. As a result, the output $z_j(t)$ in (6.63) becomes

$$z_j(t) = f\left(\sum_{i=1}^{N} w_{ji}(t)x_i(t)\right) = f\left[\boldsymbol{W}_j^T(t)\boldsymbol{X}(t)\right] \qquad (j = 1, 2, \ldots, M). \tag{6.173}$$

With this in mind, we can obtain the nonlinear versions of the PSA algorithm (6.64), the SGA algorithm (6.81), and the GHA (6.83) as follows:

(1) For the PSA algorithm (6.64):

$$\boldsymbol{W}_j(t+1) = \boldsymbol{W}_j(t) + \gamma(t)f\left[\boldsymbol{W}_j^T(t)\boldsymbol{X}(t)\right] \times \left(\boldsymbol{X}(t) - \sum_{i=1}^{M} f\left[\boldsymbol{W}_i^T(t)\boldsymbol{X}(t)\right]\boldsymbol{W}_i(t)\right) \qquad (j = 1, 2, \ldots, M). \tag{6.174}$$

(2) For the SGA algorithm (6.81):

$$\boldsymbol{W}_j(t+1) = \boldsymbol{W}_j(t) + \gamma(t)f\left[\boldsymbol{W}_j^T(t)\boldsymbol{X}(t)\right]\left(\boldsymbol{X}(t) - f\left[\boldsymbol{W}_j^T(t)\boldsymbol{X}(t)\right]\boldsymbol{W}_j(t) - 2\sum_{i=1}^{j-1} f\left[\boldsymbol{W}_i^T(t)\boldsymbol{X}(t)\right]\boldsymbol{W}_i(t)\right) \qquad (j = 1, 2, \ldots, M). \tag{6.175}$$

(3) For the GHA (6.83):

$$\boldsymbol{W}_j(t+1) = \boldsymbol{W}_j(t) + \gamma(t) f\left[\boldsymbol{W}_j^T(t)\boldsymbol{X}(t)\right] \times \left(\boldsymbol{X}(t) - \sum_{i=1}^{j} f\left[\boldsymbol{W}_i^T(t)\boldsymbol{X}(t)\right]\boldsymbol{W}_i(t)\right) \quad (j = 1, 2, \ldots, M). \tag{6.176}$$

Clearly, (6.174)–(6.176) are obtained from (6.64), (6.81), and (6.83) just by replacing the linear output $\boldsymbol{W}_j^T(t)\boldsymbol{X}(t)$ with the nonlinear output $f[\boldsymbol{W}_j^T(t)\boldsymbol{X}(t)]$. Making this same substitution, one can easily obtain the nonlinear versions of the APEX algorithm (6.107), the LEAP algorithm (6.88), and the invariant-norm PCA algorithm (6.92).

Moreover, (6.174) and (6.176) are approximative stochastic gradient algorithms for minimizing the mean-square representation error

$$\begin{aligned} E(t) &= E\left[\left\|\boldsymbol{e}_j(t)\right\|^2\right] = E\left[\left\|\boldsymbol{X}(t) - \hat{\boldsymbol{X}}(t)\right\|^2\right] \\ &= E\left[\left\|\boldsymbol{X}(t) - \sum_{i=1}^{I(j)} f\left[\boldsymbol{W}_i^T(t)\boldsymbol{X}(t)\right]\boldsymbol{W}_i(t)\right\|^2\right], \end{aligned} \tag{6.177}$$

where

$$\begin{aligned} \boldsymbol{e}_j(t) &= \boldsymbol{X}(t) - \sum_{i=1}^{I(j)} f\left[\boldsymbol{W}_i^T(t)\boldsymbol{X}(t)\right]\boldsymbol{W}_i(t) \\ &= \boldsymbol{X}(t) - \sum_{i=1}^{I(j)} z_i(t)\boldsymbol{W}_i(t) \end{aligned} \tag{6.178}$$

is the error vector and $I(j) = M$ for (6.174) and $I(j) = j$ for (6.176). Therefore, (6.174) and (6.176) can be unified as

$$\boldsymbol{W}_j(t+1) = \boldsymbol{W}_j(t) + \gamma(t) f\left[\boldsymbol{W}_j^T(t)\boldsymbol{X}(t)\right]\boldsymbol{e}_j(t) \qquad (j = 1, 2, \ldots, M). \tag{6.179}$$

A detailed derivation for (6.179) can be made in a manner similar to that for (6.64) (see Reference [31]).

In (6.177), the approximation $\hat{\boldsymbol{X}}(t)$ is linear with respect to the basis vectors $\boldsymbol{W}_j(t)$ (the weight vectors) of the expansion, but the coefficients $f[\boldsymbol{W}_j^T(t)\boldsymbol{X}(t)]$ (the outputs of neurons) of the expansion are generally nonlinear. Its main advantage seems to be that the nonlinear coefficients implicitly take higher-order statistical information into account and the outputs of the neurons become more independent than those in the linear neural networks after convergence. Extensive simulation results have shown this advantage, but a strict proof is lacking.

As mentioned in Section 6.3, the PSA algorithm (6.64) and the GHA (6.83) can be obtained by minimizing the mean-square representation error

$$E(t) = E\left[\|\boldsymbol{e}_j(t)\|^2\right] = E\left[\|\boldsymbol{X}(t) - \sum_{i=1}^{I(j)} \boldsymbol{W}_i^T(t)\boldsymbol{X}(t)\boldsymbol{W}_i(t)\|^2\right], \tag{6.180}$$

where $I(j) = M$ for (6.64) and $I(j) = j$ for (6.83). Equation (6.180) can be rewritten in a general form as

$$E(t) = \boldsymbol{L}^T E\left[g(\boldsymbol{e}_j(t))\right], \tag{6.181}$$

where $\boldsymbol{L} = [1, 1, \ldots, 1]^T$. Clearly, if $g(e) = e^2$, (6.181) coincides with (6.180). The form of Equation (6.181) indicates that we can also add the nonlinearities just by replacing $g(e) = e^2$ with more general nonlinear functions, such as

(1) the absolute value function $g(e) = |e|$,
(2) the logistic function $g(e) = \beta^2 \ln\left(\cosh(\frac{e}{\beta})\right)$,
(3) Huber's function

$$g(e) = \begin{cases} \frac{e^2}{2}, & |e| \le \beta, \\ \beta|e| - \frac{\beta^2}{2}, & |e| > \beta, \end{cases}$$

(4) Talvar's function

$$g(e) = \begin{cases} \frac{e^2}{2}, & |e| \le \beta, \\ \frac{\beta^2}{2}, & |e| > \beta, \end{cases}$$

and

(5) Andrews's function

$$g(e) = \begin{cases} \beta^2(1 - \cos(\frac{e}{\beta})), & |e| \le \pi\beta, \\ 2\beta^2, & |e| > \beta, \end{cases}$$

where β is a positive and problem-dependent parameter, called the cutoff parameter (typically, $1 \le \beta \le 3$).

Minimizing (6.181) with respect to the weight vector $\boldsymbol{W}_j$ (for $j = 1, 2, \ldots, M$) leads to a gradient algorithm as

$$\boldsymbol{W}_j(t+1) = \boldsymbol{W}_j(t) + \gamma(t)\left[\boldsymbol{W}_j^T(t) f(\boldsymbol{e}_j(t))\boldsymbol{X}(t) + \boldsymbol{X}^T(t)\boldsymbol{W}_j(t) f(\boldsymbol{e}_j(t))\right], \tag{6.182}$$

where $f(e) = dg(e)/de$. Note that in (6.182) the output neurons retain their linear input–output relationships.

As shown in [31], we can assume that the average value of $\boldsymbol{W}_j^T(t) f(\boldsymbol{e}_j(t))$ is close to zero because the error vector $\boldsymbol{e}_j(t)$ should be relatively small after the initial convergence and its sign can be either positive or negative. Hence, Equation (6.182) becomes

$$\begin{aligned} \boldsymbol{W}_j(t+1) &= \boldsymbol{W}_j(t) + \gamma(t)\boldsymbol{X}^T(t)\boldsymbol{W}_j(t) f(\boldsymbol{e}_j(t)) \\ &= \boldsymbol{W}_j(t) + \gamma(t) z_j(t) f(\boldsymbol{e}_j(t)) \qquad (j = 1, 2, \ldots, M). \end{aligned} \tag{6.183}$$

It is interesting to compare (6.183) with (6.179). They closely resemble each other; the only difference is that in (6.183) the nonlinearity $f(e)$ is applied to the error $\boldsymbol{e}_j(t)$, whereas in (6.179), the nonlinearity is applied to the output neurons. Since the output neurons in (6.183) still perform a linear operation, the weights of (6.183) will still converge to the principal components (subspace) of the input signal vector. However, (6.179) will yield different weights. In addition, in comparison with the PSA algorithm (6.64) and the GHA (6.83), the algorithm (6.183) is much more robust in the case where the input signal is corrupted by outliers or spiky impulsive noise. As a result, (6.183) is also called the robust principal component (subspace) analysis algorithm [30, 51, 52, 53].

6.6 Unsupervised Learning Algorithms for Higher-Order Statistics

An alternative method to extract adaptively the higher-order statistics of the input signal is to employ the high-order neural network given in Figure 1.8 and to generalize the above unsupervised learning algorithms of the linear neural networks for this high-order neural network.

Let us reconsider a high-order neural network with three layers – the input layer, the high-order layer, and the output layer – as shown in Figure 1.8. Similar to the input layer of the linear neural network in Figure 6.5, the input layer of the high-order neural network simply feeds the input signal vector $\boldsymbol{X}(t)$ to the second layer (the high-order layer) without any modification. The units of the high-order layer are in effect multipliers whose outputs are computed as follows:

(1) The outputs of the first-order neurons are

$$q_i^{(1)}(t) = x_i(t) \qquad (i = 1, 2, \ldots, N). \tag{6.184}$$

(2) The outputs of the second-order neurons are determined by

$$q_{ij}^{(2)}(t) = x_i(t) x_j(t) \qquad (i, j = 1, 2, \ldots, N). \tag{6.185}$$

(3) The outputs of the k'th-order neurons are determined by

$$q_{i_1 i_2 \ldots i_k}^{(k)}(t) = x_{i_1}(t) x_{i_2}(t) \ldots x_{i_k}(t) \qquad (i_j = 1, 2, \ldots, N; j = 1, 2, \ldots, k). \tag{6.186}$$

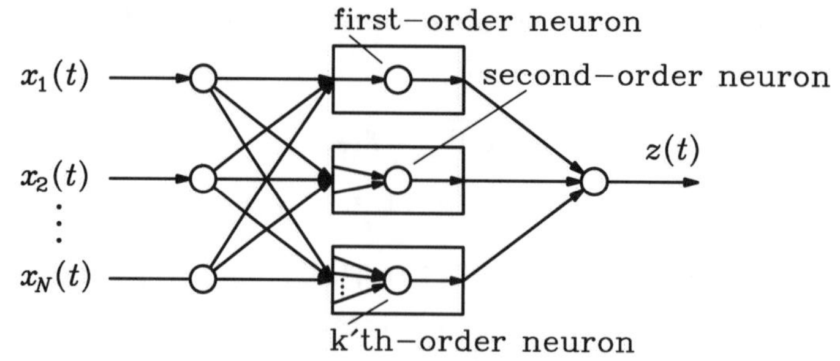

Figure 6.13. The high-order neuron network with one output neuron.

The connections between the high-order layer and the output layer are weighted, and the neurons in the output layer have a linear input–output relationship function so that they perform simple summations. For simplicity, we consider the case with only one output neuron as shown in Figure 6.13. The output of this neuron is

$$z(t) = \theta + z^{(1)}(t) + z^{(2)}(t) + \cdots + z^{(k)}(t), \tag{6.187}$$

where

$$z^{(1)}(t) = \sum_{i=1}^{N} W_i(t) q_i^{(1)}(t), \tag{6.188}$$

$$z^{(2)}(t) = \sum_{i=1}^{N}\sum_{j=1}^{N} W_{ij}(t) q_{ij}^{(2)}(t), \tag{6.189}$$

and

$$z^{(k)}(t) = \sum_{i_1=1}^{N}\sum_{i_2=1}^{N}\cdots\sum_{i_k=1}^{N} W_{i_1 i_2 \ldots i_k}(t) q_{i_1 i_2 \ldots i_k}^{(k)}(t). \tag{6.190}$$

Here θ is the threshold of the output neuron and $W_{i_1 i_2 \ldots i_k}(t)$ are the corresponding connection weights.

Based on Oja's learning rule (6.3), we obtain the following algorithm for training the weights of the high-order neural network:

(1) For the weights of the first-order neurons:

$$W_i(t+1) = W_i(t) + \gamma(t) z^{(1)}(t)\big(x_i(t) - z^{(1)}(t) W_i(t)\big) \qquad (i = 1, 2, \ldots, N). \tag{6.191}$$

Clearly, this equation is exactly the same as (6.3) except for the replacement of $z(t)$ by $z^{(1)}(t)$. Consequently, the weights of (6.191) will converge to the first principal component $\boldsymbol{S}_1$ of the input signal vector $\boldsymbol{X}(t)$.

(2) For the weights of the second-order neurons:

$$W_{ij}(t+1) = W_{ij}(t) + \gamma(t) z^{(2)}(t) \big(q_{ij}^{(2)}(t) - z^{(2)}(t) W_{ij}(t) \big) \qquad (i, j = 1, 2, \ldots, N), \tag{6.192}$$

which can be written in the matrix form as

$$\boldsymbol{W}(t+1) = \boldsymbol{W}(t) + \gamma(t) z^{(2)}(t) \big(\boldsymbol{q}^{(2)}(t) - z^{(2)}(t) \boldsymbol{W}(t) \big), \tag{6.193}$$

where $\boldsymbol{W}(t) = \{W_{ij}(t)\}$ and $\boldsymbol{q}^{(2)}(t) = \{q_{ij}^{(2)}(t)\}$. The corresponding differential equations of (6.192) are

$$\frac{dW_{ij}(t)}{dt} = \sum_{l=1}^{N} \sum_{m=1}^{N} C_{ijlm} W_{lm}(t) - \sum_{l=1}^{N} \sum_{m=1}^{N} \sum_{k=1}^{N} \sum_{s=1}^{N} W_{lm}(t) C_{kslm} W_{ks}(t) W_{ij}(t) \qquad (i, j = 1, 2, \ldots, N), \tag{6.194}$$

where C_{ijlm} and C_{kslm} are the fourth-order moments defined by

$$C_{ijlm} = E\big[x_i(t) x_j(t) x_l(t) x_m(t)\big] \tag{6.195}$$

and

$$C_{kslm} = E\big[x_k(t) x_s(t) x_l(t) x_m(t)\big] \qquad (i, j, k, l, m, s = 1, 2, \ldots, N). \tag{6.196}$$

Relabeling $(i, j) = a$, $(l, m) = b$, and $(k, s) = d$, (6.194) can be written as

$$\frac{dW_a(t)}{dt} = \sum_b C_{ab} W_b(t) - \sum_b \sum_d W_b(t) C_{bd} W_d(t) W_a(t). \tag{6.197}$$

Moreover, the differential equations (6.6) can be rewritten as

$$\frac{dw_i(t)}{dt} = \sum_j R_{ij} w_j(t) - \sum_j \sum_k w_j(t) R_{jk} w_k(t) w_i(t). \tag{6.198}$$

Comparing (6.197) with (6.198), we can envisage that the weight matrix $\boldsymbol{W}$ of (6.193) will converge to the subspace spanned by the eigenvectors of a matrix composed of the fourth-order moments, but we failed to give a rigorous proof.

(3) For the weights of the k'th-order neurons:

$$W_{i_1 i_2 \dots i_k}(t+1) = W_{i_1 i_2 \dots i_k}(t) + \gamma(t) z^{(k)}(t) \left(q^{(k)}_{i_1 i_2 \dots i_k}(t) - z^{(k)}(t) W_{i_1 i_2 \dots i_k}(t) \right)$$
$$(i_j = 1, 2, \dots, N; \quad j = 1, 2, \dots, k), \qquad (6.199)$$

and the associated differential equations are

$$\frac{dW_{i_1 i_2 \dots i_k}(t)}{dt} = \sum_{j_1 j_2 \cdots j_k} C_{i_1 i_2 \dots i_k j_1 j_2 \cdots j_k} W_{j_1 j_2 \cdots j_k}(t) - \sum_{j_1 j_2 \cdots j_k} \sum_{l_1 l_2 \dots l_k} W_{j_1 j_2 \dots j_k}(t)$$
$$\times C_{j_1 j_2 \cdots j_k l_1 l_2 \dots l_k} W_{l_1 l_2 \dots l_k}(t) W_{i_1 i_2 \dots i_k}(t)$$
$$(i_j = 1, 2, \dots, N; \quad j = 1, 2, \dots, k), \qquad (6.200)$$

where $C_{i_1 i_2 \dots i_k j_1 j_2 \dots j_k}$ and $C_{j_1 j_2 \dots j_k l_1 l_2 \dots l_k}$ are the $(2k)$th-order moments defined by

$$C_{i_1 i_2 \cdots i_k j_1 j_2 \dots j_k} = E\left[x_{i_1}(t) x_{i_2}(t) \cdots x_{i_k}(t) x_{j_1}(t) x_{j_2}(t) \cdots x_{j_k}(t) \right] \qquad (6.201)$$

and

$$C_{j_1 j_2 \cdots j_k l_1 l_2 \dots l_k} = E\left[x_{j_1}(t) x_{j_2}(t) \cdots x_{j_k}(t) x_{l_1}(t) x_{l_2}(t) \cdots x_{l_k}(t) \right]$$
$$(i_m, j_m, l_m = 1, 2, \dots, N; \quad m = 1, 2, \dots, k). \qquad (6.202)$$

The algorithm (6.199) will extract the features concerning the $(2k)$th-order moments of the input signal vector $\boldsymbol{X}(t)$, but a quantitative analysis is still lacking. In most applications, we select $k = 2$. This means that (6.192) is sufficient.

Note that the above algorithms can extract only the even-order statistics but not the odd-order (such as the third-order) statistics. To overcome this problem, we can replace the terms $z^{(1)}(t)$, $z^{(2)}(t)$, and $z^{(k)}(t)$ in (6.191), (6.192), and (6.199) by the total output $z(t)$. In this case, it will be very difficult and complicated to analyze the convergence of the corresponding learning algorithm, and more effective methods are being investigated. For more detail, one can refer to [54] and [55].

Bibliography

[1] Jolliffe, I. T., *Principal Component Analysis*, Springer-Verlag, New York, 1986.

[2] Becker, S., "Unsupervised Learning Procedures for Neural Networks," *Int. J. Neural Systems*, Vol. 2, 1991, pp. 17–33.

[3] Baldi, P. and Hornik, K., "Neural Networks and Principal Component Analysis: Learning from Examples Without Local Minima," *Neural Networks*, Vol. 2, 1989, pp. 52–58.

[4] Chauvin, Y., "Principal Component Analysis by Gradient Descent on a Constrained Linear Hebbian Cell," *Proc. Int. Joint Conf. on Neural Networks*, Washington D.C., Vol. I, 1989, pp. 373–80.

[5] Karhunen, J. and Oja, E., "Optimal Adaptive Compression of High-Dimensional Data," *Proc. Second Scandinavian Conf. on Image Analysis*, Helsinki, Finland, 1981, pp. 152–57.

[6] Karhunen, J., "Recursive Estimation of Eigenvector of Correlation Type Matrices for Signal Processing Applications," PhD Dissertation, Helsinki Univ., Finland, 1984.

[7] Dony, R. D. and Haykin, S., "Neural Networks Approaches to Image Compression," *Proc. IEEE*, Vol. 83, No. 2, 1995, pp. 288–303.

[8] Földiák, P., "Adaptive Network for Optimal Linear Feature Extraction," *Proc. Int. Joint Conf. on Neural Networks*, San Diego, Vol. I, 1989, pp. 401–6.

[9] Hornik, K. and Kuan, C. M., "Convergence Analysis for Local Feature Extraction Algorithm," *Neural Networks*, Vol. 5, 1992, pp. 229–40.

[10] Krogh, A. and Hertz, J. A., "Hebbian Learning of Principal Components," in *Parallel Processing in Neural Systems and Computers*, Eckmiller, R., et al., eds., Elsevier, Amsterdam, 1990, pp. 183–86.

[11] Linsker, R., "From Basic Network Principles to Neural Architecture," *Proceedings of the National Academy of Science, USA*, Vol. 83, 1986, pp. 7508–12.

[12] Linsker, R., "Self-Organization in a Perceptual Network," *IEEE Computer*, Vol. 21, 1988, pp. 105–17.

[13] Rubner, J. and Tavan, P., "A Self-Organization Network for Principal Analysis," *Europhysics Lett.*, Vol. 10, 1989, pp. 693–98.

[14] Yuille, A. L., Kammen, D. M., and Cohen, D. S., "Quadrature and the Development of Orientation Selective Cortical Cells by Hebb Rules," *Biol. Cybernetics*, Vol. 61, 1989, pp. 183–94.

[15] Hung, S. C. and Hung, Y. F., "Principal Component Vector Quantization for Abrupt Scene Changes," *Proc. IEEE Int. Symp. on Circuits and Systems*, 1992.

[16] Reddy, V. U., Egardt, B., and Kailath, T., "Least Squares Type Algorithm for Adaptive Implementation of Pisarenko's Harmonic Retrieval Method," *IEEE Trans. on Acoustics, Speech and Signal Processing*, Vol. 30, 1982, pp. 399–405.

[17] Banjanian, Z., Cruz, J. R., and Zrnic, D. S., "Eigen-Decomposition Methods for Frequency Estimation: A Unified Approach," *Proc. IEEE Int. Conf. on Acoustics, Speech and Signal Processing*, 1990, pp. 2595–98.

[18] Rahman, M. A. and Yu, K. B., "Total Least Squares Approach for Frequency Estimation Using Linear Prediction," *IEEE. Trans. on Acoustics, Speech and Signal Processing*, Vol. 35, No. 10, 1987, pp. 1440–54.

[19] Xu, L., Oja, E., and Suen, C. Y., "Modified Hebbian Learning for Curve and Surface Fitting," *Neural Networks*, Vol. 5, 1992, pp. 441–57.

[20] Schmidt, R. O., "Multiple Emitter Location and Signal Parameter Estimation," *IEEE Trans. on Antennas and Propagation*, Vol. 34, 1986, pp. 276–80.

[21] Griffiths, J. W. R., "Adaptive Array Processing, A Tutorial," *IEE Proc.*, Pt. F, Vol. 130, 1983, pp. 3–10.

[22] Klemm, R., "Adaptive Airborne MTI: An Auxiliary Channel Approach," *IEE Proc.*, Pt. F, Vol. 134, 1987, pp. 269–76.

[23] Barbarossa, S., Daddio, S. E., and Galati, G., "Comparison of Optimum and Linear Prediction Technique for Clutter Cancellation," *IEE Proc.*, Pt. F, Vol. 134, 1987, pp. 277–82.

[24] Oja, E., "A Simplified Neuron Model as a Principal Component Analyser," *J. Math. Biol.*, Vol. 15, 1982, pp. 267–73.

[25] Ljung, L., "Analysis of Recursive Stochastic Algorithm," *IEEE Trans. on Automatic Control*, Vol. 22, 1977, pp. 551–75.

[26] Kushner, H. J. and Clark, D. S., *Stochastic Approximation Methods for Constrained and Unconstrained Systems*, Springer-Verlag, New York, 1978.

[27] Oja, E. and Karhunen, J., "On Stochastic Approximation of the Eigenvectors and Eigenvalues of the Expectation of a Random Matrix," *J. Math. Analysis Appl.*, Vol. 106, 1985, pp. 69–84.

[28] Luo, F. L. and Li, Y. D., "Real-Time Computation of the Eigenvector Corresponding to the Largest Eigenvalue of a Positive Definite Matrix," *Neurocomputing*, Vol. 7, No.2, 1995, pp. 145–57.

[29] Oja, E., "Neural Networks, Principal Components and Subspace," *Int. J. Neural Systems*, Vol. 1, 1989, pp. 61–68.

[30] Karhunen, J. and Joutsensalo, J., "Representation and Separation of Signals Using Nonlinear PCA Type Learning," *Neural Networks*, Vol. 7, No. 1, 1994, pp. 113–28.

[31] Karhunen, J. and Joutsensalo, J., "Generalizations of Principal Component Analysis, Optimization Problems and Neural Networks," *Neural Networks*, Vol. 8, No. 4, 1995, pp. 549–62.

[32] Yan, W. Y., Helmke, U., and Moore, J. B. "Global Analysis of Oja's Flow for Neural Networks," *IEEE Trans. on Neural Networks*, Vol. 5, No. 5, 1994, pp. 674–83.

[33] Oja, E., Ogawa, H., and Wangviwattana, J., "Principal Component Analysis by Homogeneous Neural Networks, Part I: The Weight Subspace Criterion," *IEICE Trans.*, E-75D, 1992, pp. 366–75.

[34] Oja, E., Ogawa, H., and Wangviwattana, J., "Principal Component Analysis by Homogeneous Neural Networks, Part II: Analysis and Extensions of the Learning," *IEICE Trans.*, E-75D, 1992, pp. 376–82.

[35] Oja, E., *Subspace Methods of Pattern Recognition*, Research Studies Press/Wiley, Letchworth, England, 1983.

[36] Sanger, T. D., "Optimal Unsupervised Learning in a Single-Layer Linear Feedforward Neural Network," *Neural Networks*, Vol. 2, 1989, pp. 459–73.

[37] Chen, H. and Liu, R. W., "An On-Line Unsupervised Learning Machine for Adaptive Extraction," *IEEE Trans. on Circuits and Systems*, II, Vol. 41, 1994, pp. 87–98.

[38] Chen, H. and Liu, R. W., "Adaptive Distributed Orthogonalization Processing for Principal Component Analysis," *Proc. IEEE Int. Conf. on Acoustics, Speech and Signal Processing*, 1992, pp. 293–96.

[39] Luo, F. L., Unbehauen, R., and Li, Y. D., "A Principal Component Analysis Algorithm with Invariant Norm," *Neurocomputing*, Vol. 8, No. 2, 1995, pp. 213–21.

[40] Kung, S. Y. and Diammantaras, K. I., "A Neural Network Learning Algorithm for Adaptive Principal Component Extraction," *Proc. IEEE Int. Conf. on Acoustics, Speech and Signal Processing*, 1990, pp. 861–64.

[41] Kung, S. Y., Diammantaras, K. I., and Taur, J. S., "Adaptive Principal Component Extraction and Application," *IEEE Trans. on Signal Processing*, Vol. 42, No. 5, 1994, pp. 1202–17.

[42] Diammantaras, K. I. and Kung, S. Y., "An Unsupervised Neural Model for Oriented Principal Component," *Proc. IEEE Int. Conf. on Acoustics, Speech and Signal Processing*, 1991, pp. 1049–52.

[43] Luo, F. L., Unbehauen, R., and Reif, K, "Recurrent Neural Networks for Eigen-Decomposition of Positive Definite Matrix," *Proc. IEEE Workshop on Nonlinear Signal and Image Processing*, 1995, pp. 46–49.

[44] Oja, E., "Principal Components, Minor Components and Linear Neural Networks," *Neural Networks*, Vol. 5, 1992, pp. 927–35.

[45] Xu, L., Krzyzak, A., and Oja, E. "Neural Nets for Dual Subspace Pattern Recognition Method," *Int. J. Neural Systems*, Vol. 2, 1991, pp. 169–84.

[46] Robbins, H. and Monro, S., "A Stochastic Approximation Method," *Ann. Math. Stat.*, Vol. 22, 1951, pp. 400–7.

[47] Szajowski, W. J., "The Generation of Correlated Weibull Clutter for Signal Detection Problem," *IEEE Trans. on Aerospace and Electronic Systems*, Vol. 13, No. 5, 1977, pp. 536–40.

[48] Karhunen, J., "Optimization Criteria and Nonlinear PCA Neural Networks," *Proc. IEEE Int. Conf. on Neural Networks*, 1994, pp. 1241–46.

[49] Xu, L., "Theories for Unsupervised Learning: PCA and its Nonlinear Extensions," *Proc. IEEE Int. Conf. on Neural Networks*, 1994, pp. 1255–57.

[50] Palmieri, F. "Hebbian Learning and Self-Association in Nonlinear Neural Networks," *Proc. IEEE Int. Conf. on Neural Networks*, 1994, pp. 1258–63.

[51] Cichocki, A. and Unbehauen, R., "Robust Estimation of Principal Components by Using Neural Network Learning Algorithm," *Electronics Lett.*, Vol. 29, 1993, pp. 1869–70.

[52] Cichocki, A. and Unbehauen, R., *Neural Networks for Optimization and Signal Processing*, Wiley Teubner Verlag, Chichester, England, 1993.

[53] Diammantaras, K. I. and Kung, S. Y., "Cross-Correlation Neural Network Models," *IEEE Trans. on Signal Processing*, Vol. 42, No. 11, 1994, pp. 3218–23.

[54] Taylor, J. G. and Coombes, S., "Learning Higher Order Correlations," *Neural Networks*, Vol. 6, 1993, pp. 423–27.

[55] Softky, W. R. and Kammen, D. M., "Correlations in High Dimensional or Asymmetric Data Sets: Hebbian Neuronal Processing," *Neural Networks*, Vol. 4, 1993, pp. 337–47.

7

Neural Networks for Array Signal Processing

Estimating the directions of arrival (DOA) of sources is a central problem in array signal processing. Many methods for estimation of the DOA have been proposed, including the maximum likelihood (ML) technique [1, 2, 3], the minimum variance method of Capon [4], the minimum norm method of Reddi [5], the MUSIC method of Schmidt [6], and the propagator method of Marcos [7, 8]. Of all these available methods, the ML method has the best performance. Nonetheless, because of the high computational load of the multivariate nonlinear maximization problem involved, the ML technique did not become popular. Ziskind and Wax proposed an alternating projection method for computing the ML estimator that transforms the multivariate nonlinear maximization problem into a sequence of much simpler one-dimensional maximization problems [9]. However, this alternating projection method still involves costly matrix inversions. Suboptimal methods, such as the MUSIC method of Schmidt and the propagator of Marcos, are more prevalent than the ML technique when the signal-to-noise ratio and the number of samples are both not too small, because these suboptimal methods involve solving only a one-dimensional maximization problem and finding a subspace (signal subspace or noise subspace). In these suboptimal methods, the major computational burden lies in finding the signal subspace or noise subspace. For example, for the purpose of finding the noise subspace required in the MUSIC method, one has to perform the eigendecomposition. Although the propagator method of Marcos can find the noise subspace by a least-squares process from the cross-spectral matrix of the signals received on the array without eigendecomposition of matrices, this propagator method requires the computation of one matrix inversion and some matrix multiplications. As a result, its computational complexity is also intensive. In summary, each one of the aforementioned methods has certain advantages and limitations not only in terms of the estimation performance but also in terms of the computational complexity. In other words, the estimation performance and computational complexity form the central conflict in the DOA estimation methods. The main purpose of this chapter is to show how to use neural networks to overcome this conflict.

We organize this chapter as follows. Section 7.1 presents the neural network approaches for computing the ML estimator, alternating projection ML estimator, and the subspace of the propagator method. In Section 7.2, we propose two neural network models for computing the noise subspace of the MUSIC method. In Sections 7.3 and 7.4, neural network approaches for providing the DOA estimation directly from the snapshots will be given. Finally, Section 7.5 deals with another important problem of array signal processing: beamforming. We will present neural networks for computing in real time the optimal weights of a beamformer.

7.1 Real-Time Implementation of Three DOA Estimation Methods Using Neural Networks

In this section, we will show how to use the 2-D neural network proposed in Section 3.3 to perform in real time the major computation required in the ML technique, the alternating projection ML method, and the propagator method. First, we give a brief introduction to these three methods.

7.1.1 The ML and Alternating Projection ML Methods

Let us consider a uniform linear array of N omnidirectional sensors illuminated by P narrow-band signals ($P < N$). At the n'th snapshot the output of the i'th sensor can be described by

$$x_i(n) = \sum_{l=1}^{P} \alpha_l(n) \exp\left[jkd(i-1)\sin(\theta_l)\right] + \epsilon_i(n), \tag{7.1}$$

where k is the wavenumber associated with the central wavelength, d is the space between two adjacent sensors, θ_l is the angle of arrival of the l'th signal, $\alpha_l(n)$ is the complex envelope of the l'th signal of arrival, and $\epsilon_i(n)$ is the additive white noise at the i'th element.

Using vector notation, Equation (7.1) can be written as

$$\boldsymbol{X}(n) = \boldsymbol{A}(\boldsymbol{\Theta})\boldsymbol{\alpha}(n) + \boldsymbol{\epsilon}(n), \tag{7.2}$$

where the vectors $\boldsymbol{X}(n)$, $\boldsymbol{\alpha}(n)$, and $\boldsymbol{\epsilon}(n)$ are defined as

$$\boldsymbol{X}(n) = \begin{pmatrix} x_1(n) \\ x_2(n) \\ \vdots \\ x_N(n) \end{pmatrix}_{N\times 1}, \qquad \boldsymbol{\alpha}(n) = \begin{pmatrix} \alpha_1(n) \\ \alpha_2(n) \\ \vdots \\ \alpha_P(n) \end{pmatrix}_{P\times 1}, \qquad \text{and}$$

$$\epsilon(n) = \begin{pmatrix} \epsilon_1(n) \\ \epsilon_2(n) \\ \vdots \\ \epsilon_N(n) \end{pmatrix}_{N\times 1}, \tag{7.3}$$

and the $N \times P$ matrix $\boldsymbol{A}(\boldsymbol{\Theta})$ is defined as

$$\boldsymbol{A}(\boldsymbol{\Theta}) = \left[\boldsymbol{a}(\theta_1), \boldsymbol{a}(\theta_2), \ldots, \boldsymbol{a}(\theta_P)\right]. \tag{7.4}$$

Moreover,

$$\boldsymbol{a}(\theta_l) = \Big[1, \exp\left[jkd\sin(\theta_l)\right], \exp\left[j2kd\sin(\theta_l)\right], \\ \ldots, \exp\left[jkd(N-1)\sin(\theta_l)\right]\Big]^T \qquad (l = 1, 2, \ldots, P) \tag{7.5}$$

is called the l'th signal direction vector. If L snapshots are available, then we have

$$\boldsymbol{X} = \boldsymbol{A}(\boldsymbol{\Theta})\boldsymbol{\alpha} + \boldsymbol{\epsilon}, \tag{7.6}$$

where

$$\boldsymbol{X} = \left[\boldsymbol{X}(1), \boldsymbol{X}(2), \ldots, \boldsymbol{X}(L)\right]_{N\times L},$$
$$\boldsymbol{\alpha} = \left[\boldsymbol{\alpha}(1), \boldsymbol{\alpha}(2), \ldots, \boldsymbol{\alpha}(L)\right]_{P\times L},$$

and

$$\boldsymbol{\epsilon} = \left[\boldsymbol{\epsilon}(1), \boldsymbol{\epsilon}(2), \ldots, \boldsymbol{\epsilon}(L)\right]_{N\times L}.$$

Our task is to estimate the directions θ_l (for $l = 1, 2, \ldots, P$) of sources from the available matrix $\boldsymbol{X}$. Assume the noise vector $\boldsymbol{\epsilon}(n)$ is a stationary and ergodic complex-valued Gaussian process of zero mean and variance matrix $\sigma^2\boldsymbol{I}$, where σ^2 is an unknown scalar. Then the ML estimation of θ_l (for $l = 1, 2, \ldots, P$) is the solution of the optimization problem [10, 11]

$$\min_{\boldsymbol{\Theta}} \sum_{n=1}^{L} \left|\boldsymbol{X}(n) - \boldsymbol{A}(\boldsymbol{\Theta})\left(\boldsymbol{A}^H(\boldsymbol{\Theta})\boldsymbol{A}(\boldsymbol{\Theta})\right)^{-1}\boldsymbol{A}^H(\boldsymbol{\Theta})\boldsymbol{X}(n)\right|^2, \tag{7.7}$$

where $\boldsymbol{\Theta}$ denotes the set of θ_l (for $l = 1, 2, \ldots, P$) and we have used the joint density function $f(\boldsymbol{X})$ of the sampled data matrix $\boldsymbol{X}$,

$$f(\boldsymbol{X}) = \prod_{n=1}^{L} \frac{1}{\pi \det[\sigma^2 \boldsymbol{I}]} \exp\left[\frac{-1}{\sigma^2}\left|\boldsymbol{X}(n) - \boldsymbol{A}(\boldsymbol{\Theta})\boldsymbol{\alpha}(n)\right|^2\right]. \tag{7.8}$$

Equation (7.7) can be rewritten as

$$\min_{\Theta} \sum_{n=1}^{L} \left| \boldsymbol{X}(n) - \boldsymbol{P}_{A(\Theta)} \boldsymbol{X}(n) \right|^2, \tag{7.9}$$

where $\boldsymbol{P}_{A(\Theta)}$ is the projection operator onto the space spanned by the columns of the matrix $\boldsymbol{A}(\Theta)$, that is,

$$\boldsymbol{P}_{A(\Theta)} = \boldsymbol{A}(\Theta)\left(\boldsymbol{A}^H(\Theta)\boldsymbol{A}(\Theta)\right)^{-1}\boldsymbol{A}^H(\Theta). \tag{7.10}$$

With this notation, the minimization problem (7.7) can be transformed to the maximization problem

$$\max_{\Theta} \sum_{n=1}^{L} \left| \boldsymbol{P}_{A(\Theta)} \boldsymbol{X}(n) \right|^2 = \max_{\Theta} \; tr\{\boldsymbol{P}_{A(\Theta)}\boldsymbol{R}\}, \tag{7.11}$$

where $\boldsymbol{R}$ is the sample covariance matrix

$$\boldsymbol{R} = \frac{1}{L} \sum_{n=1}^{L} \boldsymbol{X}(n)\boldsymbol{X}^H(n). \tag{7.12}$$

Clearly, (7.11) is a multivariate maximization problem. In order to reduce the computational complexity, Ziskind and Wax [9] proposed an alternating projection ML method that provides the direction angles θ_l (for $l = 1, 2, \ldots, P$) of the sources by solving iteratively the following one-dimensional maximization problem:

$$\theta_i^{k+1} = arg \max_{\theta} tr\left\{\boldsymbol{P}_{[\boldsymbol{A}(\Theta_{(i)}^k), \boldsymbol{a}(\theta)]}\boldsymbol{R}\right\}. \tag{7.13}$$

Here $\boldsymbol{P}_{[\boldsymbol{A}(\Theta_{(i)}^k), \boldsymbol{a}(\theta)]}$ is the projection operator onto the space spanned by the columns of the matrix $[\boldsymbol{A}(\Theta_{(i)}^k), \boldsymbol{a}(\theta)]$, that is,

$$\boldsymbol{P}_{[\boldsymbol{A}(\Theta_{(i)}^k), \boldsymbol{a}(\theta)]} = \left[\boldsymbol{A}(\Theta_{(i)}^k), \boldsymbol{a}(\theta)\right]\left\{\left[\boldsymbol{A}(\Theta_{(i)}^k), \boldsymbol{a}(\theta)\right]^H\left[\boldsymbol{A}(\Theta_{(i)}^k), \boldsymbol{a}(\theta)\right]\right\}^{-1} \times \left[\boldsymbol{A}(\Theta_{(i)}^k), \boldsymbol{a}(\theta)\right]^H, \tag{7.14}$$

$$\boldsymbol{A}(\Theta_{(i)}^k) = \left[\boldsymbol{a}(\theta_1^k), \boldsymbol{a}(\theta_2^k), \ldots, \boldsymbol{a}(\theta_{i-1}^k), \boldsymbol{a}(\theta_{i+1}^k), \ldots, \boldsymbol{a}(\theta_P^k)\right] \quad (i = 1, 2, \ldots, P), \tag{7.15}$$

and

$$\boldsymbol{a}(\theta) = \left[1, \exp\left[jkd\sin(\theta)\right], \exp\left[j2kd\sin(\theta)\right], \ldots, \exp\left[jkd(N-1)\sin(\theta)\right]\right]^T \quad \theta \in \left[-\frac{\pi}{2}, \frac{\pi}{2}\right]. \tag{7.16}$$

Another alternative ML method involves first transforming the direction angles θ_l (for $l = 1, 2, \ldots, P$) to an AR coefficient vector $z = [z_0, z_1, \ldots, z_P]^H$. Then one finds the vector z by solving the optimization problem [1, 12, 13]

$$\min_{z} tr\{\boldsymbol{P}_Z \boldsymbol{R}\}, \tag{7.17}$$

where

$$\boldsymbol{P}_Z = \boldsymbol{Z}(\boldsymbol{Z}^H \boldsymbol{Z})^{-1} \boldsymbol{Z}^H, \tag{7.18}$$

which is the projection operator onto the space spanned by the columns of the matrix $\boldsymbol{Z}$, and

$$\boldsymbol{Z} = \begin{pmatrix} z_P^* & & 0 \\ \vdots & \ddots & z_P^* \\ z_0^* & \ddots & \vdots \\ 0 & & z_0 \end{pmatrix}. \tag{7.19}$$

Finally, one determines the roots of the polynomial constructed from the vector z; these roots correspond to the direction angles θ_l (for $l = 1, 2, \ldots, P$).

7.1.2 The Propagator Method

Consider the partitions of the vectors $\boldsymbol{a}(\theta_i)$ (for $i = 1, 2, \ldots, P$) as

$$\boldsymbol{a}(\theta_i) = \begin{pmatrix} \boldsymbol{a}_1(\theta_i) \\ \boldsymbol{a}_2(\theta_i) \end{pmatrix}$$

and the matrix $\boldsymbol{A}(\boldsymbol{\Theta})$ as

$$\boldsymbol{A}(\boldsymbol{\Theta}) = \begin{pmatrix} \boldsymbol{A}_1(\boldsymbol{\Theta}) \\ \boldsymbol{A}_2(\boldsymbol{\Theta}) \end{pmatrix}, \tag{7.20}$$

where the vector $\boldsymbol{a}_1(\theta_i)$ has P elements and the matrix $\boldsymbol{A}_1(\boldsymbol{\Theta})$ has P rows. The vector $\boldsymbol{a}_2(\theta_i)$ has $N-P$ elements and the matrix $\boldsymbol{A}_2(\boldsymbol{\Theta})$ has $N-P$ rows.

If the $P \times P$ matrix $\boldsymbol{A}_1(\boldsymbol{\Theta})$ is assumed to be nonsingular, then a unique linear operator $\boldsymbol{P}$ called the "propagator" is given by

$$\boldsymbol{P}^H \boldsymbol{A}_1(\boldsymbol{\Theta}) = \boldsymbol{A}_2(\boldsymbol{\Theta}), \tag{7.21}$$

and $\boldsymbol{P}$ defines an $N \times (N-P)$ matrix $\boldsymbol{Q}$,

$$\boldsymbol{Q} = \begin{pmatrix} \boldsymbol{P} \\ -\boldsymbol{I}_{N-P} \end{pmatrix}, \tag{7.22}$$

where $\boldsymbol{I}_{N-P}$ is the identity matrix of dimension $(N-P) \times (N-P)$. It has been shown that the subspace spanned by the columns of $\boldsymbol{Q}$ is orthogonal to the subspace spanned by the columns of $\boldsymbol{A}(\Theta)$ [7, 8]. As a result, the direction angles θ_i (for $i = 1, 2, \ldots, P$) are obtained by solving

$$\min_{\theta} \|\boldsymbol{Q}_T \boldsymbol{a}(\theta)\| \tag{7.23}$$

or

$$\min_{\theta} \|\boldsymbol{Q}^H \boldsymbol{a}(\theta)\| \qquad \theta \in \left[-\frac{\pi}{2}, \frac{\pi}{2}\right], \tag{7.24}$$

where

$$\boldsymbol{Q}_T = \boldsymbol{Q}(\boldsymbol{Q}^H \boldsymbol{Q})^{-1} \boldsymbol{Q}^H \tag{7.25}$$

and $\boldsymbol{a}(\theta)$ is the searching vector given by (7.16).

In the practical case, $\boldsymbol{A}(\Theta)$ is unknown but the sampled data matrix $\boldsymbol{X}$ is available. Under this condition, the propagator $\boldsymbol{P}$ can be estimated with the following procedures: First, we estimate the sample covariance matrix

$$\boldsymbol{R} = \frac{1}{L} \sum_{n=1}^{L} \boldsymbol{X}(n) \boldsymbol{X}^H(n). \tag{7.26}$$

Second, the matrix $\boldsymbol{R}$ is partitioned into the form

$$\boldsymbol{R} = (\boldsymbol{R}_1 \ \boldsymbol{R}_2), \tag{7.27}$$

where $\boldsymbol{R}_1$ and $\boldsymbol{R}_2$ denote $(N \times P)$- and $[N \times (N-P)]$-dimensional submatrices, respectively. Finally, the propagator $\boldsymbol{P}$ can be obtained by minimizing the cost function

$$E(\boldsymbol{P}) = \|\boldsymbol{R}_2 - \boldsymbol{R}_1 \boldsymbol{P}\|. \tag{7.28}$$

It follows from (7.28) that

$$\boldsymbol{P} = \boldsymbol{R}_1^{+} \boldsymbol{R}_2. \tag{7.29}$$

If either $\boldsymbol{R}_1^H \boldsymbol{R}_1$ or $\boldsymbol{R}_1 \boldsymbol{R}_1^H$ is nonsingular, then we have

$$\boldsymbol{P} = \left(\boldsymbol{R}_1^H \boldsymbol{R}_1\right)^{-1} \boldsymbol{R}_1^H \boldsymbol{R}_2 \tag{7.30}$$

or

$$\boldsymbol{P} = \boldsymbol{R}_1^H \left(\boldsymbol{R}_1 \boldsymbol{R}_1^H\right)^{-1} \boldsymbol{R}_2. \tag{7.31}$$

If the sources are coherent, the propagator obtained using (7.21) is not unique.

To overcome this problem, we reconstruct the sample covariance matrix as

$$\boldsymbol{R} = \begin{pmatrix} R(1) & R(2) & \cdots & R(P) \\ R(2) & R(1) & \cdots & R(P+1) \\ \vdots & \vdots & \ddots & \vdots \\ R(M) & R(M+1) & \cdots & R(N-1) \end{pmatrix}, \tag{7.32}$$

where M is selected to satisfy $M > P$ and $M = N-P$, and where

$$R(i) = \frac{1}{L}\sum_{n=1}^{L} x_1^*(n)x_{i+1}(n). \tag{7.33}$$

As we did with (7.27), we partition (7.32) into the form

$$\boldsymbol{R}^T = \begin{pmatrix} \boldsymbol{R}_3^T & \boldsymbol{R}_4^T \end{pmatrix}, \tag{7.34}$$

where $\boldsymbol{R}_3$ and $\boldsymbol{R}_4$ denote $(P \times P)$- and $[(M-P) \times P]$-dimensional submatrices, respectively. It is easy to show that the rank of $\boldsymbol{R}_3$ is P. Therefore, we can obtain a unique propagator $\boldsymbol{P}$ from

$$\min_{\boldsymbol{P}} \|\boldsymbol{R}_4^H - \boldsymbol{R}_3^H\boldsymbol{P}\| \tag{7.35}$$

or equivalently,

$$\boldsymbol{P} = \left(\boldsymbol{R}_3^H\right)^+\boldsymbol{R}_4^H, \tag{7.36}$$

which defines an $(M-P)$-dimensional noise subspace by

$$\boldsymbol{Q} = \begin{pmatrix} \boldsymbol{P} \\ -\boldsymbol{I}_{M-P} \end{pmatrix}. \tag{7.37}$$

Thus, in the coherent case, the $(N-P)$-dimensional subspace is replaced by an $(M-P)$-dimensional subspace. As a result, the searching vector $\boldsymbol{a}(\theta)$ of (7.16) becomes

$$\boldsymbol{a}(\theta) = \Big[1, \exp\big[jkd\sin(\theta)\big], \exp\big[j2kd\sin(\theta)\big], \\ \ldots, \exp\big[jkd(M-1)\sin(\theta)\big]\Big]^T. \tag{7.38}$$

7.1.3 *Real-Time Computation of the DOA Algorithms Using Neural Networks*

As pointed out in [9], in the ML technique, in the alternating projection ML method, and in the AR-model-based ML method, one has to compute repeatedly the projection operator matrices (7.10), (7.14), and (7.18), respectively. In the propagator method, the computation of the projection operator matrix given by (7.25) and the propagator given by (7.30) (or (7.31)) is required. The computational complexity of (7.30) is approximately equal to that of (7.10). Consequently, in all of these bearing estimation algorithms, the major computational burden is computing the projection operator matrix onto the space spanned by the columns of a matrix. In other words, the key to implement these algorithms in real time is to compute the projection operator matrix as fast as possible. In the following, we will present a 2-D neural network approach for computing the projection operator matrix.

Equations (7.10), (7.14), and (7.18) are unified as

$$\boldsymbol{G} = \boldsymbol{D}\boldsymbol{D}^{+} = \boldsymbol{D} \lim_{\alpha \to 0} \boldsymbol{D}^{H}(\alpha \boldsymbol{I} + \boldsymbol{D}\boldsymbol{D}^{H})^{-1}, \tag{7.39}$$

where $\boldsymbol{G}$ is the projection operator matrix onto the space spanned by the columns of the $M \times N$ matrix $\boldsymbol{D}$.

We will write (7.39) in real-valued form. It follows from the definition of the projection operator matrix that

$$\begin{aligned}
\min_{\boldsymbol{G}} \|\boldsymbol{D} - \boldsymbol{G}\boldsymbol{D}\|^2 &= \min_{\boldsymbol{G}} \|\boldsymbol{D}_r + j\boldsymbol{D}_i - \boldsymbol{G}_r\boldsymbol{D}_r + \boldsymbol{G}_i\boldsymbol{D}_i \\
&\quad - j(\boldsymbol{G}_r\boldsymbol{D}_i + \boldsymbol{G}_i\boldsymbol{D}_r)\|^2 \\
&= \min_{\boldsymbol{G}} \left\| \boldsymbol{D}_r - (\boldsymbol{G}_r \quad \boldsymbol{G}_i)\begin{pmatrix} \boldsymbol{D}_r \\ -\boldsymbol{D}_i \end{pmatrix} \right. \\
&\quad \left. + j\left[\boldsymbol{D}_i - (\boldsymbol{G}_r \quad \boldsymbol{G}_i)\begin{pmatrix} \boldsymbol{D}_i \\ \boldsymbol{D}_r \end{pmatrix}\right] \right\|^2 \\
&= \min_{\boldsymbol{G}_r, \boldsymbol{G}_i} \left\| (\boldsymbol{D}_r \quad \boldsymbol{D}_i) - (\boldsymbol{G}_r \quad \boldsymbol{G}_i)\begin{pmatrix} \boldsymbol{D}_r & \boldsymbol{D}_i \\ -\boldsymbol{D}_i & \boldsymbol{D}_r \end{pmatrix} \right\|^2 \\
&= \min_{\boldsymbol{G}_c} \|\boldsymbol{D}_{c1} - \boldsymbol{G}_c\boldsymbol{D}_c\|^2 \\
&= \min_{\boldsymbol{G}_c} tr\{(\boldsymbol{D}_{c1} - \boldsymbol{G}_c\boldsymbol{D}_c)(\boldsymbol{D}_{c1} - \boldsymbol{G}_c\boldsymbol{D}_c)^T\},
\end{aligned} \tag{7.40}$$

where

$$\boldsymbol{D}_{c1} = (\boldsymbol{D}_r \quad \boldsymbol{D}_i)_{M \times 2N}, \tag{7.41}$$

$$\boldsymbol{G}_c = (\boldsymbol{G}_r \quad \boldsymbol{G}_i)_{M \times 2M}, \tag{7.42}$$

and

$$D_c = \begin{pmatrix} D_r & D_i \\ -D_i & D_r \end{pmatrix}_{2M\times 2N} \tag{7.43}$$

are all real-valued matrices. The exact solution of (7.40) is

$$G_c = D_{c1}D_c^+ = D_{c1} \lim_{\alpha\to 0} D_c^T(\alpha I + D_c D_c^T)^{-1}. \tag{7.44}$$

Clearly, Equation (7.44) has the same form as (3.88). As a result, we can use the 2-D neural network proposed in Section 3.3 to solve (7.44) in real time by taking D_c and D_{c1} as the connection strength matrix T and bias current matrix B of the network, respectively, and selecting the other parameters to be the same as for (3.88). The network will provide the solution in the stationary state as

$$V_f = \lim_{t\to\infty} V(t) = D_{c1}D_c^T\left(\frac{1}{RK_1K_2}I + D_cD_c^T\right)^{-1}, \tag{7.45}$$

which approximates the exact solution (7.44).

Once the real-valued matrix G_c is given, the desired complex-valued G is immediately obtained from (7.42). This also shows that in the complex-valued case, two neurons are used to provide a complex value. As a result, $M \times 2M$ and $M \times 2N$ neurons in the main-network and the subnetwork, respectively, are needed, and the dimensions of the connection strength matrix and bias current matrix become $2M \times 2N$ and $M \times 2N$, respectively. With this in mind, the dynamic equations of the 2-D neural network can be written in a complex-valued form as

$$C\frac{dU(t)}{dt} = -U(t)\left(\frac{1}{R}I + K_1K_2DD^H\right) + K_2DD^H. \tag{7.46}$$

The output matrix in the stationary state can be obtained by letting $dU(t)/dt = \mathbf{0}$, that is,

$$V_f = \lim_{t\to\infty} V(t) = DD^H\left(\frac{1}{RK_1K_2}I + DD^H\right)^{-1}, \tag{7.47}$$

which is an approximation to the exact solution (7.39). The error can be made arbitrarily small by appropriately selecting the parameters R, K_1, and K_2.

In order to use the neural network to perform the computation required in the above bearing estimation algorithms, we should select the connection strength matrix T and the bias current matrix B as follows:

(1) for the ML method, $T = B = A(\Theta)$,
(2) for the alternating projection ML method, $T = B = [A(\Theta^k_{(i)}), a(\theta)]$,
(3) for the AR-model-based ML method, $T = B = Z$,
(4) for the propagator method in the incoherent-signal case, $T = R_1^H$ and $B = R_2^H$,

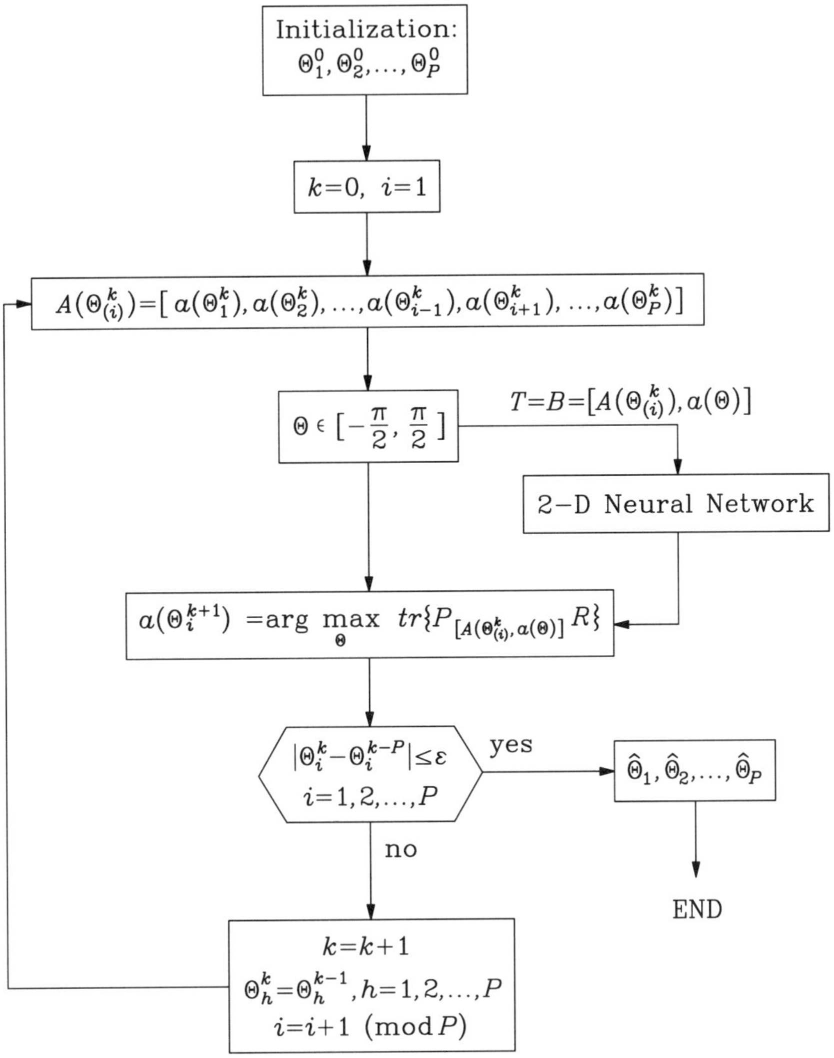

Figure 7.1. The flowchart of the alternating projection ML method using the 2-D neural network. ε is specified.

(5) for the propagator method in the coherent-signal case, $\boldsymbol{T} = \boldsymbol{R}_3$ and $\boldsymbol{B} = \boldsymbol{R}_4$,

(6) for the projection operator matrix (7.25) on the noise subspace, $\boldsymbol{T} = \boldsymbol{B} = \boldsymbol{Q}$.

For further illustration, Figure 7.1 takes the alternating projection ML method as an example to show the details of the procedure of the implementation based on the 2-D neural network.

Three sets of simulation results illustrating the above neural network approach will be presented in the following examples [14, 15]. In the simulation, we used $K_1 = K_2 = 10$, $C = 100$ pF, and $R = 10^3\ \Omega$.

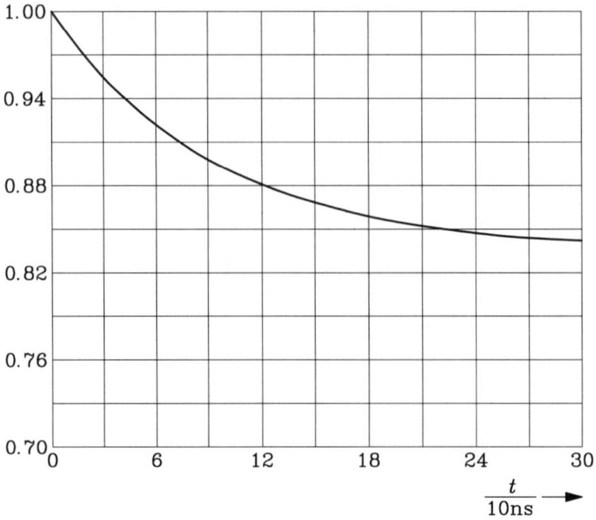

Figure 7.2. The dynamic curve of the neuron (1, 1).

Example 7.1

Given a 3×2 matrix $\boldsymbol{D}$ as

$$\boldsymbol{D} = \begin{pmatrix} 3.86539 & 3.42330 \\ 3.42330 & 2.71802 \\ 2.71802 & 1.82625 \end{pmatrix}$$

we have the exact projection matrix $\boldsymbol{G}$ for the matrix $\boldsymbol{D}$ according to (7.39) and the neural network solution $\boldsymbol{V}_f$ as

$$\boldsymbol{G} = \begin{pmatrix} 0.834656 & 0.326864 & -0.176537 \\ 0.326864 & 0.353832 & 0.348991 \\ -0.176537 & 0.348991 & 0.811512 \end{pmatrix}$$

and

$$\boldsymbol{V}_f = \begin{pmatrix} 0.834629 & 0.326869 & -0.176503 \\ 0.326869 & 0.353831 & 0.348984 \\ -0.176503 & 0.348984 & 0.811467 \end{pmatrix}.$$

The error $\|\boldsymbol{G} - \boldsymbol{V}_f\| = 7.348 \times 10^{-6}$. Figures 7.2 and 7.3 describe the dynamics of the neuron (1, 1) and the neuron (3, 3) of the main-network.

Example 7.2

Consider four planar-wave signals whose directions of arrival are $-55°$, $-45°$, $3°$, and $25°$ and whose associated signal-to-noise ratios are each 7 dB. The array is uniform and linear with 10 elements spaced a half wavelength apart. The number of snapshots is 30. $\boldsymbol{P}_{tr}$ and $\boldsymbol{P}_{ti}$ are the real and imaginary parts of the conjugate transpose of the propagator operator $\boldsymbol{P}$ computed from (7.30). $\boldsymbol{V}_{fr}$ and $\boldsymbol{V}_{fi}$ are the real and imaginary parts provided by the neural network. Obviously, $\boldsymbol{P}_{tr} \approx \boldsymbol{V}_{fr}$

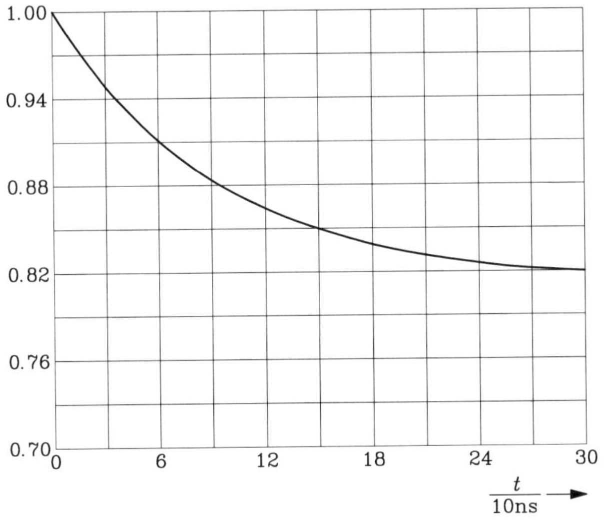

Figure 7.3. The dynamic curve of the neuron (3, 3).

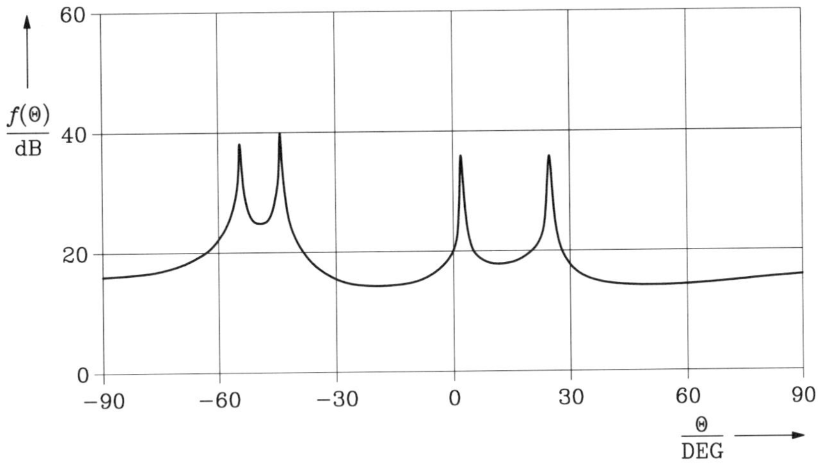

Figure 7.4. The estimated spatial spectrum obtained by using the propagator operator computed by (7.30).

and $P_{ti} \approx V_{fi}$. Figures 7.4 and 7.5 are the estimated spatial spectra obtained by use of the propagator operator computed by (7.30) and the operator provided by the neural network, respectively. We have

$$P_{tr} = \begin{pmatrix} 0.984973 & 0.184126 & 0.095016 & -0.207094 \\ -0.256845 & 0.900572 & 0.417294 & 0.126884 \end{pmatrix},$$

$$P_{ti} = \begin{pmatrix} -0.144870 & -0.237561 & 1.284330 & -0.207597 \\ -0.158476 & -0.156939 & -0.545668 & 1.377520 \end{pmatrix},$$

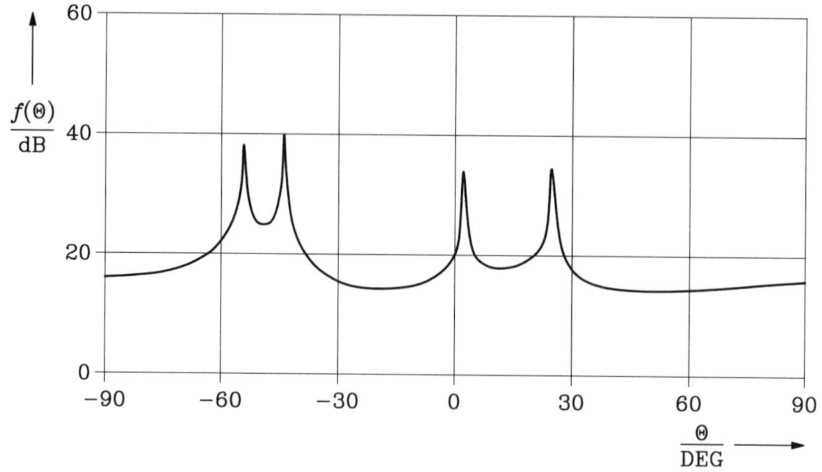

Figure 7.5. The estimated spatial spectrum obtained by using the propagator operator provided by the neural network.

Table 7.1. *Twelve Monte-Carlo runs by use of the neural network*

	1	2	3	4	5	6	7	8	9	10	11	12
θ_1	−5.5	−4.5	−5.4	−5.0	−4.8	−4.6	−4.4	−5.2	−4.8	−4.8	−5.4	−4.4
θ_2	4.6	4.3	4.8	5.4	5.6	5.4	4.8	4.8	5.2	5.0	5.2	5.4

$$V_{fr} = \begin{pmatrix} 0.984908 & 0.184149 & 0.095012 & -0.207043 \\ -0.256772 & 0.900519 & 0.417252 & 0.126874 \end{pmatrix},$$

and

$$V_{fi} = \begin{pmatrix} -0.144870 & -0.237589 & 1.284270 & -0.207548 \\ -0.158429 & -0.156931 & -0.545613 & 1.377420 \end{pmatrix}.$$

Example 7.3

Two coherent planar-wave signal sources are considered with the directions of arrival being -5° and 5°. The array is uniform and linear with 6 elements spaced a half wavelength apart. The other parameters are the same as in Example 7.2. Table 7.1 shows the results of the propagator bearing estimation method for twelve Monte-Carlo runs by use of the 2-D neural network.

7.2 Neural Networks for the MUSIC Bearing Estimation Algorithm

With the same notation as in Section 7.1, the MUSIC method of Schmidt includes essentially the following steps [6]:

(1) Estimation of the covariance matrix from the available samples

$$\boldsymbol{R} = \frac{1}{L}\sum_{n=1}^{L} \boldsymbol{X}(n)\boldsymbol{X}^H(n). \tag{7.48}$$

(2) Eigendecomposition of the matrix $\boldsymbol{R}$,

$$\boldsymbol{R} = \sum_{i=1}^{N} \lambda_i \boldsymbol{S}_i \boldsymbol{S}_i^H, \tag{7.49}$$

where $0 < \lambda_1 \leq \lambda_2 \leq \ldots \leq \lambda_N$ are the eigenvalues and $\boldsymbol{S}_1, \boldsymbol{S}_2, \ldots, \boldsymbol{S}_N$ are the corresponding orthonormal eigenvectors of the matrix $\boldsymbol{R}$.

The eigenvectors corresponding to the first P largest eigenvalues are referred to as the signal eigenvectors, and those corresponding to the minimum eigenvalues are referred to as the noise eigenvectors. The subspace spanned by the signal eigenvectors is called the signal subspace, and its orthogonal complement spanned by the noise eigenvectors is called the noise subspace.

If $\boldsymbol{R}$ is the exact covariance matrix, then we have $\lambda_1 = \lambda_2 = \ldots = \lambda_{N-P} = \sigma^2$ and

$$\boldsymbol{A}(\boldsymbol{\Theta})\boldsymbol{R}_s\boldsymbol{A}^H(\boldsymbol{\Theta}) = \sum_{i=N-P+1}^{N} (\lambda_i - \sigma^2)\boldsymbol{S}_i\boldsymbol{S}_i^H, \tag{7.50}$$

where $\boldsymbol{R}_s = E[\boldsymbol{\alpha}(n)\boldsymbol{\alpha}^H(n)]$.

Equation (7.50) shows that the signal direction vectors $\boldsymbol{a}(\theta_i)$ (for $i = 1, 2, \ldots, P$) and the signal eigenvectors $\boldsymbol{S}_i$ (for $i = N-P+1, N-P+2, \ldots, N$) span the same subspace. This implies that all signal direction vectors are orthogonal to the noise subspace.

(3) Estimation of the directions of arrival of the P signal sources by finding the values of θ corresponding to the P maxima of the function

$$f(\theta) = \frac{1}{T(\theta)} = \frac{1}{\|\boldsymbol{G}^H\boldsymbol{a}(\theta)\|^2} \qquad \theta \in \left[-\frac{\pi}{2}, \frac{\pi}{2}\right], \tag{7.51}$$

where $\boldsymbol{a}(\theta)$ is the search vector given by (7.16),

$$\boldsymbol{G} = \sum_{i=1}^{N-P} \boldsymbol{S}_i\boldsymbol{S}_i^H \tag{7.52}$$

is the orthogonal projection operator on the noise subspace, and $T(\theta) = \|\boldsymbol{G}^H\boldsymbol{a}(\theta)\|^2$ is the noise subspace projection, which can be rewritten as

$$T(\theta) = \sum_{i=1}^{N-P} |\boldsymbol{a}^H(\theta)\boldsymbol{S}_i|^2. \tag{7.53}$$

Based on the fact that all the signal direction vectors are orthogonal to the noise subspace, the P maxima of (7.51) are located at the direction angles θ_i (for $i = 1, 2, \ldots, P$).

Note that (7.49) involves computing the noise subspace. Therein lies a great computational burden making it difficult to implement the MUSIC algorithm in real time. To tackle this problem, we will present two neural network approaches for computing the noise subspace. One approach assumes that all the minimum eigenvalues are identical (that is, $\lambda_1 = \lambda_2 = \ldots = \lambda_{N-P} = \sigma^2$); the other eliminates this assumption.

7.2.1 Computation of the Noise Subspace of the Repeated Smallest Eigenvalues

A neural network for computing the noise subspace corresponding to the repeated minimum eigenvalues has been proposed by Mathew and Reddy [16]. This neural network can be represented by the differential equations

$$C_{ij}\frac{du_{ij}(t)}{dt} = -2\sum_{k=1}^{N} T_{jk}v_{ik}(t) - 4K_1\left[\psi_i(t) - 1\right]v_{ij}(t) - 2K_2 \sum_{l=1,l\neq i}^{N-P} \left[\boldsymbol{V}_i^T(t)\boldsymbol{V}_l(t)\right]v_{lj}(t), \tag{7.54}$$

$$v_{ij}(t) = g\left(u_{ij}(t)\right) \qquad (i = 1, 2, \ldots, N-P; \quad j = 1, 2, \ldots, N), \tag{7.55}$$

where the various quantities are described as follows: $u_{ij}(t)$ and $v_{ij}(t)$ are the input and output of the neuron (i, j), respectively; $g_{ij}(\cdot)$ is the input–output relationship of the neuron, which is selected to be a monotonically increasing function; C_{ij} is the input capacitor of the neuron (i, j); T_{jk} (for $j, k = 1, 2, \ldots, N$) is the connection strength between the neuron (i, j) (for $i = 1, 2, \ldots, N-P$; $j = 1, 2, \ldots, N$) and the neuron (i, k) (for $i = 1, 2, \ldots, N-P; k = 1, 2, \ldots, N$); K_1 and K_2 are positive constants; $\boldsymbol{V}_i(t) = [v_{1i}(t), v_{2i}(t), \ldots, v_{Ni}(t)]^T$ (for $i = 1, 2, \ldots, N-P$) is the i'th output vector of the network; and

$$\psi_i(t) = \sum_{j=1}^{N} \left(v_{ji}(t)\right)^2 \qquad (i = 1, 2, \ldots, N-P). \tag{7.56}$$

The structure of each neuron (i, j) is shown in Figure 7.6.

An energy function is defined as

$$E(t) = \sum_{i=1}^{N-P} \phi_i(t) + K_1 \sum_{i=1}^{N-P} \left[\psi_i(t) - 1\right]^2 + K_2 \sum_{i=1}^{N-P-1} \sum_{j=i+1}^{N-P} \left[\boldsymbol{V}_i^T(t)\boldsymbol{V}_j(t)\right]^2, \tag{7.57}$$

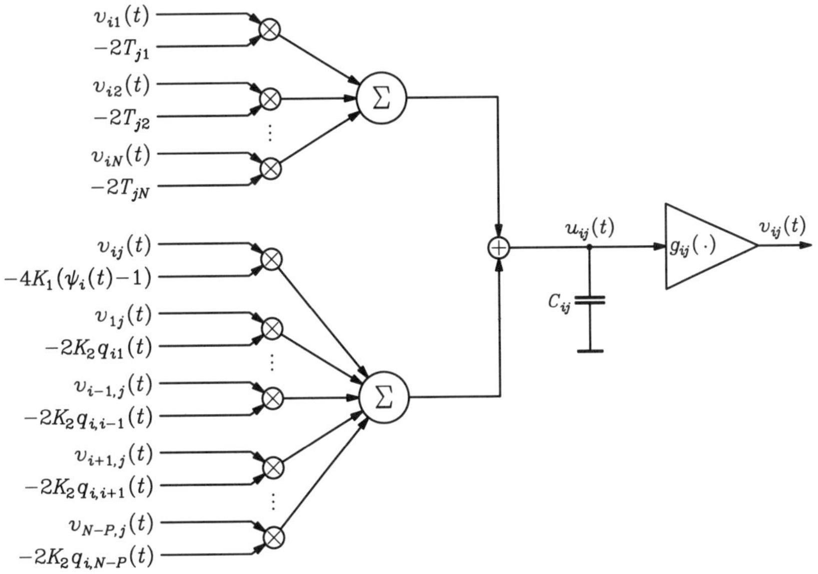

Figure 7.6. The model structure of the neuron (i, j); $q_{ij}(t) = \boldsymbol{V}_i^T(t)\boldsymbol{V}_j(t)$.

where

$$\phi_i(t) = \boldsymbol{V}_i^T(t)\boldsymbol{T}\boldsymbol{V}_i(t) \qquad (i = 1, 2, \ldots, N-P) \tag{7.58}$$

and $\boldsymbol{T} = \{T_{jk}\}$ is the $(N \times N)$-dimensional connection strength matrix, which is selected to be positive definite.

Now we will prove that this neural network is stable. This can be divided into two steps. First, because $\boldsymbol{T}$ is positive definite, $E(t)$ is positive, that is, $E(t)$ is bounded from below. Second, the time derivative of $E(t)$ is

$$\frac{dE(t)}{dt} = \sum_{i=1}^{N-P}\sum_{j=1}^{N}\frac{dv_{ij}(t)}{dt}\left(2\sum_{k=1}^{N}T_{jk}v_{ik}(t) + 4K_1\left[\psi_i(t) - 1\right]v_{ij}(t)\right.$$
$$\left. + 2K_2\sum_{l=1,l\neq i}^{N-P}\left[\boldsymbol{V}_i^T(t)\boldsymbol{V}_l(t)\right]v_{lj}(t)\right). \tag{7.59}$$

Substituting (7.54) and (7.55) into (7.59) gives

$$\frac{dE(t)}{dt} = -\sum_{i=1}^{N-P}\sum_{j=1}^{N}C_{ij}\frac{dv_{ij}(t)}{dt}\frac{du_{ij}(t)}{dt}$$
$$= -\sum_{i=1}^{N-P}\sum_{j=1}^{N}C_{ij}\left[g_{ij}^{-1}(v_{ij}(t))\right]'\left(\frac{dv_{ij}(t)}{dt}\right)^2, \tag{7.60}$$

which means that $E(t)$ is nonincreasing during the time evolution.

These two facts imply

$$\lim_{t\to\infty} \frac{dE(t)}{dt} = 0 \tag{7.61}$$

or equivalently,

$$\lim_{t\to\infty} \frac{du_{ij}(t)}{dt} = 0 \tag{7.62}$$

or

$$\lim_{t\to\infty} \frac{dv_{ij}(t)}{dt} = 0 \qquad (i = 1, 2, \ldots, N-P;\ j = 1, 2, \ldots, N). \tag{7.63}$$

Thus, the output matrix $\boldsymbol{V}(t) = [\boldsymbol{V}_1(t), \boldsymbol{V}_2(t), \ldots, \boldsymbol{V}_{N-P}(t)]$ of this network will always approach a stationary point of $E(t)$ and will provide one of the minima of $E(t)$.

In the following, we will present three theorems (proposed and proved by Mathew and Reddy in [16]). These show that each minimizer of $E(t)$ is the matrix whose columns are the eigenvectors corresponding to the repeated smallest eigenvalue σ^2 of the covariance matrix $\boldsymbol{R}$ if the matrix $\boldsymbol{R}$ is taken as the connection strength matrix $\boldsymbol{T}$ of the network and if $K_1 > \sigma^2/2$ is satisfied. For convenience, we use the following notation:

$$\bar{\boldsymbol{V}}(t) = \left[\boldsymbol{V}_1^T(t), \boldsymbol{V}_2^T(t), \ldots, \boldsymbol{V}_{N-P}^T(t)\right]^T, \tag{7.64}$$

which is an $[N(N-P)]$-dimensional column vector, and

$$\bar{\boldsymbol{R}} = bdiag[\boldsymbol{R}, \boldsymbol{R}, \ldots, \boldsymbol{R}], \tag{7.65}$$

which is an $(N-P)N \times (N-P)N$ block diagonal matrix. The energy function $E(t)$ of (7.57) can then be written as

$$\begin{aligned} E(t) = {} & \bar{\boldsymbol{V}}^T(t)\bar{\boldsymbol{R}}\bar{\boldsymbol{V}}(t) + K_1 \sum_{i=1}^{N-P} \left[\bar{\boldsymbol{V}}^T(t)\boldsymbol{E}_{ii}\bar{\boldsymbol{V}}(t) - 1\right]^2 \\ & + K_2 \sum_{i=1}^{N-P-1} \sum_{j=i+1}^{N-P} \left[\bar{\boldsymbol{V}}^T(t)\boldsymbol{E}_{ij}\bar{\boldsymbol{V}}(t)\right]^2, \end{aligned} \tag{7.66}$$

where

$$\boldsymbol{E}_{ij} = \boldsymbol{E}_i^T \boldsymbol{E}_j \qquad (i, j = 1, 2, \ldots, N-P) \tag{7.67}$$

and $\boldsymbol{E}_i$ is an $N(N-P) \times N(N-P)$ matrix with unity in the $(k, (i-1)N+k)$th location (for $k = 1, 2, \ldots, N$) and zeros elsewhere, which means that $\bar{\boldsymbol{V}}^T(t)\boldsymbol{E}_{ij}\bar{\boldsymbol{V}}(t) = \boldsymbol{V}_i^T(t)\boldsymbol{V}_j(t)$ (for $i, j = 1, 2, \ldots, N-P$).

Theorem 7.1 *An* $[N(N-P)]$*-dimensional vector* $\bar{\boldsymbol{V}}(f) = [\boldsymbol{V}_1^T(f), \boldsymbol{V}_2^T(f), \ldots, \boldsymbol{V}_{N-P}^T(f)]^T$ *is a stationary point (minimizer) of* $E(t)$ *if and only if* $\boldsymbol{V}_k(f)$ *is an eigenvector with norm* β_k *corresponding to the eigenvalue* d_k *of the matrix* $\boldsymbol{R} + K_2 \sum_{i=1}^{N-P} \boldsymbol{V}_i(f)\boldsymbol{V}_i^T(f)$*, where* $d_k = 2K_1(1-\beta_k^2) + K_2\beta_k^2$ *(for* $k = 1, 2, \ldots, N-P$*).*

Theorem 7.2 *If an* $[N(N-P)]$*-dimensional vector* $\bar{\boldsymbol{V}}(f)$ *is a stationary point of* $E(t)$*, then the corresponding vectors* $\boldsymbol{V}_1(f), \boldsymbol{V}_2(f), \ldots, \boldsymbol{V}_{N-P}(f)$ *are orthogonal to one another if and only if* $\boldsymbol{V}_k(f)$ *is an eigenvector with norm* β_k *corresponding to the eigenvalue* $2K_1(1-\beta_k^2)$ *of the matrix* $\boldsymbol{R}$ *(for* $k = 1, 2, \ldots, N-P$*).*

These results immediately follow from the gradient vector $\boldsymbol{g}_i(\boldsymbol{V}_i(t))$ of $E(t)$ given by

$$\boldsymbol{g}_i(\boldsymbol{V}_i(t)) = 2\boldsymbol{R}\boldsymbol{V}_i(t) + 4K_1\boldsymbol{V}_i(t)[\psi_i(t) - 1] + 2K_2 \sum_{l=1, l\neq i}^{N-P} \boldsymbol{V}_l(t)\boldsymbol{V}_l^T(t)\boldsymbol{V}_i(t)$$
$$(i = 1, 2, \ldots, N-P). \qquad (7.68)$$

Theorem 7.3 *The vector* $\bar{\boldsymbol{V}}(f)$ *is a global minimizer of* $E(t)$ *if and only if the corresponding vectors* $\boldsymbol{V}_1(f), \boldsymbol{V}_2(f), \ldots, \boldsymbol{V}_{N-P}(f)$ *form an orthogonal basis for the subspace spanned by the* $N-P$ *eigenvectors corresponding to the repeated smallest eigenvalue* σ^2 *of the matrix* $\boldsymbol{R}$*. The norm of the vector* $\boldsymbol{V}_i(f)$ *is* $\beta_i = \sqrt{1 - \frac{\sigma^2}{2K_1}}$ *(for* $i = 1, 2, \ldots, N-P$*).*

The proof of Theorem 7.3 is similar to that of Theorem 3.4.

Proof for the "if" part: According to the hypothesis, we have

$$\boldsymbol{R}\boldsymbol{V}_i(f) = \sigma^2\boldsymbol{V}_i(f), \qquad (7.69)$$

$$\sigma^2 = 2K_1(1 - \beta_i^2), \qquad (7.70)$$

and

$$\boldsymbol{V}_i^T(f)\boldsymbol{V}_j(f) = \beta_i^2\delta_{ij} \qquad (i, j = 1, 2, \ldots, N-P). \qquad (7.71)$$

Clearly, from Theorem 7.1, $\bar{\boldsymbol{V}}(f)$ is a stationary point of $E(t)$. To prove that $\bar{\boldsymbol{V}}(f)$ is a global minimizer we have to show that $E(\bar{\boldsymbol{V}}(t)) - E(\bar{\boldsymbol{V}}(f)) \geq 0$ for any $\bar{\boldsymbol{V}}(t)$.

Let $\bar{V}(t) = \bar{V}(f) + \Delta\bar{V}(t)$. Then we have

$$
\begin{aligned}
E(\bar{V}(t)) - E(\bar{V}(f)) \\
&= 2\sigma^2(\Delta\bar{V}(t))^T\bar{V}(f) - \sigma^2\sum_{i=1}^{N-P}\Big[2(\Delta\bar{V}(t))^T \boldsymbol{E}_{ii}\bar{V}(f) \\
&\quad + (\Delta\bar{V}(t))^T\boldsymbol{E}_{ii}\Delta\bar{V}(t)\Big] + K_2\sum_{i=1}^{N-P-1}\sum_{j=i+1}^{N-P}\Big[\bar{V}^T(f)\boldsymbol{E}_{ij}\Delta\bar{V}(t) \\
&\quad + (\Delta\bar{V}(t))^T\boldsymbol{E}_{ij}\bar{V}(f) + (\Delta\bar{V}(t))^T\boldsymbol{E}_{ij}\Delta\bar{V}(t)\Big]^2 + (\Delta\bar{V}(t))^T\bar{\boldsymbol{R}}\Delta\bar{V}(t) \\
&\quad + K_1\sum_{i=1}^{N-P}\Big[2(\Delta\bar{V}(t))^T\boldsymbol{E}_{ii}\bar{V}(f) + (\Delta\bar{V}(t))^T\boldsymbol{E}_{ii}\Delta\bar{V}(t)\Big]^2,
\end{aligned}
\tag{7.72}
$$

where we used

$$\bar{\boldsymbol{R}}\bar{V}(f) = \sigma^2\bar{V}(f). \tag{7.73}$$

Moreover, it follows that

$$(\Delta\bar{V}(t))^T\bar{\boldsymbol{R}}\Delta\bar{V}(t) \geq \sigma^2(\Delta\bar{V}(t))^T\Delta\bar{V}(t) \tag{7.74}$$

and

$$\sum_{i=1}^{N-P}(\Delta\bar{V}(t))^T\boldsymbol{E}_{ii}\bar{V}(f) = (\Delta\bar{V}(t))^T\bar{V}(f), \tag{7.75}$$

$$\sum_{i=1}^{N-P}(\Delta\bar{V}(t))^T\boldsymbol{E}_{ii}\Delta\bar{V}(t) = (\Delta\bar{V}(t))^T\Delta\bar{V}(t). \tag{7.76}$$

Substituting (7.74)–(7.76) into (7.72), we obtain

$$
\begin{aligned}
E(\bar{V}(t)) - E(\bar{V}(f)) &\geq K_1\sum_{i=1}^{N-P}\Big[2(\Delta\bar{V}(t))^T\boldsymbol{E}_{ii}\bar{V}(f) \\
&\quad + (\Delta\bar{V}(t))^T\boldsymbol{E}_{ii}\Delta\bar{V}(t)\Big]^2 \\
&\quad + K_2\sum_{i=1}^{N-P-1}\sum_{j=i+1}^{N-P}\Big[\bar{V}^T(f)\boldsymbol{E}_{ij}\Delta\bar{V}(t) \\
&\quad + (\Delta\bar{V}(t))^T\boldsymbol{E}_{ij}\bar{V}(f) + (\Delta\bar{V}(t))^T\boldsymbol{E}_{ij}\Delta\bar{V}(t)\Big]^2 \\
&\geq 0,
\end{aligned}
\tag{7.77}
$$

which means that $\bar{V}(f)$ is a global minimizer of $E(t)$.

Proof for the "only if" part: If the vector $\bar{\boldsymbol{V}}(f)$ is a global minimizer, it follows from Theorem 7.1 that

$$\left(\boldsymbol{R} + K_2 \sum_{i=1}^{N-P} \boldsymbol{V}_i(f)\boldsymbol{V}_i^T(f)\right)\boldsymbol{V}_k(f) = \left[2K_1(1-\beta_k^2) + K_2\beta_k^2\right]\boldsymbol{V}_k(f), \tag{7.78}$$

$$2K_1(1-\beta_k^2) \geq \sigma^2 + K_2 \sum_{i=1, i\neq k}^{N-P} \frac{[\boldsymbol{V}_k^T(f)\boldsymbol{V}_i(f)]^2}{\beta_k^2} \qquad (k = 1, 2, \ldots, N-P). \tag{7.79}$$

Using the notation $\boldsymbol{F} = \boldsymbol{R} + \boldsymbol{W}$ and $\boldsymbol{W} = K_2 \sum_{i=1}^{N-P} \boldsymbol{V}_i(f)\boldsymbol{V}_i^T(f)$, we have, in terms of (7.78),

$$\boldsymbol{F}\boldsymbol{W} = \boldsymbol{W}\boldsymbol{F}. \tag{7.80}$$

Together with the fact that $\boldsymbol{R}$ and $\boldsymbol{W}$ are symmetric matrices, (7.80) means that there exists an $N \times N$ orthonormal matrix $\boldsymbol{Q} = [\boldsymbol{q}_1, \boldsymbol{q}_2, \ldots, \boldsymbol{q}_N]$ such that $\boldsymbol{Q}$ simultaneously diagonalizes $\boldsymbol{F}$ and $\boldsymbol{W}$, that is,

$$\boldsymbol{F} = \sum_{i=1}^{N} d_i \boldsymbol{q}_i \boldsymbol{q}_i^T, \tag{7.81}$$

$$\boldsymbol{W} = \sum_{i\in S_w} w_i \boldsymbol{q}_i \boldsymbol{q}_i^T, \tag{7.82}$$

and

$$\boldsymbol{R} = \sum_{i=1}^{N} \lambda_i \boldsymbol{q}_i \boldsymbol{q}_i^T, \tag{7.83}$$

where d_i, w_i, and λ_i are the eigenvalues of $\boldsymbol{F}$, $\boldsymbol{W}$, and $\boldsymbol{R}$, respectively, and S_w is defined as $S_w = \{i : w_i > 0\}$. Substituting (7.82) into the Hessian matrix $\boldsymbol{H}_k(\bar{\boldsymbol{V}}(t))$ of $E(t)$ given by

$$\begin{aligned}\boldsymbol{H}_k(\bar{\boldsymbol{V}}(t)) &= 2\boldsymbol{R} + 8K_1\boldsymbol{V}_k(t)\boldsymbol{V}_k^T(t) + 4K_1\left[\psi_k(t) - 1\right]\boldsymbol{I}\\ &\quad + 2K_2 \sum_{l=1, l\neq i}^{N-P} \boldsymbol{V}_l(t)\boldsymbol{V}_l^T(t),\end{aligned} \tag{7.84}$$

we get

$$\begin{aligned}\boldsymbol{H}_k(\bar{\boldsymbol{V}}(f)) &= \sum_{i\notin S_w} 2\left[\lambda_i - 2K_1(1-\beta_k^2)\right]\boldsymbol{q}_i\boldsymbol{q}_i^T\\ &\quad + \sum_{i\in S_w} 2\left[\lambda_i + w_i - 2K_1(1-\beta_k^2)\right]\boldsymbol{q}_i\boldsymbol{q}_i^T\\ &\quad + 2(4K_1 - K_2)\boldsymbol{V}_k(f)\boldsymbol{V}_k^T(f).\end{aligned} \tag{7.85}$$

If we define the subspace $W_2 = Span\{q_i : i \in S_w\}$, then it follows that

$$H_k(\bar{V}(f)) = \sum_{i \notin S_w} 2[\lambda_i - 2K_1(1-\beta_k^2)]q_i q_i^T + 2\sum_{i=1}^{r(W)} h_i q_i q_i^T. \tag{7.86}$$

Because the vector $\bar{V}(f)$ is a global minimizer, $H_k(\bar{V}(f))$ is nonnegative definite. Hence, its eigenvalues are all nonnegative, that is,

$$\lambda_i \geq 2K_1(1-\beta_k^2), \qquad i \notin S_w, \qquad k = 1, 2, \ldots, N-P, \tag{7.87}$$

$$h_i \geq 0, \qquad i = 1, 2, \ldots, r(W). \tag{7.88}$$

Now we will show that W_2 is the noise space (that is, $W_2 = S_n = Span\{S_1, S_2, \ldots, S_{N-P}\}$) and $V_k(f)$ (for $k = 1, 2, \ldots, N-P$) are orthogonal to one another.

Suppose

$$\left(W_2 \bigcap S_n^{\perp}\right) \backslash \{O\} \neq \emptyset \tag{7.89}$$

where $S_n^{\perp}$ is the signal subspace (i.e., the orthogonal complement of S_n). Equation (7.89) implies that $\exists l \notin S_w$ such that

$$q_l \in W_2^{\perp} \bigcap S_n \tag{7.90}$$

and

$$Rq_l = \sigma^2 q_l. \tag{7.91}$$

Together with (7.89), (7.90) and (7.91) show that

$$\sigma^2 \geq 2K_1(1-\beta_k^2) \qquad (k = 1, 2, \ldots, N-P). \tag{7.92}$$

With the above supposition, it is possible to choose $k_1 \in S_w$ such that $V_{k_1}(f) \bigcap S_n^{\perp} \neq \emptyset$. But for (7.79) and (7.92) to be true simultaneously for k_1, we must have

$$V_{k_1}^T(f) V_i(f) = 0 \tag{7.93}$$

and

$$\sigma^2 = 2K_1(1-\beta_{k_1}^2) \qquad (i = 1, 2, \ldots, N-P \quad \text{and} \quad i \neq k_1). \tag{7.94}$$

Combining (7.93) with (7.78), we get

$$RV_{k_1}(f) = \sigma^2 V_{k_1}(f) = 2K_1(1-\beta_{k_1}^2) V_{k_1}(f), \tag{7.95}$$

which is a contradiction because any eigenvector of $\boldsymbol{R}$ in $\boldsymbol{S}_n^{\perp}$ must correspond to a nonminimum eigenvalue. Hence,

$$\left(\boldsymbol{W}_2 \bigcap \boldsymbol{S}_n^{\perp}\right) = \{\boldsymbol{0}\}, \quad \boldsymbol{W}_2^{\perp} \bigcap \boldsymbol{S}_n = \{\boldsymbol{0}\}. \tag{7.96}$$

Equation (7.96) shows that $\boldsymbol{W}_2 = \boldsymbol{S}_n$; consequently, $\{\boldsymbol{V}_1(f), \boldsymbol{V}_2(f), \ldots, \boldsymbol{V}_{N-P}(f)\}$ is a basis for $\boldsymbol{S}_n$. Using this result and (7.78), we have

$$\left[\sigma^2 - 2K_1(1-\beta_k^2)\right]\boldsymbol{V}_k(f) + K_2 \sum_{i=1, i\neq k}^{N-P} \boldsymbol{V}_k^T(f)\boldsymbol{V}_i(f)\boldsymbol{V}_i(f) = \boldsymbol{0}$$
$$(k = 1, 2, \ldots, N-P). \tag{7.97}$$

Since $\boldsymbol{V}_1(f), \boldsymbol{V}_2(f), \ldots, \boldsymbol{V}_{N-P}(f)$ form a basis, they are linearly independent and it follows that

$$\sigma^2 = 2K_1(1-\beta_k^2) \tag{7.98}$$

and

$$\boldsymbol{V}_k^T(f)\boldsymbol{V}_i(f) = 0, \qquad k \neq i \qquad (i, k = 1, 2, \ldots, N-P). \tag{7.99}$$

This concludes the proof of Theorem 7.3. QED

From the proof of Theorem 7.3, we know that every local minimizer of $E(t)$ is also a global minimizer because the Hessian matrix $\boldsymbol{H}_k(\bar{\boldsymbol{V}}(f))$ is at least nonnegative definite.

Theorems 7.1–7.3 guarantee that if $K_1 > \sigma^2/2$ is satisfied, the neural network will certainly converge to the noise subspace spanned by the eigenvectors corresponding to the $N-P$ repeated smallest eigenvalues of the matrix $\boldsymbol{R}$. This also means that it is necessary to know the value of σ^2 before the neural network in Figure 7.6 is used to compute the noise subspace required in the MUSIC bearing estimation algorithm. However, σ^2 will not be known a priori. Mathew and Reddy [16] suggest a practical lower bound of

$$K_1 > \frac{\text{trace}\{\boldsymbol{R}\}}{2N}. \tag{7.100}$$

In most practical uses, the noise eigenvalues of the covariance matrix $\boldsymbol{R}$ are not exactly identical. In this general case, the above neural network does not apply. In order to compute the noise subspace required in the general case that $0 < \lambda_1 \leq \lambda_2 \leq \ldots \leq \lambda_{N-P}$, we will develop an alternative neural network in the next subsection.

7.2.2 Computation of the Noise Subspace in the General Case

The neural network for computing the noise subspace $\boldsymbol{S}_n$ in the general case is represented by the differential equations

$$\frac{d\boldsymbol{V}_j(t)}{dt} = \boldsymbol{V}_j^T(t)\left[\boldsymbol{I} - \sum_{i=1}^{j-1} \boldsymbol{V}_i(t)\boldsymbol{V}_i^T(t)\right]\boldsymbol{R}\boldsymbol{V}_j(t)\boldsymbol{V}_j(t)$$

$$- \boldsymbol{V}_j^T(t)\boldsymbol{V}_j(t)\left[\boldsymbol{I} - \sum_{i=1}^{j-1} \boldsymbol{V}_i(t)\boldsymbol{V}_i^T(t)\right]\boldsymbol{R}\boldsymbol{V}_j(t)$$

$$(j = 1, 2, \ldots, N-P). \qquad (7.101)$$

$\boldsymbol{V}_j(t) = [v_{1j}(t), v_{2j}(t), \ldots, v_{Nj}(t)]^T$ is an N-dimensional vector whose element $v_{ij}(t)$ is the output of the neuron (i, j) (for $i = 1, 2, \ldots, N; j = 1, 2, \ldots, N-P$). $\boldsymbol{R}$ is the covariance matrix, which is taken as the connection strength matrix in this neural network.

If we set

$$\boldsymbol{R}_j = \boldsymbol{R} - \sum_{i=1}^{j-1} \boldsymbol{V}_i(t)\boldsymbol{V}_i^T(t)\boldsymbol{R}, \qquad (7.102)$$

$$\psi_j(t) = \boldsymbol{V}_j^T(t)\boldsymbol{V}_j(t), \qquad (7.103)$$

and

$$\phi_j(t) = \boldsymbol{V}_j^T(t)\boldsymbol{R}_j\boldsymbol{V}_j(t) \qquad (j = 1, 2, \ldots, N-P) \qquad (7.104)$$

then the differential equations (7.101) can be written as

$$\frac{d\boldsymbol{V}_j(t)}{dt} = \phi_j(t)\boldsymbol{V}_j(t) - \psi_j(t)\boldsymbol{R}_j\boldsymbol{V}_j(t) \qquad (j = 1, 2, \ldots, N-P). \qquad (7.105)$$

Figure 7.7 shows the structure of the neuron (i, j) of this network.

For the dynamics of (7.105), we have the following theorems:

Theorem 7.4 *The norm of each output vector $\boldsymbol{V}_j(t)$ is invariant during the time evolution and is equal to the norm of the initial state $\boldsymbol{V}_j(0)$, that is,*

$$\|\boldsymbol{V}_j(t)\|^2 = \|\boldsymbol{V}_j(0)\|^2, \qquad t \geq 0, \qquad (7.106)$$

or

$$\psi_j(t) = \psi_j(0), \qquad t \geq 0 \qquad (j = 1, 2, \ldots, N-P). \qquad (7.107)$$

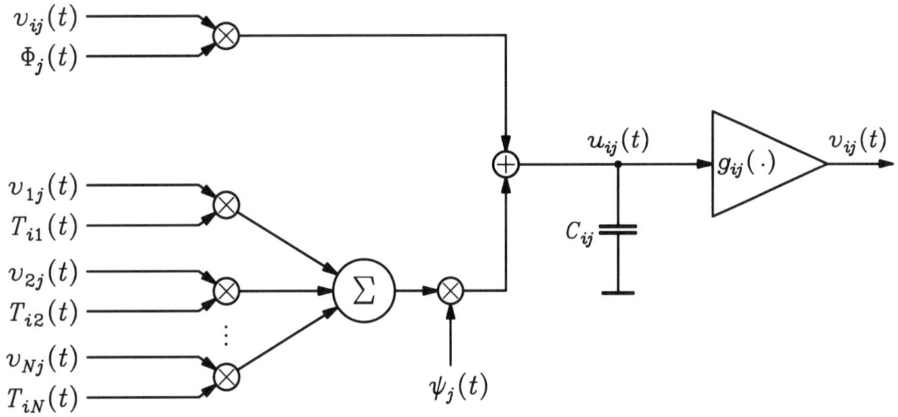

Figure 7.7. The model structure of the neuron (i, j), where $\boldsymbol{T}(t) = \boldsymbol{R}_j = \boldsymbol{R} - \sum_{i=1}^{j-1} \boldsymbol{V}_i(t)\boldsymbol{V}_i^T(t)\boldsymbol{R}$.

Proof: Multiplying (7.105) by $\boldsymbol{V}_j^T(t)$ on the left yields

$$\frac{d\|\boldsymbol{V}_j(t)\|^2}{dt} = 2\{\phi_j(t)\boldsymbol{V}_j^T(t)\boldsymbol{V}_j(t) - \psi_j(t)\boldsymbol{V}_j^T(t)\boldsymbol{R}_j\boldsymbol{V}_j(t)\} \qquad (j = 1, 2, \ldots, N-P). \tag{7.108}$$

Substituting (7.103) and (7.104) into (7.108) gives

$$\frac{d\|\boldsymbol{V}_j(t)\|^2}{dt} = 2\{\phi_j(t)\psi_j(t) - \psi_j(t)\phi_j(t)\} = 0 \qquad (j = 1, 2, \ldots, N-P), \tag{7.109}$$

which shows that (7.106) and (107) hold and concludes the proof. QED

If we set

$$\boldsymbol{V}_j(t) = \sum_{i=1}^{N} y_{ij}(t)\boldsymbol{S}_i \qquad (j = 1, 2, \ldots, N-P), \tag{7.110}$$

Theorem 7.4 shows

$$0 \le y_{ij}^2(t) \le \psi_j(0), \quad t \ge 0 \qquad (i = 1, 2, \ldots, N; \quad j = 1, 2, \ldots, N-P). \tag{7.111}$$

Theorem 7.5 *If the initial values of the output vector satisfy* $\boldsymbol{V}_j^T(0)\boldsymbol{S}_j \ne 0$ *(for* $j = 1, 2, \ldots, N-P$*), we have*

$$\lim_{t\to\infty} \boldsymbol{V}_j(t) = \lim_{t\to\infty} \sum_{i=1}^{N-P} y_{ij}(t)\boldsymbol{S}_i \qquad (j = 1, 2, \ldots, N-P). \tag{7.112}$$

Prior to the proof of Theorem 7.5, we first prove

$$y_{ij}^2(t) \leq M_{ij} \exp\left[2\psi_j(0)(\lambda_j - \lambda_i)t\right]$$
$$(j = 1, 2, \ldots, N-P; \quad i = j+1, j+2, \ldots, N), \tag{7.113}$$

where M_{ij} is a constant depending on the initial values and the eigenvalues.

For $j = 1$, we substitute (7.110) into (7.105) and obtain

$$\frac{dy_{i1}(t)}{dt} = \phi_1(t)y_{i1}(t) - \lambda_i\psi_1(0)y_{i1}(t) \qquad (i = 1, 2, \ldots, N). \tag{7.114}$$

According to the assumption of the initial value $\boldsymbol{V}_1(0)$, we may define [17, 18]

$$c_{i1}(t) = \frac{y_{i1}(t)}{y_{11}(t)} \qquad (i = 2, 3, \ldots, N). \tag{7.115}$$

Moreover,

$$\frac{dc_{i1}(t)}{dt} = \frac{y_{11}(t)\frac{dy_{i1}(t)}{dt} - y_{i1}(t)\frac{dy_{11}(t)}{dt}}{(y_{11}(t))^2}. \tag{7.116}$$

Combining Equations (7.116) and (7.114) gives

$$\frac{dc_{i1}(t)}{dt} = \psi_1(0)(\lambda_1 - \lambda_i)c_{i1}(t) \qquad (i = 2, 3, \ldots, N) \tag{7.117}$$

and

$$c_{i1}(t) = K_{i1} \exp\left[\psi_1(0)(\lambda_1 - \lambda_i)t\right] \qquad (i = 2, 3, \ldots, N), \tag{7.118}$$

where K_{i1} is a constant depending on the initial values and the eigenvalues of the matrix $\boldsymbol{R}$.

Using (7.115) yields

$$y_{i1}(t) = K_{i1}y_{11}(t) \exp\left[\psi_1(0)(\lambda_1 - \lambda_i)t\right] \qquad (i = 2, 3, \ldots, N). \tag{7.119}$$

Because $y_{11}^2(t) \leq \psi_1(0)$, we obtain

$$y_{i1}^2(t) \leq M_{i1} \exp\left[2\psi_1(0)(\lambda_1 - \lambda_i)t\right] \qquad (i = 2, 3, \ldots, N), \tag{7.120}$$

where $M_{i1} = K_{i1}^2\psi_1(0)$.

Assuming that (7.113) holds for $j = 1, 2, \ldots, p-1$, we will show that it holds for $j = p$.

For $j = p$, we have the following relationship according to (7.105) and (7.110):

$$\frac{dy_{ip}(t)}{dt} = \phi_p(t)y_{ip}(t) - \psi_p(0)\boldsymbol{S}_i^T\boldsymbol{R}_p\boldsymbol{S}_i y_{ip}(t) \qquad (i = 1, 2, \ldots, N). \tag{7.121}$$

Using (7.102) gives

$$\begin{aligned}\frac{dy_{ip}(t)}{dt} &= \phi_p(t)y_{ip}(t) - \psi_p(0)\left\{\lambda_i y_{ip}(t) - \sum_{k=1}^{p-1} \boldsymbol{S}_i^T \boldsymbol{V}_k(t)\boldsymbol{V}_k^T(t)\boldsymbol{R}\boldsymbol{S}_i y_{ip}(t)\right\}\\ &= \phi_p(t)y_{ip}(t) - \psi_p(0)\left\{\lambda_i y_{ip}(t) - \sum_{k=1}^{p-1} \lambda_i y_{ip}(t)\boldsymbol{S}_i^T \boldsymbol{V}_k(t)\boldsymbol{V}_k^T(t)\boldsymbol{S}_i\right\}\\ &= \phi_p(t)y_{ip}(t) - \psi_p(0)\lambda_i y_{ip}(t)\left[1 - \sum_{k=1}^{p-1} y_{ik}^2(t)\right]\\ &\qquad (i = 1, 2, \ldots, N). \qquad (7.122)\end{aligned}$$

In a manner similar to (7.115), we can define

$$c_{ip}(t) = \frac{y_{ip}(t)}{y_{pp}(t)} \qquad (i \neq p).$$

It follows that

$$\frac{dc_{ip}(t)}{dt} = \psi_p(0)\left\{\lambda_p\left[1 - \sum_{k=1}^{p-1} y_{pk}^2(t)\right] - \lambda_i\left[1 - \sum_{k=1}^{p-1} y_{ik}^2(t)\right]\right\} c_{ip}(t) \qquad (i \neq p) \qquad (7.123)$$

and

$$\begin{aligned}c_{ip}(t) &= K_{ip}\exp[\psi_p(0)(\lambda_p - \lambda_i)t]\\ &\quad\times \exp\left\{\psi_p(0)\int_0^t \left[\lambda_i \sum_{k=1}^{p-1} y_{ik}^2(\tau) - \lambda_p \sum_{k=1}^{p-1} y_{pk}^2(\tau)\right] d\tau\right\}\\ &\qquad (i \neq p), \qquad (7.124)\end{aligned}$$

where K_{ip} is a constant similar to K_{i1}. Moreover,

$$\begin{aligned}y_{ip}^2(t) &= K_{ip}^2 y_{pp}^2(t)\exp[2\psi_p(0)(\lambda_p - \lambda_i)t]\\ &\quad\times \exp\left\{2\psi_p(0)\int_0^t \left[\lambda_i \sum_{k=1}^{p-1} y_{ik}^2(\tau) - \lambda_p \sum_{k=1}^{p-1} y_{pk}^2(\tau)\right] d\tau\right\}\\ &\qquad (i \neq p). \qquad (7.125)\end{aligned}$$

It follows that

$$\begin{aligned}y_{ip}^2(t) &\leq K_{ip}^2 y_{pp}^2(t)\exp[2\psi_p(0)(\lambda_p - \lambda_i)t]\\ &\quad\times \exp\left(2\psi_p(0)\int_0^t \lambda_i \sum_{k=1}^{p-1} y_{ik}^2(\tau)d\tau\right) \qquad (i \neq p). \qquad (7.126)\end{aligned}$$

For $i = p+1, p+2, \ldots, N$, substituting $y_{ik}^2(t) \le M_{ik} \exp[2\psi_k(0)(\lambda_k - \lambda_i)t]$ (for $k = 1, 2, \ldots, p-1$) and $y_{pp}^2(t) \le \psi_p(0)$ into the above, we obtain

$$y_{ip}^2(t) \le K_{ip}^2 \psi_p(0) \exp\big[2\psi_p(0)(\lambda_p - \lambda_i)t\big] \\ \times \exp\left\{ 2\psi_p(0) \int_0^t \lambda_i \sum_{k=1}^{p-1} M_{ik} \exp\big[2\psi_k(0)(\lambda_k - \lambda_i)\tau\big] d\tau \right\} \\ (i = p+1, p+2, \ldots, N). \tag{7.127}$$

Because

$$2\psi_p(0) \int_0^t \lambda_i \sum_{k=1}^{p-1} M_{ik} \exp\big[2\psi_k(0)(\lambda_k - \lambda_i)\tau\big] d\tau \\ = \lambda_i \sum_{k=1}^{p-1} \frac{M_{ik}}{(\lambda_k - \lambda_i)} \frac{\psi_p(0)}{\psi_k(0)} \Big[\exp(2\psi_k(0)(\lambda_k - \lambda_i)t) - 1 \Big] \\ \le -\lambda_i \sum_{k=1}^{p-1} \frac{M_{ik}}{\lambda_k - \lambda_i} \frac{\psi_p(0)}{\psi_k(0)}, \tag{7.128}$$

we obtain

$$y_{ip}^2(t) \le M_{ip} \exp\big[2\psi_p(0)(\lambda_p - \lambda_i)t\big] \qquad (i = p+1, p+2, \ldots, N), \tag{7.129}$$

where

$$M_{ip} = K_{ip}^2 \psi_p(0) \exp\left(-\lambda_i \sum_{k=1}^{p-1} \frac{M_{ik}}{\lambda_k - \lambda_i} \frac{\psi_p(0)}{\psi_k(0)} \right). \tag{7.130}$$

This completes the proof of (7.113).

Since the eigenvalues of the matrix $\boldsymbol{R}$ satisfy $\lambda_{N-P+1} > \lambda_{N-P} \ge \lambda_{N-P-1} \ge \ldots \ge \lambda_1 > 0$, it is easy to get from (7.113) that

$$\lim_{t\to\infty} y_{ij}(t) = 0 \\ (j = 1, 2, \ldots, N-P; \quad i = N-P+1, N-P+2, \ldots, N). \tag{7.131}$$

Combining (7.110) with (7.131), we obtain

$$\lim_{t\to\infty} \boldsymbol{V}_j(t) = \lim_{t\to\infty} \sum_{i=1}^{N-P} y_{ij}(t)\boldsymbol{S}_i \qquad (j = 1, 2, \ldots, N-P). \tag{7.132}$$

This concludes the proof of Theorem 7.5.

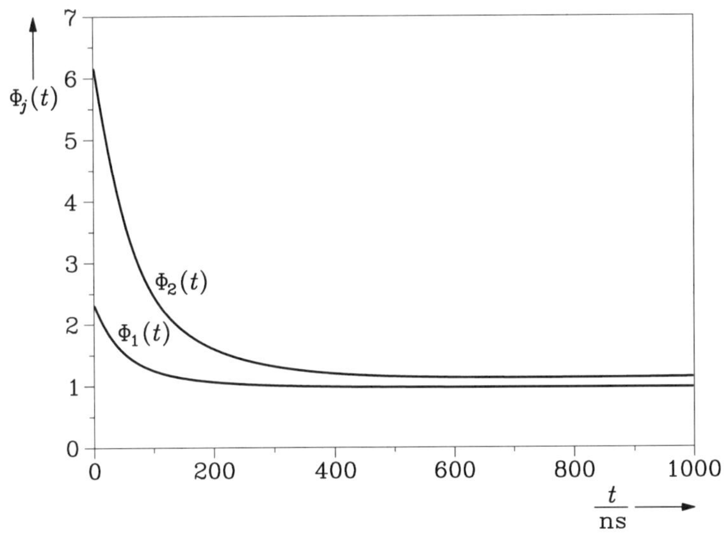

Figure 7.8. The dynamic curves of the variables $\phi_j(t)$ (for $j = 1, 2$) in Example 7.4.

Theorems 7.4 and 7.5 guarantee that the neural network with the differential equations (7.101) will converge to the noise subspace $\boldsymbol{S}_n$ as required in the MUSIC algorithm. In fact, this neural network can be applied to any positive definite matrix. In other words, this network can be used to provide the subspace spanned by the eigenvectors corresponding to the M smallest eigenvalues of any positive definite matrix. To illustrate this, we will give two sets of simulation examples. In Example 7.4, the smallest eigenvalues of the given positive definite matrix $\boldsymbol{R}$ are distinct; in Example 7.5, the smallest eigenvalues of $\boldsymbol{R}$ are not distinct. $\boldsymbol{V}_j(f)$ and $\boldsymbol{V}_j(0)$ (for $j = 1, 2, \ldots, M$) are the output vectors of the network in the stationary state and the initial state, respectively. $\lambda_j(f)$ are the values computed by $\boldsymbol{V}_j(f)$,

$$\lambda_j(f) = \frac{\boldsymbol{V}_j^T(f)\boldsymbol{R}\boldsymbol{V}_j(f)}{\boldsymbol{V}_j^T(f)\boldsymbol{V}_j(f)}$$

(for $j = 1, 2, \ldots, M$); $\lambda_j(a)$ (for $j = 1, 2, \ldots, M$) are the exact eigenvalues of the given matrix. For further illustrations, Figures 7.8 and 7.9 describe the time evolution of the variables $\phi_j(t)$ (for $j = 1, 2, \ldots, M$).

Example 7.4

$$\boldsymbol{R} = \begin{pmatrix} 2.3279 & 1.0825 & 0.8574 & -3.2503 \\ 1.0825 & 6.2087 & 4.3588 & -3.4051 \\ 0.8574 & 4.3588 & 4.8211 & -2.7944 \\ -3.2503 & -3.4051 & -2.7944 & 8.8905 \end{pmatrix}$$

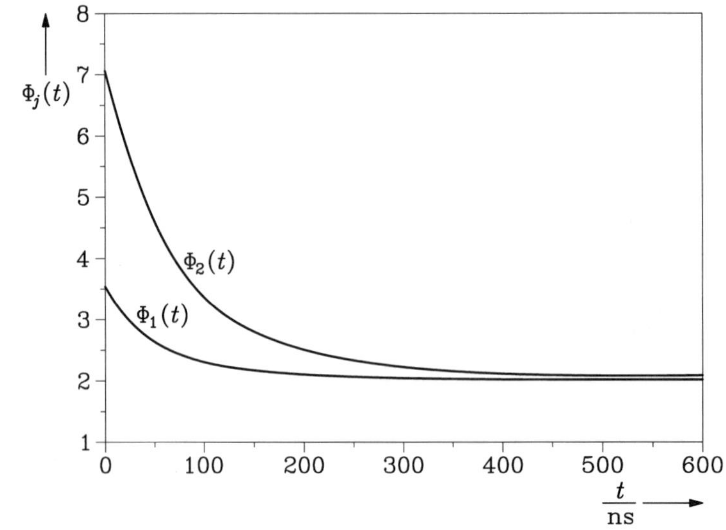

Figure 7.9. The dynamic curves of the variables $\phi_j(t)$ (for $j = 1, 2$) in Example 7.5.

$$\boldsymbol{V}_1(f) = \begin{pmatrix} 0.9152 \\ 0.0391 \\ 0.0397 \\ 0.4049 \end{pmatrix}, \qquad \boldsymbol{V}_2(f) = \begin{pmatrix} 0.0522 \\ 0.6723 \\ -0.7593 \\ 0.0463 \end{pmatrix}$$

$$\boldsymbol{V}_1(0) = \begin{pmatrix} 1.0000 \\ 0.0000 \\ 0.0000 \\ 0.0000 \end{pmatrix}, \qquad \boldsymbol{V}_2(0) = \begin{pmatrix} 0.0000 \\ 1.0000 \\ 0.0000 \\ 0.0000 \end{pmatrix}$$

$$\lambda_1(f) = 0.9681, \qquad \lambda_2(f) = 1.0997$$

$$\lambda_1(a) = 0.9680, \qquad \lambda_2(a) = 1.1000.$$

Example 7.5

$$\boldsymbol{R} = \begin{pmatrix} 3.5678 & 1.1626 & 0.9302 & -3.5269 \\ 1.1626 & 7.1180 & 4.3627 & -3.4317 \\ 0.9302 & 4.3627 & 5.7218 & -2.7970 \\ -3.5269 & -3.4317 & -2.7970 & 10.0905 \end{pmatrix}$$

$$\boldsymbol{V}_1(f) = \begin{pmatrix} 0.9213 \\ 0.0409 \\ 0.0451 \\ 0.4333 \end{pmatrix}, \qquad \boldsymbol{V}_2(f) = \begin{pmatrix} 0.0847 \\ 0.7279 \\ -0.8198 \\ 0.0654 \end{pmatrix}$$

$$V_1(0) = \begin{pmatrix} 1.0000 \\ 0.0000 \\ 0.0000 \\ 0.0000 \end{pmatrix}, \qquad V_2(0) = \begin{pmatrix} 0.0000 \\ 1.0000 \\ 0.0000 \\ 0.0000 \end{pmatrix}$$

$$\lambda_1(f) = 2.0001, \qquad \lambda_2(f) = 2.0003$$

$$\lambda_1(a) = 2.0000, \qquad \lambda_2(a) = 2.0000.$$

We note that $\lambda_j(f) = \lambda_j(a)$ (for $j = 1, 2, \ldots, M$) in the above simulation results means that

$$\lim_{t\to\infty} y_{ij}(t) = 0 \qquad (i = 1, 2, \ldots, j-1; \quad j = 1, 2, \ldots, M) \tag{7.133}$$

and

$$\lim_{t\to\infty} y_{jj}(t) = \pm\sqrt{\psi_j(0)} \qquad (j = 1, 2, \ldots, M). \tag{7.134}$$

Moreover, (7.112) in Theorem 7.5 becomes

$$\lim_{t\to\infty} V_j(t) = \pm\sqrt{\psi_j(0)} S_i \qquad (j = 1, 2, \ldots, M). \tag{7.135}$$

That is, the output vector $V_j(t)$ (for $j = 1, 2, \ldots, M$) will converge to the eigenvector corresponding to the j'th eigenvalue λ_j of the given matrix R. Although many other simulation results have also exhibited this property, we failed to find a rigorous proof.

The following examples present two simulation results of the MUSIC bearing estimation method using the above neural network [19].

Example 7.6

Consider four planar-wave signals whose directions of arrival are $-55°$, $-45°$, $3°$, and $25°$, and whose associated signal-to-noise ratios are each 7 dB. The array is uniform and linear with 10 elements spaced a half wavelength apart. The number of snapshots is 30. P_{tr} and P_{ti} are the real and imaginary parts of the exact orthogonal projection operator on the noise subspace computed from (7.52). V_r and V_i are the real and imaginary parts provided by the neural network. Obviously, $P_{tr} \approx V_r$ and $P_{ti} \approx V_i$. Figures 7.10 and 7.11 show the estimated spatial spectrum by use of the exact noise subspace S_n and the solution provided by the neural network, respectively. We have

$$P_{tr} = \begin{pmatrix} 0.258489 & 0.062819 & 0.006841 & -0.046045 & -0.252478 & -0.029564 \\ 0.062819 & 0.270058 & -0.036132 & -0.001520 & -0.094764 & -0.254078 \\ 0.006841 & -0.036132 & 0.474691 & -0.126503 & 0.000633 & -0.040808 \\ -0.046045 & -0.001520 & -0.126503 & 0.475472 & -0.037899 & 0.012248 \\ -0.252478 & -0.094764 & 0.000633 & -0.037899 & 0.266104 & 0.061587 \\ -0.029564 & -0.254078 & -0.040808 & 0.012248 & 0.061587 & 0.255185 \end{pmatrix},$$

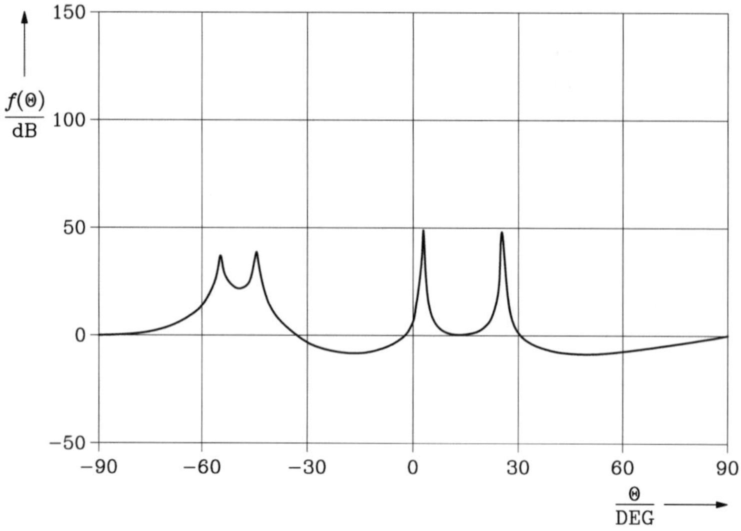

Figure 7.10. The estimated spatial spectrum obtained by use of the exact noise subspace $\boldsymbol{S}_n$.

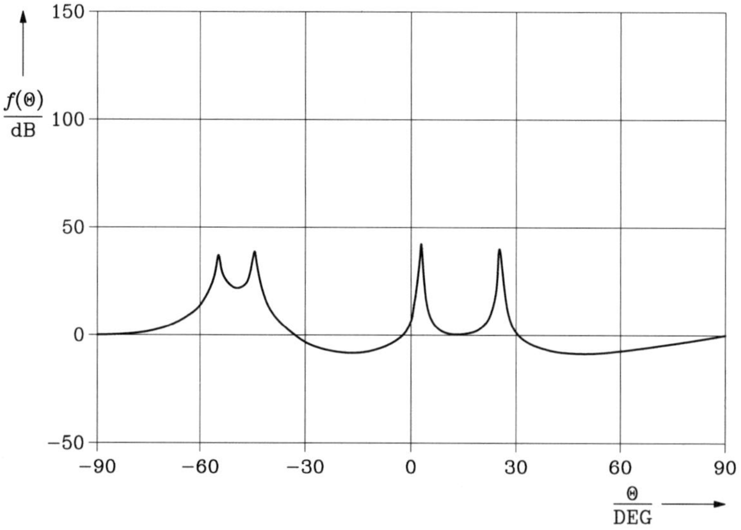

Figure 7.11. The estimated spatial spectrum obtained by use of the neural network solution.

$$\boldsymbol{P}_{ti} = \begin{pmatrix} 0.000000 & -0.056791 & 0.341879 & -0.028773 & -0.002339 & -0.000028 \\ 0.056791 & 0.000000 & 0.021180 & 0.318210 & -0.115776 & 0.001007 \\ -0.341879 & -0.021180 & 0.000000 & 0.112655 & 0.315792 & -0.024351 \\ 0.028733 & -0.318210 & -0.112655 & 0.000000 & 0.018426 & 0.338482 \\ 0.002339 & 0.115776 & -0.315792 & -0.018426 & 0.000000 & -0.062159 \\ 0.000028 & -0.001007 & 0.024351 & -0.338482 & 0.062159 & 0.000000 \end{pmatrix},$$

Table 7.2. *Ten Monte-Carlo Runs of the MUSIC method using the neural network*

	1	2	3	4	5	6	7	8	9	10
θ_1	−5.4	−4.8	−4.8	−4.8	−5.24	−5.2	−5.2	−5.0	−5.4	−5.1
θ_2	5.2	5.1	5.0	4.8	4.8	4.7	5.2	5.4	4.8	5.2

$$V_r = \begin{pmatrix} 0.258663 & 0.062867 & 0.006519 & -0.045759 & -0.252443 & -0.029731 \\ 0.062930 & 0.269792 & -0.035787 & -0.001851 & -0.094887 & -0.253722 \\ 0.006703 & -0.036042 & 0.479740 & -0.126505 & 0.000751 & -0.040946 \\ -0.045850 & -0.001595 & -0.126765 & 0.4758835 & -0.038196 & 0.012388 \\ -0.252352 & -0.094861 & 0.000851 & -0.038142 & 0.266230 & 0.061501 \\ -0.029766 & -0.253845 & -0.040997 & 0.0123716 & 0.061532 & 0.255080 \end{pmatrix},$$

and

$$V_i = \begin{pmatrix} 0.000204 & -0.056920 & 0.341844 & -0.0287862 & -0.002527 & 0.000300 \\ 0.056486 & 0.000095 & 0.0215070 & 0.317870 & -0.115518 & 0.000821 \\ -0.341868 & -0.0211394 & -0.000210 & 0.112851 & 0.315791 & -0.024392 \\ 0.028954 & -0.318522 & -0.112338 & -0.000277 & 0.018402 & 0.338665 \\ 0.002206 & 0.115919 & -0.315778 & -0.018465 & 0.000122 & -0.062194 \\ 0.000014 & -0.001025 & 0.0242153 & -0.338823 & 0.061930 & 0.000066 \end{pmatrix}.$$

Example 7.7
Two planar-wave signal sources whose directions of arrival are -5° and 5° are considered. The array is uniform and linear and has 5 elements spaced a half wavelength apart. The other parameters are the same as in Example 7.6. Table 7.2 shows results of the MUSIC bearing estimation method for ten Monte-Carlo runs by use of the neural network with the differential equations (7.105).

7.3 Neural Networks for the ML Bearing Estimation

In Sections 7.1 and 7.2, neural networks are only used to perform the major computation required in the ML method, the alternating projection ML method, the propagator method, and the MUSIC method. The outputs of these neural networks are thus the projection operators or noise subspace but not the desired bearing estimation values. In the following sections, we will develop neural networks that can provide the bearing estimation directly from the sampled data. This section will present a neural network approach whose performance is equal to that of the ML method. In addition, the envelopes and initial phases of arrival of the signals are also given in this approach. In Section 7.4, a new neural network approach is given whose direction resolution depends primarily on the number of neurons.

For simplicity, let us consider the real-valued case of (7.1). At the n'th snapshot the output of the i'th sensor can be described by

$$x_i(n) = \sum_{l=1}^{P} \alpha_l(n) \cos\left[kd(i-1)\sin(\theta_l) + \varphi_l\right] + \epsilon_i(n) \qquad (i = 1, 2, \ldots, N), \tag{7.136}$$

where φ_l is the initial phase of arrival of the l'th signal source and the various other quantities are the same as those in (7.1) except that they all take real values. In addition, the $\epsilon_i(n)$ $(i = 1, 2, \ldots, N)$ are assumed to be the additive, white, zero-mean noises with identical Gaussian probability distribution functions and to be independent of one another.

Our problem is: Given L snapshots (i.e., $x_i(1), x_i(2), \ldots, x_i(L)$; $i = 1, 2, \ldots, N$), estimate θ_l, $\alpha_l(n)$, and φ_l (for $l = 1, 2, \ldots, P$).

With the above assumption, the joint probability density function of the received vectors $\boldsymbol{X}(n) = [x_1(n), x_2(n), \ldots, x_N(n)]^T$ (for $n = 1, 2, \ldots, L$) is

$$f(\boldsymbol{X}) = \prod_{n=1}^{L} \frac{1}{\pi \det[\sigma^2 \boldsymbol{I}]} \times \exp\left\{\frac{-1}{\sigma^2} \sum_{i=1}^{N} \left(x_i(n) - \sum_{l=1}^{P} \alpha_l(n) \cos\left[kd(i-1)\sin(\theta_l) + \varphi_l\right]\right)^2\right\}. \tag{7.137}$$

The corresponding log-likelihood function (omitting the constant term) is

$$lf = -LN \log(\sigma^2) - \frac{1}{\sigma^2} \sum_{n=1}^{L} \sum_{i=1}^{N} \left(x_i(n) - \sum_{l=1}^{P} \alpha_l(n) \cos\left[kd(i-1)\sin(\theta_l) + \varphi_l\right]\right)^2. \tag{7.138}$$

The estimation of σ^2 under the ML criterion is

$$\hat{\sigma}^2 = \frac{1}{LN} \sum_{n=1}^{L} \sum_{i=1}^{N} \left(x_i(n) - \sum_{l=1}^{P} \alpha_l(n) \cos\left[kd(i-1)\sin(\theta_l) + \varphi_l\right]\right)^2. \tag{7.139}$$

Substituting this result into the log-likelihood function (7.138) and ignoring constant terms, we get the log-likelihood function for the parameters θ_l, $\alpha_l(n)$, and φ_l (for $l = 1, 2, \ldots, P$; $n = 1, 2, \ldots, L$):

$$lf(\theta_l, \alpha_l(n), \varphi_l) = -LN \log\left[\frac{1}{LN} \sum_{n=1}^{L} \sum_{i=1}^{N} \left(x_i(n) - \sum_{l=1}^{P} \alpha_l(n) \cos\left[kd(i-1)\sin(\theta_l) + \varphi_l\right]\right)^2\right]. \tag{7.140}$$

That is, the parameters θ_l , $\alpha_l(n)$, and φ_l can be obtained by solving the following maximization problem:

$$\max_{\theta_l,\alpha_l(n),\varphi_l} \; -LN \log\left[\frac{1}{LN}\sum_{n=1}^{L}\sum_{i=1}^{N}\left(x_i(n)-\sum_{l=1}^{P}\alpha_l(n)\right.\right.$$
$$\left.\left.\times\cos\left[kd(i-1)\sin(\theta_l)+\varphi_l\right]\right)^2\right]. \tag{7.141}$$

Since the logarithm function is a monotonic function, this problem can be expressed in the following form:

$$\min_{\theta_l,\alpha_l(n),\varphi_l} \sum_{n=1}^{L}\sum_{i=1}^{N}\left(x_i(n)-\sum_{l=1}^{P}\alpha_l(n)\cos\left[kd(i-1)\sin(\theta_l)+\varphi_l\right]\right)^2, \tag{7.142}$$

which is a typical nonlinear least-squares problem. We will now show how to use neural networks to solve (7.142) in real time.

For mathematical convenience, we assume that $\alpha_l(n)$ is a slowly changing function during the time of L snapshots, that is, $\alpha_l(n) = \alpha_l$ (for $n = 1, 2, \ldots, L$). This assumption affects only the number of unknown variables without affecting the form of (7.142). In addition, without loss of generality, let us choose $d = 1/k$. With these conditions, (7.142) becomes

$$\min_{\theta_l,\alpha_l,\varphi_l} \sum_{n=1}^{L}\sum_{i=1}^{N}\left(x_i(n)-\sum_{l=1}^{P}\alpha_l\cos\left[(i-1)\sin(\theta_l)+\varphi_l\right]\right)^2. \tag{7.143}$$

A neural network for solving (7.143) can be represented by the following differential equations:

$$C_{1i}\frac{du_{1i}(t)}{dt} = -\frac{u_{1i}(t)}{R_{1i}} + v_{3i}(t)\cos\left(v_{1i}(t)\right)$$
$$\times\sum_{j=1}^{N}q_j(t)(j-1)\sin\Big[(j-1)\sin\left(v_{1i}(t)\right)+v_{2i}(t)\Big], \tag{7.144}$$

$$C_{2i}\frac{du_{2i}(t)}{dt} = -\frac{u_{2i}(t)}{R_{2i}} + v_{3i}(t)\sum_{j=1}^{N}q_j(t)\sin\Big((j-1)\sin\left(v_{1i}(t)\right)+v_{2i}(t)\Big), \tag{7.145}$$

$$C_{3i}\frac{du_{3i}(t)}{dt} = -\frac{u_{3i}(t)}{R_{3i}} - \sum_{j=1}^{N}q_j(t)\Big(\cos\left[(j-1)\sin\left(v_{1i}(t)\right)+v_{2i}(t)\right]\Big), \tag{7.146}$$

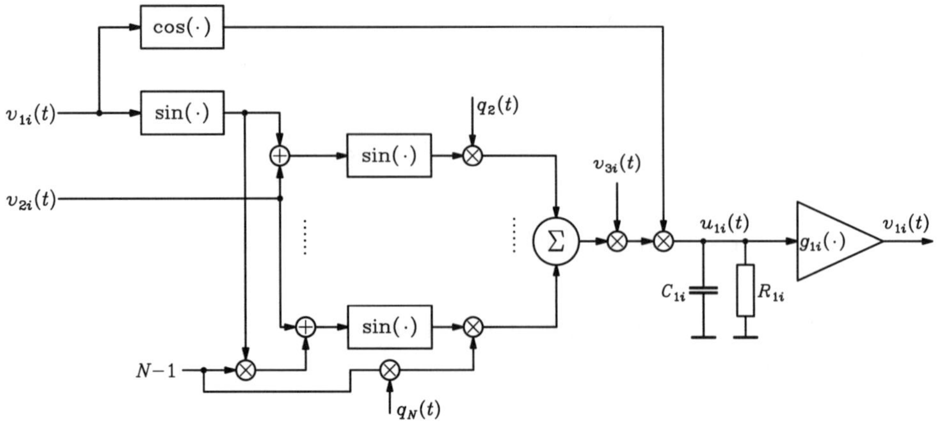

Figure 7.12. The neuron represented by the differential equations (7.144).

and

$$q_j(t) = \sum_{n=1}^{L}\left(\sum_{i=1}^{P} v_{3i}(t)\cos\left[(j-1)\sin\left(v_{1i}(t)\right) + v_{2i}(t)\right] - b_j(n)\right)$$
$$(i = 1, 2, \ldots, P). \qquad (7.147)$$

This neural network is composed of three kinds of neurons whose inputs and outputs are denoted by $u_{ki}(t)$ and $v_{ki}(t)$ (for $k = 1, 2, 3$ and $i = 1, 2, \ldots, P$), respectively. R_{ki} and C_{ki} are the input resistor and input capacitor, respectively, of the neuron (k, i). The input–output relationship function of the neuron (k, i) is denoted by $g_{ki}(\cdot)$ and selected to guarantee that

(1) $\int_0^{v_{ki}(t)} g_{ki}^{-1}(v)dv$ (for $k = 1, 2, 3$; $i = 1, 2, \ldots, P$) is bounded from below and that

(2) $g_{ki}(u)$ (for $k = 1, 2, 3$; $i = 1, 2, \ldots, P$) is a monotonically increasing function.

$b_j(n)$ (for $j = 1, 2, \ldots, N; n = 1, 2, \ldots, L$) is a kind of bias current. The structures of the three different kinds of neurons are shown in Figures 7.12–7.14.

In order to guarantee that this neural network is stable, we define an energy function as follows:

$$E(t) = \frac{1}{2}\sum_{n=1}^{L}\sum_{j=1}^{N}\left(\sum_{i=1}^{P} v_{3i}(t)\cos\left[(j-1)\sin\left(v_{1i}(t)\right) + v_{2i}(t)\right] - b_j(n)\right)^2$$
$$+ \sum_{i=1}^{P}\sum_{k=1}^{3}\frac{1}{R_{ki}}\int_0^{v_{ki}(t)} g_{ki}^{-1}(v)dv. \qquad (7.148)$$

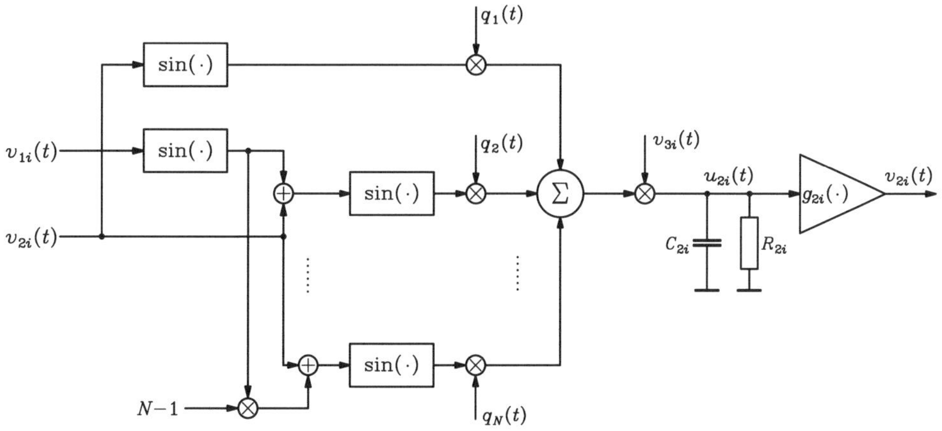

Figure 7.13. The neuron represented by the differential equations (7.145).

Because the first term on the right-hand side of (7.148) is nonnegative and because

$$\sum_{i=1}^{P}\sum_{k=1}^{3}\frac{1}{R_{ki}}\int_{0}^{v_{ki}(t)} g_{ki}^{-1}(v)dv$$

is bounded from below, $E(t)$ is bounded from below.

Moreover, from (7.148), we can differentiate $E(t)$ with respect to time t to get

$$\begin{aligned}\frac{dE(t)}{dt} = &\sum_{n=1}^{L}\sum_{j=1}^{N}\left(\sum_{k=1}^{P} v_{3k}(t)\cos\left\{(j-1)\sin\left[v_{1k}(t)\right]+v_{2k}(t)\right\}-b_j(n)\right)\\ &\times\left\{\sum_{i=1}^{P}\cos\left[(j-1)\sin\left(v_{1i}(t)\right)+v_{2i}(t)\right]\frac{dv_{3i}(t)}{dt}\right.\\ &-\sum_{i=1}^{P} v_{3i}(t)(j-1)\cos\left(v_{1i}(t)\right)\sin\left[(j-1)\sin\left(v_{1i}(t)\right)+v_{2i}(t)\right]\\ &\times\left.\frac{dv_{1i}(t)}{dt}-\sum_{i=1}^{P} v_{3i}(t)\sin\left[(j-1)\sin\left(v_{1i}(t)\right)+v_{2i}(t)\right]\frac{dv_{2i}(t)}{dt}\right\}\\ &+\sum_{i=1}^{P}\sum_{k=1}^{3}\frac{u_{ki}(t)}{R_{ki}}\frac{dv_{ki}(t)}{dt}. \qquad (7.149)\end{aligned}$$

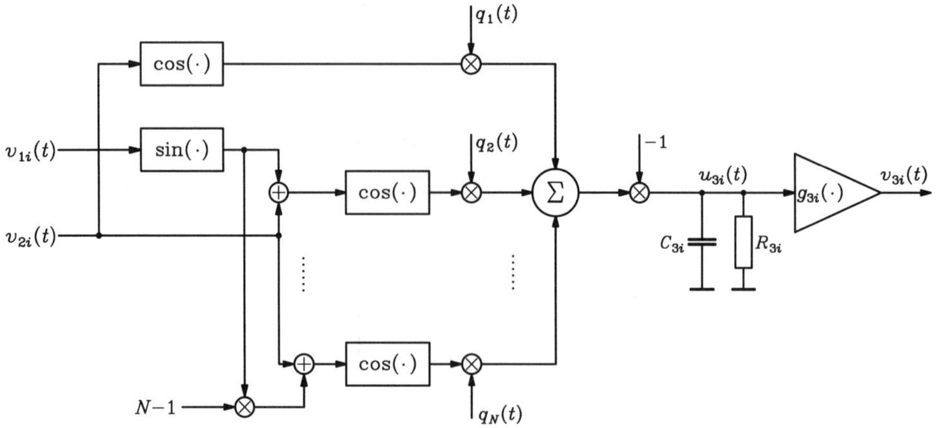

Figure 7.14. The neuron represented by the differential equations (7.146).

From (7.147) and (7.149) it follows that

$$\begin{aligned}\frac{dE(t)}{dt} &= \sum_{i=1}^{P}\left(\frac{u_{1i}(t)}{R_{1i}} - \sum_{j=1}^{N} q_j(t)v_{3i}(t)(j-1)\cos\left(v_{1i}(t)\right)\right.\\ &\quad\left.\times \sin\left[(j-1)\sin\left(v_{1i}(t)\right) + v_{2i}(t)\right]\right)\frac{dv_{1i}(t)}{dt}\\ &\quad+\sum_{i=1}^{P}\left(\frac{u_{2i}(t)}{R_{2i}} - \sum_{j=1}^{N} q_j(t)v_{3i}(t)\sin\left[(j-1)\sin(v_{1i}(t)) + v_{2i}(t)\right]\right)\\ &\quad\times\frac{dv_{2i}(t)}{dt} + \sum_{i=1}^{P}\left(\frac{u_{3i}(t)}{R_{3i}} + \sum_{j=1}^{N} q_j(t)\right.\\ &\quad\left.\times\cos\left[(j-1)\sin\left(v_{1i}(t)\right) + v_{2i}(t)\right]\right)\frac{dv_{3i}(t)}{dt}. \end{aligned} \tag{7.150}$$

Moreover, substituting (7.144)–(7.146) into (7.150), we obtain

$$\begin{aligned}\frac{dE(t)}{dt} &= -\sum_{k=1}^{3}\sum_{i=1}^{P} C_{ki}\frac{dv_{ki}(t)}{dt}\frac{du_{ki}(t)}{dt}\\ &= -\sum_{k=1}^{3}\sum_{i=1}^{P} C_{ki}\left[g_{ki}^{-1}\left(v_{ki}(t)\right)\right]'\left(\frac{dv_{ki}(t)}{dt}\right)^2. \end{aligned} \tag{7.151}$$

Since C_{ki} is positive and $[g_{ki}^{-1}(v_{ki}(t))]'$ is a monotonically increasing function, each term in (7.151) is nonnegative and

$$\frac{dE(t)}{dt} \leq 0, \qquad \frac{dE(t)}{dt} = 0 \quad \rightarrow \quad \frac{dv_{ki}(t)}{dt} = 0$$
$$(k = 1, 2, 3; \quad i = 1, 2, \ldots, P). \tag{7.152}$$

Together with the fact that $E(t)$ is bounded from below, (7.152) shows that the time evolution of this neural network is a motion in the state space that seeks out the minimum of the energy function $E(t)$ and comes to a stop at such a point. One of the local minimum points of $E(t)$ is provided by $dv_{ki}(t)/dt = 0$ (for $k = 1, 2, 3$; $i = 1, 2, \ldots, P$).

Now we will show how to use this network to solve the optimization problem (7.143) of the bearing estimation. If we choose the outputs of the sensors $x_j(n)$ (for $j = 1, 2, \ldots, N; n = 1, 2, \ldots, L$) as the bias currents $b_j(n)$ of the network and all the neurons to be linear amplifiers with gains K (i.e., $v_{ki}(t) = Ku_{ki}(t)$), then we know that such a constructed neural network is stable and provides a minimum of

$$E(t) = \frac{1}{2}\sum_{n=1}^{L}\sum_{j=1}^{N}\left\{\sum_{i=1}^{P} v_{3i}(t)\cos\left[(j-1)\sin\left(v_{1i}(t)\right) + v_{2i}(t)\right] - x_j(n)\right\}^2$$
$$+\sum_{i=1}^{P}\sum_{k=1}^{3}\frac{(v_{ki}(t))^2}{2R_{ki}K}. \tag{7.153}$$

Comparing (7.153) and (7.143) shows that the minimum of (7.153) is approximately equal to the minimum of (7.143) with arbitrarily small error if we choose large enough K and R_{ki}. Hence, the outputs $v_{1i}(f)$, $v_{2i}(f)$, and $v_{3i}(f)$ (for $i = 1, 2, \ldots, P$) in the stationary state of the neural network will provide the direction angles θ_i, the initial phases φ_i, and the envelopes α_i of arrival of P signal sources, respectively.

We can make several additional comments regarding this neural network approach:

(1) Because the available data $x_j(n)$ (for $j = 1, 2, \ldots, N; n = 1, 2, \ldots, L$) are directly taken as the bias currents of the network without any computations and the other parameters are fixed, this network is very suitable for real-time applications.

(2) Choosing large enough K and R_{ki} makes (7.153) arbitrarily close to (7.143). This means that the estimation performance of this network is the same as that of the ML method.

(3) This network is easily generalized for the case in which the data take complex values. In this case, the number of the neurons is increased by a factor of two.

(4) For the time-varying envelopes $\alpha_i(n)$ (for $i = 1, 2, \ldots, N$), the number of

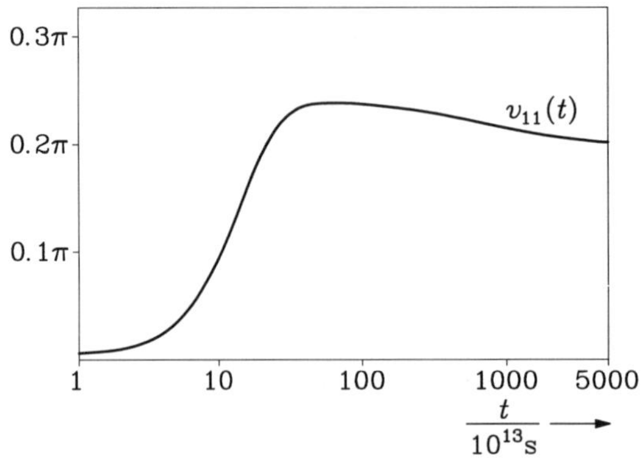

Figure 7.15. The dynamic curve of the neuron (1, 1) in the first simulation example.

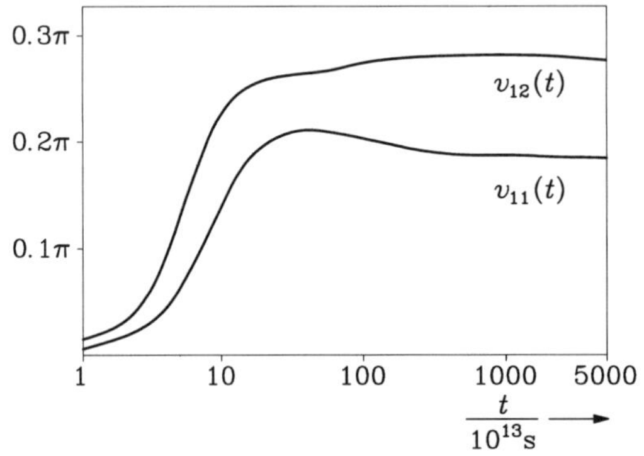

Figure 7.16. The dynamic curves of the neurons (1, 1) and (1, 2) in the second simulation example.

neurons of the second kind (corresponding to $v_{2i}(t)$) will increase, but the architecture of the network will not change.

We have simulated this proposed neural network for bearing estimation [20]. Three sets of simulation results are given in Figures 7.15–7.17 and Tables 7.3–7.5. In these figures, the curves $v_{1i}(t)$ (for $i = 1, 2, \ldots, P$) describe the time evolutions of the neurons of the first kind. In Tables 7.3–7.5, numerical results corresponding to $v_{1i}(t)$ are given. The errors between the solutions in the stationary state of the network and the real directions of the sources are also presented in the tables. From the tables, we can also find the time evolution of the energy function $E(t)$.

In these simulation results, we used $R_{ki} = 100\ \text{k}\Omega$, $C_{ki} = 100$ pF (for $k = 1, 2, 3$; $i = 1, 2, \ldots, P$), and $K = 1$. The sensor number and snapshot number are 5 and 22, respectively.

In the first simulation example, only one source is considered. It has a direction of 54° and a signal-to-noise ratio of 20 dB. In the second example, there are two

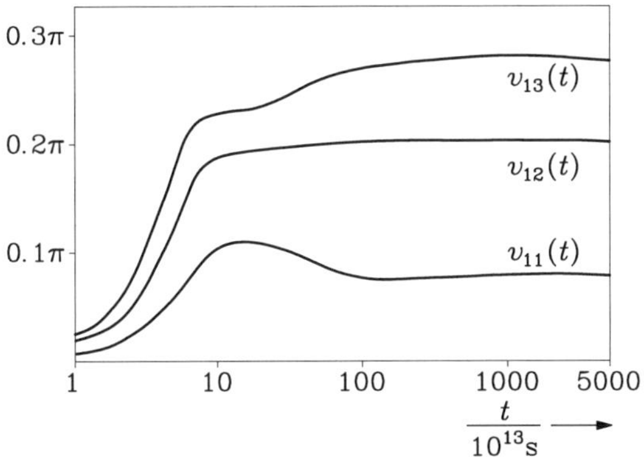

Figure 7.17. The dynamic curves of the neurons (1, 1), (1, 2), and (1, 3) in the third simulation example.

Table 7.3. *The time evolution of the output $v_{11}(t)$ and the energy function $E(t)$*

$t \times 10^{-13}$	$E(t)$	$v_{11}(t)$ (rad)
1	53.7964	0.017453
2	47.7512	0.029197
3	42.8004	0.046268
⋮	⋮	⋮
4940	0.0020115	0.629626
4945	0.0020114	0.629615
4950	0.0020113	0.629605
Estimation	error	0.076280°

Table 7.4. *The time evolution of $E(t)$, $v_{11}(t)$, and $v_{12}(t)$*

$t \times 10^{-13}$	$E(t)$	$v_{11}(t)$ (rad)	$v_{12}(t)$ (rad)
1	234.421	0.017453	0.034907
2	182.155	0.041758	0.086624
3	143.549	0.083185	0.178499
⋮	⋮	⋮	⋮
4940	0.0020137	0.603602	0.897712
4945	0.0020137	0.603602	0.897712
4950	0.0020137	0.603602	0.897712
Estimation	error	1.41615°	2.56489°

Table 7.5. *The time evolution of $E(t)$, $v_{11}(t)$, $v_{12}(t)$ and $v_{13}(t)$*

$t\times 10^{-13}$	$E(t)$	$v_{11}(t)$ (rad)	$v_{12}(t)$ (rad)	$v_{13}(t)$ (rad)
1	263.590	0.017453	0.034907	0.052360
2	191.934	0.041604	0.086520	0.134579
3	139.868	0.079826	0.171968	0.274347
⋮	⋮	⋮	⋮	⋮
4940	0.0021042	0.244283	0.624450	0.872154
4945	0.0021042	0.244289	0.624440	0.872159
4950	0.0021042	0.244295	0.624434	0.872163
Estimation	error	4.00293°	0.22368°	4.00156°

sources; their directions are 36° and 54° and the signal-to-noise ratios are 20 dB. In the third example, three sources are considered. Their directions of arrival are 18°, 36°, 54°, respectively, and their associated signal-to-noise ratios are all 20 dB.

Finally, it should be noted that this neural network approach suffers from the local minimum problem as encountered in the ML method [11]. Although some methods such as the continuous-time analog of the simulated annealing algorithm [21] have been proposed so as to deliver the global solution, this remains an unsolved problem.

7.4 Hypothesis-Based Bearing Estimation Using Neural Networks

As pointed out in Sections 1.1 and 1.2, one of the important properties of Hopfield-type neural networks is their ability to simultaneously test a large number of alternative hypotheses and make decisions on them using the given data at a remarkable speed. In effect, the bearing estimation of arrival of the sources can be mapped to this kind of hypothesis-decision problem [22, 23, 24]. Hypothesize M sources to impinge on a uniform linear array of N omnidirectional sensors. Then our task is to indicate which among the M hypotheses are true (i.e., which are the signals present in the given data). For each hypothesis, we have a vector

$$Y_i = \alpha_i e^{j\varphi_i}\Big[1, \exp\big[jkd\sin(\theta_i)\big], \ldots, \exp\big[jkd(N-1)\sin(\theta_i)\big]\Big]^T \qquad (i = 1, 2, \ldots, M), \qquad (7.154)$$

where θ_i, α_i, and φ_i are the direction, envelope, and initial phase of the i'th hypothesized source, respectively.

With the received vector $\boldsymbol{X}(n)$ at the n'th snapshot, we have the following optimization function:

$$E = \left\| \boldsymbol{X}(n) - [\boldsymbol{Y}_1, \boldsymbol{Y}_2, \ldots, \boldsymbol{Y}_M]\boldsymbol{V} \right\|^2, \tag{7.155}$$

where $\boldsymbol{V} = [v_1, v_2, \ldots, v_M]^T$ is an M-dimensional vector. Clearly, the global optimization solution of (7.155) is

$$\begin{cases} v_i = 1, & i = i_1, i_2, \ldots, i_P, \\ v_i = 0, & i \neq i_1, i_2, \ldots, i_P, \end{cases} \tag{7.156}$$

where the i_j (for $j = 1, 2, \ldots, P$) index the P true signal sources.

Now we will show how to use the continuous-time Hopfield neural network in Figure 1.2 to solve the problem (7.155). Equation (7.155) can be written as

$$E = \boldsymbol{X}^H(n)\boldsymbol{X}(n) + \boldsymbol{V}^T\boldsymbol{Y}^H\boldsymbol{Y}\boldsymbol{V} - \boldsymbol{X}^H(n)\boldsymbol{Y}\boldsymbol{V} - \boldsymbol{V}^T\boldsymbol{Y}^H\boldsymbol{X}, \tag{7.157}$$

where $\boldsymbol{Y} = [\boldsymbol{Y}_1, \boldsymbol{Y}_2, \ldots, \boldsymbol{Y}_M]$. Eliminating the constant term of (7.157), we obtain

$$E = \boldsymbol{V}^T\boldsymbol{Y}^H\boldsymbol{Y}\boldsymbol{V} - 2\,\mathrm{real}\big(\boldsymbol{X}^H(n)\boldsymbol{Y}\boldsymbol{V}\big). \tag{7.158}$$

Comparing (7.158) with (1.26) implies that the problem (7.155) can be solved by using the continuous-time Hopfield neural network with M neurons, provided we choose the connection strength T_{ij} and bias current b_i as

$$T_{ij} = -2\,\mathrm{real}\big(\boldsymbol{Y}_i^H\boldsymbol{Y}_j\big) \tag{7.159}$$

and

$$b_i = \mathrm{real}\big(\boldsymbol{X}^H(n)\boldsymbol{Y}_i\big) \qquad (i, j = 1, 2, \ldots, M). \tag{7.160}$$

In addition, the activation function of each neuron is selected as the sigmoid function $g_i(u) = \frac{1}{2}(1 + \tanh(\frac{u}{u_0}))$ so that the neuron outputs $v_i(t)$ (for $i = 1, 2, \ldots, M$) are allowed to take on the values in the range of zero to one. According to Theorem 1.2, such a constructed network is stable. Furthermore, from the output vector $\boldsymbol{V}_f$ in the stationary state, we can find that certain v_{i_j} (for $j = 1, 2, \ldots, P$) are (or close to) one, and hence we infer that the corresponding $\boldsymbol{Y}_{i_j}$ are the ones that comprise the sampled data $\boldsymbol{X}(n)$. In other words, once v_{i_j} is found, we can immediately determine θ_i, α_i, and φ_i from the corresponding hypothesized vector $\boldsymbol{Y}_{i_j}$ (for $j = 1, 2, \ldots, P$).

In practical hardware implementation, T_{ii} (self-connections) (for $i = 1, 2, \ldots, M$) should be zeros so as to avoid unexpected oscillations [25, 26]. This can be achieved by adding the term $-\sum_{i=1}^{M}(\boldsymbol{Y}_i^H\boldsymbol{Y}_i)v_i(v_i - 1)$ to (7.158), which gives

$$E_1 = \boldsymbol{V}^T\boldsymbol{Y}^H\boldsymbol{Y}\boldsymbol{V} - 2\,\mathrm{real}\big(\boldsymbol{X}^H(n)\boldsymbol{Y}\boldsymbol{V}\big) - \sum_{i=1}^{M}(\boldsymbol{Y}_i^H\boldsymbol{Y}_i)v_i(v_i - 1). \tag{7.161}$$

Hopfield and Tank have shown that E and E_1 have the same minima [25, 26]. As a result, the network for solving (7.158) is the same as that for (7.161) except that the connection strengths and bias currents for (7.161) become

$$T_{ij} = -2\,\mathrm{real}(\boldsymbol{Y}_i^H \boldsymbol{Y}_j) \quad (i \neq j), \tag{7.162}$$

$$b_i = \mathrm{real}\left(\boldsymbol{X}^H(n)\boldsymbol{Y}_i - \frac{1}{2}(\boldsymbol{Y}_i^H \boldsymbol{Y}_i) \right), \tag{7.163}$$

and

$$T_{ii} = 0 \qquad (i, j = 1, 2, \ldots, M). \tag{7.164}$$

Resolution

It is easy to see that the direction resolution of this approach depends mainly on the number M of the hypothesized vectors (consequently, on the number of neurons). For the purpose of high resolution, M will be very large, which will make hardware implementation difficult. To decrease the number of neurons, we could eliminate the phase and amplitude variables (in general, they need not be estimated) by minimizing the objective function [22]

$$E = \left\| \boldsymbol{X}(n) - \left[\boldsymbol{P}_{Y_1}\boldsymbol{X}(n), \boldsymbol{P}_{Y_2}\boldsymbol{X}(n), \ldots, \boldsymbol{P}_{Y_M}\boldsymbol{X}(n) \right] \boldsymbol{V} \right\|^2, \tag{7.165}$$

where $\boldsymbol{P}_{Y_i} = \boldsymbol{Y}_i(\boldsymbol{Y}_i^H \boldsymbol{Y}_i)^{-1}\boldsymbol{Y}_i^H$ is the projection matrix of the i'th hypothesized vector $\boldsymbol{Y}_i$ and

$$\boldsymbol{Y}_i = \left[1, \exp\left[jkd\sin(\theta_i)\right], \ldots, \exp\left[jkd(N-1)\sin(\theta_i)\right] \right]^T \qquad (i = 1, 2, \ldots, M). \tag{7.166}$$

Since the phase and amplitude variables have been eliminated in (7.166), the number of neurons in the network for minimizing (7.165) will be greatly reduced in comparison with that for (7.158). For (7.165), the connection strength and the bias currents of the Hopfield neural network are

$$T_{ij} = -2\mathrm{real}\left[\boldsymbol{X}^H(n)\boldsymbol{P}_{Y_i}^H \boldsymbol{P}_{Y_j}\boldsymbol{X}(n) \right] \qquad (i \neq j), \tag{7.167}$$

$$b_i = \frac{1}{2}\mathrm{real}\left[\boldsymbol{X}^H(n)\boldsymbol{P}_{Y_i}\boldsymbol{X}(n) \right], \tag{7.168}$$

and

$$T_{ii} = 0 \qquad (i, j = 1, 2, \ldots, M). \tag{7.169}$$

Note that the cost for eliminating the phase and amplitude variables (that is, the reduction of the number of neurons) is that the connection strengths are dependent on the sampled data $\boldsymbol{X}(n)$, which gives rise to a programming-complexity problem.

Programming-Complexity Problem

From Equations (7.159), (7.160), (7.167), and (7.168), we see that it is necessary to invest additional computations to find the right connection strengths and the bias currents from the sampled data and the hypothesized vectors before the network begins to work. This programming-complexity problem can be overcome to a certain extent by using the neural network in Figure 2.2 [27]. If we choose the activation functions of the neurons on the right and left parts of the network as $f(u) = u$ and $g_i(u) = \frac{1}{2}(1+\tanh(\frac{u}{u_0}))$, respectively, then the corresponding neural network in Figure 2.2 provides one of the minima of the energy function

$$E(t) = \frac{1}{2}\sum_{j=1}^{N}\left(\sum_{i=1}^{M} T_{ji}v_i(t) - b_j\right)^2 + \sum_{i=1}^{M}\frac{1}{R}\int_0^{v_i(t)} g^{-1}(v)dv. \tag{7.170}$$

If the steepness of $g_i(u)$ is large enough, then the second term on the right-hand side of (7.170) can be ignored. Thus,

$$E(t) = \frac{1}{2}\sum_{j=1}^{N}\left(\sum_{i=1}^{M} T_{ji}v_i(t) - b_j\right)^2, \tag{7.171}$$

which has the same form as (7.155) or (7.165).

To solve (7.155), the connection strength matrix and the bias current vector of the neural network in Figure 2.2 must be $\boldsymbol{T} = \boldsymbol{Y}$ and $\boldsymbol{B} = \boldsymbol{X}(n)$, respectively, that is, the hypothesized matrix and the sampled data are directly taken as the parameters of the network without any computations. This means that the programming complexity is zero if the neural network in Figure 2.2 is used to solve Equation (7.155).

To solve (7.165), the connection strength matrix $\boldsymbol{T}$ and the bias current vector $\boldsymbol{B}$ are given as

$$\boldsymbol{T} = \left[\boldsymbol{P}_{Y_1}\boldsymbol{X}(n), \boldsymbol{P}_{Y_2}\boldsymbol{X}(n), \ldots, \boldsymbol{P}_{Y_M}\boldsymbol{X}(n)\right] \tag{7.172}$$

and

$$\boldsymbol{B} = \boldsymbol{X}(n). \tag{7.173}$$

Equation (7.172) shows that the programming complexity is not zero but has been greatly reduced in comparison with (7.167).

The total number of connections (conductances) in Figures 2.2 and 1.2 required to solve the problem (7.155) (or (7.165)) are $O(MN)$ and $O(M^2)$, respectively. In general, $N \ll M$, which means that the network of Figure 2.2 is much easier to implement in hardware than the network of Figure 1.2, although the former needs $M+N$ neurons and the latter needs only M neurons. In addition, we can prove that these two networks have the same time evolution in finding the solution of (7.155) (or (7.165)). This can be achieved by analyzing the differential equations of the networks. The corresponding differential equations of the network of Figure 2.2 for solving (7.155) are

$$C\frac{du_i(t)}{dt} = -\sum_{j=1}^{N} Y_{ji}\left(\sum_{k=1}^{M} Y_{jk}v_k(t) - x_j(n)\right) - \frac{u_i(t)}{R}, \tag{7.174}$$

and those of the other network are

$$C\frac{du_i(t)}{dt} = \sum_{j=1}^{M} T_{ij}v_j(t) - \frac{u_i(t)}{R_i} + b_i \qquad (i = 1, 2, \ldots, M). \tag{7.175}$$

Note that in (7.174) we used $\boldsymbol{T} = \boldsymbol{Y}$. Substituting (7.159) and (7.160) into (7.175), we see that (7.175) becomes

$$C\frac{du_i(t)}{dt} = -\sum_{j=1}^{N}\sum_{k=1}^{M} Y_{ji}Y_{jk}v_k(t) - \frac{u_i(t)}{R_i} + \sum_{j=1}^{N} Y_{ji}x_j(n)$$
$$(i = 1, 2, \ldots, M), \tag{7.176}$$

which is equal to (7.174). Thus, the network in Figure 2.2 is more suitable than that in Figure 1.2 in solving this bearing estimation problem.

Local Minimum Problem

These two neural networks are likely to find a local minimum rather than the desired global one. To overcome this problem, global optimization methods, such as gain annealing [25], iterated descent [23], and stochastic networks [28], can be used. These methods will, however, have costs in terms of programming complexity and time consumption.

Several Snapshots

Thus far we have dealt only with the one-snapshot case. In the case that there are L ($L > 1$) snapshots, we can define two objective functions corresponding to (7.155) and (7.165), respectively,

$$E = \frac{1}{L}\sum_{n=1}^{L} \left\|\boldsymbol{X}(n) - [\boldsymbol{Y}_1, \boldsymbol{Y}_2, \ldots, \boldsymbol{Y}_M]\boldsymbol{V}\right\|^2 \tag{7.177}$$

and

$$E = \frac{1}{L}\sum_{n=1}^{L} \left\|\boldsymbol{X}(n) - \left[\boldsymbol{P}_{Y_1}\boldsymbol{X}(n), \boldsymbol{P}_{Y_2}\boldsymbol{X}(n), \ldots, \boldsymbol{P}_{Y_M}\boldsymbol{X}(n)\right]\boldsymbol{V}\right\|^2. \tag{7.178}$$

If the Hopfield neural network in Figure 1.2 is used to solve these two problems, the corresponding connection strengths and bias currents are

$$T_{ij} = -2\,\mathrm{real}(\boldsymbol{Y}_i^H \boldsymbol{Y}_j), \tag{7.179}$$

$$b_i = \frac{1}{L}\,\mathrm{real}\left(\sum_{n=1}^{L} \boldsymbol{X}^H(n)\boldsymbol{Y}_i\right) \tag{7.180}$$

and

$$T_{ij} = -2\,\mathrm{real}\left(\frac{1}{L}\sum_{n=1}^{L} \boldsymbol{X}^H(n)\boldsymbol{P}_{\boldsymbol{Y}_i}^H \boldsymbol{P}_{\boldsymbol{Y}_j}\boldsymbol{X}(n)\right) \qquad (i \neq j), \tag{7.181}$$

$$b_i = \frac{1}{2L}\,\mathrm{real}\left(\sum_{n=1}^{L} \boldsymbol{X}^H(n)\boldsymbol{P}_{\boldsymbol{Y}_i}\boldsymbol{X}(n)\right) \qquad (i, j = 1, 2, \ldots, M). \tag{7.182}$$

Equations (7.180)–(7.182) can also be computed using the following recursive expressions:

$$b_i(n+1) = \frac{n}{n+1} b_i(n) + \frac{n}{n+1}\,\mathrm{real}\left[\boldsymbol{X}^H(n+1)\boldsymbol{Y}_i\right] \tag{7.183}$$

and

$$T_{ij}(n+1) = \frac{n}{n+1} T_{ij}(n) - \frac{2n}{n+1}\,\mathrm{real}\left[\boldsymbol{X}^H(n+1)\boldsymbol{P}_{\boldsymbol{Y}_i}^H \boldsymbol{P}_{\boldsymbol{Y}_j}\boldsymbol{X}(n+1)\right] \qquad (i \neq j), \tag{7.184}$$

$$b_i(n+1) = \frac{n}{n+1} b_i(n) + \frac{n}{2(n+1)}\,\mathrm{real}\left[\boldsymbol{X}^H(n+1)\boldsymbol{P}_{\boldsymbol{Y}_i}\boldsymbol{X}(n+1)\right] \qquad (i, j = 1, 2, \ldots, M). \tag{7.185}$$

Thus, from Equations (7.183)–(7.185) we see that the programming complexity for several snapshots can be reduced to the same order as that for only one snapshot.

Wideband Signal Sources

This method based on hypotheses and neural networks can be generalized for the case of wideband signal sources. In this case, we have *MK* hypothesized vectors

$$\boldsymbol{Y}_i(l) = \alpha_i(l)e^{j\varphi_i(l)}\left[1, \exp\left[jk_l d\sin(\theta_i)\right], \ldots, \exp\left[jk_l d(N-1)\sin(\theta_i)\right]\right]^T \qquad (i = 1, 2, \ldots, M; \quad l = 1, 2, \ldots, K), \tag{7.186}$$

where we have divided the frequency band of the hypothesized signals into K bands. Consequently, $\alpha_i(l)$, $\varphi_i(l)$, and k_l are, respectively, the envelope, initial phase, and wavenumber of the i'th hypothesized source in the l'th frequency band, and θ_i is the direction of the i'th hypothesized source.

With this in mind, we have a cost function similar to (7.155), namely

$$E = \frac{1}{K}\sum_{l=1}^{K}\left\|\boldsymbol{X}(n) - \left[\boldsymbol{Y}_1(l), \boldsymbol{Y}_2(l), \ldots, \boldsymbol{Y}_M(l)\right]\boldsymbol{V}\right\|^2. \tag{7.187}$$

If the neural network in Figure 1.2 is used to solve (7.187), then the connection strengths and bias currents are computed by use of

$$T_{ij} = -2\,\mathrm{real}\left(\frac{1}{K}\sum_{l=1}^{K}\boldsymbol{Y}_i^H(l)\boldsymbol{Y}_j(l)\right) \qquad (i \neq j) \tag{7.188}$$

and

$$b_i = \frac{1}{K}\,\mathrm{real}\left(\sum_{l=1}^{K}\boldsymbol{X}^H(n)\boldsymbol{Y}_i(l)\right) \qquad (i, j = 1, 2, \ldots, M). \tag{7.189}$$

To reduce the number of the neurons, we define a cost function similar to (7.165):

$$E = \frac{1}{K}\sum_{l=1}^{K}\left\|\boldsymbol{X}(n) - \left[\boldsymbol{P}_{\boldsymbol{Y}_1(l)}\boldsymbol{X}(n), \boldsymbol{P}_{\boldsymbol{Y}_2(l)}\boldsymbol{X}(n), \ldots, \boldsymbol{P}_{\boldsymbol{Y}_M(l)}\boldsymbol{X}(n)\right]\boldsymbol{V}\right\|^2, \tag{7.190}$$

where $\boldsymbol{P}_{\boldsymbol{Y}_i(l)} = \boldsymbol{Y}_i(l)(\boldsymbol{Y}_i^H(l)\boldsymbol{Y}_i(l))^{-1}\boldsymbol{Y}_i^H(l)$. Consequently, the expressions to compute the connection strengths and bias currents of the network in Figure 1.2 are

$$T_{ij} = -2\,\mathrm{real}\left(\frac{1}{K}\sum_{l=1}^{K}\boldsymbol{X}^H(n)\boldsymbol{P}^H_{\boldsymbol{Y}_i(l)}\boldsymbol{P}_{\boldsymbol{Y}_j(l)}\boldsymbol{X}(n)\right) \qquad (i \neq j) \tag{7.191}$$

and

$$b_i = \frac{1}{2K}\,\mathrm{real}\left(\sum_{l=1}^{K}\boldsymbol{X}^H(n)\boldsymbol{P}_{\boldsymbol{Y}_i(l)}\boldsymbol{X}(n)\right) \qquad (i, j = 1, 2, \ldots, M). \tag{7.192}$$

In the case that there are several ($L > 1$) snapshots, (7.189) becomes

$$b_i = \mathrm{real}\left(\frac{1}{KL}\sum_{l=1}^{K}\sum_{n=1}^{L}\boldsymbol{X}^H(n)\boldsymbol{Y}_i(l)\right) \qquad (i = 1, 2, \ldots, M), \tag{7.193}$$

Equation (7.188) remains unchanged, and (7.191) and (7.192) become

$$T_{ij} = -2\,\mathrm{real}\left(\frac{1}{KL}\sum_{l=1}^{K}\sum_{n=1}^{L}\boldsymbol{X}^H(n)\boldsymbol{P}^H_{\boldsymbol{Y}_i(l)}\boldsymbol{P}_{\boldsymbol{Y}_j(l)}\boldsymbol{X}(n)\right) \qquad (i \neq j) \tag{7.194}$$

and

$$b_i = \text{real}\left(\frac{1}{2KL}\sum_{l=1}^{K}\sum_{n=1}^{L} \boldsymbol{X}^H(n)\boldsymbol{P}_{\boldsymbol{Y}_i(l)}\boldsymbol{X}(n)\right) \qquad (i, j = 1, 2, \ldots, M). \tag{7.195}$$

Recursive alternatives of (7.194) and (7.195) can easily be obtained using the same approach as for (7.181) and (7.182).

Two-Dimensional Signal Sources

The above method can also be generalized to the two-dimensional case. For more detail, see Reference [29].

7.5 Beamforming Using Neural Networks

Beamforming is another important problem in array signal processing [30, 31]. Beamforming is in effect a spatial-domain filtering that forms multiple beams by applying appropriate delay and weighting elements to the signals received by the sensors of an array so as to suppress unwanted jamming interferences. Similar to the design of the coefficients of the temporal-domain filtering discussed in Chapter 2, the design of the weights on the sensor outputs is the key problem in beamforming. The most common technique for computing the weights is based on a closed-loop gradient descent algorithm [32]. The disadvantage of this algorithm is the poor convergence for a broad dynamic range signal environment. Several modified methods have been proposed [30], but they suffer from the signal-cancellation problem. This problem can be overcome through the application of linear constraints on the weights. The basic concept of this linear-constraint-based beamformer is to constrain the response of the beamformer such that the desired signals are passed with specified gain and phase. The weights are chosen to minimize the output power subject to the response constraint. In a rapid time-varying environment, the weights of this linear-constraint-based beamformer should be updated in real time. However, the evaluation of the weights is computationally intensive and can hardly meet the real-time requirement. In this section, we will develop neural network approaches for computing in real time the weights of this linear-constraint-based beamformer [33, 34].

Let us consider a linear array composed of N isotropic antenna elements that receive signals from sources of variation frequency f_0 located far from the array. The output of the i'th sensor at the n'th sampling time can be described by

$$x_i(n) = \alpha(n)\exp\Big[j2\pi f_0\big(n + \tau_i(\theta, \varphi)\big)\Big] + \epsilon_i(n) + x_{Ii}(n) \qquad (i = 1, 2, \ldots, N), \tag{7.196}$$

where

$$\tau_i(\theta,\varphi) = \frac{\boldsymbol{r}_i \boldsymbol{a}(\theta,\varphi)}{c} \tag{7.197}$$

is the time delay of the i'th element relative to a reference point. The vector $\boldsymbol{r}_i$ describes the position of the i'th element, $\boldsymbol{a}(\theta,\varphi)$ is a unit vector in the direction (θ,φ) of the signal source, and c is the propagation speed of the plane wave in free space. $x_{Ii}(n)$ is the component of the directional interferences received by the i'th element and possesses the same statistics as the signal source. $\alpha(n)$ is the complex envelope of the signal and $\epsilon_i(n)$ is the zero-mean additive white noise at the i'th element. With these, the output $y(n)$ of the beamformer is

$$y(n) = \sum_{i=1}^{N} W_i x_i(n) = \boldsymbol{W}^T \boldsymbol{X}(n), \tag{7.198}$$

where $\boldsymbol{W} = [W_1, W_2, \ldots, W_N]^T$ and $\boldsymbol{X}(n) = [x_1(n), x_2(n), \ldots, x_N(n)]^T$ are the weight vector and the input vector, respectively, of the beamformer. The mean output power of the beamformer is given by

$$E[y(n)y^*(n)] = \boldsymbol{W}^H \boldsymbol{R} \boldsymbol{W}, \tag{7.199}$$

where $\boldsymbol{R} = E[\boldsymbol{X}^*(n)\boldsymbol{X}^T(n)]$.

For the purpose of achieving the optimal utilization of the mean output power of the beamformer, the weights are chosen such that the output contains minimal influence due to noise and interference signals arriving from other directions. There exist many different criteria and algorithms for choosing the optimum beamformer weights [35]. The optimum weights of the linear-constraint-based beamformer mentioned above are obtained by solving the following constrained optimization problem:

$$\begin{cases} \min_W \boldsymbol{W}^H \boldsymbol{R} \boldsymbol{W} \\ s.t. \quad \boldsymbol{W}^T \boldsymbol{a} = 1, \end{cases} \tag{7.200}$$

where

$$\boldsymbol{a} = \Big[1, \exp\big[jkd\cos(\theta_0)\big], \exp\big[j2kd\cos(\theta_0)\big], \\ \ldots, \exp\big[jkd(N-1)\cos(\theta_0)\big]\Big]^T \tag{7.201}$$

is called the constraint vector. Here d is the element spacing, k is the wavenumber, and θ_0 is the look direction angle (the angle between the axis of the linear array and the direction of the desired signal source).

Using Lagrange's method we can define a cost function

$$E(\boldsymbol{W}) = \boldsymbol{W}^H \boldsymbol{R} \boldsymbol{W} + \lambda(1 - \boldsymbol{W}^T \boldsymbol{a}), \tag{7.202}$$

where λ is an arbitrary constant. Differentiating with respect to the weight vector $\boldsymbol{W}$ and equating to zero give the optimum weight vector $\boldsymbol{W}_{\text{opt}}$ of (7.200) as

$$\boldsymbol{W}_{\text{opt}} = \frac{\boldsymbol{R}^{-1}\boldsymbol{a}^*}{\boldsymbol{a}^T\boldsymbol{R}^{-1}\boldsymbol{a}^*}. \tag{7.203}$$

Clearly, the computational complexity of (7.203) is very intensive.

In some practical applications, we can use a constant λ instead of $1/(\boldsymbol{a}^T\boldsymbol{R}^{-1}\boldsymbol{a}^*)$, that is,

$$\boldsymbol{W}_{\text{opt}} = \lambda\boldsymbol{R}^{-1}\boldsymbol{a}^*. \tag{7.204}$$

For simplicity, we let $\lambda = 1$; then (7.204) becomes

$$\boldsymbol{W}_{\text{opt}} = \boldsymbol{R}^{-1}\boldsymbol{a}^*. \tag{7.205}$$

Equation (7.205) has the same form as (2.67) and can be written in the real-valued form

$$\boldsymbol{W}_{c\text{opt}} = \boldsymbol{R}_c^{-1}\boldsymbol{a}_c, \tag{7.206}$$

where

$$\boldsymbol{W}_{c\text{opt}} = \begin{pmatrix} \boldsymbol{W}_{r\text{opt}} \\ \boldsymbol{W}_{i\text{opt}} \end{pmatrix}, \qquad \boldsymbol{R}_c = \begin{pmatrix} \boldsymbol{R}_r & -\boldsymbol{R}_i \\ \boldsymbol{R}_i & \boldsymbol{R}_r \end{pmatrix}, \quad \text{and} \quad \boldsymbol{a}_c = \begin{pmatrix} \boldsymbol{a}_r \\ -\boldsymbol{a}_i \end{pmatrix}$$

and $\boldsymbol{W}_{\text{opt}} = \boldsymbol{W}_{r\text{opt}} + j\boldsymbol{W}_{i\text{opt}}$, $\boldsymbol{R} = \boldsymbol{R}_r + j\boldsymbol{R}_i$, and $\boldsymbol{a}^* = \boldsymbol{a}_r - j\boldsymbol{a}_i$. Therefore, (7.206) can be solved by the neural network in Figure 2.2 by letting the known matrix $\boldsymbol{R}_c$ be the connection strength matrix $\boldsymbol{T}$, the constraint vector $\boldsymbol{a}_c$ be the bias current vector $\boldsymbol{B}$ of the network, and the other parameters of the network be selected as for (2.1). According to the analysis given in Section 2.1, this network gives the solution

$$V_f = \lim_{t\to\infty} V(t) = \left(\frac{1}{RK_1K_2}\boldsymbol{I} + \boldsymbol{R}_c^T\boldsymbol{R}_c\right)^{-1}\boldsymbol{R}_c^T\boldsymbol{a}_c \approx \boldsymbol{R}_c^{-1}\boldsymbol{a}_c^*, \tag{7.207}$$

which is the approximation to the optimum solution (7.206).

In the general case, using the penalty function method [36] for (7.200), it is sufficient to solve the unconstrained problem of the form

$$\min_{\boldsymbol{W}} \left[\boldsymbol{W}^H\boldsymbol{R}\boldsymbol{W} + k(\boldsymbol{W}^T\boldsymbol{a} - 1)^2\right], \tag{7.208}$$

where k a positive constant.

From [36] we know that (7.200) and (7.208) have the same solution if k tends to infinity. However, if k is large enough, the solution of (7.208) can approximate that of (7.200).

Equation (7.208) can be written in the real-valued form

$$\min_{\boldsymbol{W}_c} \left[\boldsymbol{W}_c^T\boldsymbol{R}_c\boldsymbol{W}_c + k(\boldsymbol{A}_c\boldsymbol{W}_c - \boldsymbol{e}_c)^T(\boldsymbol{A}_c\boldsymbol{W}_c - \boldsymbol{e}_c)\right], \tag{7.209}$$

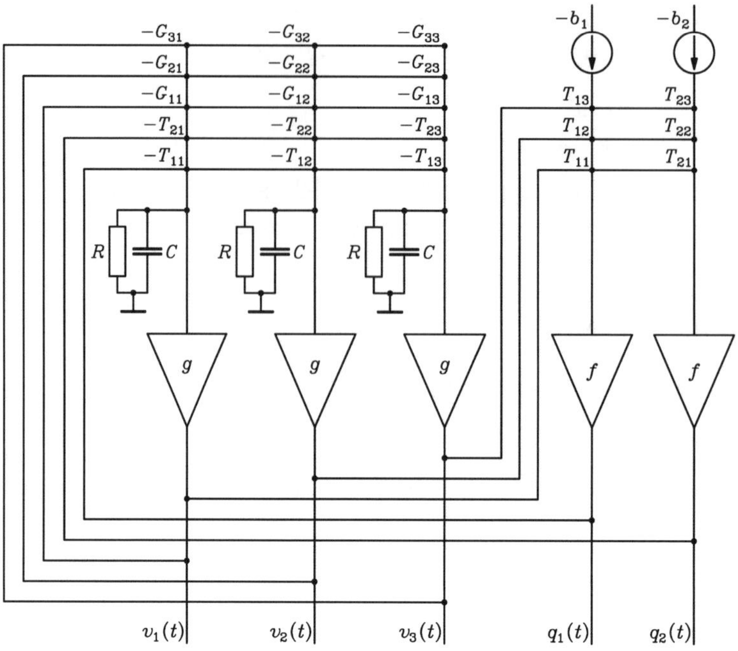

Figure 7.18. A neural network for solving Equation (7.209). $N = 3$ and $M = 2$.

where

$$A_c = \begin{pmatrix} a_r^T & -a_i^T \\ a_i^T & a_r^T \end{pmatrix}, \qquad e_c = \begin{pmatrix} 1 \\ 0 \end{pmatrix},$$

and the other variables are the same as those in (7.206).

The neural network to solve (7.209), shown in Figure 7.18, is similar to the neural network in Figure 2.2. The main difference is that the input currents of the neurons in the left part come from not only the outputs of the neurons in the right part but also from its own outputs. The dynamic equations of the neural network in Figure 7.18 can be obtained in terms of Kirchhoff's laws as

$$C\frac{du_i(t)}{dt} = -\sum_{j=1}^{M} T_{ji} f\left(\sum_{k=1}^{N} T_{jk} v_k(t) - b_j\right) - \frac{u_i(t)}{R} - \sum_{i=1}^{N}\sum_{j=1}^{N} G_{ij} v_j(t)$$

$$(i = 1, 2, \ldots, N), \qquad (7.210)$$

where various quantities are the same as those of (2.4) except that G_{ij} (for $i, j = 1, 2, \ldots, N$) denotes the connection strength between the i'th neuron and the j'th neuron in the left part.

By defining and analyzing an energy function as

$$E(t) = \sum_{j=1}^{M} F\left(\sum_{i=1}^{N} T_{ji} v_i(t) - b_j\right) + \frac{1}{2}\sum_{i=1}^{N}\sum_{j=1}^{N} G_{ji} v_i(t) v_j(t)$$
$$+ \sum_{i=1}^{N} \frac{1}{R}\int_0^{v_i(t)} g^{-1}(v)dv \tag{7.211}$$

we can prove that this network is stable and provides one minimum of $E(t)$ in the stationary state.

To solve (7.209), we choose $\boldsymbol{T} = \boldsymbol{A}_c, \boldsymbol{G} = \{G_{ij}\} = \boldsymbol{R}_c, \boldsymbol{B} = \boldsymbol{e}_c$, and $f(u) = ku$. Consequently, the energy function becomes

$$E(t) = \frac{1}{2}k\|\boldsymbol{e}_c - \boldsymbol{A}_c \boldsymbol{V}(t)\|^2 + \frac{1}{2}\boldsymbol{V}^T(t)\boldsymbol{R}_c\boldsymbol{V}^T(t) + \sum_{i=1}^{N}\frac{1}{R}\int_0^{v_i(t)} g^{-1}(v)dv, \tag{7.212}$$

where $\boldsymbol{V}(t) = [v_1(t), v_2(t), \ldots, v_N(t)]^T$ is the output vector of the left part of the network. Since (7.212) approximates (7.209), the output vector in the stationary state of such a constructed neural network will provide the approximation of the minimum of (7.209).

In most practical applications, there are only L sampled vectors $\boldsymbol{X}(n)$ available but the covariance matrix $\boldsymbol{R}$ is unknown. There are two ways to deal with this case. One way is first to estimate the covariance matrix $\boldsymbol{R}$ from the available sampled vectors by

$$\hat{\boldsymbol{R}} = \frac{1}{L}\sum_{n=1}^{L} \boldsymbol{X}^*(n)\boldsymbol{X}^T(n)$$

and then to use the same method as in the above to find the weights of the beamformer except that $\boldsymbol{R}$ is replaced by $\hat{\boldsymbol{R}}$. The other way is to reconstruct the objective function (7.200) as

$$\begin{cases} \min_{\boldsymbol{W}} \sum_{n=1}^{L} |\boldsymbol{X}^T(n)\boldsymbol{W}|^2 \\ s.t. \quad \boldsymbol{W}^T\boldsymbol{a} = 1 \end{cases} \tag{7.213}$$

Clearly, (7.213) is exactly the same as the problem described by (2.62). As a result, the neural networks for the constrained LS algorithm given in Section 2.3 can be used immediately to solve (7.213).

Bibliography

[1] Blesler, Y. and Macovski, A., "Exact Maximum Likelihood Parameter Estimation of Superimposed Exponential Signals in Noise," *IEEE Trans. on Acoustics, Speech and Signal Processing*, Vol. 34, 1986, pp. 1081–89.

[2] Liggt, W. S., "Passive Sonar: Fitting Models to Multiple Time-Series," *Nato ASI on Signal Processing*, Academic, New York, 1973, pp. 327–45.

[3] Schweppe, F. C., "Sensor Array Data Processing for Multiple Signal Sources," *IEEE Trans. on Information Theory*, Vol. 14, 1968, pp. 294–305.

[4] Capon, J., "High-Resolution Frequency-Wavenumber Spectrum Analysis," *Proc. IEEE*, Vol. 57, 1969, pp. 1408–18.

[5] Reddi, S. S., "Multiple Source Location-a Digital Approach," *IEEE Trans. on Aerospace Electronics Systems*, Vol. 15, 1979, pp. 95–105.

[6] Schmidt, R. O., "Multiple Emitter Location and Signal Parameter Estimation," *IEEE Trans. on Antennas and Propagation*, Vol. 34, No. 3, 1986, pp. 276–80.

[7] Marcos, S. and Munier, J., "Source Localization Using a Distorted Antenna," *Proc. IEEE Int. Conf. on Acoustics, Speech and Signal Processing*, 1989, pp. 2756–59.

[8] Marcos, S. and Benidier, M., "On High Resolution Array Processing Non-Based on the Eigenanalysis Approach," *Proc. IEEE Int. Conf. on Acoustics, Speech and Signal Processing*, 1990, pp. 2956–59.

[9] Ziskind, I. and Wax, M., "Maximum Likelihood Localization of Multisources by Alternating Projection," *IEEE Trans. on Acoustics, Speech and Signal Processing*, Vol. 36, No. 10, 1988, pp. 1553–60.

[10] Wax, M., "Detection and Estimation of Superimposed Signals," PhD Dissertation, Stanford Univ., 1985.

[11] Ziskind, I. and Wax, M., "Maximum Likelihood Localization of Diversed Polarized Sources by Simulated Annealing," *IEEE Trans. on Antennas and Propagation*, Vol. 38, No. 3, 1990.

[12] Boheme, J., "Estimating the Source Parameters by Maximum Likelihood and Nonlinear Regression," *Proc. IEEE Int. Conf. on Acoustics, Speech and Signal Processing*, 1980, pp. 307–10.

[13] Hwang, J. K. and Chen, Y. C., "Real-Time Computation of the Maximum Likelihood Criterion for Some Superimposed Signal Problems," *Electronics Lett.*, Vol. 26, No. 23, 1990, pp. 1969–71.

[14] Luo, F. L. and Bao, Z., "Real-Time Neural Computation of Maximum Likelihood Criterion for Bearing Estimation Problems," *Neural Networks*, Vol. 5, 1992, pp. 765–69.

[15] Luo, F. L. and Bao, Z., "Real-Time Implementation of Propagator Bearing Estimation Method by Use of a Neural Network," *IEEE J. Oceanic Engineering*, Vol. 17, No. 4, 1992, pp. 320–25.

[16] Mathew, G. and Reddy, V. U., "Orthogonal Eigensubspace Estimation Using Neural Networks," *IEEE Trans. on Acoustics, Speech and Signal Processing*, Vol. 42, No. 7, 1994. pp. 1803–11.

[17] Karhunen, J., "Recursive Estimation of Eigenvectors of Correlation Type Matrices for Signal Processing Applications," PhD Dissertation, Helsinki Univ., Finland, 1984.

[18] Oja, E. and Karhunen, J., "On Stochastic Approximation of the Eigenvectors and Eigenvalues of the Expectation of a Random Matrix," *J. Math. Analysis Appl.*, Vol. 106, 1985, pp. 69–84.

[19] Luo, F. L. and Li, Y. D., "Real-Time Neural Computation of the Noise Subspace for the MUSIC Algorithm," *Proc. IEEE Int. Conf. on Acoustics, Speech and Signal Processing*, 1993, pp. 1485–88.

[20] Luo, F. L., Ji, H. B., and Zhao, X. P., "Neural Network Approach to ML Bearing Estimation," *J. Electronics*, Vol. 10, No. 1, 1993, pp. 1–8.

[21] Jelonek, T. M., Reilly, J. P., and Wu, Q., "Real-Time Analog Global Optimization with Constraints: Application to the Direction of Arrival Estimation Problem," *IEEE Trans. on Circuits and Systems*, II, Vol. 42, No. 4, 1995, pp. 223–44.

[22] Rastogi, R., Gupta, P. K., and Kumaresan, R., "Array Signal Processing with Interconnected Neuron-Like Elements," *Proc. IEEE Int. Conf. on Acoustics, Speech and Signal Processing*, 1987, pp. 2328–31.

[23] Jha, S., Chapman, R. and Durrai, T. S., "Bearing Estimation Using Neural Networks," *Proc. IEEE Int. Conf. on Acoustics, Speech and Signal Processing*, 1988, pp. 2156–59.

[24] Goryn, D. and Kaveh, M., "Neural Networks for Narrowband and Wideband Direction Finding," *Proc. IEEE Int. Conf. on Acoustics, Speech and Signal Processing*, 1988, pp. 2164–67.

[25] Hopfield, J. J. and Tank, D. W., "Neural Computation of Decision Optimization Problems," *Biol. Cybernetics*, Vol. 52, 1985, pp. 141–52.

[26] Tank, D. W. and Hopfield, J. J., "Simple Neural Optimization Networks: A/D Converter, Signal Decision Circuit, and Linear Programming Circuit," *IEEE Trans. on Circuits and Systems*, Vol. 33, 1986, pp. 533–41.

[27] Luo, F. L., Li, Y. D., and Bao, Z., "A Modified Hopfield Neural Network and its Applications," *Int. J. Neural Networks*, Vol. 3, No. 4, 1992, pp. 135–41.

[28] Jha, S. and Durrai, T. S., "Bearing Estimation Using Neural Optimization Methods," *Proc. IEEE Int. Conf. on Acoustics, Speech and Signal Processing*, 1990, pp. 889–92.

[29] Yang, Z. K., "Two Dimensional Direction Finding by Use of Neural Networks," *Proc. Int. China Conf. of Circuits and Systems*, ShenZheng, 1991.

[30] Van Veen, B. D. and Buckley, K. M., "Beamforming: A Versatile Approach to Spatial Filtering," *IEEE ASSP Magazine*, 1988, pp. 4–24.

[31] Luo, F. L., "Real-Time Neural Network Implementation of Total Least Squares Adaptive Beamforming," *Proc. 24th URSI Assembly*, Kyoto, Japan, 1993.

[32] Widrow, B. and Stearns, S., *Adaptive Signal Processing*, Prentice-Hall, Englewood Cliffs, NH, 1985.

[33] Chang, P. R., Yang, W. H., and Chan, K. K., "A Neural Network Approach to MVDR Beamforming Problem," *IEEE Trans. on Antennas and Propagation*, Vol. 40, No. 3, 1992, pp. 313–22.

[34] Lo, K. W., "Comments on a Neural Network Approach to MVDR Beamforming Problem," *IEEE Trans. on Antennas and Propagation*, Vol. 41, No. 9, 1993, pp. 1344–45.

[35] Godara, L. C., "Error Analysis of the Optimal Antenna Array Processors," *IEEE Trans. on Aerospace Electronics Systems*, Vol. 22, 1986, pp. 395–409.

[36] Luenberger, D. G., *Linear and Nonlinear Programming*, Addision-Wesley, Reading, MA, 1978, pp. 366–69.

8

Neural Networks for System Identification

System identification is a fundamental problem in many fields of signal processing such as automatic control, communication systems, and seismic processing [1, 2]. System identification is mainly concerned with the determination of the input–output mappings of systems. Such an identification allows one to predict the system outputs, and as a result this problem has considerable impact in the applications mentioned above where forecasting is of extreme importance. The current emphasis is on real-time identification, nonlinear identification, and blind identification. Based on asynchronous parallel and distributed processing, nonlinear dynamics, global interconnection of network elements, self-organization, and the high-speed computational capability of neural networks; neural network approaches for system identification offer many advantages over traditional ones. In this chapter, we will present architectures and algorithms of neural networks for system identification. Section 8.1 will first give a brief introduction of the system identification problem. Sections 8.2–8.4 will deal mainly with nonlinear system identification by use of various kinds of neural networks such as MLP networks, RBF networks, and recurrent neural networks. Real-time identification based on neural networks will be reported in Section 8.5. Blind identification with neural networks will be discussed in Section 8.6.

8.1 Fundamentals of System Identification

Let us consider a system with N inputs (denoted by a vector $\boldsymbol{X}(t) = [x_1(t), x_2(t), \ldots, x_N(t)]^T$) and M outputs (denoted by a vector $\boldsymbol{Y}(t) = [y_1(t), y_2(t), \ldots, y_M(t)]^T$) as shown in Figure 8.1. The common method for representing this system is to use vector differential equations [3]

$$\begin{cases} \frac{d\boldsymbol{Z}(t)}{dt} = \boldsymbol{F}\big[\boldsymbol{Z}(t), \boldsymbol{X}(t)\big], \\ \boldsymbol{Y}(t) = \boldsymbol{G}\big[\boldsymbol{Z}(t)\big], \end{cases} \tag{8.1}$$

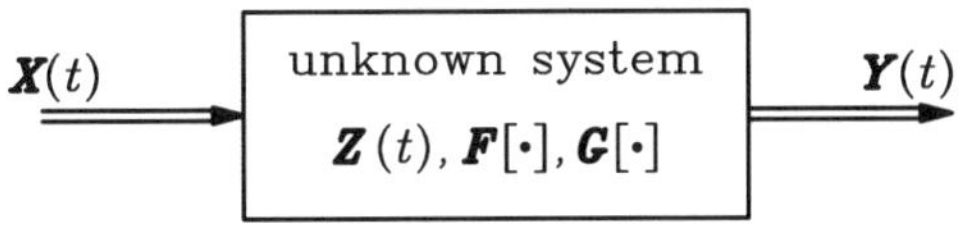

Figure 8.1. A system to be identified.

where the L-dimensional vector $\boldsymbol{Z}(t) = [z_1(t), z_2(t), \ldots, z_L(t)]^T$ denotes the state of the system at time t, and where $\boldsymbol{F}[\cdot]$ and $\boldsymbol{G}[\cdot]$ are nonlinear maps. The state vector $\boldsymbol{Z}(t)$ is determined by the state at time $t_0 < t$ and the input vector $\boldsymbol{X}(t)$ defined over the interval $[t_0, t)$. The output vector $\boldsymbol{Y}(t)$ is completely determined by the state vector $\boldsymbol{Z}(t)$ at time t. The discrete-time version of (8.1) is

$$\begin{cases} \boldsymbol{Z}(n+1) = \boldsymbol{F}\big[\boldsymbol{Z}(n), \boldsymbol{X}(n)\big], \\ \boldsymbol{Y}(n) = \boldsymbol{G}\big[\boldsymbol{Z}(n)\big]. \end{cases} \tag{8.2}$$

Equation (8.2) is referred to as the input-state-output representation of the system as shown in Figure 8.2.

If the system to be considered is linear and time invariant, then (8.2) becomes

$$\begin{cases} \boldsymbol{Z}(n+1) = \boldsymbol{A}\boldsymbol{Z}(n) + \boldsymbol{B}\boldsymbol{X}(n), \\ \boldsymbol{Y}(n) = \boldsymbol{C}\boldsymbol{Z}(n), \end{cases} \tag{8.3}$$

where $\boldsymbol{A}$, $\boldsymbol{B}$, and $\boldsymbol{C}$ are $L \times L$, $L \times N$, and $M \times L$ matrices, respectively.

The task of identifying an unknown system (in which the maps $\boldsymbol{F}[\cdot]$ and $\boldsymbol{G}[\cdot]$ are unknown) is to construct an operator $\boldsymbol{P}[\cdot]$ for which we have

$$\big\|\boldsymbol{P}\big[\boldsymbol{X}(n)\big] - \boldsymbol{Y}(n)\big\| < \epsilon, \tag{8.4}$$

where $\|\cdot\|$ is a suitably defined norm and ϵ is the specified error. For linear time-invariant systems, this problem can be transformed to finding the matrices $\boldsymbol{A}$, $\boldsymbol{B}$, and $\boldsymbol{C}$. This is called parameter identification.

Using available samples of the input, output, and state of the system, we can obtain

$$\begin{aligned} &\boldsymbol{Y}(n) = \boldsymbol{G}\big[\boldsymbol{Z}(n)\big], \\ &\boldsymbol{Y}(n+1) = \boldsymbol{G}\Big[\boldsymbol{F}\big[\boldsymbol{Z}(n), \boldsymbol{X}(n)\big]\Big], \\ &\vdots \\ &\boldsymbol{Y}(n+k-1) = \boldsymbol{G}\Bigg[\boldsymbol{F}\Bigg[\cdots \boldsymbol{F}\Big[\boldsymbol{F}\big[\boldsymbol{Z}(n), \boldsymbol{X}(n)\big], \\ &\qquad\qquad \boldsymbol{X}(n+1)\Big], \ldots, \boldsymbol{X}(n+k-2)\Bigg]\Bigg]. \end{aligned}$$

This set of nonlinear equations can be used as the constraints of (8.4) to construct the desired identification operator $\boldsymbol{P}[\cdot]$.

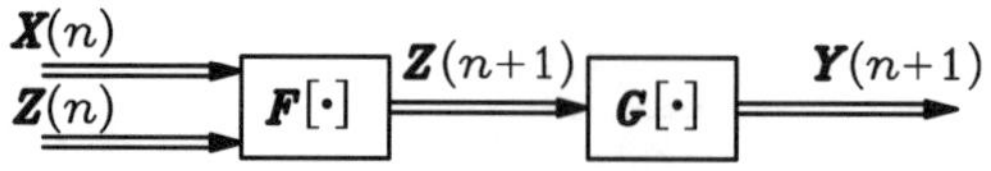

Figure 8.2. Input-state-output representation of a system.

Although the input-state-output representation (8.2) of a system is quite general, all system states $\boldsymbol{Z}(n)$ are not usually available for measurement. In this case, the representation without using system states is desirable. Under some mild assumptions, it has been shown that a wide class of discrete-time systems can be represented by the following nonlinear difference equation model [4, 5]:

$$\begin{aligned}\boldsymbol{Y}(n) = \boldsymbol{F}\big[&\boldsymbol{Y}(n-1), \boldsymbol{Y}(n-2), \ldots, \boldsymbol{Y}(n-n_y), \boldsymbol{X}(n-1),\\ &\boldsymbol{X}(n-2), \ldots, \boldsymbol{X}(n-n_x)\big],\end{aligned} \tag{8.5}$$

where n_y and n_x are the maximum lags in the output vector and the input vector, respectively. Note that $\boldsymbol{F}$ is different from that in (8.2). Clearly, in (8.5), the output vector $\boldsymbol{Y}(n)$ is determined by the past values of the system input vector and output vector. There are four special cases for (8.5):

(1) $$\boldsymbol{Y}(n) = \sum_{i=1}^{n_y} \boldsymbol{A}_i \boldsymbol{Y}(n-i) + \sum_{j=1}^{n_x} \boldsymbol{B}_j \boldsymbol{X}(n-j), \tag{8.6}$$

where $\boldsymbol{A}_i$ and $\boldsymbol{B}_j$ (for $i = 1, 2, \ldots, n_y$; $j = 1, 2, \ldots, n_x$) are $M \times M$ and $M \times N$ matrices, respectively. Equation (8.6) means that the system is linear and time invariant. Equation (8.6) is usually called the ARMAX (autoregressive moving average with exogenous inputs) model.

(2) $$\begin{aligned}\boldsymbol{Y}(n) = \boldsymbol{F}\big[\boldsymbol{Y}(n-1), \boldsymbol{Y}(n-2), \ldots, \boldsymbol{Y}(n-n_y)\big] + \boldsymbol{G}\big[&\boldsymbol{X}(n-1),\\ &\boldsymbol{X}(n-2), \ldots, \boldsymbol{X}(n-n_x)\big].\end{aligned} \tag{8.7}$$

(3) $$\boldsymbol{Y}(n) = \sum_{i=1}^{n_y} \boldsymbol{A}_i \boldsymbol{Y}(n-i) + \boldsymbol{G}\big[\boldsymbol{X}(n-1), \boldsymbol{X}(n-2), \ldots, \boldsymbol{X}(n-n_x)\big]. \tag{8.8}$$

(4) $$\boldsymbol{Y}(n) = \boldsymbol{F}\big[\boldsymbol{Y}(n-1), \boldsymbol{Y}(n-2), \ldots, \boldsymbol{Y}(n-n_y)\big] + \sum_{j=1}^{n_x} \boldsymbol{B}_i \boldsymbol{X}(n-j). \tag{8.9}$$

In Cases (2)–(4) $\boldsymbol{F}[\cdot]$ and $\boldsymbol{G}[\cdot]$ are nonlinear maps. The systems represented by (8.6)–(8.9) are also shown in Figures 8.3–8.6, respectively.

If we incorporate the system noise, then (8.5) becomes

$$\begin{aligned}\boldsymbol{Y}(n) = \boldsymbol{F}\big[&\boldsymbol{Y}(n-1), \boldsymbol{Y}(n-2), \ldots, \boldsymbol{Y}(n-n_y),\\ &\boldsymbol{X}(n-1), \boldsymbol{X}(n-2), \ldots, \boldsymbol{X}(n-n_x),\\ &\boldsymbol{e}(n-1), \boldsymbol{e}(n-2), \ldots, \boldsymbol{e}(n-n_e)\big] + \boldsymbol{e}(n),\end{aligned} \tag{8.10}$$

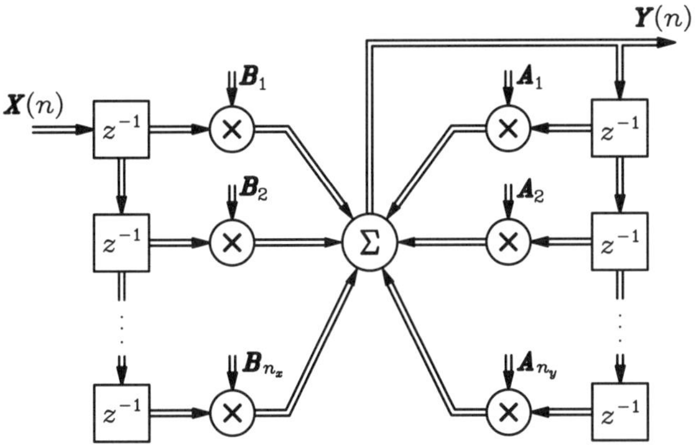

Figure 8.3. The system represented by Equation (8.6).

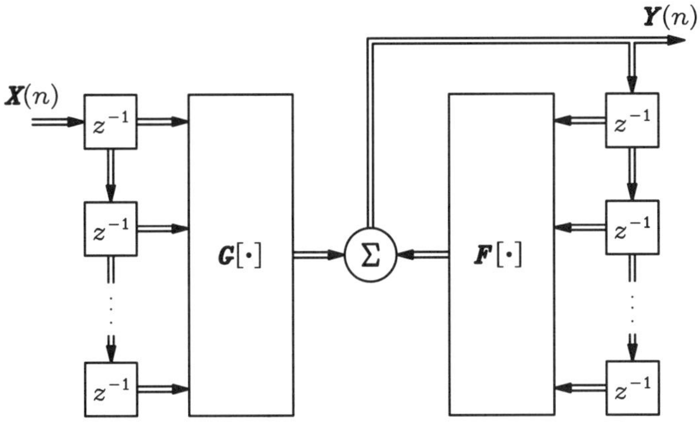

Figure 8.4. The system represented by Equation (8.7).

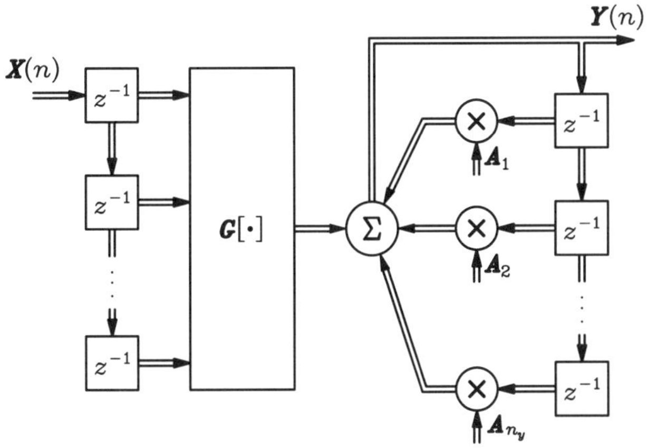

Figure 8.5. The system represented by Equation (8.8).

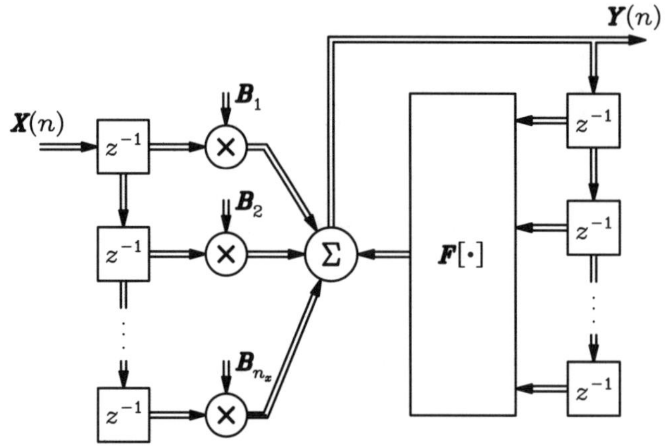

Figure 8.6. The system represented by Equation (8.9).

where $\boldsymbol{e}(n) = [e_1(n), e_2(n), \ldots, e_M(n)]^T$ is the noise vector and n_e is the maximum lag in the noise vector. In fact, one can obtain Equation (8.10) by extending the ARMAX model to the nonlinear case. Consequently, (8.10) is referred to as the nonlinear ARMAX (NARMAX) representation of systems.

With these models in mind, system identification becomes the construction of the related nonlinear maps $\boldsymbol{F}[\cdot]$ and $\boldsymbol{G}[\cdot]$ and parameters $\boldsymbol{A}_i$, and $\boldsymbol{B}_j$. In the following sections, we will show how to use neural networks to attack this problem.

8.2 System Identification Using MLP Networks

As pointed out in Section 1.4, the input–output relationship of an MLP network can approximate an arbitrary nonlinear map and is completely determined by the network parameters such as the connection weights and thresholds. This suggests that MLP networks can be used to construct the nonlinear maps related to the system identification operator $\boldsymbol{P}[\cdot]$. Without loss of generality, we consider a system represented by the model

$$\begin{aligned}\boldsymbol{Y}(n) = \boldsymbol{F}\big[&\boldsymbol{Y}(n-1), \boldsymbol{Y}(n-2), \ldots, \boldsymbol{Y}(n-n_y),\\ &\boldsymbol{X}(n-1), \boldsymbol{X}(n-2), \ldots, \boldsymbol{X}(n-n_x)\big] + \boldsymbol{e}(n). \qquad (8.11)\end{aligned}$$

This means that n_e in the general NARMAX model (8.10) is equal to zero. As a matter of fact, the system identification techniques based on the simple system (8.11) can readily be extended to the general case (8.10) [6].

The MLP network for constructing the system mapping $\boldsymbol{F}[\cdot]$ of (8.11) is shown in Figure 8.7. Three layers are assumed in this section but more layers are a direct generalization. The input layer has $n_i = n_y M + n_x N$ neurons whose input

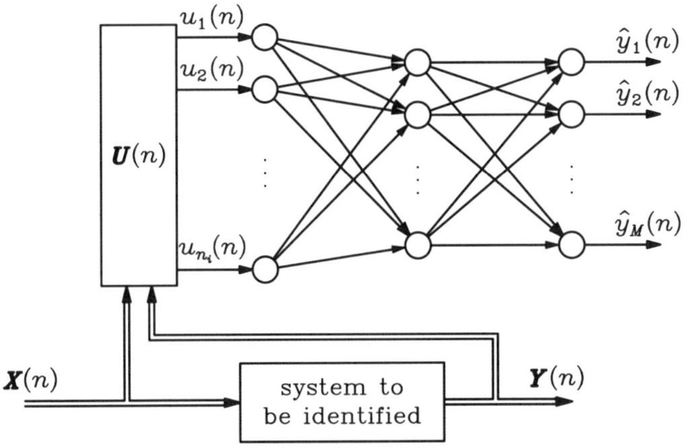

Figure 8.7. The MLP network for nonlinear system identification.

vector is

$$\begin{aligned} U(n) &= \left[u_1(n), u_2(n), \ldots, u_{n_i}(n)\right]^T \\ &= \left[Y^T(n-1), Y^T(n-2), \ldots, Y^T(n-n_y),\right. \\ &\qquad \left. X^T(n-1), X^T(n-2), \ldots, X^T(n-n_x)\right]^T. \end{aligned} \tag{8.12}$$

Thus, the input vector of the network consists of the past values of the input vector and output vector of the system. The input layer simply feeds the vector $U(n)$ to the hidden layer without any modification. The hidden layer has n_h neurons with nonlinear transfer functions (such as the sigmoid function). The output layer has M neurons, which correspond to the M outputs of the unknown system.

Let $W_{ij}^{(1)}$ denote the connection weight between the i'th neuron in the input layer and the j'th neuron in the hidden layer (for $i = 1, 2, \ldots, n_i$; $j = 1, 2, \ldots, n_h$); let $q_j(n)$ and $f_j(\cdot)$ (for $j = 1, 2, \ldots, n_h$) be the output and activation function of the j'th neuron in the hidden layer, respectively; let $W_{jk}^{(2)}$ denote the connection weight between the j'th neuron in the hidden layer and the k'th neuron in the output layer; let $\hat{y}_k(n)$, and $g_k(\cdot)$ be the output and activation function of the k'th output neuron, respectively. Then we have

$$\begin{aligned} \hat{y}_k(n) &= g_k\left(\sum_{j=1}^{n_h} W_{jk}^{(2)} q_j(n)\right) \\ &= g_k\left(\sum_{j=1}^{n_h} W_{jk}^{(2)} f_j\left[\sum_{i=1}^{n_i} W_{ij}^{(1)} u_i(n)\right]\right) \qquad (k = 1, 2, \ldots, M). \end{aligned} \tag{8.13}$$

Note that in (8.13) the threshold of each neuron has been selected to be zero and the activation function $g_k(\cdot)$ of the k'th output neuron is usually no longer the sigmoid function so that it can provide values greater than unity. For simplicity, we select $g_k(\cdot)$ to be a linear function, that is, all the output neurons perform only simple summations. With this in mind, (8.13) becomes

$$\hat{y}_k(n) = \sum_{j=1}^{n_h} W_{jk}^{(2)} f_j\left(\sum_{i=1}^{n_i} W_{ij}^{(1)} u_i(n)\right). \tag{8.14}$$

Clearly, the output vector provided by the network is

$$\begin{aligned}\hat{\boldsymbol{Y}}(n) &= \left[\hat{y}_1(n), \hat{y}_2(n), \ldots, \hat{y}_M(n)\right]^T \\ &= \hat{\boldsymbol{F}}\left[\boldsymbol{U}(n)\right],\end{aligned} \tag{8.15}$$

where the nonlinear mapping $\hat{\boldsymbol{F}}$ is completely determined by the connection weights $W_{ij}^{(1)}$ and $W_{jk}^{(2)}$.

Moreover, according to (8.4), we obtain an optimization problem as

$$\min_{\hat{\boldsymbol{F}}} \|\boldsymbol{Y}(n) - \hat{\boldsymbol{Y}}(n)\| = \min_{\boldsymbol{W}} \|\boldsymbol{Y}(n) - \hat{\boldsymbol{Y}}(n)\|, \tag{8.16}$$

where $\boldsymbol{W}$ denotes the set of all connection weights. Inspection of Equation (8.16) shows us that we have changed the determination of the nonlinear mapping of the unknown system to the determination of the parameters (connection weights) of the MLP network.

In most practical cases, K samples of the input vector $\boldsymbol{X}(n)$ and the output vector $\boldsymbol{Y}(n)$ are available. These available samples can be used as the training data to find the parameter set $\boldsymbol{W}$. With these training data and by using the Euclidean norm in (8.16), we have

$$\min_{\boldsymbol{W}} \sum_{n=1}^{I} \boldsymbol{E}^T(n)\boldsymbol{E}(n) = \min_{\boldsymbol{W}} \sum_{n=1}^{I} \|\boldsymbol{Y}(n) - \hat{\boldsymbol{Y}}(n)\|^2, \tag{8.17}$$

where

$$\boldsymbol{E}(n) = \boldsymbol{Y}(n) - \hat{\boldsymbol{Y}}(n) \tag{8.18}$$

and where I denotes the data length in which the values of the corresponding vector $\boldsymbol{U}(n)$ of the MLP network are known. Clearly, $I < K$. It is easy to see that (8.17) is exactly the same as (1.53). Consequently, any algorithm for solving (1.53) can immediately be used to find the weight set $\boldsymbol{W}$ of (8.17). As an example, the version of the BP algorithm for solving (8.17) is

$$W_{jk}^{(2)}(t+1) = W_{jk}^{(2)}(t) + \gamma_1 \sum_{n=1}^{I} \delta_k^{(2)}(n) q_j(n)$$

$$(j = 1, 2, \ldots, n_h; \quad k = 1, 2, \ldots, M) \tag{8.19}$$

and

$$W_{ij}^{(1)}(t+1) = W_{ij}^{(1)}(t) + \gamma_2 \sum_{n=1}^{I} \delta_j^{(1)}(n) u_i(n)$$

$$(i = 1, 2, \ldots, n_i; \quad j = 1, 2, \ldots, n_h), \tag{8.20}$$

where the various variables are

$$\delta_k^{(2)}(n) = y_k(n) - \hat{y}_k(n), \tag{8.21}$$

$$\delta_j^{(1)}(n) = \left[f_j\left(r_j(n)\right)\right]' \sum_{k=1}^{M} \delta_k^{(2)}(n) W_{jk}^{(2)}(t), \tag{8.22}$$

$$\left[f_j\left(r_j(n)\right)\right]' = \left.\frac{\partial f_j(r)}{\partial r}\right|_{r=r_j(n)}, \tag{8.23}$$

and

$$r_j(n) = \sum_{i=1}^{n_i} W_{ij}^{(1)}(t) u_i(n) \qquad (n = 1, 2, \ldots, I). \tag{8.24}$$

The quantities γ_1 and γ_2 are the learning-rate parameters. The details of the procedures are shown in Figure 8.8. Note that $W_{ij}^{(1)}(t)$ and $W_{jk}^{(2)}(t)$ denote the connection weights at the t'th iteration.

This is in effect the batch-processing version of the BP algorithm that applies to the case in which all the training data are available when the training of the connection weights is initiated. If each new output vector $\boldsymbol{Y}(n)$ and input vector $\boldsymbol{X}(n)$ of the unknown system become available during the training, then (8.19) and (8.20) become

$$W_{jk}^{(2)}(n+1) = W_{jk}^{(2)}(n) + \gamma_1 \delta_k^{(2)}(n) q_j(n)$$

$$(j = 1, 2, \ldots, n_h; \quad k = 1, 2, \ldots, M) \tag{8.25}$$

and

$$W_{ij}^{(1)}(n+1) = W_{ij}^{(1)}(n) + \gamma_2 \delta_j^{(1)}(n) u_i(n)$$

$$(i = 1, 2, \ldots, n_i; \quad j = 1, 2, \ldots, n_h), \tag{8.26}$$

where $W_{ij}^{(1)}(n)$ and $W_{jk}^{(2)}(n)$ denote the connection weights at the n'th iteration. Note that the corresponding connection weights in (8.22) and (8.24) should be replaced by $W_{ij}^{(1)}(n)$ and $W_{jk}^{(2)}(n)$.

As is known, the BP algorithm is one of the gradient-descent-based techniques. Now, we will present a more general algorithm based on gradient-descent techniques for solving (8.17) [5, 6]. For convenience, we arrange the weight set $\boldsymbol{W}$ into

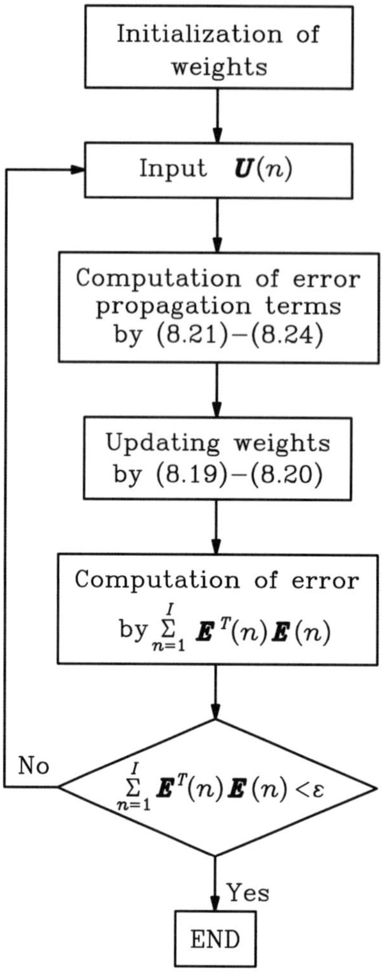

Figure 8.8. The flowchart of the BP algorithm for solving Equation (8.17). ϵ is specified.

an n_w-dimensional parameter vector as $\boldsymbol{W} = [W_1, W_2, \ldots, W_{n_w}]^T$, where $n_w = n_i \times n_h + n_h \times M$ and the gradient of $\hat{\boldsymbol{Y}}(n)$ with respect to W_i is denoted as $\boldsymbol{a}_i(n)$, which is a $1 \times M$ row vector. Consequently, we have

$$\boldsymbol{\Psi}(n, t) = \left[\frac{d\hat{\boldsymbol{Y}}(n)}{d\boldsymbol{W}}\right]^T = \left[\boldsymbol{a}_1^T(n), \boldsymbol{a}_2^T(n), \ldots, \boldsymbol{a}_{n_w}^T(n)\right]^T. \tag{8.27}$$

With these in mind, the general gradient-descent-based algorithm is

$$\boldsymbol{W}(t+1) = \boldsymbol{W}(t) + \gamma \boldsymbol{Q}(t), \tag{8.28}$$

where $\boldsymbol{Q}(t)$ is a search direction based on the gradient of $\sum_{n=1}^{I} \boldsymbol{E}^T(n)\boldsymbol{E}(n)$ with respect to $\boldsymbol{W}(t)$ and γ is a search-rate parameter, which may vary with t. The most commonly used search direction is the modified negative gradient direction defined by

$$Q(t) = M(W)\left(\sum_{n=1}^{I} \Psi(n, t)E(n)\right), \tag{8.29}$$

where $M(W)$ is an $(n_w \times n_w)$-dimensional positive definite matrix and $\sum_{n=1}^{I} \Psi(n, t)$ $E(n)$ is the gradient of $-\sum_{n=1}^{I} E^T(n)E(n)$ with respect to $W(t)$.

If $M(W)$ is selected to be the identity matrix, then (8.28) is in effect another version of the BP algorithm. To speed up the convergence of the training, we can use the Gauss–Newton search direction in which the matrix $M(W)$ is selected as the inverse of the approximate Hessian matrix of $\sum_{n=1}^{I} E^T(n)E(n)$, that is,

$$(M(W))^{-1} = H(W) = \sum_{n=1}^{I} \Psi(n, t)\Psi^T(n, t). \tag{8.30}$$

Clearly, this algorithm based on the Gauss–Newton direction requires much more computation than the BP algorithm. One way to reduce the computational complexity is to choose $(M(W))^{-1}$ to be the near-diagonal matrix

$$(M(W))^{-1} = \sum_{n=1}^{I} \begin{pmatrix} \Psi_1(n, t)\Psi_1^T(n, t) & \mathbf{0} & \cdots & \mathbf{0} \\ \mathbf{0} & \Psi_2(n, t)\Psi_2^T(n, t) & \cdots & \mathbf{0} \\ \vdots & \vdots & \ddots & \vdots \\ \mathbf{0} & \mathbf{0} & \cdots & \Psi_p(n, t)\Psi_p^T(n, t) \end{pmatrix}, \tag{8.31}$$

where $p = n_h + M$ and $\Psi_i(n, t)$ (for $i = 1, 2, \ldots, p$) is an $(m_i \times M)$-dimensional matrix; m_i denotes the number of the related weights of the i'th neuron if all the neurons in the hidden layer and the output layer are numbered from 1 to p. If the related connection weights of the i'th neuron are denoted by the vector W_i, then $\Psi_i(n, t)$ is the gradient of $\hat{Y}(n)$ with respect to W_i. By using (8.29)–(8.31), the search direction $Q(t)$ can be decomposed into p smaller vectors:

$$Q_i(t) = \left[\sum_{n=1}^{I} \Psi_i(n, t)\Psi_i^T(n, t)\right]^{-1} \sum_{n=1}^{I} \Psi_i(n, t)E(n). \tag{8.32}$$

Consequently, Equation (8.28) can be decomposed into p parallel subalgorithms as

$$W_i(t + 1) = W_i(t) + \gamma_i Q_i(t) \qquad (i = 1, 2, \ldots, p). \tag{8.33}$$

Because the matrices involved in (8.32) have fewer dimensions than those in (8.30), the computational complexity of (8.33) is less than that of (8.28). In addition, a diagonal matrix KI may be added to the right-hand side of (8.31) to guarantee the positive definiteness of $(M(W))^{-1}$ without affecting the decomposition.

If each new output vector $\boldsymbol{Y}(n)$ and input vector $\boldsymbol{X}(n)$ become available during the training, then (8.28) and (8.33) become [6, 7]

$$\boldsymbol{W}(n+1) = \boldsymbol{W}(n) + \gamma \boldsymbol{M}(n)\boldsymbol{\Psi}(n)\boldsymbol{E}(n) \tag{8.34}$$

and

$$\boldsymbol{W}_i(n+1) = \boldsymbol{W}_i(n) + \gamma \boldsymbol{M}_i(n)\boldsymbol{\Psi}_i(n)\boldsymbol{E}(n) \qquad (i = 1, 2, \ldots, p). \tag{8.35}$$

Moreover, $\boldsymbol{M}(n)$ and $\boldsymbol{M}_i(n)$ are computed by the recursive algorithms

$$\begin{aligned} \boldsymbol{M}(n) = \frac{1}{\lambda}\Big\{&\boldsymbol{M}(n-1) - \boldsymbol{M}(n-1)\boldsymbol{\Psi}(n) \\ &\times \big[\lambda \boldsymbol{I} + \boldsymbol{\Psi}^T(n)\boldsymbol{M}(n-1)\boldsymbol{\Psi}(n)\big]^{-1}\boldsymbol{\Psi}^T(n)\boldsymbol{M}(n-1)\Big\} \end{aligned} \tag{8.36}$$

and

$$\begin{aligned} \boldsymbol{M}_i(n) = \frac{1}{\lambda_i}\Big\{&\boldsymbol{M}_i(n-1) - \boldsymbol{M}_i(n-1)\boldsymbol{\Psi}_i(n) \\ &\times \big[\lambda_i \boldsymbol{I} + \boldsymbol{\Psi}_i^T(n)\boldsymbol{M}_i(n-1)\boldsymbol{\Psi}_i(n)\big]^{-1}\boldsymbol{\Psi}_i^T(n)\boldsymbol{M}_i(n-1)\Big\} \\ &\qquad (i = 1, 2, \ldots, p), \end{aligned} \tag{8.37}$$

where λ and λ_i are positive constants and are usually less than unity.

In some practical applications, a phenomenon known as "covariance wind-up" may occur, that is, $\boldsymbol{M}(n)$ and $\boldsymbol{M}_i(n)$ may become explosive [7, 8]. Many numerical measures have been developed to overcome this problem. As an example, a technique called constant trace adjustment is presented as follows:

$$\begin{aligned} \hat{\boldsymbol{M}}(n) &= \boldsymbol{M}(n-1) - \boldsymbol{M}(n-1)\boldsymbol{\Psi}(n) \\ &\quad \times \big[\lambda \boldsymbol{I} + \boldsymbol{\Psi}^T(n)\boldsymbol{M}(n-1)\boldsymbol{\Psi}(n)\big]^{-1}\boldsymbol{\Psi}^T(n)\boldsymbol{M}(n-1), \\ \boldsymbol{M}(n) &= \frac{K_0}{trace[\hat{\boldsymbol{M}}(n)]}\hat{\boldsymbol{M}}(n) \end{aligned} \tag{8.38}$$

and

$$\begin{aligned} \hat{\boldsymbol{M}}_i(n) &= \boldsymbol{M}_i(n-1) - \boldsymbol{M}_i(n-1)\boldsymbol{\Psi}_i(n) \\ &\quad \times \big[\lambda_i \boldsymbol{I} + \boldsymbol{\Psi}_i^T(n)\boldsymbol{M}_i(n-1)\boldsymbol{\Psi}_i(n)\big]^{-1}\boldsymbol{\Psi}_i^T(n)\boldsymbol{M}_i(n-1), \\ \boldsymbol{M}_i(n) &= \frac{K_0}{trace[\hat{\boldsymbol{M}}_i(n)]}\hat{\boldsymbol{M}}_i(n) \qquad (i = 1, 2, \ldots, p), \end{aligned} \tag{8.39}$$

where K_0 is a positive constant.

Clearly, (8.25) and (8.26) in the BP algorithm can be obtained by simply letting $\boldsymbol{M}(n)$ and $\boldsymbol{M}_i(n)$ in (8.38) and (8.39) be the identity matrix.

In (8.34) and (8.35), $\boldsymbol{\Psi}(n)\boldsymbol{E}(n)$ and $\boldsymbol{\Psi}_i(n)\boldsymbol{E}(n)$ are known as stochastic gradients. These can be replaced by so-called smoothed stochastic gradients calculated using [9, 10]

$$\boldsymbol{\Delta}(n) = \gamma_3\boldsymbol{\Delta}(n-1) + \gamma_4\boldsymbol{\Psi}(n)\boldsymbol{E}(n) \tag{8.40}$$

and

$$\boldsymbol{\Delta}_i(n) = \gamma_5\boldsymbol{\Delta}_i(n-1) + \gamma_6\boldsymbol{\Psi}_i(n)\boldsymbol{E}(n) \qquad (i = 1, 2, \ldots, p), \tag{8.41}$$

where γ_3, γ_4, γ_5, and γ_6 are gain parameters. With these parameters, (8.34) and (8.35) become

$$\boldsymbol{W}(n+1) = \boldsymbol{W}(n) + \gamma\boldsymbol{M}(n)\boldsymbol{\Delta}(n) \tag{8.42}$$

and

$$\boldsymbol{W}_i(n+1) = \boldsymbol{W}_i(n) + \gamma\boldsymbol{M}_i(n)\boldsymbol{\Delta}_i(n) \qquad (i = 1, 2, \ldots, p). \tag{8.43}$$

In summary, we have presented three algorithms in the batch-processing version and recursive-processing version for solving (8.17) of the system identification problem: the BP algorithm ((8.19)–(8.20) and (8.25)–(8.26)), the Gauss–Newton algorithm ((8.28) and (8.34)), and the simplified Gauss–Newton algorithm ((8.33) and (8.35)). The Gauss–Newton algorithm has the most extensive computational complexity and the fastest convergence of the three; the BP algorithm has the least computation but the slowest convergence. In practical applications, the simplified Gauss–Newton algorithm may be the most attractive because it can be integrated into a fully parallel computational structure like the BP algorithm and can deliver a fast convergence compared to that of the Gauss–Newton algorithm.

In addition, other loss functions can be used to replace $\sum_{n=1}^{I} \boldsymbol{E}^T(n)\boldsymbol{E}(n)$ of (8.17), for example,

$$\min_{\boldsymbol{W}} \frac{1}{2} \log\left[\det \sum_{n=1}^{I} \boldsymbol{E}(n)\boldsymbol{E}^T(n)\right] \tag{8.44}$$

and

$$\min_{\boldsymbol{W}} \sum_{n=1}^{I} \left[\boldsymbol{E}^T(n)\boldsymbol{\Lambda}\boldsymbol{E}(n)\right], \tag{8.45}$$

where $\boldsymbol{\Lambda}$ is a given $M \times M$ symmetric positive matrix. More detailed discussions of loss functions and their effects on performance can be found in References [10] and [11].

We make three further comments concerning this kind of system identification based on the MLP network and related algorithms:

Comment 1 Because the input vector $\boldsymbol{U}(n)$ is determined by (8.12), this MLP-based scheme only applies to the identification of a system that can be represented by the NARMAX model (8.11). However, for the special case of the input-state-output representation

$$\begin{cases} \boldsymbol{Z}(n+1) = \boldsymbol{F}\big[\boldsymbol{Z}(n), \boldsymbol{X}(n)\big], \\ \boldsymbol{Y}(n) = \boldsymbol{Z}(n) \end{cases} \tag{8.46}$$

the MLP-based scheme still applies as long as we select the input vector $\boldsymbol{U}(n) = [\boldsymbol{Y}^T(n-1), \boldsymbol{X}^T(n-1)]^T$. In fact, in this case, the output vector and the state vector of the unknown system are identical.

Comment 2 Because the solutions provided by these algorithms cannot be guaranteed to be the global minimum, it is necessary to test the model validity after determining the connection weights of the MLP network. If modeling is adequate, $\boldsymbol{E}(n)$ in (8.18) will be unpredictable from (uncorrelated with) all linear and nonlinear combinations of the past inputs and outputs. Billings and Coworkers [12, 13, 14] have proposed five conditions that an adequately identified model should satisfy. As a brief illustration, these five conditions in the single-input and single-output case are:

$$r_{ee}(k) = \delta(k),$$

$$r_{ex}(k) = 0,$$

$$r_{e(ex)}(k) = 0 \quad (k \geq 0),$$

$$r_{\hat{x}e}(k) = 0,$$

and

$$r_{\hat{x}(e^2)}(k) = 0,$$

where $\delta(k)$ is the impulse function, $r_{uv}(k)$ indicates the cross-correlation between the two time series $u(n)$ and $v(n)$, $e(n) = y(n) - \hat{y}(n)$, $x(n)$ and $y(n)$ are the input and output of the system, and $ex(n) = e(n+1)x(n+1)$, $\hat{x}(n) = x^2(n) - (E[x(n)])^2$. Clearly, these five conditions can be used to test the validity of the MLP-based scheme. If the network after training fails to satisfy these five conditions, then we have to select other values for the initial weights and restart the training until the identified model is adequate.

Alternatively, a statistical test known as the chi-squared test can be employed to validate the identified model [15]. In this test, we first set a confidence level α such as 95%, then use the available error vector $\boldsymbol{E}(n)$, the input vector $\boldsymbol{X}(n)$, and the output vector $\boldsymbol{Y}(n)$ of the system to calculate the chi-squared statistic, and finally compare the statistic with $\chi^2(1-\alpha)$, where $\chi^2(1-\alpha)$ is the critical value of the chi-squared distribution for the

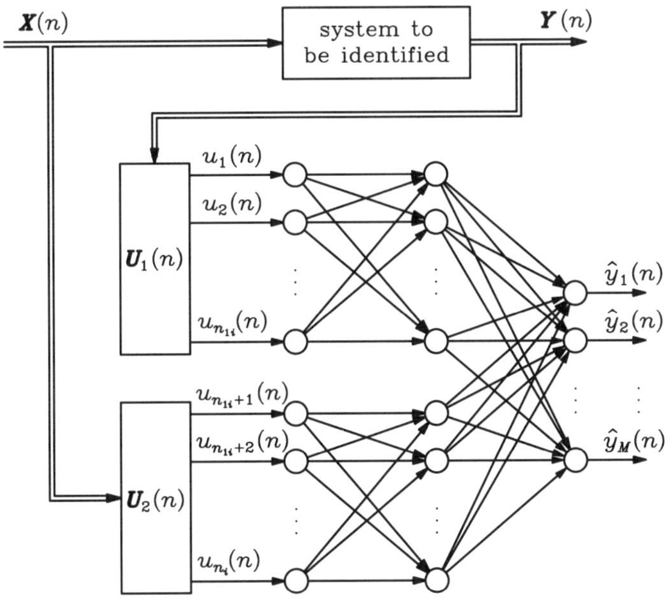

Figure 8.9. The MLP network for identifying the system represented by (8.7).

given level $(1-\alpha)$. If the statistic is less than $\chi^2(1-\alpha)$, then the model is regarded as adequate; otherwise, the identified model is invalid.

Comment 3 For the systems represented by the four special cases ((8.6)–(8.9)) of the NARMAX model, the related identification algorithms will become much simpler.

In Case 1, the system is linear and time invariant and hence many available parameter identification algorithms [1, 2] can immediately be used to find the coefficient vectors $\boldsymbol{A}_i$ and $\boldsymbol{B}_j$. Moreover, some explicit expressions to calculate $\boldsymbol{A}_i$ and $\boldsymbol{B}_j$ have been reported [1, 2]. The current key problems concern how to implement the available identification algorithms in real time and how to identify the system under the condition that the input of the system is not known a priori. We will deal with these two aspects in Section 8.5 and Section 8.6, respectively.

In Case 2, the nonlinear dependence of $\boldsymbol{Y}(n)$ on $\boldsymbol{Y}(n-i)$ and $\boldsymbol{X}(n-j)$ is assumed to be separable. For this case, the input layer and hidden layer of the MLP network in Figure 8.7 can be separated into two independent parts as shown in Figure 8.9. Consequently, the input vector $\boldsymbol{U}(n)$ and the output vector $\boldsymbol{q}(n) = [q_1(n), q_2(n), \ldots, q_{n_h}(n)]^T$ of the hidden neuron are each partitioned into two parts:

$$\boldsymbol{U}(n) = \begin{pmatrix} \boldsymbol{U}_1(n) \\ \boldsymbol{U}_2(n) \end{pmatrix}, \qquad \boldsymbol{q}(n) = \begin{pmatrix} \boldsymbol{q}_1(n) \\ \boldsymbol{q}_2(n) \end{pmatrix}, \tag{8.47}$$

where

$$\boldsymbol{U}_1(n) = \left[u_1(n), u_2(n), \ldots, u_{n_{1i}}(n)\right]^T$$
$$= \left[\boldsymbol{Y}^T(n-1), \boldsymbol{Y}^T(n-2), \ldots, \boldsymbol{Y}^T(n-n_y)\right]^T, \tag{8.48}$$
$$\boldsymbol{U}_2(n) = \left[u_{n_{1i}+1}(n), u_{n_{1i}+2}(n), \ldots, u_{n_i}(n)\right]^T,$$
$$= \left[\boldsymbol{X}^T(n-1), \boldsymbol{X}^T(n-2), \ldots, \boldsymbol{X}^T(n-n_x)\right]^T, \tag{8.49}$$
$$\boldsymbol{q}_1(n) = \left[q_1(n), q_2(n), \ldots, q_{n_{h_1}}(n)\right]^T, \tag{8.50}$$

and

$$\boldsymbol{q}_2(n) = \left[q_{n_{h_1}+1}(n), q_{n_{h_1}+2}(n), \ldots, q_{n_h}(n)\right]^T. \tag{8.51}$$

In (8.48)–(8.51), $n_{1i} = n_y M$, and n_{h_1} is the hidden neuron number corresponding to $\boldsymbol{U}_1(n)$; n_{h_1} is adjustable; $n_h - n_{h_1}$ is the hidden neuron number corresponding to $\boldsymbol{U}_2(n)$. Note that there are no connection weights among the following neurons: (1) the i'th neuron (for $i = 1, 2, \ldots, n_{1i}$) in the input layer with the j'th neuron (for $j = n_{h_1} + 1, n_{h_1} + 2, \ldots, n_h$) in the hidden layer and (2) the i'th neuron (for $i = n_{1i} + 1, n_{1i} + 2, \ldots, n_i$) in the input layer with the j'th neuron (for $j = 1, 2, \ldots, n_{h_1}$) in the hidden layer. In other words, the weights $W_{ij}^{(1)} = 0$ (for $i = 1, 2, \ldots, n_{1i}; j = n_{h_1}+1, n_{h_1}+2, \ldots, n_h$ and $i = n_{1i}+1, n_{1i}+2, \ldots, n_i; j = 1, 2, \ldots, n_{h_1}$). Thus, we can partition the parameter vector $\boldsymbol{W}$ into two subvectors, which results in fewer computations when the identification algorithms given above are used to find the weights.

Cases 3 and 4 can be considered as special cases of Case 2. As a result, on the basis of (8.48)–(8.51), we can further simplify the architecture of the MLP network for Case 3 and Case 4 as follows: For Case 3, the hidden neurons (for $i = 1, 2, \ldots, n_{h_1}$) are selected to have linear activation functions instead of nonlinear ones, and the number n_{h_1} is the same $(n_y M)$ as that of the first part of the input layer. In addition,

$$W_{ij}^{(1)} = 1, \qquad i = j,$$

and

$$W_{ij}^{(1)} = 0, \qquad i \neq j \qquad (i, j = 1, 2, \ldots, n_{1i}). \tag{8.52}$$

This means that $q_j(n) = u_j(n)$ (for $j = 1, 2, \ldots, n_{1i}$) and the weights $W_{jk}^{(2)}$ (for $j = 1, 2, \ldots, n_{1i}; k = 1, 2, \ldots, M$) after training will approximate the corresponding element of the coefficient vector $\boldsymbol{A}_j$ in (8.8).

For Case 4, the hidden neurons (for $i = n_{h_1} + 1, n_{h_1} + 2, \ldots, n_h$) are selected to have a linear activation function such as that in Case 3, and the number $n_h - n_{h_1}$

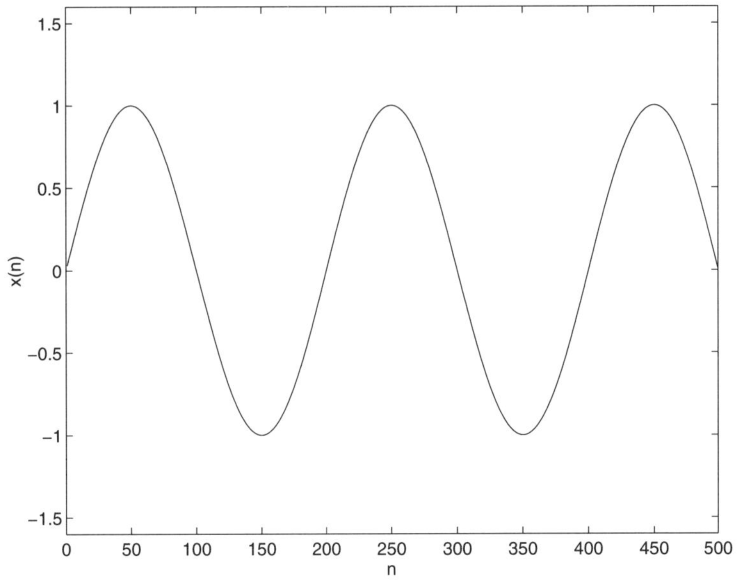

Figure 8.10. The input $x(n)$ of the system.

is the same $(n_x N)$ as that of the second part of the input layer. In addition,

$$W_{ij}^{(1)} = 1, \qquad i = j,$$

and

$$W_{ij}^{(1)} = 0 \qquad i \neq j \qquad (i, j = n_{h_1} + 1, n_{h_1} + 2, \ldots, n_h). \tag{8.53}$$

This means that $q_j(n) = u_j(n)$ (for $j = n_{h_1} + 1, n_{h_1} + 2, \ldots, n_h$) and the weights $W_{jk}^{(2)}$ (for $j = n_{h_1} + 1, n_{h_1} + 2, \ldots, n_h$; $k = 1, 2, \ldots, M$) after training will approximate the corresponding element of the coefficient vector $\boldsymbol{B}_j$ in (8.9).

In the final part of this section, we give a set of simulation results for the MLP-based system identification scheme. We considered the single-input single-output system

$$\begin{aligned} y(n) = {} & \left(0.8 - 0.5e^{-y^2(n-1)}\right) y(n-1) \\ & - \left(0.3 + 0.9e^{-y^2(n-2)}\right) y(n-2)x(n-1) \\ & + 0.2x(n-2) + 0.1x(n-1)x(n-2) + e(n). \end{aligned} \tag{8.54}$$

In the simulation, we used $n_i = 4$, $n_h = 20$, $I = 500$, and the BP algorithm (8.19)–(8.20). The system noise $e(n)$ is a zero-mean Gaussian white sequence with variance 0.01; the system input $x(n) = \sin(0.1\pi n)$. Figures 8.10–8.13 show $x(n)$, $y(n)$, $\hat{y}(n)$, and the error $y(n) - \hat{y}(n)$.

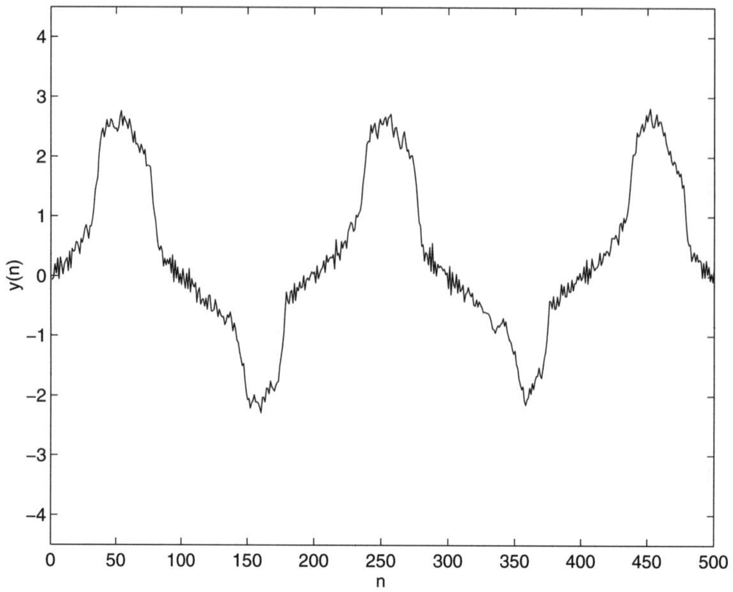

Figure 8.11. The output $y(n)$ of the system.

These simulation results demonstrate further the effectiveness of the MLP-based system identification scheme. However, this MLP-based system identification technique is still very preliminary and additional investigations are strongly recommended.

8.3 System Identification Using RBF Networks

In Section 1.6 and Section 5.3, it was pointed out that the RBF network is an alternative of the MLP network for performing a nonlinear mapping. As a result, the RBF network can immediately be employed to find the nonlinear maps related with the system identification operator $\boldsymbol{P}[\cdot]$. For simplicity, but without loss of generality, we continue to consider the system represented by (8.11).

The RBF network for finding the system mapping $\boldsymbol{F}[\cdot]$ is exactly the same as that given in Figure 1.7. This network consists of three layers. The input layer has $n_i = n_y M + n_x N$ neurons with a linear function that simply feed the input signals to the hidden layer. The input vector is

$$\begin{aligned}\boldsymbol{U}(n) &= \left[u_1(n), u_2(n), \ldots, u_{n_i}(n)\right]^T \\ &= \left[\boldsymbol{Y}^T(n-1), \boldsymbol{Y}^T(n-2), \ldots, \boldsymbol{Y}^T(n-n_y),\right. \\ &\qquad \left.\boldsymbol{X}^T(n-1), \boldsymbol{X}^T(n-2), \ldots, \boldsymbol{X}^T(n-n_x)\right]^T.\end{aligned}$$

This equation is the same as (8.12) and means that the input of the network is

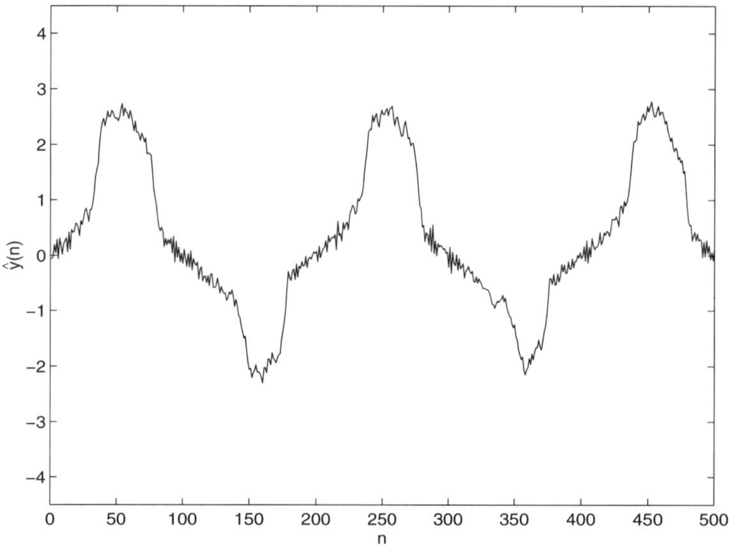

Figure 8.12. The output $\hat{y}(n)$ of the neural network (after convergence).

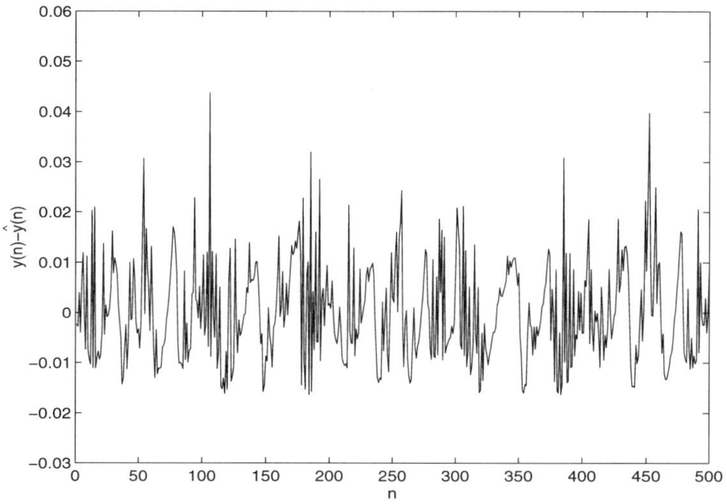

Figure 8.13. The difference between the output $y(n)$ of the system and the output $\hat{y}(n)$ of the neural network.

given by the past values of the inputs and the outputs of the system to be identified. The connection between the input layer and the hidden layer are not weighted: Each hidden neuron receives each corresponding input value unaltered. The hidden neurons are processing units that perform the radial basis function. The transfer function $f_j(\cdot)$ (for $j = 1, 2, \ldots, n_h$) of each hidden neuron in this RBF network can be the Gaussian function, multiquadratic function, inverse multiquadratic function, thin-plate-spline function, piecewise-linear function, or the cubic approximation function, where n_h is the hidden neuron number. Each hidden neuron has a center vector denoted by $\boldsymbol{C}_j$. If we use the Euclidean norm to measure the distance between the input vector $\boldsymbol{U}(n)$ with the center vector $\boldsymbol{C}_j$, then the output $h_j(n)$ of the j'th neuron in the hidden layer is

$$h_j(n) = f_j\left(\|\boldsymbol{U}(n) - \boldsymbol{C}_j\|\right) \qquad (j = 1, 2, \ldots, n_h). \tag{8.55}$$

The connections between the hidden layer and the output layer are weighted. Each neuron of the output layer is a linear combinator, that is, the output neuron performs simple summations. With this in mind, the output $\hat{y}_k(n)$ of the k'th neuron in the output layer at time n is

$$\hat{y}_k(n) = \sum_{j=1}^{n_h} W_{kj} h_j(n) = \sum_{j=1}^{n_h} W_{kj} f_j\left(\|\boldsymbol{U}(n) - \boldsymbol{C}_j\|\right) \qquad (k = 1, 2, \ldots, M), \tag{8.56}$$

where W_{kj} is the connection weight between the j'th neuron in the hidden layer and the k'th neuron in the output layer. Note that the neuron number of the output layer must be equal to the output number (M) of the system to be identified. The output vector provided by the network is

$$\begin{aligned} \hat{\boldsymbol{Y}}(n) &= \left[\hat{y}_1(n), \hat{y}_2(n), \ldots, \hat{y}_M(n)\right]^T \\ &= \hat{\boldsymbol{F}}\left[\boldsymbol{U}(n)\right], \end{aligned} \tag{8.57}$$

where $\hat{\boldsymbol{F}}$ denotes the nonlinear mapping given by the network. Based on (8.56), the mapping $\hat{\boldsymbol{F}}$ is completely determined by the three sets of network parameters – the center vectors, the connection weights, and the scaling gains (σ_i in (1.71)).

According to (8.4), we obtain

$$\min_{\hat{\boldsymbol{F}}} \|\boldsymbol{Y}(n) - \hat{\boldsymbol{Y}}(n)\| = \min_{\boldsymbol{W},\boldsymbol{C},\boldsymbol{\sigma}} \|\boldsymbol{Y}(n) - \hat{\boldsymbol{Y}}(n)\|, \tag{8.58}$$

where $\boldsymbol{W}$, $\boldsymbol{C}$, and $\boldsymbol{\sigma}$ denote the set of the connection weights, the center vectors, and the scaling gains, respectively. Like (8.17), Equation (8.58) means that we have changed the determination of the nonlinear mapping of unknown systems to the determination of the parameters of the RBF network. Now we will show how to determine the parameters $\boldsymbol{W}$, $\boldsymbol{C}$, and $\boldsymbol{\sigma}$ from a set of the available input vectors and output vectors of the system to be identified, which generally includes the three steps mentioned in Section 1.6:

(1) determining the center vectors $\boldsymbol{C}_j$ (for $j = 1, 2, \ldots, n_h$),
(2) determining the scalar gain $\boldsymbol{\sigma}_j$ (for $j = 1, 2, \ldots, n_h$), and
(3) determining the connection weights $\boldsymbol{W}_{ij}$ (for $i = 1, \ldots, M$; $j = 1, \ldots, n_h$).

For the center vectors the simplest technique is to choose them randomly from a subset of the available input vectors $\boldsymbol{U}(n)$ (for $n = 1, 2, \ldots, I$). A more appropriate alternative is the "k-means clustering algorithm," which has two versions: a batch version when all data are available initially and an "on-line" version, which does not require all the data at one time. The two versions both minimize the optimization problem [16, 17]

$$E_{k\text{-means}} = \sum_{j=1}^{n_h} \sum_{n=1}^{I} D_{jn} \|\boldsymbol{U}(n) - \boldsymbol{C}_j\|^2, \tag{8.59}$$

where D_{jn} is the cluster partition or membership function, which forms an $n_h \times I$ matrix. Each column represents an available sample vector and each row represents a cluster. Each column has a single "1" in the row corresponding to the cluster nearest to that training point and zeros elsewhere.

Note that this is an unsupervised procedure using only the network input vector $\boldsymbol{U}(n)$. This has similarities to the Kohonen self-organizing algorithm ((1.62)–(1.64)).

If the hidden neurons take the Gaussian function, multiquadratic function, or inverse multiquadratic function as their activation functions, that is,

$$f_j(r) = e^{-\frac{r^2}{\sigma_j^2}}, \tag{8.60}$$

$$f_j(r) = (r^2 + \sigma_j^2)^{\frac{1}{2}}, \tag{8.61}$$

or

$$f_j(r) = (r^2 + \sigma_j^2)^{-\frac{1}{2}} \qquad (j = 1, 2, \ldots, n_h), \tag{8.62}$$

then it is necessary to determine the scaling gain σ_j. An appropriate method for doing this is based on the P-nearest neighbor heuristic, that is,

$$\sigma_j = \frac{1}{P} \sum_{i=1}^{P} \|\boldsymbol{C}_j - \boldsymbol{C}_i\|^2 \qquad (j = 1, 2, \ldots, n_h), \tag{8.63}$$

where $\boldsymbol{C}_i$ (for $i = 1, 2, \ldots, P$) are the P-nearest neighbors of $\boldsymbol{C}_j$. However, it has been shown that the scaling gains are much less influential in comparison with the center vectors and weights [18]. As a result, for simplicity, we can set all scaling gains σ_j to be identical and on the order of the power of the inputs of the system to be identified.

In the case that the Gaussian function is taken as the activation function, Tan et al. [19] proposed another effective method. In this method one first determines the scaling gains and then determines the centers. This method can be presented as follows: Assume the operating region of a system (defined by the bounds on all the coordinates in the vector $\boldsymbol{U}(n)$) to be fixed within a hypercube with the minimum length $2r_{\min}$. The bound $r_{\min}$ can be determined either by considering the physical nature of the system to be identified or by applying situations in which the system needs to operate within a prescribed operating range of interest. Then the scaling gain σ_j of the Gaussian function is chosen to be within $[\frac{1}{10}r_{\min}, r_{\min}]$. For simplicity, all σ_j are often selected to be a single σ. After the determination of σ, the center vectors will be selected by the following steps:

(1) Determine $\boldsymbol{U}(1)$ according to (8.12) by use of the available input vectors and the output vectors of the system, and set $\boldsymbol{U}(1)$ to be the first center vector $\boldsymbol{C}_1$.

(2) At time n, compare $\boldsymbol{U}(n)$ with all the existing center vectors. If the distances between $\boldsymbol{U}(n)$ and the existing center vectors are all greater than σ, then set $\boldsymbol{U}(n)$ to be a new center vector.

(3) Repeat this process until the number of center vectors stabilizes.

Observe that the idea underlining the above algorithm is to perceive a hidden neuron of the RBF network as a disk with a data point as its center and the scaling gain σ as its effective radius. These disks are used to cover the operating region for the unknown system. The algorithm proceeds in such a way that a new hidden neuron will be generated when the current data point (input vector $\boldsymbol{U}(n)$) falls outside the area covered by the existing neurons. Properly initialized, the algorithm will lay down a set of hidden neurons recursively on the incoming data sequences; these will gradually cover the whole operating region of the system to be identified.

Tan et al. [19] have shown that the above algorithm can effectively avoid the supposed center vectors often incurred by the k-means clustering algorithm.

We now need to determine the connection weights W_{ij} (for $i = 1, \ldots, M$; $j = 1, \ldots, n_h$). Because an RBF network has only one layer of weighted connections and the output neurons are simple summation units, (8.58) will become a linear least-squares problem once the center vectors and the scaling parameter have been determined, that is,

$$\min_{\boldsymbol{W}} \sum_{n=1}^{I} \left\|\boldsymbol{Y}(n) - \hat{\boldsymbol{Y}}(n)\right\|^2 = \min_{\boldsymbol{W}} \left\|\boldsymbol{W}\boldsymbol{H} - \boldsymbol{Y}\right\|^2, \tag{8.64}$$

where $\boldsymbol{W} = \{W_{ij}\}$ is an $M \times n_h$ matrix of the connection weights, $\boldsymbol{H}$ is an $n_h \times I$ matrix consisting of the outputs of the hidden neurons and whose elements are computed with

$$H_{in} = f_i\left(\left\|\boldsymbol{U}(n) - \boldsymbol{C}_i\right\|\right) \qquad (i = 1, 2, \ldots, n_h; \quad n = 1, 2, \ldots, I), \tag{8.65}$$

and $\boldsymbol{Y} = [\boldsymbol{Y}(1), \boldsymbol{Y}(2), \ldots, \boldsymbol{Y}(I)]$ is the $M \times I$ matrix consisting of the output vectors of the system to be identified.

We can find the connection weight matrix $\boldsymbol{W}$ from (8.64) in an explicit form as

$$\boldsymbol{W} = \boldsymbol{Y}\boldsymbol{H}^{+} = \boldsymbol{Y} \lim_{\alpha \to 0} \boldsymbol{H}^T(\boldsymbol{H}\boldsymbol{H}^T + \alpha \boldsymbol{I})^{-1}, \tag{8.66}$$

where $\boldsymbol{H}^{+}$ is the pseudoinverse of $\boldsymbol{H}$. Therefore, if all the related data are available at one time, the weight matrix $\boldsymbol{W}$ can be obtained directly by performing (8.66). A number of numerical algorithms for solving (8.66) have been proposed [20, 21], including the Gaussian elimination and Cholesky decomposition, Gram–Schmidt transformation, Householder transformation, Givens method, and singular value decomposition method. If either $\boldsymbol{H}\boldsymbol{H}^T$ or $\boldsymbol{H}^T\boldsymbol{H}$ is of full rank, then the Gaussian elimination and Cholesky decomposition are the most economical way of computing $\boldsymbol{W}$. With these methods the computational complexity is at about half of that of the orthogonal decomposition and about one quarter to one eighth of that of the singular value decomposition. Because the matrix dimensions involved in nonlinear system identification are generally very large, the ill-conditioning problem is always a possibility when the Gaussian elimination and Cholesky decomposition are used to solve (8.66). In this case, the methods based on an orthogonal decomposition (Gram–Schmidt transformation, Householder transformation, and Givens method) are more accurate. It has been shown that the orthogonal-decomposition-based methods have much better numerical properties in comparison with the methods based on the Gaussian elimination and Cholesky decomposition [20, 21]. When the rank of the matrix $\boldsymbol{H}$ is unknown, the singular value decomposition method is particularly useful. However, this method is computationally the most expensive.

In addition to these numerical methods, we can employ the high-computational capability of analog neural networks to perform (8.66) in real time. In effect, if we compare (8.66) with (3.88), we see that (8.66) can be performed in real time by the 2-D neural network of Figure 3.6 as long as we let the matrix $\boldsymbol{H}$ and $\boldsymbol{Y}$ be the connection matrix $\boldsymbol{T}$ and the bias current matrix $\boldsymbol{B}$ of the network, respectively, and other parameters be the same as those for (3.88). The network will provide the solution

$$\boldsymbol{V}_f = \boldsymbol{Y}\boldsymbol{H}^T\left(\frac{1}{RK_1K_2}\boldsymbol{I} + \boldsymbol{H}\boldsymbol{H}^T\right)^{-1}, \tag{8.67}$$

which is obviously an approximation of the exact solution (8.66).

Thus far we have only dealt with the case in which all the related data are available at one time (i.e., batch-processing). If each new output vector $\boldsymbol{Y}(n)$ and input vector $\boldsymbol{X}(n)$ of the system to be identified become available during the training, then the following recursive procedure can be used to determine the right connection weights:

(1) Initialize randomly all connection weights.
(2) Compute the output vector $\hat{\boldsymbol{Y}}(n)$ by use of (8.55)–(8.57).

(3) Compute the error term $e_i(n)$ of each output neuron

$$e_i(n) = y_i(n) - \hat{y}_i(n) \qquad (i = 1, 2, \ldots, M). \tag{8.68}$$

(4) Adjust the connection weights according to

$$W_{ij}(n+1) = W_{ij}(n) + \gamma e_i(n) f_j\left(\left\|\boldsymbol{U}(n) - \boldsymbol{C}_j\right\|\right)$$
$$(i = 1, 2, \ldots, M; \quad j = 1, 2, \ldots, n_h), \tag{8.69}$$

where γ is the learning-rate parameter.

(5) Compute the total error

$$\epsilon = \left\|\boldsymbol{Y}(n) - \hat{\boldsymbol{Y}}(n)\right\|^2 \tag{8.70}$$

and iterate the computation by returning to Step (2) until this error is less than the specified one.

Note that (8.70) can also be replaced by

$$\epsilon = \frac{1}{n}\sum_{k=1}^{n} \left\|\boldsymbol{Y}(k) - \hat{\boldsymbol{Y}}(k)\right\|^2, \tag{8.71}$$

where $\hat{\boldsymbol{Y}}(k)$ should be computed on the basis of the weights at the n'th iteration instead of those at the k'th iteration, that is,

$$\hat{y}_i(k) = \sum_{j=1}^{n_h} W_{ij}(n) f_j\left(\left\|\boldsymbol{U}(k) - \boldsymbol{C}_j\right\|\right) \qquad (i = 1, 2, \ldots, M). \tag{8.72}$$

Usually, the value of ϵ obtained by (8.71) is less than that found by (8.70).

This recursive procedure is similar to a 2-D LMS algorithm that can be derived from (8.64) by use of the gradient-descent method. The Gauss–Newton search direction mentioned in the last section can also be used to speed up the convergence. The algorithm based on the Gauss–Newton search direction is in effect a kind of recursive-least-squares algorithm [20, 21].

Because (8.64) is a linear LS problem, we do not encounter the local minimum problem in finding the connection weights of the network. This is one of the advantages of RBF networks over MLP networks for system identification. However, the local minimum problem persists in finding the center vectors of the hidden neurons of RBF networks. In other words, one must still test the model validity after determining the connection weights, center vectors, and scaling gains of RBF networks. In Reference [22] a sufficient condition for the global convergence has been proposed. More details can be found in this reference.

In addition, when the RBF network is used to identify the systems represented by the four special cases ((8.6)–(8.9)) of (8.5), it is possible to develop simpler structures and algorithms in the same way as given in Comment 3 of the last section.

Finally, we should mention that another effective way to identify nonlinear systems is to use alternative nonlinear mapping neural networks, such as functional link neuron networks [6] or higher-order neural networks [23, 24].

8.4 Recurrent Neural Networks for System Identification

In Sections 8.2 and 8.3, the input vector of the neural networks for identifying the unknown system consists of the past values of the input vector and the output vector of the system. This is called the series-parallel identification method [3]. An alternative method is the parallel identification method. Here one replaces the past values of the output vector of the system by the past values of the output vector of the neural network in setting the input vector of the neural network, that is,

$$\begin{aligned} \boldsymbol{U}(n) &= \left[u_1(n), u_2(n), \ldots, u_{n_i}(n)\right]^T \\ &= \left[\hat{\boldsymbol{Y}}^T(n-1), \hat{\boldsymbol{Y}}^T(n-2), \ldots, \hat{\boldsymbol{Y}}^T(n-n_{\hat{y}}),\right. \\ &\qquad \left.\boldsymbol{X}^T(n-1), \boldsymbol{X}^T(n-2), \ldots, \boldsymbol{X}^T(n-n_x)\right]^T, \end{aligned} \tag{8.73}$$

where $n_{\hat{y}}$ is the maximum lag in the output vectors of the network whose value can be different from that of n_y. This means that feedback is included in the above feedforward neural networks (MLP networks and RBF networks) as shown in Figure 8.14. The network described by Figure 8.14 is thus called a recurrent neural network [25].

The output vector of the recurrent neural network in Figure 8.14 can be written as

$$\begin{aligned} \hat{\boldsymbol{Y}}(n) &= \left[\hat{y}_1(n), \hat{y}_2(n), \ldots, \hat{y}_M(n)\right]^T \\ &= \hat{\boldsymbol{F}}\left[\hat{\boldsymbol{Y}}^T(n-1), \hat{\boldsymbol{Y}}^T(n-2), \ldots, \hat{\boldsymbol{Y}}^T(n-n_{\hat{y}}), \boldsymbol{X}^T(n-1),\right. \\ &\qquad \left.\boldsymbol{X}^T(n-2), \ldots, \boldsymbol{X}^T(n-n_x)]^T\right], \end{aligned} \tag{8.74}$$

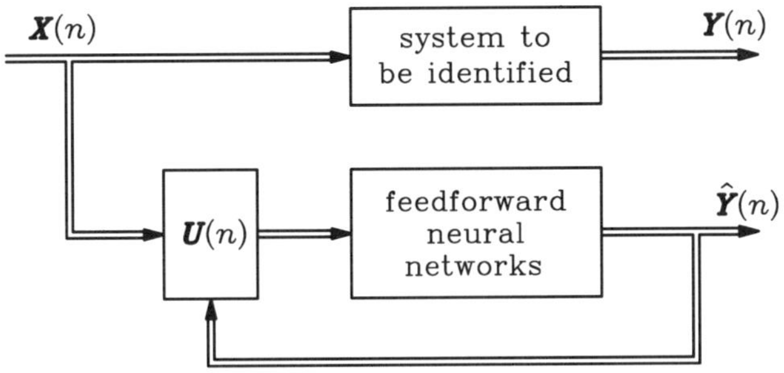

Figure 8.14. A neural network with feedback.

where $\hat{\boldsymbol{F}}$ is an operator depending on the connection weights (and other parameters) of the network. The key problem of recurrent neural networks for system identification remains the determination of the network parameters by use of the known input vectors and output vectors of the system to be identified. Because the input vector of recurrent neural networks depends on the past values of its output vector, the algorithms presented in Sections 8.2 and 8.3 will no longer apply to determining the parameters of recurrent networks. In the following, we will take a recurrent neural network based on the MLP network in Figure 8.7 as an example to present the algorithms for determining these parameters. Clearly, in this case, the concerned parameters are the connection weights. For simplicity, we arrange all the connection weights into an n_w-dimensional parameter vector as $\boldsymbol{W} = [W_1, W_2, \ldots, W_{n_w}]^T$, where $n_w = n_i n_h + n_h M$ and all other variables are exactly the same as those used in Section 8.2 except that $n_i = n_{\hat{y}} M + n_x N$.

According to (8.4), the weight vector $\boldsymbol{W}$ can be obtained by solving the following optimization:

$$\min_{\boldsymbol{W}} \sum_{n=1}^{I} \boldsymbol{E}^T(n)\boldsymbol{E}(n) = \min_{\boldsymbol{W}} \sum_{n=1}^{I} \left\|\boldsymbol{Y}(n) - \hat{\boldsymbol{Y}}(n)\right\|^2. \tag{8.75}$$

Using the gradient-descent method, we get a recursive algorithm

$$\boldsymbol{W}(t+1) = \boldsymbol{W}(t) + \gamma \boldsymbol{\Phi}(t), \tag{8.76}$$

where $\boldsymbol{\Phi}(t)$ is the gradient vector of $\sum_{n=1}^{I} \boldsymbol{E}^T(n)\boldsymbol{E}(n)$ with respect to the vector $\boldsymbol{W}(t)$, that is,

$$\boldsymbol{\Phi}(t) = \frac{d(\sum_{n=1}^{I} \boldsymbol{E}^T(n)\boldsymbol{E}(n))}{d\boldsymbol{W}(t)}. \tag{8.77}$$

Note that $\boldsymbol{\Phi}(t)$ is no longer equal to $\sum_{n=1}^{I} \boldsymbol{\Psi}(n,t)\boldsymbol{E}(n)$ mainly because the input vector $\boldsymbol{U}(n)$ depends on the past values of the output vector of the network. Therefore, we should use the ordered derivatives, that is,

$$\boldsymbol{\Phi}(t) = \frac{d(\sum_{n=1}^{I} \boldsymbol{E}^T(n)\boldsymbol{E}(n))}{d\boldsymbol{W}(t)} = \sum_{n=1}^{I} \left(\frac{d\hat{\boldsymbol{Y}}(n)}{d\boldsymbol{W}(t)}\right)^T \frac{\partial^+ \sigma}{\partial \hat{\boldsymbol{Y}}(n)}, \tag{8.78}$$

where $\sigma = \sum_{n=1}^{I} \boldsymbol{E}^T(n)\boldsymbol{E}(n)$ and the partial derivatives with a plus sign are the ordered derivatives. For further illustration of the ordered derivatives, we give an example. For a function

$$L = L(x_1, x_2, \ldots, x_n) \tag{8.79}$$

and

$$x_j = l_j(x_1, x_2, \ldots, x_{j-1}) \qquad (j = 2, 3, \ldots, n), \tag{8.80}$$

the ordered derivatives are calculated as

$$\frac{\partial^+ L}{\partial x_j} = \frac{\partial L}{\partial x_j} + \sum_{k=j+1}^{n} \frac{\partial^+ L}{\partial x_k} \frac{\partial l_k}{\partial x_j}. \tag{8.81}$$

Similarly, $\partial^+\sigma/\partial\hat{\boldsymbol{Y}}(n)$ is calculated as

$$\begin{aligned}\frac{\partial^+\sigma}{\partial\hat{\boldsymbol{Y}}(n)} &= \frac{\partial\sigma}{\partial\hat{\boldsymbol{Y}}(n)} + \sum_{k=n+1}^{n+n_{\hat{y}}} \frac{\partial^+\sigma}{\partial\hat{\boldsymbol{Y}}(k)} \frac{\partial\hat{\boldsymbol{Y}}(k)}{\partial\hat{\boldsymbol{Y}}(n)} \\ &= \frac{\partial\sigma}{\partial\hat{\boldsymbol{Y}}(n)} + \sum_{k=1}^{n_{\hat{y}}} \frac{\partial^+\sigma}{\partial\hat{\boldsymbol{Y}}(n+k)} \frac{\partial\hat{\boldsymbol{Y}}(n+k)}{\partial\hat{\boldsymbol{Y}}(n)}.\end{aligned} \tag{8.82}$$

Substituting (8.82) into (8.78), we obtain

$$\boldsymbol{\Phi}(t) = \sum_{n=1}^{I} \left[\left(\frac{d\hat{\boldsymbol{Y}}(n)}{d\boldsymbol{W}(t)}\right)^T \frac{\partial\sigma}{\partial\hat{\boldsymbol{Y}}(n)} + \left(\frac{d\hat{\boldsymbol{Y}}(n)}{d\boldsymbol{W}(t)}\right)^T \sum_{k=1}^{n_{\hat{y}}} \frac{\partial^+(\sigma)}{\partial\hat{\boldsymbol{Y}}(n+k)} \frac{\partial\hat{\boldsymbol{Y}}(n+k)}{\partial\hat{\boldsymbol{Y}}(n)} \right]. \tag{8.83}$$

It is easy to get

$$\left(\frac{d\hat{\boldsymbol{Y}}(n)}{d\boldsymbol{W}(t)}\right)^T \frac{\partial\sigma}{\partial\hat{\boldsymbol{Y}}(n)} = \boldsymbol{\Psi}(n,t)\boldsymbol{E}(n). \tag{8.84}$$

Clearly, this is the same result as from (8.29) if we let the matrix $\boldsymbol{M}(\boldsymbol{W})$ be the identity matrix. To speed up the convergence, we can replace $\boldsymbol{\Psi}(n,t)\boldsymbol{E}(n)$ by $\boldsymbol{M}(\boldsymbol{W})\boldsymbol{\Psi}(n,t)\boldsymbol{E}(n)$ as used in Section 8.2.

With this in mind, we can summarize the procedures for determining the connection weights of the recurrent neural network for system identification as follows:

(1) Initialize randomly all connection weights and the $n_{\hat{y}}$ output vectors of the neural network, that is, $\hat{\boldsymbol{Y}}(0), \hat{\boldsymbol{Y}}(-1), \ldots, \hat{\boldsymbol{Y}}(1-n_{\hat{y}})$. For simplicity, all these output vectors can be set to zero.

(2) Compute the output vector $\hat{\boldsymbol{Y}}(n)$ (for $n = 1, 2, \ldots, I$) by use of (8.14) and by taking (8.73) as the input vector of the network. Note that I input vectors of the system are required in this step.

(3) Compute the related gradient terms in terms of Equations (8.82)–(8.84).

(4) Adjust the connection weights according to (8.76).

(5) Compute the total error σ according to $\sigma = \sum_{n=1}^{I} \boldsymbol{E}^T(n)\boldsymbol{E}(n)$ and iterate the computation by returning to Step (2) until this error is less than the specified one.

This procedure is also a kind of batch-processing. For the case that each new input vector and output vector of the system become available recursively, (8.76)

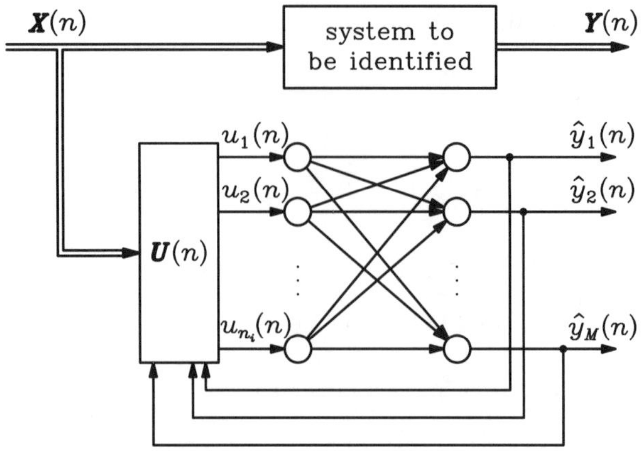

Figure 8.15. A simplified recurrent neural network.

will become

$$\boldsymbol{W}(n+1) = \boldsymbol{W}(n) + \gamma \boldsymbol{\Phi}(n), \tag{8.85}$$

where $\boldsymbol{\Phi}(n)$ is the gradient vector of $\boldsymbol{E}^T(n)\boldsymbol{E}(n)$ with respect to the vector $\boldsymbol{W}(n)$, that is,

$$\boldsymbol{\Phi}(n) = \boldsymbol{\Psi}(n)\boldsymbol{E}(n). \tag{8.86}$$

$\boldsymbol{\Psi}(n)$ is the gradient of $\hat{\boldsymbol{Y}}(n)$ with respect to $\boldsymbol{W}(n)$, a relationship similar to that in (8.27).

Qin et al. [26] showed both analytically and by simulation results that the weights by (8.85) will be different from those by (8.76), but the difference is small and can be controlled by the learning parameter γ. This difference will converge to zero as fast as γ goes to zero.

It is interesting to note that we can simplify the structure of the above recurrent network for system identification as shown in Figure 8.15. This simplified recurrent network has only two layers. The input layer has $n_i = n_{\hat{y}}M + n_x N$ neurons whose input vector is calculated by using (8.73) and simply feeds the vector $\boldsymbol{U}(n)$ to the second layer (the output layer) without any modification. Thus the input layer of the simplified network is exactly the same as that in Figure 8.14. The output layer has M neurons, which correspond to M outputs of the unknown system. The activation function $g_j(\cdot)$ of each output neuron should be a nonlinear function such as the sigmoid function $K\frac{1-e^{-\alpha x}}{1+e^{-\alpha x}}$ so that the network has the necessary nonlinear dynamics for identifying nonlinear systems. In addition, the constant K is used to provide values greater than unity. Note that the output neurons in Figure 8.7 and Figure 8.14 take a linear function as their activation function for simplicity.

Let W_{ij} denote the connection weight between the i'th neuron in the input layer and the j'th neuron in the output layer (for $i = 1, 2, \ldots, n_i$; $j = 1, 2, \ldots, M$).

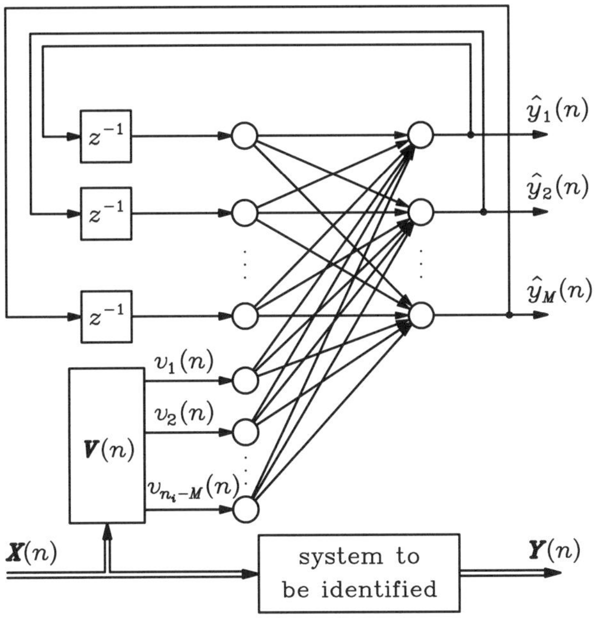

Figure 8.16. A simplified recurrent neural network with the one-step delayed vector.

Then the output of the network is

$$\hat{y}_j(n) = g_j\left(\sum_{i=1}^{n_i} W_{ij}u_i(n)\right) \qquad (j = 1, 2, \ldots, M). \tag{8.87}$$

In applying this simplified network to system identification, our task is still to determine the connection weights. Steps (1)–(5) listed above still apply to the determination of these weights. However, because the dimensions of the weights in the simplified network are much less than those in Figure 8.14, the computational complexity required in Figure 8.15 is much less than that in Figure 8.14.

In (8.73), $n_{\hat{y}}$-step delayed output vectors of the network are used. To simplify further the structure and algorithm of the network for system identification we can use only the one-step delayed vector (see Figure 8.16). In this case, the output of the network can be written as

$$\hat{y}_j(n) = g_j\left(\sum_{i=1}^{M} W_{ij}\hat{y}_i(n-1)\right) + \sum_{i=1}^{n_i-M} W_{ij}v_i(n) \qquad (j = 1, 2, \ldots, M), \tag{8.88}$$

where $v_i(n)$ is the i'th element of the vector $[\boldsymbol{X}^T(n-1), \boldsymbol{X}^T(n-2), \ldots, \boldsymbol{X}^T(n-n_x)]^T$.

Moreover,

$$\hat{y}_j(n) = g_j\left(\sum_{i=1}^{n_i} W_{ij}u_i(n)\right), \tag{8.89}$$

where

$$u_i(n) = \begin{cases} \hat{y}_i(n-1), & 1 \leq i \leq M, \\ v_{i-M}(n), & M+1 \leq i \leq n_i. \end{cases} \tag{8.90}$$

With these, we can simplify the algorithm (8.76) to train the network connection weights as [27]

$$W_{ij}(t+1) = W_{ij}(t) + \gamma \sum_{n=1}^{I}\sum_{l=1}^{M}(y_l(n) - \hat{y}_l(n))\xi_{ij}^{l}(n)$$
$$(i = 1, 2, \ldots, n_i; \quad j = 1, 2, \ldots, M), \tag{8.91}$$

where

$$\xi_{ij}^{l}(n+1) = g_l'\left(\sum_{i=1}^{n_i} W_{il}u_i(n)\right)\left(\delta_{lj}u_i(n) + \sum_{k=1}^{M} W_{kl}\xi_{ij}^{k}(n)\right) \tag{8.92}$$

with initial conditions

$$\xi_{ij}^{l}(0) = 0. \tag{8.93}$$

In (8.92), δ_{lj} is the Kronecker delta, which equals 1 when $l = j$ and zero otherwise. When each new input vector and output vector of the unknown system become available, recursively, (8.91) will become

$$W_{ij}(n+1) = W_{ij}(n) + \gamma \sum_{l=1}^{M}(y_l(n) - \hat{y}_l(n))\xi_{ij}^{l}(n)$$
$$(i = 1, 2, \ldots, n_i; \quad j = 1, 2, \ldots, M), \tag{8.94}$$

where $\xi_{ij}^{l}(n)$ is still computed by (8.92).

When these recurrent neural networks are used to identify the system represented by (8.6)–(8.9), we can also use the method of Section 8.2 (see Comment 3) to further simplify the architecture and algorithm. More important is that for Case 2 and Case 3 the coefficient matrices $\boldsymbol{A}_j$ and $\boldsymbol{B}_j$ can immediately be obtained by the available algorithms [3] without recurrent neural networks. Hence, recurrent neural networks are only used to determine the nonlinear parts of the system to be identified.

Finally, we should mention that in addition to the series-parallel identification method based on feedforward neural networks and the parallel identification method based on recurrent neural networks, fully parallel identification methods are worthy of investigation. In fully parallel identification methods, the input vector of neural networks may consist of three parts: the past values of the output vector of the network, the input vector, and the output vector of the system to be identified. This may offer more flexibility and better performance for identifying certain complicated nonlinear systems.

8.5 Neural Networks for Real-Time System Identification

In some high-speed applications such as radar and navigation systems, it is desirable to perform system identification as fast as possible. However, reaching this goal is not easy if the conventionally digital and sequentially computational techniques are employed mainly because the computational complexity involved in system identification is very extensive. Although many methods have been proposed to reduce the computational complexity, it remains difficult to deliver the desired real-time performance. In this section, we will show how to employ the high-speed computational capability of neural networks to attack this problem. As shown in Equation (8.4), system identification is in effect a multidimensional nonlinear optimization problem. This problem can in principle be solved in real time by analog neural networks for nonlinear programming [28]. However, because the operations involving nonlinear system identification are so complicated, the architectures and dynamic equations of analog neural networks for performing nonlinear system identification in real time are not well known and remain an open problem. Based on this, we emphasize in this section the presentation and development of neural network approaches for identifying linear systems. In the first subsection, we will deal with single-input single-output (SISO) systems. The generalization for multi-input multi-output (MIMO) systems will be given in the second subsection.

8.5.1 Neural Networks for Real-Time Identification of SISO Systems

Let us consider a general linear time-invariant SISO system represented by

$$y(n) = \sum_{i=1}^{n_y} a_i y(n-i) + \sum_{j=1}^{n_x} b_j x(n-j) + e(n), \tag{8.95}$$

where $y(n)$ and $x(n)$ are the output and the input of the system, respectively; $e(n)$ is the system noise; n_y and n_x are the lags in the output and input of the system, respectively; and a_i (for $i = 1, 2, \ldots, n_y$) and b_j (for $j = 1, 2, \ldots, n_x$) are the model coefficients.

It is worth mentioning that

(1) if $x(n)$ is white noise, then (8.95) is an ARMA model,
(2) if $n_x = 0$, then (8.95) is an AR model,
(3) if $n_y = 0$ and $x(n)$ is white noise, then (8.95) is an MA model, and
(4) if $n_y = 0$, then (8.95) is an FIR model.

As pointed out in Section 8.1, the identification problem of the system (8.95) consists of estimating the system parameters a_i (for $i = 1, 2, \ldots, n_y$) and b_j (for $j = 1, 2, \ldots, n_x$) from I known data of the input $x(n)$ and output $y(n)$ (for $n = 1, 2, \ldots, I$).

With these known data, we have the following equation:

$$\begin{pmatrix} y(I-1) & \cdots & y(I-n_y) & x(I-1) & \cdots & x(I-n_x) \\ y(I-2) & \cdots & y(I-n_y-1) & x(I-2) & \cdots & x(I-n_x-1) \\ \vdots & \ddots & \vdots & \vdots & \ddots & \vdots \\ y(n_m-1) & \cdots & y(n_m-n_y) & x(n_m-1) & \cdots & x(n_m-n_x) \end{pmatrix}$$

$$\times \begin{pmatrix} a_1 \\ a_2 \\ \vdots \\ a_{n_y} \\ b_1 \\ b_2 \\ \vdots \\ b_{n_x} \end{pmatrix} + \begin{pmatrix} e(I) \\ e(I-1) \\ \vdots \\ e(n_m) \end{pmatrix} = \begin{pmatrix} y(I) \\ y(I-1) \\ \vdots \\ y(n_m) \end{pmatrix}, \tag{8.96}$$

where $n_m = \max\{n_y, n_x\}$. Equation (8.96) can be rewritten in a matrix form

$$\boldsymbol{Y} = \boldsymbol{Z}\boldsymbol{W} + \boldsymbol{e}, \tag{8.97}$$

where

$$\boldsymbol{Y} = \begin{pmatrix} y(I) \\ y(I-1) \\ \vdots \\ y(n_m) \end{pmatrix}_{(I-n_m+1)\times 1}, \qquad \boldsymbol{e} = \begin{pmatrix} e(I) \\ e(I-1) \\ \vdots \\ e(n_m) \end{pmatrix}_{(I-n_m+1)\times 1}, \tag{8.98}$$

$$\boldsymbol{W} = \begin{pmatrix} a_1 \\ a_2 \\ \vdots \\ a_{n_y} \\ b_1 \\ b_2 \\ \vdots \\ b_{n_x} \end{pmatrix}_{(n_y+n_x)\times 1},$$

and

$$
\boldsymbol{Z} = \begin{pmatrix} y(I-1) & \cdots & y(I-n_y) & x(I-1) & \cdots & x(I-n_x) \\ y(I-2) & \cdots & y(I-n_y-1) & x(I-2) & \cdots & x(I-n_x-1) \\ \vdots & \ddots & \vdots & \vdots & \ddots & \vdots \\ y(n_m-1) & \cdots & y(n_m-n_y) & x(n_m-1) & \cdots & x(n_m-n_x) \end{pmatrix}. \tag{8.99}
$$

The dimension of $\boldsymbol{Z}$ is $(I-n_m+1)\times(n_y+n_x)$. $\boldsymbol{Z}$ and $\boldsymbol{Y}$, which are both known are referred to as the data matrix and data vector, respectively. $\boldsymbol{W}$ is the unknown coefficient vector, and the noise vector $\boldsymbol{e}$ is also unknown.

Note that the data matrix $\boldsymbol{Z}$ and data vector $\boldsymbol{Y}$ have other different forms if we take different data-window processing as shown in Section 3.1. For example,

$$
\boldsymbol{Y} = \begin{pmatrix} y(I) \\ y(I-1) \\ \vdots \\ y(2) \end{pmatrix}_{(I-1)\times 1}, \qquad \boldsymbol{e} = \begin{pmatrix} e(I) \\ e(I-1) \\ \vdots \\ e(2) \end{pmatrix}_{(I-1)\times 1},
$$

and

$$
\boldsymbol{Z} = \begin{pmatrix} y(I-1) & \cdots & y(I-n_y) & x(I-1) & \cdots & x(I-n_x) \\ y(I-2) & \cdots & y(I-n_y-1) & x(I-2) & \cdots & x(I-n_x-1) \\ \vdots & \ddots & \vdots & \vdots & \ddots & \vdots \\ y(1) & \cdots & 0 & x(1) & \cdots & 0 \end{pmatrix} \tag{8.100}
$$

where the dimension of $\boldsymbol{Z}$ is $(I-1)\times(n_y+n_x)$. In (8.100), all the values of $y(n)$ and $x(n)$ for $n \leq 0$ are set to zero. Different forms of the data matrix $\boldsymbol{Z}$ and data vector $\boldsymbol{Y}$ will give a different estimation performance.

The most common way to solve for the coefficient vector $\boldsymbol{W}$ from (8.97) is to use the least-squares criterion, that is,

$$
\min_{\boldsymbol{W}} \|\boldsymbol{Y} - \boldsymbol{Z}\boldsymbol{W}\|^2. \tag{8.101}
$$

The solution of (8.101) is

$$
\boldsymbol{W} = (\boldsymbol{Z}^T\boldsymbol{Z})^{-1}\boldsymbol{Z}^T\boldsymbol{Y}. \tag{8.102}
$$

Clearly, the computational complexity invested in (8.102) is very extensive.

Comparing (8.102) with (2.3) and (2.14), we see that if we let the data matrix $\boldsymbol{Z}$ and vector $\boldsymbol{Y}$ be taken as the connection strength matrix $\boldsymbol{T}$ and bias current vector $\boldsymbol{B}$ of the neural network of Figure 2.2, respectively (this means that the programming complexity is zero), then such a constructed network is stable and provides the solution

$$V_f = \left(\frac{1}{RK_1K_2}I + Z^T Z\right)^{-1} Z^T Y, \tag{8.103}$$

which is an approximation of (8.102). The error can be made arbitrarily small by appropriately selecting the parameters K_1, K_2, and R. As shown in Section 2.1, the time taken by this network to provide the solution is within an elapsed time of only a few characteristic time constants of the network (on the order of hundreds of nanoseconds). These guarantee that the neural network in Figure 2.2 can perform the computations required in (8.102) at a very high speed and with arbitrarily small error. As a result, this neural network approach is suitable for real-time identification of linear time-invariant SISO systems.

So far we have limited ourselves to the case in which $\boldsymbol{Z}$ and $\boldsymbol{Y}$ take real values. However, this neural network approach can immediately be generalized for the case in which $\boldsymbol{Z}$ and $\boldsymbol{Y}$ take complex values. In the complex-valued case, (8.102) becomes

$$W = (Z^H Z)^{-1} Z^H Y \tag{8.104}$$

and can be written in the real-valued form as

$$W_c = (Z_c^T Z_c)^{-1} Z_c^T Y_c, \tag{8.105}$$

where

$$W_c = \begin{pmatrix} W_r \\ W_i \end{pmatrix}, \quad Z_c = \begin{pmatrix} Z_r & -Z_i \\ Z_i & Z_r \end{pmatrix}, \quad \text{and} \quad Y_c = \begin{pmatrix} Y_r \\ Y_i \end{pmatrix}$$

and where $\boldsymbol{W}$, $\boldsymbol{Z}$, and $\boldsymbol{Y}$ are broken down into real and imaginary parts

$$W = W_r + jW_i, \quad Z = Z_r + jZ_i, \quad \text{and} \quad Y = Y_r + jY_i. \tag{8.106}$$

With these in mind, we let the data matrix $\boldsymbol{Z}_c$ and vector $\boldsymbol{Y}_c$ be taken as the connection strength matrix $\boldsymbol{T}$ and bias current vector $\boldsymbol{B}$ of the neural network of Figure 2.2, respectively. Then such a constructed network is stable and provides the solution

$$V_f = \left(\frac{1}{RK_1K_2}I + Z_c^T Z_c\right)^{-1} Z_c^T Y_c, \tag{8.107}$$

which is an approximation to (8.105).

It is easy to see that the number of neurons required for a complex-valued system is twice that required for a real-valued system.

Note that to obtain a better performance, we can also use the total-least-squares criterion and the related neural network given in Section 2.4 to find the system parameters $\boldsymbol{W}$. The price incurred by the TLS method is that the architecture of the network is more complicated than that of the LS method.

8.5.2 Neural Networks for Real-Time Identification of MIMO Systems

Consider a linear time-invariant system with N inputs and M outputs represented by (8.6), that is,

$$\boldsymbol{Y}(n) = \sum_{i=1}^{n_y} \boldsymbol{A}_i \boldsymbol{Y}(n-i) + \sum_{j=1}^{n_x} \boldsymbol{B}_j \boldsymbol{X}(n-j) + \boldsymbol{e}(n). \tag{8.108}$$

Note that the system noise vector is included in (8.108). The various quantities in (8.108) are exactly the same as those in (8.6).

Assume that I samples of the input vector $\boldsymbol{X}(n)$ and the output vector $\boldsymbol{Y}(n)$ of the system to be identified are available, that is, $\boldsymbol{X}(n)$ and $\boldsymbol{Y}(n)$ (for $n = 1, 2, \ldots, I$) are known. Our task is to find $(M \times M)$-dimensional matrices $\boldsymbol{A}_i$ and $(M \times N)$-dimensional matrices $\boldsymbol{B}_j$ (for $i = 1, 2, \ldots, n_y$; $j = 1, 2, \ldots, n_x$) using these available data.

Similar to the matrix form (8.97) we have

$$\boldsymbol{Y} = \boldsymbol{W}\boldsymbol{Z} + \boldsymbol{E}, \tag{8.109}$$

where

$$\boldsymbol{Y} = \left[\boldsymbol{Y}(I), \boldsymbol{Y}(I-1), \ldots, \boldsymbol{Y}(n_m)\right]_{M\times(I-n_m+1)}, \tag{8.110}$$

$$\boldsymbol{E} = \left[\boldsymbol{e}(I), \boldsymbol{e}(I-1), \ldots, \boldsymbol{e}(n_m)\right]_{M\times(I-n_m+1)}, \tag{8.111}$$

$$\boldsymbol{W} = \left[\boldsymbol{A}_1, \boldsymbol{A}_2, \ldots, \boldsymbol{A}_{n_y}, \boldsymbol{B}_1, \boldsymbol{B}_2, \ldots, \boldsymbol{B}_{n_x}\right]_{M\times(n_y M + n_x N)}, \tag{8.112}$$

and

$$\boldsymbol{Z} = \begin{pmatrix} \boldsymbol{Y}(I-1) & \boldsymbol{Y}(I-2) & \cdots & \boldsymbol{Y}(n_m-1) \\ \boldsymbol{Y}(I-2) & \boldsymbol{Y}(I-3) & \cdots & \boldsymbol{Y}(n_m-2) \\ \vdots & \vdots & \ddots & \vdots \\ \boldsymbol{Y}(I-n_y) & \boldsymbol{Y}(I-n_y-1) & \cdots & \boldsymbol{Y}(n_m-n_y) \\ \boldsymbol{X}(I-1) & \boldsymbol{X}(I-2) & \cdots & \boldsymbol{X}(n_m-1) \\ \boldsymbol{X}(I-2) & \boldsymbol{X}(I-3) & \cdots & \boldsymbol{X}(n_m-2) \\ \vdots & \vdots & \ddots & \vdots \\ \boldsymbol{X}(I-n_x) & \boldsymbol{X}(I-n_x-1) & \cdots & \boldsymbol{X}(n_m-n_x) \end{pmatrix}_{(n_y M + n_x N)\times(I-n_m+1)} \tag{8.113}$$

Note that $\boldsymbol{Y}$, $\boldsymbol{E}$, and $\boldsymbol{Z}$ can take other forms such as

$$\boldsymbol{Y} = \left[\boldsymbol{Y}(I), \boldsymbol{Y}(I-1), \ldots, \boldsymbol{Y}(2)\right]_{M\times(I-1)}, \tag{8.114}$$

$$\boldsymbol{E} = \left[\boldsymbol{e}(I), \boldsymbol{e}(I-1), \ldots, \boldsymbol{e}(2)\right]_{M\times(I-1)}, \tag{8.115}$$

and

$$\boldsymbol{Z} = \begin{pmatrix} \boldsymbol{Y}(I-1) & \boldsymbol{Y}(I-2) & \cdots & \boldsymbol{Y}(1) \\ \boldsymbol{Y}(I-2) & \boldsymbol{Y}(I-3) & \cdots & \boldsymbol{0} \\ \vdots & \vdots & \ddots & \vdots \\ \boldsymbol{Y}(I-n_y) & \boldsymbol{Y}(I-n_y-1) & \cdots & \boldsymbol{0} \\ \boldsymbol{X}(I-1) & \boldsymbol{X}(I-2) & \cdots & \boldsymbol{X}(1) \\ \boldsymbol{X}(I-2) & \boldsymbol{X}(I-3) & \cdots & \boldsymbol{0} \\ \vdots & \vdots & \ddots & \vdots \\ \boldsymbol{X}(I-n_x) & \boldsymbol{X}(I-n_x-1) & \cdots & \boldsymbol{0} \end{pmatrix}_{(n_y M + n_x N)\times(I-1)}. \tag{8.116}$$

In (8.109), $\boldsymbol{Z}$ and $\boldsymbol{Y}$ are known but $\boldsymbol{W}$ and $\boldsymbol{E}$ are not. The matrix $\boldsymbol{W}$ can be obtained by solving the optimization problem

$$\min_{\boldsymbol{W}} tr\{(\boldsymbol{Y} - \boldsymbol{W}\boldsymbol{Z})(\boldsymbol{Y} - \boldsymbol{W}\boldsymbol{Z})^T\}. \tag{8.117}$$

Its solution is

$$\boldsymbol{W} = \boldsymbol{Y}\boldsymbol{Z}^{+}. \tag{8.118}$$

Equation (8.118) has exactly the same form as (3.88). As a result, we can use the 2-D neural network proposed in Section 3.3 to solve (8.118) in real time by taking $\boldsymbol{Z}$ and $\boldsymbol{Y}$ as the connection strength matrix $\boldsymbol{T}$ and bias current matrix $\boldsymbol{B}$ of the network, respectively, and selecting the other parameters to be the same as for (3.88). The network will provide the solution in the stationary state as

$$\boldsymbol{V}_f = \lim_{t\to\infty} \boldsymbol{V}(t) = \boldsymbol{Y}\boldsymbol{Z}^T\left(\frac{1}{RK_1K_2}\boldsymbol{I} + \boldsymbol{Z}\boldsymbol{Z}^T\right)^{-1}, \tag{8.119}$$

which approximates the exact solution (8.118). The error between (8.118) and (8.119) can be made arbitrarily small by appropriately selecting the other parameters of the network such as R, K_1, and K_2.

For a complex-valued system, the connection strength matrix $\boldsymbol{T}$ and bias current matrix $\boldsymbol{B}$ of the 2-D network will be selected as $\boldsymbol{Z}_c$ and $\boldsymbol{Y}_c$, respectively, where

$$\boldsymbol{Y}_c = \begin{pmatrix} \boldsymbol{Y}_r & \boldsymbol{Y}_i \end{pmatrix} \tag{8.120}$$

and

$$\boldsymbol{Z}_c = \begin{pmatrix} \boldsymbol{Z}_r & \boldsymbol{Z}_i \\ -\boldsymbol{Z}_i & \boldsymbol{Z}_r \end{pmatrix}. \tag{8.121}$$

The matrices Z_r, Y_r and Z_i, Y_i are the real and imaginary parts of Z and Y, respectively. The network will provide the solution

$$V_f = Y_c Z_c^T \left(\frac{1}{RK_1K_2} I + Z_c Z_c^T \right)^{-1}, \tag{8.122}$$

which is a very close approximation of the exact solution $W_c = [W_r \; W_i]$, where W_r and W_i are the real and imaginary parts of W with complex values given in (8.118); that is,

$$W = YZ^+ = \lim_{\alpha \to 0} YZ^H (\alpha I + ZZ^H)^{-1}. \tag{8.123}$$

Because the available samples of the input vectors and output vectors of the system to be identified are directly taken as the connection strength matrix and basis current matrix of the network without any computation, the programming complexity for this network is zero. Together with the high-speed computational capability of the 2-D neural network depicted in Section 3.3, the above argument shows that this 2-D neural network approach is very suitable for real-time identification of linear MIMO systems.

8.6 Blind System Identification and Neural Networks

In the previous sections of this chapter, knowledge of the inputs and outputs of the system to be identified is available. However, in many practical applications, knowledge of the inputs of the system is not available or not sufficient [29]. For example, in seismic signal processing, we use the received signal to estimate the sequence of reflection coefficients corresponding to the various layers of the earth model. The received signal is itself made up of echos produced at the different layers of the model and is considered to result from the convolution of the reflection coefficients with the excitation, which is ordinarily in the form of a short-duration pulse. The problem is that the exact waveform of the excitation responsible for the generation of the received signal is usually unknown. A similar problem arises in image restoration. In image restoration, we have an unknown system that represents blurring effects caused by photographic or electronic imperfections or both. An original image of interest constitutes the system input. The system output is a blurred version of the original image. Given the blurred image, the task is to restore the original image. To achieve this purpose, we first have to identify the system without knowledge of the original image. Another example is the binary signal reconstruction without transmitting the training data (see Section 5.5). In these cases, blind identification techniques are required. Hardly a new topic, blind system identification has been extensively studied during the two decades since Sato first proposed the innovative idea of self-recovering adaptive identification

[30]. The available blind identification methods can be divided into two groups: statistical methods and deterministic methods [31, 32]. Statistical methods identify systems by assuming the inputs of systems to be random signals. Although these statistical methods do not require explicit knowledge of the system inputs, they do require certain statistical assumptions on the system inputs (e.g, the inputs are assumed to be white noise). In deterministic methods, system inputs are treated as arbitrary deterministic signals. Because these blind identification methods work under the condition that prior knowledge is either unavailable or insufficient, they can be considered as a self-organizing learning process (or say, unsupervised learning methods) of neural networks. Nonetheless, much may be gained for blind identification from neural networks, in particular, that part of the subject that deals with self-organization and unsupervised learning [33]. This may include two parts. One part is to use the principles of the available self-organization and unsupervised learning algorithms of neural networks to develop new methods for blind identification. The other part is to employ the high-speed computational capability and nonlinear mapping capability of neural networks to implement blind identification algorithms. However, this very preliminary idea merits much more investigation. As a further illustration concerning the applications of neural networks to blind identification, we will now present an example.

Let us consider the blind identification approach of multichannel linear FIR filters proposed in Reference [31]. This blind identification approach mainly includes the following steps:

(1) Estimation of the autocorrelation matrix related to the output vector of the system to be identified by

$$\boldsymbol{R} = \frac{1}{I}\sum_{n=1}^{I} \boldsymbol{G}(n)\boldsymbol{G}^H(n), \tag{8.124}$$

where $\boldsymbol{G}(n)$ (for $n = 1, 2, \ldots, I$) is an L-dimensional vector and can be immediately obtained by the available samples of the output vector $\boldsymbol{Y}(n)$ of the unknown system.

(2) Eigendecomposition of the matrix $\boldsymbol{R}$, that is,

$$\boldsymbol{R} = \sum_{i=1}^{L} \lambda_i \boldsymbol{S}_i \boldsymbol{S}_i^H, \tag{8.125}$$

where $0 < \lambda_1 \leq \lambda_2 \leq \cdots \leq \lambda_L$ are the eigenvalues and $\boldsymbol{S}_1, \boldsymbol{S}_2, \ldots, \boldsymbol{S}_L$ are the corresponding orthonormal eigenvectors of the matrix $\boldsymbol{R}$.

Under appropriate conditions (see [31]), any column vector of the matrix consisting of the coefficient set $\boldsymbol{W}$ of the system is orthogonal to the subspace spanned by the eigenvectors corresponding to the P smallest eigenvalues of the matrix $\boldsymbol{R}$.

(3) Estimation of the coefficient set $\boldsymbol{W}$ by finding the values corresponding to the minimization of the function

$$f(\boldsymbol{W}) = \boldsymbol{W}^H \boldsymbol{Q} \boldsymbol{W}, \tag{8.126}$$

where $\boldsymbol{Q}$ is a matrix determined completely by the eigenvectors corresponding to the P smallest eigenvalues of the matrix $\boldsymbol{R}$.

Clearly, this algorithm is similar to the MUSIC algorithm presented in Section 7.2, and the key problem is to perform the eigendecomposition so as to find the desired eigenvectors. This task can be accomplished by using the neural networks proposed in Figure 7.6 or in Figure 7.7, where we take the matrix $\boldsymbol{R}$ as the connection strength matrix $\boldsymbol{T}$ of these networks. In contrast, if each sample of the output vector of the system becomes available recursively, the MCA networks proposed in Section 6.4 can be used to extract adaptively the necessary eigenvectors from the vector $\boldsymbol{G}(n)$. For example, we can use the invariant-norm MCA algorithm as

$$\begin{aligned}\boldsymbol{W}_{jc}(n+1) = \boldsymbol{W}_{jc}(n) - \gamma(n)\Bigg\{ & \boldsymbol{W}_{jc}^T(n)\boldsymbol{W}_{jc}(n)\Bigg[\boldsymbol{G}_c(n)\boldsymbol{z}_{jc}(n) \\ & - \sum_{i=1}^{j-1} \boldsymbol{z}_{jc}^T(n)\boldsymbol{z}_{ic}(n)\boldsymbol{W}_{ic}(n)\Bigg] - \boldsymbol{z}_{jc}^T(n)\boldsymbol{z}_{jc}(n)\boldsymbol{W}_{jc}(n) \\ & + \sum_{i=1}^{j-1} \boldsymbol{z}_{jc}^T(n)\boldsymbol{z}_{ic}(n)\boldsymbol{W}_{jc}^T(n)\boldsymbol{W}_{ic}(n)\boldsymbol{W}_{jc}(n)\Bigg\} \\ & (j = 1, 2, \ldots, P),\end{aligned} \tag{8.127}$$

where

$$\boldsymbol{z}_{jc}(n) = \begin{pmatrix} z_{jr}(n) \\ z_{ji}(n) \end{pmatrix}, \qquad \boldsymbol{W}_{jc}(n) = \begin{pmatrix} \boldsymbol{W}_{jr}(n) \\ \boldsymbol{W}_{ji}(n) \end{pmatrix}$$

and

$$z_{jr}(n) = \boldsymbol{W}_{jr}^T(n)\boldsymbol{G}_r(n) + \boldsymbol{W}_{ji}^T(n)\boldsymbol{G}_i(n), \tag{8.128}$$

$$z_{ji}(n) = -\boldsymbol{W}_{jr}^T(n)\boldsymbol{G}_i(n) + \boldsymbol{W}_{ji}^T(n)\boldsymbol{G}_r(n) \qquad (j = 1, 2, \ldots, P). \tag{8.129}$$

Here $\boldsymbol{G}_r(n)$ and $\boldsymbol{G}_i(n)$ are the real and imaginary parts of the vector $\boldsymbol{G}(n)$, respectively.

With the conclusions given in Section 6.4, the weight vector $\boldsymbol{W}_{jc}$ will converge to the desired eigenvector.

It is easy to see that the MCA approaches have avoided the computations required in the first step (the estimation of an autocorrelation matrix) of the blind identification algorithm. However, the third step is still necessary for finding the coefficient set $\boldsymbol{W}$. The authors are currently developing a neural network approach that can provide the desired system coefficient set $\boldsymbol{W}$ adaptively and directly from the sampled data of the system outputs without any programming complexity, that is, this neural network can perform all computations (the estimation of the autocorrelation matrix, eigendecomposition, and the computations of (8.126)) required in the above blind identification algorithm. This envisaged neural network will have two layers: The first layer will perform the desired eigendecomposition with the sampled data as the inputs; the second layer will provide the coefficients of the system to be identified. This neural network can also be used to perform simultaneously all the computations required in the MUSIC and other eigenstructure-based algorithms.

Because eigendecomposition is also a key step in the deterministic approach for blind identification proposed in Reference [32], the above eigendecomposition neural networks can immediately be applied to this deterministic approach.

As pointed out in Section 8.5, if the inputs are white noise, then the general model (8.108) will become an ARMA model. Consequently, the system identification becomes the ARMA model parameter estimation. Usually, the model parameters can be determined by using the sampled data and the second-order (for minimum phase systems) or higher-order statistics (for nonminimum phase systems) of the system outputs. In this sense, the related parameter estimation can be considered as blind identification. From Chapter 3, we learned that the main computation required in the ARMA model parameter estimation is in solving linear equations under some criteria (such as the LS criterion and the TLS criterion). This suggests that we can use the neural network of Figure 2.2 and the 2-D neural network of Figure 3.6 to perform the major computations in the related ARMA model parameter estimation methods.

Bibliography

[1] Pillai, S. U. and Shim, T. I., *Spectrum Estimation and System Identification*, Springer–Verlag, New York, 1993.

[2] Soederstroem, T. and Storica, P., *System Identification,* Prentice Hall, Englewood Cliffs, NJ, 1989.

[3] Narendra, K. S. and Parthasarathy, K., "Identification and Control of Dynamical Systems Using Neural Networks," *IEEE Trans. on Neural Networks*, Vol. 1, No. 1, 1990, pp. 4–26.

[4] Chen, S. and Billings, S. A., "Modelling and Analysis of Nonlinear Time Series," *Int. J. Control*, Vol. 50, No. 6, 1989, pp. 2151–71.

[5] Chen, S., Billings, S. A., and Grant, P. M., "Nonlinear System Identification Using Neural Networks," *Int. J. Control*, Vol. 51, No. 6, 1990, pp. 1191–214.

[6] Chen, S. and Billings, S. A., "Neural Networks for Nonlinear Dynamic System Modelling and Identification," *Int. J. Control*, Vol. 56, No. 2, 1992, pp. 319–46.

[7] Chen, S., Billings, S. A., and Grant, P. M., "Parallel Recursive Prediction Error Algorithm for Training Layered Neural Networks," *Int. J. Control*, Vol. 51, No. 6, 1990, pp. 1215–28.

[8] Salgado, M. E., Goodwin, G. C., and Middleton, R. H., "Modified Least Squares Algorithm Incorporating Exponential Resetting and Forgetting," *Int. J. Control*, Vol. 47, No. 7, 1988, pp. 477–91.

[9] Ljung, L., "Analysis of Recursive Stochastic Algorithm," *IEEE Trans. on Automatic Control*, Vol. 22, 1977, pp. 551–75.

[10] Ljung, L., "Convergence Analysis of Parametric Identification Methods," *IEEE Trans. on Automatic Control*, Vol. 23, 1978, pp. 770–83.

[11] Goodwin, G. C. and Payne, R. L., *Dynamic System Identification: Experiment Design and Data Analysis*, Academic Press, New York, 1977.

[12] Billings, S. A. and Voon, W. S. F., "Correlation Based Model Validity Tests for Nonlinear Models," *Int. J. Control*, Vol. 44, 1986, pp. 235–44.

[13] Billings, S. A. and Chen, S., "Identification of Nonlinear Rational System Using a Prediction-Error Estimation Algorithm," *Int. J. System Science*, Vol. 20, 1989, pp. 467–94.

[14] Leontarities, I. J. and Billings, S. A., "Model Selection and Validation Methods for Nonlinear System," *Int. J. of Control*, Vol. 45, 1987, pp. 311–41.

[15] Bohlin, T., "Maximum-Power Validation of Models Without Higher-Order Fitting," *Automatica,* Vol. 4, 1978, pp. 137–46.

[16] Leonard, J. A., and Kramer, M. A., "Radial Basis Function Networks for Classifying Process Faults," *IEEE Control Systems Magazine*, April 1991, pp. 31–38.

[17] Moody, J. E. and Darken, C. J., "Fast Learning in Networks of Locally Tuned Processing Units," *Neural Computation*, Vol. 1, 1989, pp. 281–94.

[18] Chen, S., Billings, S. A., Cowan, C. F. N., and Grant, P. M., "Practical Identification of NARMAX Models Using Radial Basis Function Networks," *Int. J. Control*, Vol. 52, No. 6, 1990, pp. 1327–50.

[19] Tan, S., Hao, J., and Vandewalle, J., "Efficient Identification of RBF Neural Net Models for Nonlinear Discrete-Time Multivariable Dynamical Systems," *Neurocomputing*, Vol. 9, No. 1, 1995, pp. 11–26.

[20] Chen, S., Cowan, C. F. N., and Grant, P. M., "Orthogonal Least Squares Learning Algorithm for Radial Basis Function Networks," *IEEE Trans. on Neural Networks*, Vol. 2, 1991, pp. 302–9.

[21] Chen, S., Billings, S. A., and Luo, W., "Orthogonal Least Squares Methods and Their Applications to Nonlinear System Identification," *Int. J. Control*, Vol. 50, No. 5, 1989, pp. 1873–96.

[22] Gorinevsky, D., "On the Persistency of Excitation in Radial Basis Function Network Identification of Nonlinear Systems," *IEEE Trans. on Neural Networks*, Vol. 6, No. 5, 1995, pp. 1237–44.

[23] Kosmatopoulos, E. B., Polycarpou, M. M., Christodoulo, M. A., and Ioannou, P. A., "High-Order Neural Network Structure for Identification of Dynamic Systems," *IEEE Trans. on Neural Networks*, Vol. 6, No. 2, 1995, pp. 422–31.

[24] Kuschewski, J. G., Hui, S., and Zak, S. H., "Application of Feedforward Neural Networks to Dynamical System Identification and Control," *IEEE Trans. on Control Systems Technology*, Vol. 1, No. 1, 1993, pp. 37–49.

[25] Giles, C. L., Kuhn, G. M., and Williams, R. J., "Dynamic Recurrent Neural Networks," *IEEE Trans. on Neural Networks*, Vol. 5, No. 2, 1994, pp. 153–55.

[26] Qin, S. Z., Su, H. T., and McAvoy, T. J., "Comparison of Four Neural Network Learning

Methods for Dynamic System Identification," *IEEE Trans. on Neural Networks*, Vol. 3. No. 1, 1992, pp. 123–30.

[27] Haykin, S., *Neural Networks, A Comprehensive Foundation*, IEEE Press, New York, 1994.

[28] Cichocki, A. and Unbehauen, R., *Neural Networks for Optimization and Signal Processing*, Wiley Teubner Verlag, Chichester, England, 1993.

[29] Haykin, S., *Blind Deconvolution*, Prentice Hall, Englewood Cliffs, NJ, 1994.

[30] Sato, Y., "A Method of Self-Recovering Equalization for Multilevel Amplitude-Modulations," *IEEE Trans. on Communications*, Vol. 23, No. 6, 1975, pp. 679–82.

[31] Moulines, E., Duhamel, P., Cardoso, J., and Mayrargue, S., "Subspace Methods for the Blind Identification of Multi-Channel FIR Filters," *Proc. IEEE Int. Conf. on Acoustics, Speech and Signal Processing*, Vol. 4, 1994, pp. 573–77.

[32] Liu, H., Xu, G., and Tong, L., "A Deterministic Approach to Blind Identification of Multi-Channel FIR Systems," *Proc. IEEE Int. Conf. on Acoustics, Speech and Signal Processing*, Vol. 4, 1994, pp. 581–84.

[33] Haykin, S., "Blind Equalization Formulated as a Self-Organized Learning Process," *Proc. Twenty-Sixth Asilomar Conf. on Signals, Systems and Computers*, Pacific Grove, CA, 1992, pp. 346–50.

9

Neural Networks for Signal Compression

Signal compression is used to achieve a low bit rate in the digital representation of signals with a minimum loss of signal quality. The function of compression is usually referred to as low bit rate coding or coding, for short. Signal compression has found a wide use and is playing more and more important roles in many respects of signal communications, signal storage, and message encryption [1, 2, 3, 4]. However, despite the efforts of many well-established scientists working in this field, a good compromise involving quality, complexity, and compression ratio has not yet been reached. Recently, there has been a tremendous interest in applying neural networks to signal compression, and some promising results have been reported [5, 6, 7]. The characteristics of neural networks involved in signal compression include massively parallel structures; a high-degree of interconnection; the propensity for storing experiential knowledge; and the capabilities of high-speed computation, nonlinear mapping, and self-organization [5]. The high-speed computational capability of neural networks can be employed to implement real-time signal compression. Furthermore, the nonlinear nature of neural networks can be exploited to design nonlinear predictors for predictive coding. In addition, self-organizing systems are very suitable for adaptive coding and can offer many advantages over traditional coding methods.

In this chapter, we intend to give a relatively detailed presentation of applications of neural networks to signal compression. Section 9.1 deals with the real-time implementation of the optimal linear predictive coding algorithm. In Sections 9.2 and 9.3, we will discuss MLP networks and high-order networks for nonlinear predictive coding systems, respectively. Section 9.4 shows how to use neural networks to perform the eigendecomposition required in the optimal Karhunen–Loève transform coding. Section 9.5 focuses on neural network approaches for wavelet transform coding. In Section 9.6, we will show how to apply self-organizing systems to vector quantization.

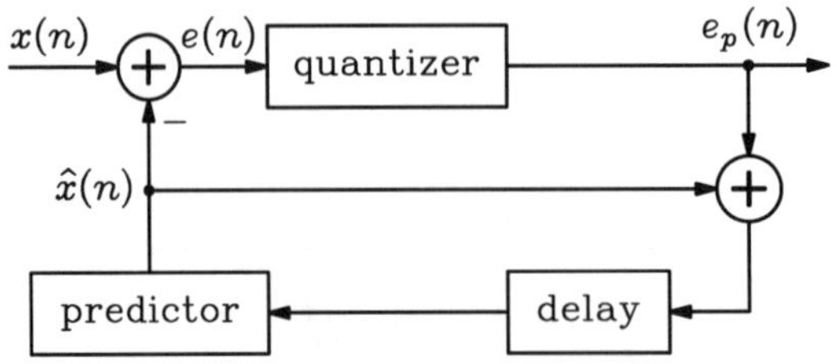

Figure 9.1. A predictive coding system.

9.1 Neural Networks for Linear Predictive Coding

Predictive coding systems have been commonly used for the encoding of speech, image, and video signals. Predictive coding systems primarily employ the correlation property of signals to achieve the desired compression, because correlation implies a redundancy in the raw data [8, 9]. Figure 9.1 shows a block diagram of a predictive coding system. The predictor uses P past samples $x(n-1)$, $x(n-2), \ldots, x(n-P)$ of a signal to estimate the value of the current sample $x(n)$. If we denote the estimation of $x(n)$ by $\hat{x}(n)$, then the difference is $e(n) = \hat{x}(n) - x(n)$, which is used for storage and transmission. It is obvious that the bit rate required for representing the difference $e(n)$ is lower than that for the original signal $x(n)$. As the accuracy of the predictor increases, the variance of the difference will decrease so as to deliver a higher compression ratio. Thus the key problems for predictive coding involve designing and implementing the predictor.

The simplest predictor is a linear predictor:

$$\hat{x}(n) = \sum_{i=1}^{P} a_i x(n-i) \tag{9.1}$$

and

$$e(n) = \sum_{i=0}^{P} a_i x(n-i), \tag{9.2}$$

where a_i (for $i = 0, 1, \ldots, P$) is the prediction coefficient and P is the order. Note that $a_0 = -1$. Under the criterion of minimizing the mean squared error $E[e^2(n)]$, the coefficients satisfy

$$\sum_{k=0}^{P} a_k R(m-k) = 0 \qquad (m = 1, 2, \ldots, P), \tag{9.3}$$

where $R(k) = E[x(n)x^*(n+k)]$ is the autocorrelation function of $x(n)$. This can be written in a matrix form

$$\begin{pmatrix} R(0) & R(-1) & \cdots & R(1-P) \\ R(1) & R(0) & \cdots & R(2-P) \\ \vdots & \vdots & \ddots & \vdots \\ R(P-1) & R(P-2) & \cdots & R(0) \end{pmatrix} \begin{pmatrix} a_1 \\ a_2 \\ \vdots \\ a_P \end{pmatrix} = \begin{pmatrix} R(1) \\ R(2) \\ \vdots \\ R(P) \end{pmatrix}. \qquad (9.4)$$

Once the $P+1$ autocorrelation values $R(k)$ (for $k = 0, 1, \ldots, P$) are known (note that $R(-k) = R^*(k)$), the coefficients a_i (for $i = 1, 2, \ldots, P$) can be computed by solving (9.4), and the linear predictor is thereby determined. However, in most practical applications, the required autocorrelation values are not available, but the samples $x(0), x(1), \ldots, x(N-1)$ are available. In this case, the procedure to design the predictor is as follows:

(1) Estimation of the $P+1$ autocorrelation values from the available samples $x(0), x(1), \ldots, x(N-1)$,

$$\hat{R}(k) = \frac{1}{N} \sum_{n=0}^{N-k-1} x^*(n+k)x(n) \qquad (k = 0, 1, \ldots, P). \qquad (9.5)$$

(2) Construction of the equations corresponding to (9.3) by using (9.5) and the relationship $\hat{R}(-k) = \hat{R}^*(k)$, that is,

$$\sum_{k=1}^{P} a_k \hat{R}(m-k) = \hat{R}(m) \qquad (m = 1, 2, \ldots, P) \qquad (9.6)$$

or, in a matrix form,

$$\begin{pmatrix} \hat{R}(0) & \hat{R}(-1) & \cdots & \hat{R}(1-P) \\ \hat{R}(1) & \hat{R}(0) & \cdots & \hat{R}(2-P) \\ \vdots & \vdots & \ddots & \vdots \\ \hat{R}(P-1) & \hat{R}(P-2) & \cdots & \hat{R}(0) \end{pmatrix} \begin{pmatrix} a_1 \\ a_2 \\ \vdots \\ a_P \end{pmatrix} = \begin{pmatrix} \hat{R}(1) \\ \hat{R}(2) \\ \vdots \\ \hat{R}(P) \end{pmatrix}. \qquad (9.7)$$

(3) Determination of the coefficients a_i (for $i = 1, 2, \ldots, P$) by solving Equation (9.7).

Substituting (9.5) into (9.7), we get

$$\boldsymbol{X}^H \boldsymbol{X} \boldsymbol{A} = \boldsymbol{X}^H \boldsymbol{X}_1, \qquad (9.8)$$

where

$$X = \begin{pmatrix} x(0) & 0 & \cdots & 0 \\ x(1) & x(0) & \cdots & 0 \\ \vdots & \vdots & \ddots & \vdots \\ x(P-1) & x(P-2) & \cdots & x(0) \\ \vdots & \vdots & \ddots & \vdots \\ x(N-1) & x(N-2) & \cdots & x(N-P) \\ \vdots & \vdots & \ddots & \vdots \\ 0 & \cdots & x(0) & x(1) \\ 0 & \cdots & 0 & x(0) \end{pmatrix}_{(N+P)\times P}, \tag{9.9}$$

$$A = \begin{pmatrix} a_1 \\ a_2 \\ \vdots \\ a_P \end{pmatrix}_{P\times 1}, \quad \text{and} \quad X_1 = \begin{pmatrix} x(1) \\ x(2) \\ \vdots \\ x(N-1) \\ 0 \\ \vdots \\ 0 \end{pmatrix}_{(N+P)\times 1}. \tag{9.10}$$

X and X_1 are referred to as the data matrix and data vector, respectively, and A is the coefficient vector. It is easy to show that (9.8) can also be obtained under the LS criterion, that is,

$$\min_{A} \|XA - X_1\|^2. \tag{9.11}$$

This is a special case (because X and X_1 are correlated) of the LS problem (2.2). Consequently, the neural network in Figure 2.2 can immediately be used to solve (9.8) provided that we let the connection strength matrix T be the data matrix X, the bias current vector B be the data vector X_1, and the other parameters of the network be the same as for (2.2). More details of the dynamic and stationary performance of the neural network for solving (9.8) can be found in Section 2.1. Although a large number of effective and fast algorithms to solve (9.8) are available (for example, the Levinson–Durbin algorithm shown in Figure 3.1), using the neural network in Figure 2.2 can provide the solution at a much higher speed. This is desirable in high-speed communication systems such as videophones and videoconferencing.

As pointed out in Section 2.1, the neural network of Figure 2.2 does not apply to the case in which the number of samples involving the predictor increases as time progresses. In this case, we should use the neural network for the RLS algorithm (presented in Section 2.2): At time n ($n > P$), (9.11) becomes

$$\min_{A_n} \sum_{i=P}^{n} (x(i) - \hat{x}(i))^2 = \min_{A_n} \|\boldsymbol{X}_n \boldsymbol{A}_n - \boldsymbol{X}_{1,n}\|^2, \tag{9.12}$$

where $\boldsymbol{A}_n$ is the coefficient vector of the predictor at time n and

$$\boldsymbol{X}_n = \begin{pmatrix} x(P-1) & x(P-2) & \cdots & x(0) \\ x(P) & x(P-1) & \cdots & x(1) \\ \vdots & \vdots & \ddots & \vdots \\ x(n-1) & x(n-2) & \cdots & x(n-P) \end{pmatrix}_{(n-P+1)\times P}, \tag{9.13}$$

$$\boldsymbol{A}_n = \begin{pmatrix} a_{1,n} \\ a_{2,n} \\ \vdots \\ a_{P,n} \end{pmatrix}_{P\times 1}, \quad \text{and} \quad \boldsymbol{X}_{1,n} = \begin{pmatrix} x(P) \\ x(P+1) \\ \vdots \\ x(n) \end{pmatrix}_{(n-P+1)\times 1}. \tag{9.14}$$

The solution of (9.12) is

$$\boldsymbol{A}_n = \boldsymbol{X}_n^{+} \boldsymbol{X}_{1,n} = \lim_{\alpha \to 0} \left(\boldsymbol{X}_n^H \boldsymbol{X}_n + \alpha \boldsymbol{I}\right)^{-1} \boldsymbol{X}_n^H \boldsymbol{X}_{1,n}. \tag{9.15}$$

Consequently, at time $n+1$, we have

$$\boldsymbol{X}_{n+1} = \begin{pmatrix} x(P-1) & x(P-2) & \cdots & x(0) \\ x(P) & x(P-1) & \cdots & x(1) \\ \vdots & \vdots & \ddots & \vdots \\ x(n-1) & x(n-2) & \cdots & x(n-P) \\ x(n) & x(n-1) & \cdots & x(n-P+1) \end{pmatrix}_{(n-P+2)\times P}, \tag{9.16}$$

$$\boldsymbol{A}_{n+1} = \begin{pmatrix} a_{1,n+1} \\ a_{2,n+1} \\ \vdots \\ a_{P,n+1} \end{pmatrix}_{P\times 1}, \quad \text{and} \quad \boldsymbol{X}_{1,n+1} = \begin{pmatrix} x(P) \\ x(P+1) \\ \vdots \\ x(n) \\ x(n+1) \end{pmatrix}_{(n-P+2)\times 1}. \tag{9.17}$$

The solution of (9.12) at time $n+1$ is

$$\boldsymbol{A}_{n+1} = \boldsymbol{X}_{n+1}^{+}\boldsymbol{X}_{1,n+1} = \lim_{\alpha\to 0}\left(\boldsymbol{X}_{n+1}^{H}\boldsymbol{X}_{n+1} + \alpha\boldsymbol{I}\right)^{-1}\boldsymbol{X}_{n+1}^{H}\boldsymbol{X}_{1,n+1}. \tag{9.18}$$

We use the continuous-time Hopfield neural network in Figure 1.2 by selecting the neurons with a linear input–output relationship function $g(u) = K_1 u$ instead of the sigmoid function $\frac{1}{2}(1+\tanh(\frac{u}{u_0}))$. In addition, at time n, the connection strengths and bias currents of the network are computed from the expressions

$$\boldsymbol{T}_n = \boldsymbol{X}_n^H\boldsymbol{X}_n \quad \text{and} \quad \boldsymbol{B}_n = -\boldsymbol{X}_n^H\boldsymbol{X}_{1,n}. \tag{9.19}$$

Consequently, at time $n+1$,

$$\boldsymbol{T}_{n+1} = \boldsymbol{X}_{n+1}^H\boldsymbol{X}_{n+1} \quad \text{and} \quad \boldsymbol{B}_{n+1} = -\boldsymbol{X}_{n+1}^H\boldsymbol{X}_{1,n+1}. \tag{9.20}$$

Note that in (9.19) and (9.20) we have considered the complex-valued case ((2.49) and (2.56) describe only the real-valued case). According to the analysis given in Section 2.2, this network will provide

$$\boldsymbol{V}_{n,f} = \lim_{t\to\infty}\boldsymbol{V}(t) = \left(\frac{1}{RK_1}\boldsymbol{I} + \boldsymbol{X}_n^H\boldsymbol{X}_n\right)^{-1}\boldsymbol{X}_n^H\boldsymbol{X}_{1,n} \tag{9.21}$$

and

$$\boldsymbol{V}_{n+1,f} = \lim_{t\to\infty}\boldsymbol{V}(t) = \left(\frac{1}{RK_1}\boldsymbol{I} + \boldsymbol{X}_{n+1}^H\boldsymbol{X}_{n+1}\right)^{-1}\boldsymbol{X}_{n+1}^H\boldsymbol{X}_{1,n+1}. \tag{9.22}$$

Equations (9.21) and (9.22) approximate the exact solutions (9.15) and (9.18), respectively.

Although the computation complexity invested in computing the connection strengths and bias currents by use of (9.19) and (9.20) is $O(n^2P^2)$, which is much less than the complexity required in (9.15) and (9.18), the computational complexity needed in (9.19) and (9.20) is on the same order as that of the complexity of the fast RLS algorithm [10]. As a result, it is necessary to decrease the complexity for computing the connection strengths and bias currents before this neural network is applied to real-time problems [11].

Equations (9.19) can be rewritten as

$$T_n(i,j) = \sum_{k=0}^{n-P} x^*(P-i+k)x(P-j+k) \quad \text{and}$$

$$B_n(i) = -\sum_{k=0}^{n-P} x^*(P-i+k)x(P+k) \qquad (i,j = 1,2,\ldots,P), \tag{9.23}$$

where $T_n(i,j)$ and $B_n(i)$ are the corresponding elements of the matrix $\boldsymbol{T}_n$ and the vector $\boldsymbol{B}_n$ at time n, respectively.

Similarly, (9.20) can be rewritten as

$$T_{n+1}(i,j) = \sum_{k=0}^{n-P+1} x^*(P-i+k)x(P-j+k) \quad \text{and}$$

$$B_{n+1}(i) = -\sum_{k=0}^{n-P+1} x^*(P-i+k)x(P+k) \qquad (i,j=1,2,\ldots,P), \tag{9.24}$$

where $T_{n+1}(i,j)$ and $B_{n+1}(i)$ are the corresponding elements of the matrix $\boldsymbol{T}_{n+1}$ and the vector $\boldsymbol{B}_{n+1}$ at time $n+1$, respectively.

Moreover,

$$\begin{aligned} T_{n+1}(i,j) &= \sum_{k=0}^{n-P} x^*(P-i+k)x(P-j+k) \\ &\quad + x^*(n-i+1)x(n-j+1) \\ &= T_n(i,j) + x^*(n-i+1)x(n-j+1) \end{aligned} \tag{9.25}$$

and

$$\begin{aligned} B_{n+1}(i) &= -\sum_{k=0}^{n-P} x^*(P-i+k)x(P+k) - x^*(n-i+1)x(n+1) \\ &= B_n(i) - x^*(n-i+1)x(n+1) \qquad (i,j=1,2,\ldots,P). \end{aligned} \tag{9.26}$$

Equations (9.25) and (9.26) mean that only $x^*(n)x(n)$ and $x^*(n-i+1)x(n+1)$ (for $i=1,2,\ldots,P$) need to be computed at time $n+1$ if the connection strengths and bias currents at time n are available. Note that $x^*(n-i+1)x(n-j+1)$ (for $i \neq 1$ or $j \neq 1$) in (9.25) has been available at time $n+1$. From (9.25) and (9.26), we see that the complexity for recursively computing the connection strengths and bias currents is $O(P)$, which is much less than the complexity of the available fast RLS algorithms. In addition, no division computations are needed if this proposed neural network is used. Once the connection strengths and bias currents have been determined, the neural network can provide the desired coefficients of the linear predictor during an elapsed time of only a few characteristic time constants of the network with an arbitrarily small error. These properties demonstrate the effectiveness of the above neural network approaches for finding the coefficients of the linear predictive coding system in real time.

9.2 MLP Networks for Nonlinear Predictive Coding

It has been shown that nonlinear predictive coding systems can achieve a better performance than linear systems [12, 13, 14, 15]. However, the design and implementation of nonlinear predictors are not as mathematically tractable as those of linear predictors and hence comprise the key problems facing nonlinear coding systems. Recently, nonlinear mapping neural networks such as MLP networks, RBF networks, and high-order networks have been successfully applied. The preliminary idea for such neural networks was presented in Subsection 2.6.2. In this section and the next section, we will give details of the neural network approaches for nonlinear predictors.

Using the same notation as that in the last section, the output of a nonlinear predictor is

$$\hat{x}(n) = g\big(x(n-1), x(n-2), \ldots, x(n-P)\big), \tag{9.27}$$

where $g(\cdot)$ is a nonlinear mapping. The prediction error $e(n)$ at time n is given by

$$e(n) = x(n) - \hat{x}(n) = x(n) - g\big(x(n-1), x(n-2), \ldots, x(n-P)\big). \tag{9.28}$$

When designing a nonlinear predictor, the goal is to find the mapping $g(\cdot)$ so that the mean squared value of the prediction error, $E[e^2(n)]$, is minimized.

We know that MLP networks can approximate nonlinear mappings by determining the connection weights, thresholds, and other network parameters. This suggests that the determination of the mapping $g(\cdot)$ of (9.27) can be transformed to the determination of the network parameters when an MLP network is used as a nonlinear predictor (this is shown in Figure 9.2). This network consists of three layers: the input layer, the hidden layer, and the output layer. The input layer has

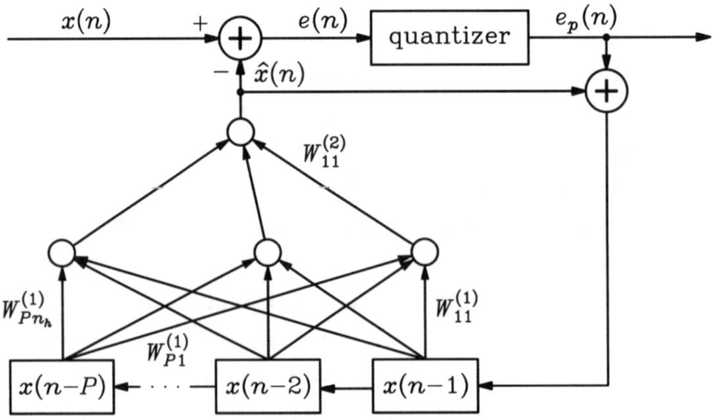

Figure 9.2. A nonlinear predictive coding system based on an MLP network.

P neurons whose input vector is

$$\boldsymbol{X}(n) = \left[x(n-1), x(n-2), \ldots, x(n-P)\right]^T. \tag{9.29}$$

That is, the input vector of the network consists of the P past values of the signal to be encoded. The input layer simply feeds the vector $\boldsymbol{X}(n)$ to the hidden layer without any modification. The hidden layer has n_h neurons with nonlinear transfer functions (such as the sigmoid function). The output layer has only one neuron, which provides the prediction value of the signal at time n.

Let $W_{ij}^{(1)}$ denote the connection weight between the ith neuron in the input layer and the j'th neuron in the hidden layer (for $i = 1, 2, \ldots, P$; $j = 1, 2, \ldots, n_h$); let $q_j(n)$ and $f_j(\cdot)$ (for $j = 1, 2, \ldots, n_h$) be the output and activation function of the j'th neuron in the hidden layer, respectively; let $W_{j1}^{(2)}$ denote the connection weight between the j'th neuron in the hidden layer and the output neuron; and let $o(\cdot)$ be the activation function of the output neuron. Then we have the output $\hat{x}(n)$ of this neural network predictor as

$$\begin{aligned}\hat{x}(n) &= o\left(\sum_{j=1}^{n_h} W_{j1}^{(2)} q_j(n)\right) \\ &= o\left[\sum_{j=1}^{n_h} W_{j1}^{(2)} f_j\left(\sum_{i=1}^{P} W_{ij}^{(1)} x(n-i)\right)\right].\end{aligned} \tag{9.30}$$

Note that in (9.30) the thresholds of all related neurons are selected to be zero and the activation function $o(\cdot)$ of the output neuron is usually no longer the sigmoid function so that it can provide values greater than unity. For simplicity, we select $o(\cdot)$ to be a linear function (i.e., the output neuron performs only simple summations). With this in mind, Equation (9.30) becomes

$$\hat{x}(n) = \sum_{j=1}^{n_h} W_{j1}^{(2)} f_j\left(\sum_{i=1}^{P} W_{ij}^{(1)} x(n-i)\right). \tag{9.31}$$

Under the criterion of minimizing the mean squared error $E[e^2(n)]$, we have

$$\begin{aligned}\min_{\mathbf{W}} E\left[e^2(n)\right] &= \min_{\mathbf{W}} E\left[\left(x(n) - \hat{x}(n)\right)^2\right] \\ &= \min_{\mathbf{W}} E\left[\left(\sum_{j=1}^{n_h} W_{j1}^{(2)} f_j\left(\sum_{i=1}^{P} W_{ij}^{(1)} x(n-i)\right) - x(n)\right)^2\right],\end{aligned} \tag{9.32}$$

where $\boldsymbol{W}$ stands for the set of all the connection weights. Equation (9.32) poses a very complicated nonlinear optimization problem. Its solution requires the values of the autocorrelation function and higher-order statistics of the signal $x(n)$, which are unknown in practical cases. In practical cases, only N samples, $(x(0), x(1), \ldots, x(N-1))$ are available. With this in mind, (9.32) can be replaced by

$$\begin{aligned}\min_{\mathbf{W}} \sum_{n=P}^{N-1} e^2(n) &= \min_{\mathbf{W}} \sum_{n=P}^{N-1} [x(n) - \hat{x}(n)]^2 \\ &= \min_{\mathbf{W}} \sum_{n=P}^{N-1} \left[\sum_{j=1}^{n_h} W_{j1}^{(2)} f_j \left(\sum_{i=1}^{P} W_{ij}^{(1)} x(n-i) \right) - x(n) \right]^2 .\end{aligned} \tag{9.33}$$

This is a nonlinear LS problem similar to (1.53). With the gradient-descent techniques used in the BP algorithm, we can determine the connection weights by use of

$$W_{j1}^{(2)}(t+1) = W_{j1}^{(2)}(t) + \gamma_1 \sum_{n=P}^{N-1} \delta_1^{(2)}(n) q_j(n) \tag{9.34}$$

and

$$W_{ij}^{(1)}(t+1) = W_{ij}^{(1)}(t) + \gamma_2 \sum_{n=P}^{N-1} \delta_j^{(1)}(n) x(n-i) \qquad (i = 1, 2, \ldots, P; \quad j = 1, 2, \ldots, n_h), \tag{9.35}$$

where

$$q_j(n) = f_j \left(\sum_{i=1}^{P} W_{ij}^{(1)} x(n-i) \right), \tag{9.36}$$

$$\delta_1^{(2)}(n) = x(n) - \hat{x}(n), \tag{9.37}$$

$$\delta_j^{(1)}(n) = \left[f_j \left(u_j^{(1)}(n) \right) \right]' \delta_1^{(2)}(n) W_{j1}^{(2)}(t), \tag{9.38}$$

$$\left[f_j \left(u_j^{(1)}(n) \right) \right]' = \left. \frac{\partial f_j(u)}{\partial u} \right|_{u=u_j^{(1)}(n)}, \tag{9.39}$$

and

$$u_j^{(1)}(n) = \sum_{i=1}^{P} W_{ij}^{(1)}(t) x(n-i) \qquad (n = P, P+1, \ldots, N-1). \tag{9.40}$$

The quantities γ_1 and γ_2 are the learning-rate parameters; $W_{ij}^{(1)}(t)$ and $W_{j1}^{(2)}(t)$ denote the related connection weights at the tth iteration. Equations (9.34) and (9.35) will be iterated until the error $\sum_{n=P}^{N-1} e^2(n)$ is less than a specified one.

For the case in which each new sample of the signal $x(n)$ becomes available recursively, (9.34) and (9.35) become

$$W_{j1}^{(2)}(n+1) = W_{j1}^{(2)}(n) + \gamma_1 \sum_{t=P}^{n} \delta_1^{(2)}(t) q_j(t) \tag{9.41}$$

and

$$W_{ij}^{(1)}(n+1) = W_{ij}^{(1)}(n) + \gamma_2 \sum_{t=P}^{n} \delta_j^{(1)}(t) x(t-i)$$
$$(i = 1, 2, \ldots, P; \quad j = 1, 2, \ldots, n_h), \tag{9.42}$$

where

$$q_j(t) = f_j\left(\sum_{i=1}^{P} W_{ij}^{(1)}(t) x(t-i)\right), \tag{9.43}$$

$$\delta_1^{(2)}(t) = x(t) - \hat{x}(t), \tag{9.44}$$

$$\delta_j^{(1)}(t) = \left[f_j(u_j^{(1)}(t))\right]' \delta_1^{(2)}(t) W_{j1}^{(2)}(t), \tag{9.45}$$

$$\left[f_j(u_j^{(1)}(t))\right]' = \left.\frac{\partial f_j(u)}{\partial u}\right|_{u=u_j^{(1)}(t)}, \tag{9.46}$$

and

$$u_j^{(1)}(t) = \sum_{i=1}^{P} W_{ij}^{(1)}(t) x(t-i) \qquad (t = P, P+1, \ldots, n). \tag{9.47}$$

In these expressions, the n available samples are all used, which is suitable for a stationary signal. However, if the signal is nonstationary, we prefer the window-shift processing in which only the $M + P - 1$ latest samples are used to find the connection weights. Then (9.41) and (9.42) become

$$W_{j1}^{(2)}(n+1) = W_{j1}^{(2)}(n) + \gamma_1 \sum_{t=n-M+1}^{n} \delta_1^{(2)}(t) q_j(t) \tag{9.48}$$

and

$$W_{ij}^{(1)}(n+1) = W_{ij}^{(1)}(n) + \gamma_2 \sum_{t=n-M+1}^{n} \delta_j^{(1)}(t) x(t-i)$$
$$(i = 1, 2, \ldots, P; \quad j = 1, 2, \ldots, n_h). \tag{9.49}$$

For simplicity (also for the case in which the statistics of the signal $x(n)$ vary rapidly with time), we let $M = 1$. Then (9.48) and (9.49) become

$$W_{j1}^{(2)}(n+1) = W_{j1}^{(2)}(n) + \gamma_1 \delta_1^{(2)}(n) q_j(n) \tag{9.50}$$

and

$$W_{ij}^{(1)}(n+1) = W_{ij}^{(1)}(n) + \gamma_2 \delta_j^{(1)}(n) x(n-i)$$

$$(i = 1, 2, \ldots, P; \quad j = 1, 2, \ldots, n_h). \qquad (9.51)$$

Clearly, Equations (9.50) and (9.51) are similar to the algorithm (1.60).

Additionally, because in (9.50) and (9.51) only one new sample is involved in each iteration, which is different from the general case in which P new samples (a new sample vector) are involved, it is possible to develop a simpler algorithm that in principle will be similar to (9.25) and (9.26). This is one of the topics of research currently being carried out by the authors.

Dianat et al. [13] simulated image compression by using an MLP network with a three-neuron input layer, one hidden layer of 30 neurons, and one output neuron. The input consists of the three immediate causal neighbors. Their experiments showed that, under the condition of the same input configuration, one can achieve a 4.1 dB improvement in signal-to-noise ratio (SNR) over the optimal linear predictive coding system and an improvement in error entropy from 4.7 bits per pixel (BPP) to 3.9 BPP.

9.3 High-Order Neural Networks for Nonlinear Predictive Coding

With the above in mind, we can easily design nonlinear predictors based on other nonlinear mapping neural networks such as RBF networks and high-order neural networks. As an example, Figure 9.3 depicts the architecture of a high-order neural network for nonlinear predictive coding. This neural network consists of three layers: the input layer, the high-order layer, and the output layer. The input layer of the high-order neural network simply feeds the observation vector $X(n) = [x(n-1), x(n-2), \ldots, x(n-P)]^T$ to the second layer (the high-order layer) without modification. Thus, there are P neurons in the input layer. The units of the high-order layer are in effect multipliers whose outputs are computed as follows:

(1) The outputs of the first-order neurons are

$$q_i^{(1)}(n) = x(n-i) \qquad (i = 1, 2, \ldots, P). \qquad (9.52)$$

(2) The outputs of the second-order neurons are determined by

$$q_{ij}^{(2)}(n) = x(n-i)x(n-j) \qquad (i, j = 1, 2, \ldots, P). \qquad (9.53)$$

(3) The outputs of the k'th-order neurons are determined by

$$q_{i_1 i_2 \ldots i_k}^{(k)}(n) = x(n-i_1)x(n-i_2)\ldots x(n-i_k)$$

$$(i_j = 1, 2, \ldots, P; j = 1, 2, \ldots, k). \qquad (9.54)$$

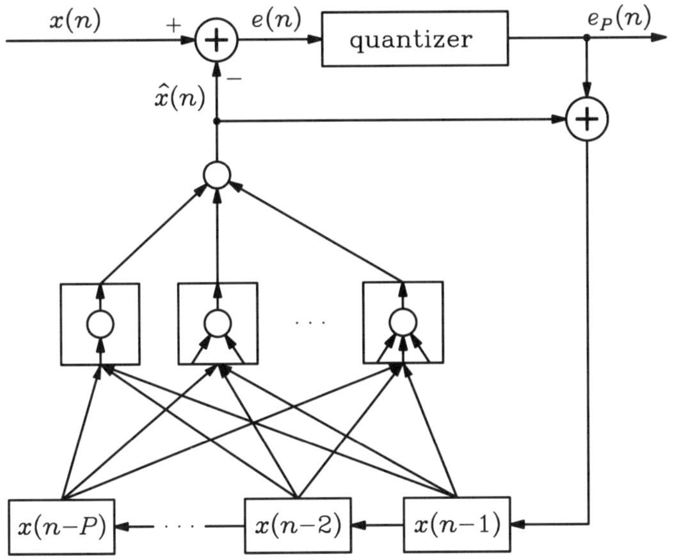

Figure 9.3. A nonlinear predictive coding system based on high-order neural networks.

The output layer has only one neuron with the activation function $f(\cdot)$. The connections between the neurons in the high-order layer and the output neuron are weighted. The output $\hat{x}(n)$ of the output neuron is

$$\begin{aligned}\hat{x}(n) = f(u(n)) &= f\Bigg(\sum_{i=1}^{P} W(i) q_i^{(1)}(n) + \sum_{i=1}^{P}\sum_{j=1}^{P} W(ij) q_{ij}^{(2)}(n) \\ &\quad + \cdots + \sum_{i_1=1}^{P}\sum_{i_2=1}^{P}\cdots\sum_{i_k=1}^{P} W(i_1 i_2 \ldots i_k) q_{i_1 i_2 \ldots i_k}^{(k)}(n)\Bigg) \\ &= f\Bigg(\sum_{i=1}^{P} W(i) x(n-i) + \sum_{i=1}^{P}\sum_{j=1}^{P} W(ij) x(n-i) \\ &\quad \times x(n-j) + \cdots + \sum_{i_1=1}^{P}\sum_{i_2=1}^{P}\cdots\sum_{i_k=1}^{P} W(i_1 i_2 \ldots i_k) \\ &\quad \times x(n-i_1) x(n-i_2) \ldots x(n-i_k)\Bigg),\end{aligned} \tag{9.55}$$

where $u(n)$ is the input of the output neuron and the $W(\cdot)$ are the corresponding connection weights between the high-order neurons with the output neuron. Note that in (9.55) we let the threshold of the output neuron be zero.

Using the available samples of the signal $x(n)$, we can determine all the related connection weights of the network by

$$\Delta W(i) = \gamma_1 \sum_{n=P}^{N-1} \delta(n)x(n-i) \tag{9.56}$$

for the weights related to the first-order neurons, by

$$\Delta W(ij) = \gamma_2 \sum_{n=P}^{N-1} \delta(n)x(n-i)x(n-j) \tag{9.57}$$

for the weights related to the second-order neurons, and by

$$\Delta W(i_1 i_2 \ldots i_k) = \gamma_k \sum_{n=P}^{N-1} \delta(n)x(n-i_1)x(n-i_2)\ldots x(n-i_k) \tag{9.58}$$

for the weights related to the k'th-order neurons, where

$$\delta(n) = \left(x(n) - \hat{x}(n)\right)\left[f\left(u(n)\right)\right]', \tag{9.59}$$

$$\left[f\left(u(n)\right)\right]' = \left.\frac{\partial f(u)}{\partial u}\right|_{u=u(n)}, \tag{9.60}$$

and

$$\begin{aligned} u(n) = {} & \sum_{i=1}^{P} W(i)x(n-i) + \sum_{i=1}^{P}\sum_{j=1}^{P} W(ij)x(n-i)x(n-j) \\ & + \cdots + \sum_{i_1=1}^{P}\sum_{i_2=1}^{P}\cdots\sum_{i_k=1}^{P} W(i_1 i_2 \cdots i_k)x(n-i_1) \\ & \times x(n-i_2)\cdots x(n-i_k) \qquad (n = P, P+1, \ldots, N-1). \end{aligned} \tag{9.61}$$

The quantities γ_1, γ_2, and γ_k are the learning-rate parameters. They can take the same values. Corresponding to (9.41)–(9.42) and (9.48)–(9.49), respectively, are the following expressions:

$$\Delta W(i) = \gamma_1 \sum_{t=P}^{n} \delta(t)x(t-i), \tag{9.62}$$

$$\Delta W(ij) = \gamma_2 \sum_{t=P}^{n} \delta(t)x(t-i)x(t-j), \tag{9.63}$$

$$\Delta W(i_1 i_2 \ldots i_k) = \gamma_k \sum_{t=P}^{n} \delta(t)x(t-i_1)x(t-i_2)\ldots x(t-i_k) \tag{9.64}$$

and

$$\Delta W(i) = \gamma_1 \sum_{t=n-M+1}^{n} \delta(t)x(t-i), \tag{9.65}$$

$$\Delta W(ij) = \gamma_2 \sum_{t=n-M+1}^{n} \delta(t)x(t-i)x(t-j), \tag{9.66}$$

$$\Delta W(i_1 i_2 \cdots i_k) = \gamma_k \sum_{t=n-M+1}^{n} \delta(t)x(t-i_1)x(t-i_2)\cdots x(t-i_k). \tag{9.67}$$

Readers can immediately obtain expressions for $M = 1$ from (9.65)–(9.67) that will be similar to (1.88) (note that $M + P - 1$ in (9.65)–(9.67) is the sample number involved in the current iteration).

It is worth mentioning that if we select the activation function $f(\cdot)$ of the output neuron as a linear function, then (9.55) will become the discrete-time Volterra expansion:

$$\begin{aligned} \hat{x}(n) = {} & \sum_{i=1}^{P} W(i)x(n-i) + \sum_{i=1}^{P}\sum_{j=1}^{P} W(ij)x(n-i)x(n-j) \\ & + \cdots + \sum_{i_1=1}^{P}\sum_{i_2=1}^{P}\cdots\sum_{i_k=1}^{P} W(i_1 i_2 \cdots i_k) \\ & \times x(n-i_1)x(n-i_2)\cdots x(n-i_k). \end{aligned} \tag{9.68}$$

Comparing (9.68) with (2.98), we see that this high-order network acts in effect as a Volterra filter (as shown in Figure 2.9). Consequently, the connection weights $\boldsymbol{W}$ of the network are the coefficients of the Volterra filter. As a result, the available methods for designing and implementing Volterra filters can immediately be applied to this high-order network. We can also use the neural network in Figure 2.2 to compute in real time the connection weights as given in Section 2.5.

Manikopoulos [16] has given simulation results of high-order neural networks for nonlinear predictive coding of image signals. In the simulation, the neuron number P in the input layer is 4 and the activation function $f(\cdot)$ of the output neuron is a linear function. The simulation results showed that significant improvements in both SNR and BPP over the optimal linear predictive coding systems were achieved.

We should also point out that there has recently been an interest in signal compression based on models of human perception [17, 18, 19]. One main aspect of this topic is to replace the mean squared error cost function used above by some perception-based cost functions such as the p-Hölder ($p \neq 2$) norms [20]. The motivation behind this is that the mean squared error is an objective measure that is not always perfectly correlated with subjective quality assessments, and furthermore, it spreads degradations all over the signal. However, no matter what the cost function is, the above mapping neural networks are still suitable and the gradient-descent techniques can still be employed to find the connection weights and other network parameters that define the mapping $g(\cdot)$ of the nonlinear predictor in a coding system.

9.4 Neural Networks for the Karhunen–Loève Transform Coding

Another important approach to signal compression uses transformations that operate on a signal to produce a set of coefficients. According to the redundancies, only a subset of these coefficients are chosen and quantized for transmission or for storage. The goal of this technique is to choose a transformation for which such a subset of coefficients is adequate to reconstruct a signal under a specified criterion. There have been many transformation forms. With respect to minimizing the mean squared error, the Karhunen–Loève transformation (KLT) is the best of all linear transformations [21]. In the past, the KLT was interesting theoretically, but it was generally not used in practice. One major problem is that the KLT is signal dependent, that is, for different classes of inputs, the basis functions of the transform change. In addition, these basis functions are, in general, not known analytically. Furthermore, even if the transform matrix is known, the number of operations needed to perform the transform is too large for practical signal compression purposes. Owing to these difficulties, fixed-basis transforms such as the discrete cosine transform (DCT) are used. The DCT is the suboptimal transform used most often in practice and has been adopted by the Joint Photographic Experts' Group (JPEG) and the Moving Picture Experts' Group (MPEG) as the industrial standard [22, 23]. However, at very low bit rates the DCT-based coding techniques suffer from numerous problems [4]. For example, in image compression, the DCT-based coding will generate blocking and mosquito artifacts in the transmitted and reconstructed images. In order to attack these problems, wavelet transform coding has been proposed [24, 25]. It has been shown that wavelet transforms are more suitable for the reduction of the redundancies than DCT in many cases [4, 26]. This property is brought about by the fact that the related basis functions in the wavelet transform overlap one another and decay smoothly to zero at their end points. Unfortunately, wavelet transform coding is very complicated in the algorithm and architecture. The main purpose of the next two sections is to show how to use neural networks to implement these transform-based coding techniques. In this section, we will deal with the PCA networks and their algorithms for the KLT coding. Section 9.5 will give the architecture and learning rules for wavelet transform coding.

Let us consider a signal vector $\boldsymbol{X}(n) = [x(n-1), x(n-2), \ldots, x(n-P)]^T$ (for $n = P, P+1, \ldots, N-1$). A linear transformation, which can be written as an $(M \times P)$-dimensional matrix $\boldsymbol{W} = \{w_{ij}\}$ with $M \leq P$, is performed on each block with the M rows of $\boldsymbol{W}$. $\boldsymbol{W}_j = [w_{j1}, w_{j2}, \ldots, w_{jP}]^T$ (for $j = 1, 2, \ldots, M$) is the basis vector of the transformation. The resulting M-dimensional coefficient vector $\boldsymbol{Z}$ is calculated as

$$\boldsymbol{Z}(n) = \boldsymbol{W}\boldsymbol{X}(n). \tag{9.69}$$

If the basis vectors are orthonormal, that is,

$$\boldsymbol{W}_i^T \boldsymbol{W}_j = \begin{cases} 1, & i = j, \\ 0, & i \neq j \end{cases} \qquad (i, j = 1, 2, \ldots, M), \tag{9.70}$$

then the inverse transformation is given by the transpose of the forward transformation matrix, resulting in the reconstructed vector $\hat{\boldsymbol{X}}(n)$ as

$$\hat{\boldsymbol{X}}(n) = \boldsymbol{W}^T \boldsymbol{Z}(n) = \boldsymbol{W}^T \boldsymbol{W}\boldsymbol{X}(n). \tag{9.71}$$

The matrix $\boldsymbol{W}$ of the KLT consists of M rows of the eigenvectors corresponding to the largest eigenvalues of the autocorrelation matrix

$$\boldsymbol{R} = E\left[\boldsymbol{X}(n)\boldsymbol{X}^T(n)\right], \tag{9.72}$$

that is, $\boldsymbol{W} = [\boldsymbol{S}_1, \boldsymbol{S}_2, \ldots, \boldsymbol{S}_M]^T$. Note that $\lambda_1 \geq \lambda_2 \geq \cdots \geq \lambda_P \geq 0$ denote the eigenvalues and $\boldsymbol{S}_1, \boldsymbol{S}_2, \ldots, \boldsymbol{S}_P$ denote the corresponding orthonormal eigenvectors of the matrix $\boldsymbol{R}$. This is obtained by minimizing the mean squared error

$$E\left[\left\|\boldsymbol{X}(n) - \hat{\boldsymbol{X}}(n)\right\|^2\right]. \tag{9.73}$$

Thus, the KLT gives the optimal solution of (9.73).

As pointed out in the beginning of this section, the KLT is signal dependent. For different signals, $\boldsymbol{R}$ and $\boldsymbol{W}$ will be different. In practical applications, $\boldsymbol{R}$ is not known exactly but can be estimated by

$$\boldsymbol{R} = \frac{1}{N-P} \sum_{n=P}^{N-1} \boldsymbol{X}(n)\boldsymbol{X}^T(n). \tag{9.74}$$

Furthermore, the determination of the KLT matrix $\boldsymbol{W}$ is related to the PCA. In other words, all of the PCA algorithms presented in Section 6.3 can be employed to find the KLT matrix $\boldsymbol{W}$. As an example, the invariant-norm PCA algorithm [27] will be used in the following.

If the matrix $\boldsymbol{R}$ has been obtained from (9.74), then we can construct an analog neural network whose dynamic differential equations are defined by (6.94), that is,

$$\frac{d\boldsymbol{W}_j(t)}{dt} = \boldsymbol{W}_j^T(t)\boldsymbol{W}_j(t)\left[\boldsymbol{I} - \sum_{i=1}^{j-1} \boldsymbol{W}_i(t)\boldsymbol{W}_i^T(t)\right]\boldsymbol{R}\boldsymbol{W}_j(t)$$
$$-\boldsymbol{W}_j^T(t)\left[\boldsymbol{I} - \sum_{i=1}^{j-1} \boldsymbol{W}_i(t)\boldsymbol{W}_i^T(t)\right]\boldsymbol{R}\boldsymbol{W}_j(t)\boldsymbol{W}_j(t)$$
$$(j = 1, 2, \ldots, M) \tag{9.75}$$

or

$$\frac{d\boldsymbol{W}_j(t)}{dt} = \psi_j(t)\boldsymbol{R}_j\boldsymbol{W}_j(t) - \phi_j(t)\boldsymbol{W}_j(t) \qquad (j = 1, 2, \ldots, M), \tag{9.76}$$

where

$$\boldsymbol{R}_j = \boldsymbol{R} - \sum_{i=1}^{j-1} \boldsymbol{W}_i(t)\boldsymbol{W}_i^T(t)\boldsymbol{R}, \tag{9.77}$$

$$\psi_j(t) = \boldsymbol{W}_j^T(t)\boldsymbol{W}_j(t), \tag{9.78}$$

and

$$\phi_j(t) = \boldsymbol{W}_j^T(t)\boldsymbol{R}_j\boldsymbol{W}_j(t) \qquad (j = 1, 2, \ldots, M). \tag{9.79}$$

According to the analysis given in Section 6.3, such a constructed neural network will be stable and will provide M eigenvectors corresponding to the M largest eigenvalues of the matrix $\boldsymbol{R}$. Thus,

$$\lim_{t\to\infty} \boldsymbol{W}_j(t) = \pm\sqrt{\psi_j(0)}\boldsymbol{S}_j \qquad (j = 1, 2, \ldots, M). \tag{9.80}$$

Moreover, the norm of each vector $\boldsymbol{W}_j(t)$ is invariant during the time evolution of the network and is equal to the norm of the initial vector $\boldsymbol{W}_j(0)$. Clearly, for the KLT, we select $\boldsymbol{W}_j^T(0)\boldsymbol{W}_j(0) = 1$, and then

$$\lim_{t\to\infty} \boldsymbol{W}_j(t) = \pm\boldsymbol{S}_j \qquad (j = 1, 2, \ldots, M). \tag{9.81}$$

Note that in [28] we have proved that (9.75) is exponentially stable, which means that $\boldsymbol{W}_j(t)$ will converge exponentially to $\pm\boldsymbol{S}_j$.

One possible discrete-time version of (9.75) is

$$\boldsymbol{W}_j(t+1) = \boldsymbol{W}_j(t) + \gamma(t)\left[\boldsymbol{I} - \sum_{i=1}^{j-1} \boldsymbol{W}_i(t)\boldsymbol{W}_i^T(t)\right]\boldsymbol{R}\boldsymbol{W}_j(t)$$
$$-\boldsymbol{W}_j^T(t)\left[\boldsymbol{I} - \sum_{i=1}^{j-1} \boldsymbol{W}_i(t)\boldsymbol{W}_i^T(t)\right]\boldsymbol{R}\boldsymbol{W}_j(t)\boldsymbol{W}_j(t)$$
$$(j = 1, 2, \ldots, M), \tag{9.82}$$

where $\gamma(t)$ is the step parameter. It is easy to see that although (9.82) has a parallel processing structure, it is not too attractive in comparison with the available direct eigendecomposition methods [21] because many iterations are required in (9.82) from the initiation to the convergence. The number of iterations depends on the choice of the initialized values for each vector $\boldsymbol{W}_j(t)$ and the step parameter $\gamma(t)$. However, if the signal varies slowly and if each new sample of the signal $x(n)$ becomes available as time progresses, then the above problem will become less important because we can use (9.82) to reconstruct an adaptive algorithm as

$$\boldsymbol{W}_j(n+1) = \boldsymbol{W}_j(n) + \gamma(n)\left[\boldsymbol{I} - \sum_{i=1}^{j-1} \boldsymbol{W}_i(n)\boldsymbol{W}_i^T(n)\right]\boldsymbol{R}(n)\boldsymbol{W}_j(n)$$
$$-\boldsymbol{W}_j^T(n)\left[\boldsymbol{I} - \sum_{i=1}^{j-1} \boldsymbol{W}_i(n)\boldsymbol{W}_i^T(n)\right]\boldsymbol{R}(n)\boldsymbol{W}_j(n)\boldsymbol{W}_j(n)$$
$$(j = 1, 2, \ldots, M), \tag{9.83}$$

where

$$\boldsymbol{R}(n) = \frac{1}{n-P+1}\sum_{t=P}^{n} \boldsymbol{X}(t)\boldsymbol{X}^T(t). \tag{9.84}$$

Moreover, for window-shift processing (i.e., only the $M+P-1$ latest samples are used), Equation (9.84) becomes

$$\boldsymbol{R}(n) = \frac{1}{M}\sum_{t=n-M+1}^{n} \boldsymbol{X}(t)\boldsymbol{X}^T(t). \tag{9.85}$$

Using the same method as for (9.25), we obtain a recursive expression to compute each element of $\boldsymbol{R}(n)$ as

$$R_{ij}(n) = \frac{n-P}{n-P+1}R_{ij}(n-1) + \frac{1}{n-P+1}x(n-i)x(n-j)$$
$$(i, j = 1, 2, \ldots, P), \tag{9.86}$$

where $R_{ij}(\cdot)$ stands for the (i, j)th element of the matrix $\boldsymbol{R}(\cdot)$. Equation (9.86) means that only $x(n-1)x(n-i)$ (for $i = 1, 2, \ldots, P$) needs to be computed at time n if the matrix $\boldsymbol{R}(n-1)$ at time $n-1$ is available. Note that $x(n-i)x(n-j)$ (for $i > 1$ and $j > 1$) in (9.86) has been available at time n. This greatly reduces the computational complexity of (9.83). It is easy to see that the storage of $\boldsymbol{R}(n)$ is necessary in the algorithm (9.86). To avoid the storage of the past values of the matrix $\boldsymbol{R}(n)$, we select $M = 1$, and then (9.85) becomes

$$\boldsymbol{R}(n) = \boldsymbol{X}(n)\boldsymbol{X}^T(n). \tag{9.87}$$

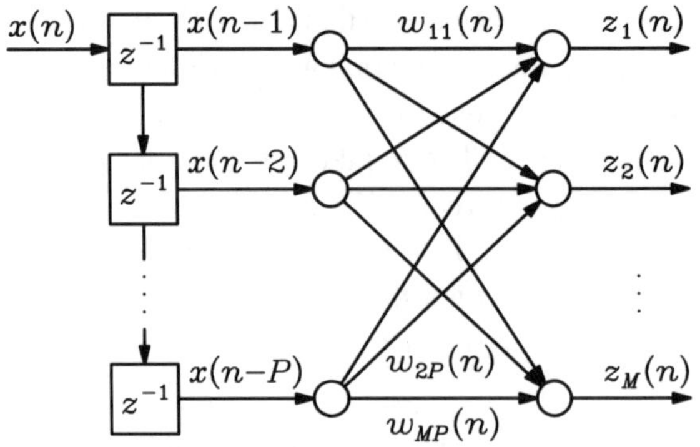

Figure 9.4. The neural network architecture corresponding to the algorithm (9.89).

Substituting (9.87) into (9.83), we have

$$\boldsymbol{W}_j(n+1) = \boldsymbol{W}_j(n) + \gamma(n)\left[\boldsymbol{I} - \sum_{i=1}^{j-1} \boldsymbol{W}_i(n)\boldsymbol{W}_i^T(n)\right]\boldsymbol{X}(n)\boldsymbol{X}^T(n)\boldsymbol{W}_j(n)$$

$$-\boldsymbol{W}_j^T(n)\left[\boldsymbol{I} - \sum_{i=1}^{j-1} \boldsymbol{W}_i(n)\boldsymbol{W}_i^T(n)\right]\boldsymbol{X}(n)\boldsymbol{X}^T(n)\boldsymbol{W}_j(n)\boldsymbol{W}_j(n)$$

$$(j = 1, 2, \ldots, M). \qquad (9.88)$$

Let us denote $z_j(n) = \boldsymbol{W}_j^T(n)\boldsymbol{X}(n)$. Then it follows that

$$\boldsymbol{W}_j(n+1) = \boldsymbol{W}_j(n) + \gamma(n)z_j(n)\left\{\left[\boldsymbol{X}(n) - \sum_{i=1}^{j-1} z_i(n)\boldsymbol{W}_i(n)\right]\right.$$

$$\left. - z_j(n)\boldsymbol{W}_j(n) + \sum_{i=1}^{j-1} z_i(n)\boldsymbol{W}_j^T(n)\boldsymbol{W}_i(n)\boldsymbol{W}_j(n)\right\}$$

$$(j = 1, 2, \ldots, M). \qquad (9.89)$$

Clearly, this equation is exactly the same as (6.92). The corresponding network architecture for (9.89) is shown in Figure 9.4. Like the neural network in Figure 6.5, this network consists of two layers. The first layer is the input layer and has P neurons that simply feed the input vector $\boldsymbol{X}(n)$ to the second layer without any modification. The second layer is the output layer and has M neurons. The input–output relationship of each neuron in the output layer is a linear function so that they perform simple summations. The connections between the input layer and output layer are weighted. The connection weight vector is denoted as $\boldsymbol{W}_j(n) = [w_{j1}(n), w_{j2}(n), \ldots, w_{jP}(n)]^T$ (for $j = 1, 2, \ldots, M$), where $w_{ji}(n)$ is the connection strength between the i'th input neuron and the j'th output neuron. Furthermore,

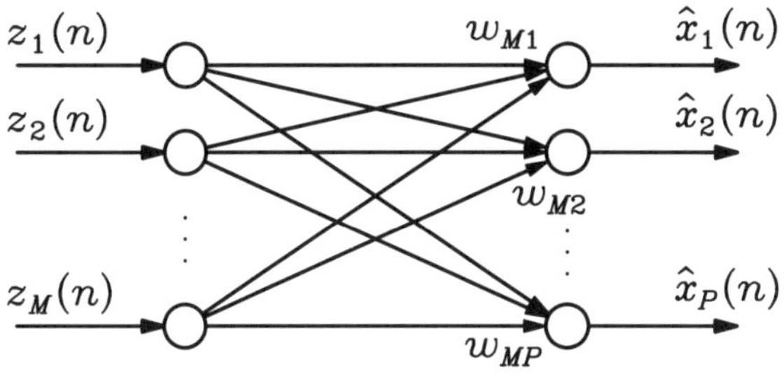

Figure 9.5. The neural network architecture for the inverse KLT.

$z_j(n) = \boldsymbol{W}_j^T(n)\boldsymbol{X}(n)$ is the output of the j'th neuron in the output layer. This also means that after convergence the output vector $\boldsymbol{Z}(n) = [z_1(n), z_2(n), \ldots, z_M(n)]^T$ provides the desired KLT transform $\boldsymbol{Z}(n) = \boldsymbol{W}\boldsymbol{X}(n)$. Based on (9.71), the inverse KLT can be implemented by the network shown in Figure 9.5. Obviously, the connection weight matrix in Figure 9.5 is the transpose of that in Figure 9.4.

In comparison with (9.83), Equation (9.89) is computationally more efficient but has slower convergence. As a result, one has to make a trade-off in practical applications of the KLT transform coding.

9.5 Neural Networks for Wavelet Transform Coding

Coding techniques based on wavelet transforms are considered to be promising in signal compression, especially in image compression, where they can provide high rates of compression while maintaining good quality [24, 25]. Like Fourier transforms, wavelet transforms fall into two categories: discrete wavelet transforms (DWTs) and continuous wavelet transforms (CWTs). In this section, we deal only with the DWT. The discrete wavelet transform is generally defined as

$$W(a, b) = \sum_n \frac{1}{\sqrt{a}} x(n) h\left(\frac{n-b}{a}\right), \tag{9.90}$$

where $\frac{1}{\sqrt{a}}h(\frac{n-b}{a})$ is called a wavelet, b is a shift, a is a scale factor, and $\sqrt{a}$ is a normalization factor. The term $h(n)$ ($b = 0$, $a = 1$) is referred to as the mother wavelet and usually has the form [29]

$$h(n) = w(n) f(n), \tag{9.91}$$

where $w(n)$ is a window function (often Gaussian) and $f(n)$ is a modulation term. $w(n)$ and $f(n)$ are both scaled and shifted. The term $\frac{1}{\sqrt{a}}h(\frac{n-b}{a})$ (for $a \neq 1$ or $b \neq 0$) is also called the daughter wavelet.

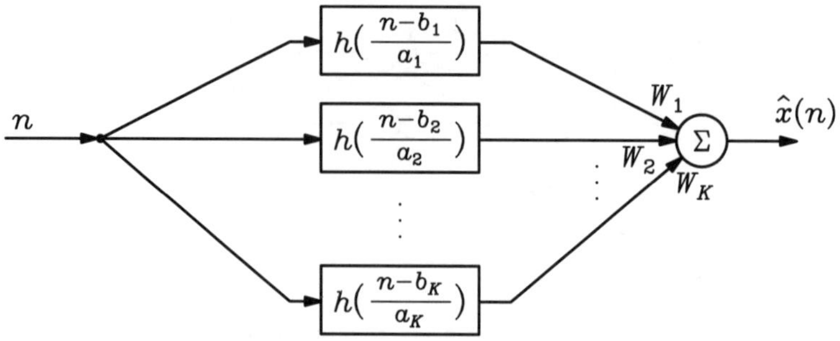

Figure 9.6. The neural network architecture for Equation (9.93).

The inverse discrete wavelet transform (IDWT) is

$$\hat{x}(n) = \sum_a \sum_b \frac{1}{\sqrt{a}} W(a, b) h\left(\frac{n - b}{a}\right). \tag{9.92}$$

One simple version of (9.92) can be written as [30]

$$\hat{x}(n) = \sum_{k=1}^{K} W_k h\left(\frac{n - b_k}{a_k}\right), \tag{9.93}$$

where the W_k, b_k, and a_k (for $k = 1, 2, \ldots, K$) are the weight coefficients, shifts, and dilations for each daughter wavelet.

The neural network for implementing (9.93) is shown in Figure 9.6. This network consists of two layers. The first layer is the input layer and has K neurons. The daughter wavelet function $h(\frac{n-b_k}{a_k})$ (for $k = 1, 2, \ldots, K$) serves as the activation function of the corresponding neuron in the input layer. The second layer is the output layer and has only one neuron that performs simple summations. The weight coefficient W_k (for $k = 1, 2, \ldots, K$) in (9.93) is taken as the connection weight of the k'th input neuron with the output neuron. Note that the input of the network is the time index n. With these in mind, the output neuron will provide $\hat{x}(n)$ as given in Equation (9.93).

One major task in wavelet transform coding of the signal $x(n)$ is to determine the parameters W_k, b_k, and a_k (for $k = 1, 2, \ldots, K$). If N samples of $x(n)$, $(x(0), x(1), \ldots, x(N-1))$ are available, then these parameters can be determined by solving the optimization problem

$$\min_{\mathbf{W}} \sum_{n=0}^{N-1} (\hat{x}(n) - x(n))^2 = \min_{\mathbf{W}} \sum_{n=0}^{N-1} \left[x(n) - \sum_{k=1}^{K} W_k h\left(\frac{n - b_k}{a_k}\right)\right]^2, \tag{9.94}$$

where $\mathbf{W} = \{W_k, b_k, a_k;\ k = 1, 2, \ldots, K\}$ is the parameter set. Clearly, this is a typical nonlinear LS problem. Using gradient-descent techniques, we have the algorithm to update the set $\boldsymbol{W}$ as

$$\boldsymbol{W}(t+1) = \boldsymbol{W}(t) + \gamma \boldsymbol{Q}(t), \tag{9.95}$$

where $\boldsymbol{Q}(t)$ is a search direction based on the gradients of the total representation error squares

$$\sum_{n=0}^{N-1} e^2(n) = \sum_{n=0}^{N-1} \left[x(n) - \sum_{k=1}^{K} W_k h\left(\frac{n-b_k}{a_k}\right)\right]^2 \tag{9.96}$$

with respect to $\boldsymbol{W}(t)$, and γ is a search-rate parameter that can vary with t. Note that $\boldsymbol{W}(t)$ denotes the parameter set at the tth iteration. Usually, we use the negative gradient defined by

$$\boldsymbol{Q}(t) = -\sum_{n=0}^{N-1} \frac{de^2(n)}{d\boldsymbol{W}(t)} = 2\sum_{n=0}^{N-1} e(n) \frac{d\hat{x}(n)}{d\boldsymbol{W}(t)} \tag{9.97}$$

as the search direction.

It is easy to see that Equation (9.95) is exactly the same as (8.28). Consequently, all the modified versions of (8.28) given in Section 8.2 can immediately be used to replace (9.95) so as to improve the performance or to reduce the complexity.

If each new sample of the signal $x(n)$ becomes available as time progresses, then the parameter set at time $n+1$ is computed by

$$\boldsymbol{W}(n+1) = \boldsymbol{W}(n) + \gamma \boldsymbol{Q}(n), \tag{9.98}$$

where $\boldsymbol{Q}(n)$ can be taken as

$$\boldsymbol{Q}(n) = -\sum_{t=0}^{n} \frac{de^2(t)}{d\boldsymbol{W}(n)} = 2\sum_{t=0}^{n} e(t) \frac{d\hat{x}(t)}{d\boldsymbol{W}(n)}. \tag{9.99}$$

Note that (9.99) corresponds to (9.97) and that other modified search directions can also be used.

For the case of window-shift processing (where only M samples are involved for each iteration), (9.99) becomes

$$\boldsymbol{Q}(n) = -\sum_{t=n-M+1}^{n} \frac{de^2(t)}{d\boldsymbol{W}(n)} = 2\sum_{t=n-M+1}^{n} e(t) \frac{d\hat{x}(t)}{d\boldsymbol{W}(n)}. \tag{9.100}$$

Compared with (9.97), (9.99) and (9.100) are more suitable for adaptive wavelet transform coding in which the related signal $x(n)$ varies with time. In addition, it is necessary to mention that the parameter W_k (for $k = 1, 2, \ldots, K$) in batch-processing (corresponding to (9.95)) can also be computed directly by (9.90) if a_k and b_k are available.

In effect, the above network and the related algorithms can easily be extended to a more general form of (9.93) as

$$\hat{x}(n) = \sum_{k=1}^{K} \sum_{i=1}^{I} W_{k,i} h\left(\frac{n - b_k}{a_i}\right). \tag{9.101}$$

In (9.101), two-dimensional weight coefficients are involved; hence, the structure of the network will be more complicated than that in Figure 9.6. The advantage of (9.101) over (9.93) is that because there are more adjustable parameters a better performance could be achieved when to code the signal $x(n)$.

Because wavelets are usually not orthogonal, we are unable to present a network and an algorithm similar to those for the KLT. The development and implementation of neural network approaches for wavelet transforms remain an interesting research topic [31, 32].

9.6 Neural Networks for Vector Quantization

As is well known, the process of quantization maps a signal $x(n)$ into a series of L discrete messages. For the k'th message, there exist a pair of thresholds θ_k and θ_{k+1} and an output value $\hat{x}(n)$ such that $\theta_k < \hat{x}(n) \leq \theta_{k+1}$. For a given set of quantization values, the optimal thresholds are equidistant from the values. This concept can be extended to vector quantization (VQ). In VQ, a P-dimensional data vector $\boldsymbol{X}(n)$ is represented as one of a finite set of L symbols. Associated with each symbol is a P-dimensional vector $\boldsymbol{C}_i$ (for $i = 1, 2, \ldots, L$) called a codeword [33]. The complete set of the L codewords is called the codebook. Each element of the codeword vector can be, for example, a successive sample value of $\boldsymbol{X}(n)$ or a parameter extracted from the signal. VQ mainly consists of two parts: the codebook design and the vector encoding. The codebook is usually obtained through a training process using a large set of training data that is typical of the data encountered in practice. The most popular algorithm for VQ codebook design is the Linde, Buzo, and Gray (LBG) algorithm (also called the generalized Lloyd algorithm because it is a generalized version of the scalar quantization algorithm proposed by Lloyd in 1957 [34]). In the LBG algorithm, L codewords are initially set to random values. On each iteration, each data vector is classified based on its nearest codeword. Each codeword is then replaced by the mean of its resulting class. The iterations continue until a minimum acceptable error is achieved. This algorithm minimizes the mean squared error over the training set. During the encoding process, each P-dimensional vector $\boldsymbol{X}(n)$ that is to be encoded is compared to each of the L codewords in the codebook, and the distance $d(\boldsymbol{X}(n), \boldsymbol{C}_i)$ (for $i = 1, 2, \ldots, L$) (Euclidean distance is usually used) between the vector $\boldsymbol{X}(n)$ and the codeword $\boldsymbol{C}_i$ is computed. The vector $\boldsymbol{X}(n)$ is then encoded as the index j of the codeword

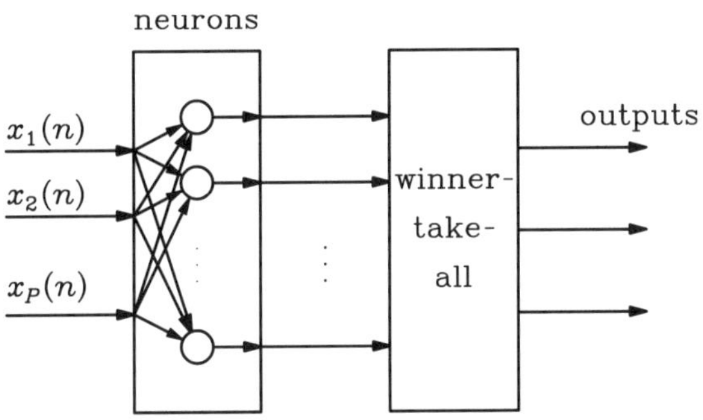

Figure 9.7. The neural network architecture for the VQ encoding.

that yields the minimum distance. The receiver, which is assumed to have a copy of the codebook, uses this index to look up the corresponding codeword $\boldsymbol{C}_j$. The codeword $\boldsymbol{C}_j$ is then used as the encoded value of the vector $\boldsymbol{X}(n)$. Although VQ can provide a large reduction in the bit rate, its application has been limited primarily because of the prohibitive amounts of computation associated with both the vector encoding and the codebook design stages mentioned above. In this section, we will show how to use neural networks to perform VQ [35, 36, 37].

First, let us consider the encoding process in VQ using neural networks. In effect, it is straightforward to formulate a neural network structure for the encoding step in VQ, as shown in Figure 9.7. This network consists of L neurons. The i'th (for $i = 1, 2, \ldots, L$) neuron is associated with a P-dimensional weight vector $\boldsymbol{W}_i$, which is selected to be the i'th codeword, that is, $\boldsymbol{W}_i = \boldsymbol{C}_i$. Given any P-dimensional input vector $\boldsymbol{X}(n)$ that is to be encoded, $\boldsymbol{X}(n)$ is fed in parallel to all the L neurons. The output $h_i(n)$ of each neuron is the distance $d(\boldsymbol{X}(n), \boldsymbol{C}_i)$ of the input vector $\boldsymbol{X}(n)$ from the weight vector $\boldsymbol{W}_i$, that is,

$$h_i(n) = d\big(\boldsymbol{X}(n), \boldsymbol{W}_i\big). \tag{9.102}$$

In structure this configuration is similar to the input layer and hidden layer of RBF networks. The distance $d(\cdot)$ corresponds to the activation function of the hidden neurons of the RBF networks. The major difference is that in (9.102) all the central vectors of the hidden neurons are zero.

The neuron with the minimum distance is called the winning neuron, and the input vector $\boldsymbol{X}(n)$ is encoded as the index of the winning neuron. Clearly, all computations except for the determination of the winning neuron are performed in parallel. In fact, we can also use some available neuron networks such as those proposed in [34, 36] to determine effectively the winning neuron.

Comparing the process of the VQ codebook design with the self-organizing systems presented in Section 1.5, we see that the Kohonen self-organizing feature

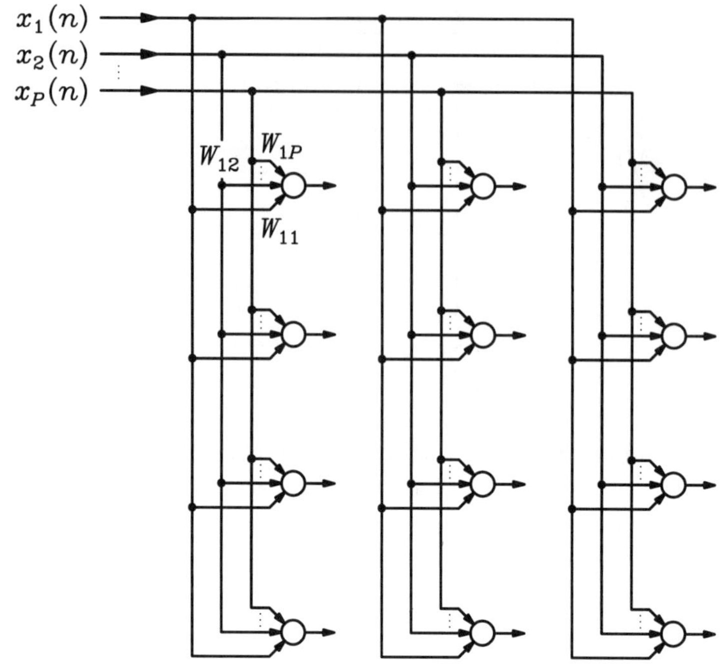

Figure 9.8. The neural network architecture for the VQ codebook design.

map (KSFM) algorithm ((1.62)–(1.64)) can immediately be applied to the codebook design from the available training data vectors.

Let the training vector $\boldsymbol{X}(n)$ be connected in parallel to all neurons in the network (as shown in Figure 9.8). The connection weight vector between the input vector $\boldsymbol{X}(n)$ and the i'th neuron is denoted by $\boldsymbol{W}_i(n) = [W_{i1}(n), W_{i2}(n), \ldots, W_{iP}(n)]^T$.

For the purpose of making the weight vectors in the stable state form the codewords required in VQ, the following steps can be used [38]:

(1) Initialize randomly all connection weight vectors and select the size (width or radius) of the neighborhood set N_c (which is defined by all the neurons neighboring a so-called central neuron).

The size N_c can also be time variable. Experimentally it has turned out to be advantageous to let N_c be very wide at the beginning and to shrink it monotonically with time.

(2) Compute the distance $d(\boldsymbol{X}(n), \boldsymbol{W}_i(n))$ $(i \in N_c)$.

(3) Find the central neuron i^* satisfying

$$d\big(\boldsymbol{X}(n), \boldsymbol{W}_{i^*}(n)\big) = \min_i d\big(\boldsymbol{X}(n), \boldsymbol{W}_i(n)\big). \tag{9.103}$$

If the Euclidean distance measure is used, then (9.103) becomes

$$\big\|\boldsymbol{X}(n) - \boldsymbol{W}_{i^*}(n)\big\| = \min_i \big\|\boldsymbol{X}(n) - \boldsymbol{W}_i(n)\big\|. \tag{9.104}$$

The central neuron i^* is in effect the winning neuron whose weight vector in the stable state will be taken as the codeword.

(4) Adjust the connection weights by

$$\begin{cases} \boldsymbol{W}_i(n+1) = \boldsymbol{W}_i(n) + \gamma(n)\big(\boldsymbol{X}(n) - \boldsymbol{W}_i(n)\big), & i \in N_c, \\ \boldsymbol{W}_i(n+1) = \boldsymbol{W}_i(n), & i \notin N_c, \end{cases} \tag{9.105}$$

where $\gamma(n)$ is an adaptive parameter that should satisfy $0 < \gamma(n) < 1$, $\gamma(n) \to 0$, $\sum_n \gamma^p(n) < \infty$ for some p, and $\sum_n \gamma(n) = \infty$.

(5) Iterate the computation by presenting a new input vector and returning to Step (2) until the weight vectors stabilize their values.

This algorithm for codebook design has been shown to be closely related to the LBG algorithm. In fact, the LBG algorithm can be considered as the batch version of the above KSFM algorithm, and both result in a minimum mean squared error codebook [37, 39]. However, the KSFM algorithm offers a number of advantages over the LBG algorithm. These advantages include less sensitivity to initialization of the codebook, better rate distortion performance, and fast convergence. It has also been shown [34] that the codebook obtained by the KSFM algorithm is optimal in the sense that the average distortion is minimized. More important is that the KSFM algorithm applies to the case that each new input vector becomes available as time progresses. This aspect can be used to design the codebook adaptively and proves to be crucial in those applications where the signal statistics are changing over time.

Another problem with the LBG algorithm is that the frequency of use of the words in the codebook can be quite uneven with some codewords being underutilized. This problem can be overcome to some extent by the so-called frequency-sensitive competitive learning (FSCL) algorithm [40]. The FSCL algorithm updates the weight of the winning neuron by

$$\boldsymbol{W}_i(n+1) = \boldsymbol{W}_i(n) + S(n)Q_i(n)\big(\boldsymbol{X}(n) - \boldsymbol{W}_i(n)\big), \tag{9.106}$$

where

$$Q_i(n) = \begin{cases} 1, & d\big(\boldsymbol{X}(n), \boldsymbol{W}_i(n)\big) < d\big(\boldsymbol{X}(n), \boldsymbol{W}_j(n)\big), \quad i \neq j, \\ 0, & \text{otherwise}, \end{cases} \tag{9.107}$$

$$S(n) = \begin{cases} \frac{1}{F_i(n)}, & 1 \leq F_i(n) \leq F_{th}, \\ 0, & \text{otherwise}, \end{cases} \tag{9.108}$$

and

$$F_i(n+1) = F_i(n) + Q_i(n). \tag{9.109}$$

$S(n)$ and F_{th} are called the frequency-sensitive training rate and the upper-threshold frequency, respectively. Note that according to (9.106) only the weight vector of

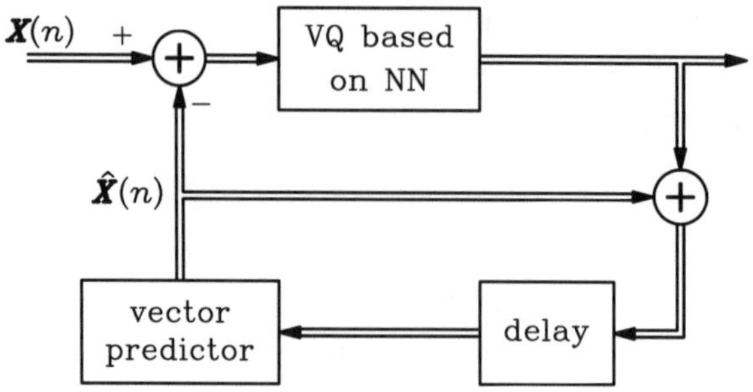

Figure 9.9. Predictive VQ based on neural networks.

the winning neuron is updated. If $F_i(n)$ is larger than F_{th}, then $S(n)$ is set to be zero and no further training will be performed for this winning neuron so as to avoid overutilization and underutilization. The selection of the upper-threshold frequency is heuristic and depends on the statistics of the signal to be encoded. Empirically, an adequate F_{th} is chosen to be two or three times larger than the average winning frequency.

An alternative FSCL algorithm involves replacing the distance $d(\mathbf{X}(n), \mathbf{W}_i(n))$ by $F(u_i)d(\mathbf{X}(n), \mathbf{W}_i(n))$, where u_i is the count of how many times the neuron i has won and $F(\cdot)$ is a nondecreasing function called the "fairness function." This function can take many forms, for example, $F(u) = u$ and $F(u) = u^{\exp(-n/T)}$, where n is the training iteration number and T is a constant. The choice of $F(u) = u^{\exp(-n/T)}$ means that the "fairness function" depends on the training iteration. This choice initially emphasizes uniformity of codeword usage, but later emphasizes minimizing the distortion as training progresses. It has been shown that the fairness function has a significant effect on the performance of the resulting codebook [36].

The VQ techniques described here do not take into account the correlations that exist among the adjacent codewords. These correlations can be exploited for further reducing the coding rate [41]. One direct way to achieve this is to use a predictor. The output vector of the predictor is first subtracted from the original signal and then the residual vector is encoded by the above VQ techniques based on neural networks. This is shown in Figure 9.9. Clearly, any of the techniques given in Section 9.1 and Section 9.2 can be used to deal with the design and implementation of the predictor.

Finally, we should point out that these neural network approaches for VQ are still very preliminary. Further investigations are desirable.

Bibliography

[1] Jayant, N., "Signal Compression: Technology Targets and Research Directions," *IEEE J. Selected Areas in Communications*, Vol. 10, No. 5, 1992, pp. 796–818.

[2] Storer, J. A., "Scanning the Special Section on Data Compression," *Proc. IEEE*, Vol. 82, No. 6, 1994, p. 856.

[3] Zhang, Y. Q., Li, W., and Liou, M. L. (eds.), *Proc. IEEE*, Special issue on Advances in Image and Video Compression, Vol. 83, No. 2, 1995.

[4] Tzou, K. H., Musmann, H. G., and Aizawa, K. (eds.), *IEEE Trans. on Circuits and Systems for Video Technology*, Special issue on Very-Low-Bit-Rate Video Coding, Vol. 4, No. 3, 1994.

[5] Dony, R. D. and Haykin, S., "Neural Network Approaches to Image Compression," *Proc. IEEE*, Vol. 83, No. 2, 1995, pp. 288–303.

[6] Niemann, H. and Wu, J. K., "Neural Network Adaptive Image Coding," *IEEE Trans. on Neural Networks*, Vol. 4, No. 4, 1993, pp. 615–27.

[7] Stark, J., "A Neural Network to Compute the Hutchinson Metric in Fractal Image Processing," *IEEE Trans. on Neural Networks*, Vol. 2, No. 1, 1991, pp. 156–58.

[8] Kroon, P. and Swaminathan, K., "A High Quality Multi–Rate Real-Time CELP Coder," *IEEE J. Selected Areas in Communications*, Vol. 10, No. 5, 1992, pp. 850–57.

[9] Tremain, T. E., "The Government Standard Linear Predictive Coding Algorithm: LPC-10," *Speech Technology*, Vol. 1, No. 2, 1982, pp. 40–49.

[10] Gioffi, J. M. and Kailath, T., "Fast RLS Transversal Filters for Adaptive Filtering," *IEEE Trans. on Acoustics, Speech and Signal Processing*, Vol. 32, No. 2, 1984, pp. 304–37.

[11] Luo, F. L. and Li, Y. D., "Neural Networks for the Exact RLS Algorithm," *Appl. Math. Computation*, Vol. 60, No. 2, 1994, pp. 103–12.

[12] Li, J. and Manikopoulos, C. N., "Nonlinear Prediction in Image Coding with DPCM," *Electronics Lett.*, Vol. 26, No. 17, 1990, pp. 1357–59.

[13] Dianat, S. A., Nasrabadi, N. M., and Venkataraman, S., "A Nonlinear Predictor for Differential Pulse-Code Encoder (DPCM) Using Neural Networks," *Proc. IEEE Int. Conf. on Acoustics, Speech and Signal Processing*, 1991, pp. 2793–96.

[14] Sicuranzi, G. L., Ramponi, G., and Marsi, S., "Artificial Neural Network for Image Compression," *Electronics Lett.*, Vol. 26, 1990, pp. 477–79.

[15] Namphol, A., Arozullah, M., and Chin, S., "Higher Order Data Compression with Neural Networks," *Proc. Int. Joint Conf. on Neural Networks*, Baltimore, Vol. I, 1991, pp. 55–59.

[16] Manikopoulos, C. N., "Neural Network Approach to DPCM System Design for Image Coding," *IEE Proc.*, Part I, Vol. 139, No. 5, 1992, pp. 501–7.

[17] Jayant, N., Johnston, J., and Safranek, R., "Signal Compression Based on Models of Human Perception," *Proc. IEEE*, Vol. 81, No. 10, 1993, pp. 1385–422.

[18] Passaggio, F., Anguita, D., and Zunino, R., "Human Visual System for Image Compression by BP," *Proc. IEEE Int. Conf. on Neural Networks*, 1994, pp. 1221–24.

[19] Parodi, G. and Passaggio, F., "Size-Adaptive Neural Network for Image Compression," *Proc. IEEE Int. Conf. on Neural Networks*, 1994, pp. 945–47.

[20] Mougeot, M., Azencott, R., and Angeniol, B., "Image Compression with Back Propagation: Improvement of the Visual Restoration Using Different Cost Functions," *Neural Networks*, Vol. 4, 1991, pp. 467–76.

[21] Reed, I. S. and Lan, L. S., "A Fast Approximate KLT Transform for Data Compression," *J. Visual Communication Image Representation*, Vol. 5, No. 4, 1994, pp. 304–16.

[22] Girod, B., Gall, D. J. L, Sezan, M. I., Vetterli, M., and Yasuda, H. (eds.), *IEEE Trans. on Image Processing*, Special issue on Image Sequence Compression, Vol. 3, No. 5, 1994.

[23] Kuo, C. C. J. and Storer, J. A. (eds.), *J. Visual Communication Image Representation*, Special issue on Still and Video Image Compression, Vol. 5, No. 4, 1994.

[24] DeVore, R., Jawerth, B., and Lucier, B., "Image Compression Through Wavelet Transform Coding," *IEEE Trans. on Information Theory*, Vol. 38, No. 2, 1992, pp. 719–46.

[25] Rioul, O. and Duhamel, P., "Fast Algorithm for Discrete and Continuous Wavelet Transforms," *IEEE Trans. on Information Theory*, Vol. 38, No. 2, 1992, pp. 569–85.

[26] Katto, J., Ohki, J. I., Nogaki, S., and Ohta, M., "A Wavelet Codec with Overlapped Motion Compensation for Very Low Bit-Rate Environment," *IEEE Trans. on Circuits and Systems for Video Technology*, Vol. 4, No. 3, 1994, pp. 329–38.

[27] Luo, F. L., Unbehauen, R., and Li, Y. D., "A Principal Component Analysis Algorithm with Invariant Norm," *Neurocomputing*, Vol. 8, No. 2, 1995, pp. 213–21.

[28] Reif, K., Luo, F. L., and Unbehauen, R., "The Exponential Stability of the Invariant-Norm PCA Algorithm," Accepted by *IEEE Trans. on Circuits and Systems*, Part II. 1996.

[29] Szu, H. H. and Gaulfield, H. J., "Wavelet Transforms," *Optical Engineering*, Vol. 31, No. 9, 1992, pp. 1823–24.

[30] Szu, H. H., Telfer, B., and Kadambe, S., "Neural Network Adaptive Wavelets for Signal Representation and Classification," *Optical Engineering*, Vol. 31, No. 9, 1992, pp. 1907–15.

[31] Daugman, J. G., "Complete Discrete 2-D Gabor Transforms by Neural Networks for Image Analysis and Compression," *IEEE Trans. on Acoustics, Speech and Signal Processing*, Vol. 36, No. 7, 1988, pp. 1169–79.

[32] Zhang, Q. and Benveniste, A., "Approximation by Nonlinear Wavelet Networks," *Proc. IEEE Int. Conf. on Acoustics, Speech and Signal Processing*, 1991, pp. 3417–20.

[33] Gray, R. M., "Vector Quantization," *IEEE Acoustics, Speech and Signal Processing Magazine*, Vol. 1, No. 2, 1984, pp. 4–29.

[34] Lee, T. S. and Peterson, A. M., "Adaptive Vector Quantization Using a Self-Development Neural Network," *IEEE J. Selected Areas in Communications*, Vol. 8, No. 8, 1990, pp. 1458–71.

[35] Nasrabadi, N. M. and Feng, Y., "Vector Quantization of Image Based on Kohonen Self-Organizing Map," *Proc. IEEE Conf. on Neural Networks*, San Diego, Vol. I, 1988, pp. 101–5.

[36] Krishnamurthy, A. K., Ahalt, S. C., Melton, D. E., and Chen, P., "Neural Network for Vector Quantization of Speech and Images," *IEEE J. Selected Areas in Communications*, Vol. 8, No. 8, 1990, pp. 1449–57.

[37] Luttrell, S. P., "Self-Organization: A Derivation from First Principle of a Class of Learning Algorithms," *Proc. IEEE Conf. on Neural Networks*, Washington DC, 1989, pp. 495–98.

[38] Kohonen, T., "The Self-Organizing Map," *Proc. IEEE*, Vol. 78, No. 9, 1990, pp. 1464–80.

[39] McAuliffe, J. D., Atlas, L. E., and Rivera, C., "A Comparison of the LBG Algorithm and Kohonen Neural Network Paradigm for Image Quantization," *Proc. IEEE Int. Conf. on Acoustics, Speech and Signal Processing*, 1990, pp. 2293–96.

[40] Fang, W. C., Sheu, B. J., Chen, O. T. C., and Choi, J., "A VLSI Neural Processor for Image Data Compression Using Self-Organization Networks," *IEEE Trans. on Neural Networks*, Vol. 3, No. 3, 1992, pp. 507–15.

[41] Mohsenian, N., Rizvi, S. A., and Nasrabadi, N. M., "Predictive Vector Quantization Using a Neural Network Approach," *Optical Engineering*, Vol. 32, No. 7, 1993, pp. 1503–13.

Index